Biogeography

Biogeography

An Ecological and Evolutionary Approach

C. Barry Cox
*Former Head of Biological Sciences
at King's College London, UK*

Peter D. Moore
*Emeritus Reader in Ecology,
King's College London, UK*

Richard J. Ladle
*Titular Professor of Conservation Biogeography,
Federal University of Alagoas, Brazil*

NINTH EDITION

WILEY Blackwell

This ninth edition first published 2016 © 2016 by John Wiley & Sons, Ltd

Edition history: 2010, 2005, 2000, 1993, 1985, 1980, 1976, 1973 published by John Wiley & Sons, Inc.

Registered office: John Wiley & Sons, Ltd, The Atrium, Southern Gate, Chichester, West Sussex, PO19 8SQ, UK

Editorial offices: 9600 Garsington Road, Oxford, OX4 2DQ, UK
The Atrium, Southern Gate, Chichester, West Sussex, PO19 8SQ, UK
111 River Street, Hoboken, NJ 07030-5774, USA

For details of our global editorial offices, for customer services and for information about how to apply for permission to reuse the copyright material in this book please see our website at www.wiley.com/wiley-blackwell.

The right of the author to be identified as the author of this work has been asserted in accordance with the UK Copyright, Designs and Patents Act 1988.

Library of Congress Cataloging-in-Publication Data

Names: Cox, C. Barry (Christopher Barry), 1931- author. | Moore, Peter D.,
    author. | Ladle, Richard J., author.
Title: Biogeography : an ecological and evolutionary approach / C. Barry Cox,
    Peter D. Moore, Richard Ladle.
Description: Ninth edition. | Chichester, UK; Hoboken, NJ : John Wiley &
    Sons, 2016. | Includes bibliographical references and index.
Identifiers: LCCN 2016000665 | ISBN 9781118968574 (cloth) | ISBN 9781118968581
    (pbk.)
Subjects: LCSH: Biogeography.
Classification: LCC QH84 .C65 2016 | DDC 577.2/2—dc23 LC record available at
http://lccn.loc.gov/2016000665

A catalogue record for this book is available from the British Library.

Wiley also publishes its books in a variety of electronic formats. Some content that appears in print may not be available in electronic books.

Cover image: © gettyimages / Chad Ehlers

Set in 9/11.5 Trump Mediaeval LT Std Roman by Aptara Inc., New Delhi, India

Printed in Singapore by C.O.S. Printers Pte Ltd

1   2016

Contents

Entry numbers in **bold type** in the index indicate the pages on which the concept involved is defined. These words or concepts are shown in bold type on the page in question, and also appear in the Glossary if the concept is used in a variety of pages.

Preface

To interpret the phenomena of biogeography, we need to understand many different areas of science – for example, evolution, taxonomy, ecology, geology, palaeontology and climatology. Although each area makes its own individual contribution, a textbook such as this therefore has to cover a similar scope and must be suitable for students with a variety of different backgrounds. This is particularly necessary today, when the coming-together of molecular methods of demonstrating relationships, and of the cladistic technique of imposing pattern upon the resulting data, is revolutionizing our understanding of biogeography.

Many changes have taken place in the study of biogeography over the 43 years and nine editions that have now passed in the life of this textbook. Back in 1973, the depth of the problem that our species poses for our planet's biota and climate was hardly appreciated, and the 'greenhouse effect' was still mainly a concern for horticulturalists rather than a worry for the whole planet. It was not until the 1980s that the evidence that the Earth's climate is changing, and that this is increasingly the result of human activities, became steadily greater. In interpreting the interaction between the physical world and the living world, and of the human impact on each of these, biogeography clearly has a major role to play in assessing the likely results of climatic change and in suggesting how best to counter them. As climatic change renders old crop areas less fertile, will it be possible to find new areas to replace them – and, if so, where? Or shall we have to find new varieties of plant, adapted to these new conditions – and, if so, where are we likely to find them? The increasing urgency of these questions led to a great increase in the amount of research on biogeography during the 1990s.

But it is not only our foodstuffs that are threatened by climate change; it is the diversity of life that inhabits environments that is shrinking and disappearing. This is not just a concern for the curators of museums and herbaria, for we are also becoming aware of the extent to which we rely on this diversity for new drugs as well as for new food-plants. So we became aware of the need to census this diversity, in order to appreciate where it is greatest and where it is under particular threat. Which habitats are threatened, where, and how should we attempt to preserve them?

Until recently, biologists were unable to document the dates at which new species appeared and diverged from one another. As a result, it was impossible to be sure of the relationship between these biological processes and events such as the separation of units of land by plate tectonics or climatic change. The rise of molecular methods of investigation that provide reliable dates for the times of appearance and divergence of species has given us new confidence in the accuracy of our biogeographical analyses, based on rigorous techniques of analysis of relationships in time and space, using increasingly complex and sophisticated computer algorithms. At last it seems that biogeographical research is revealing, with increasing scope and detail, a single, consistent story of the history of the biogeography of the world today and of the processes that led to it.

This understanding may have come just in time, for it is clear that it is imperative that we conserve what is left of the natural world of our planet. In this new edition, we welcome the contributions of Richard Ladle, who has not only helped in the revision of some previously existing chapters, but also contributed a new chapter on 'Conservation Biogeography' (Chapter 14). In this chapter he outlines the startling new techniques that are now becoming available for gathering and integrating information on the distribution of species. It is up to the new generations of biogeographers to find ways to use this increasing wealth of data to construct clear arguments that will convince the (sometimes reluctant) politicians and businessmen of the validity of their case. Only then can the vital step of transforming scientific knowledge into potential action take place. We can only hope that this will happen sufficiently soon to save the living world of our planet as we know it today.

After all these years and editions, this is probably the last time that Peter Moore and I will contribute to this book; it has been a long and happy collaboration. We welcome Richard Ladle as the first of a new group of biogeographers who will, we hope, take the book into the future. It is also appropriate now to remember that the first edition was the work, not only of Dr Peter Moore and myself, but also of our colleague Dr Ian Healey, who sadly died before that work was published.

Barry Cox

Acknowledgements

Our first thanks must go to Ward Cooper of Wiley-Blackwell for making this new edition possible, and also to Kelvin Matthews, Emma Strickland and Jane Andrew for all their hard work in getting it through the production process.

As noted earlier, biogeography involves the study of a very great range of data in the fields of both earth sciences and biological sciences, and it is nowadays impossible for any one person to cover all the literature in such a huge area. Our task in trying to identify the significant new references has been greatly aided by many people, but we would like to thank in particular the following:

Professor David Bellwood, School of Marine and Tropical Biology, James Cook University, Queensland, Australia

Professor Alex Rogers, Department of Zoology, Oxford University, UK

Dr Isabel Sanmartín, Reál Jardín Botánico, Madrid, Spain.

We also thank Professor Robert Hall, Earth Sciences Department, Royal Holloway College, University of London, UK for providing the special set of palaeomaps and giving us permission to use them.

The History of Biogeography

Chapter 1

This introductory chapter begins with an explanation of why the study of the history of a subject is important, and highlights some of the important lessons that students may gain from it. This is followed by a review of the ways in which each of the areas of research in biogeography developed from its foundation to today.

Lessons from the Past

One of the best reasons for studying history is to learn from it; otherwise, it becomes merely a catalogue of achievement. So, for example, it is often valuable to think about why and when a particular advance was made. Was it the result of personal courage in confronting the current orthodoxy of religion or science? Was it the result of the mere accumulation of data, or was it allowed by the development of new techniques in the field of research, or in a neighbouring field, or by a new intellectual permissiveness? But the study of history also gives us the opportunity to learn other lessons – and the first of these is humility. We must be wary, when considering the ideas of earlier workers, not to fall into the trap of arrogantly dismissing them as in some way inferior to ourselves, simply because they did not perceive the 'truths' that we now see so clearly. In studying their ideas and suggestions, one soon realizes that their intellect was no less penetrating than those that we can see at work today. However, compared to the scientists of today, they were handicapped by lack of knowledge and by living in a world in which, explicitly or implicitly, it was difficult or impossible to ask some questions.

Firstly, less was known and understood. When Isaac Newton, who originated the theory of gravitational attraction, wrote that he had 'stood on the shoulders of giants', he was acknowledging that in his own work he was building upon that of generations of earlier thinkers, and was taking their ideas and perceptions as the foundations of his own. So, the further we go back in time, the more we see intellects that had to start afresh, with a page that was either blank or contained little in the way of earlier ideas or syntheses.

Secondly, we must be very aware that, for every generation, the range of theories that might be suggested was (and is!) limited by what contemporary society or science views as permissible or respectable. Attitudes towards the ideas of evolution (see Chapter 6) and continental drift (this chapter) are good examples of such inhibitions in the 19th and 20th centuries. The history of scientific debate is rarely, if ever, one of dispassionate, unemotional evaluation of new ideas, particularly if they conflict with one's own. Scientists, like all men and women, are the product of their upbringing and experience, affected by their political and religious beliefs (or disbeliefs), by their position in society, by their own previous judgments and publicly expressed opinions and by their ambitions – just as 'there's no business like show business', there's no interest like self-interest! Very good examples of this, discussed further in this chapter, are the use of the concept of evolution by the rising middle-class scientists of England as a weapon against the 19th-century establishment and, at the individual level, the history of Leon Croizat.

Biogeography: An Ecological and Evolutionary Approach, Ninth Edition. Edited by C. Barry Cox, Peter D. Moore, Richard J. Ladle.
© 2016 John Wiley & Sons, Ltd. Published 2016 by John Wiley & Sons, Ltd.

In our survey of the history of biogeography, we shall therefore see people who, like most of us, grew up accepting the intellectual and religious ideas current in their time, but who also had the curiosity to ask questions of the world of nature around them. Sometimes the only answers that they could find contradicted or challenged the current ideas, and it was only natural then to seek ways to circumvent the problem. Could these ideas be reinterpreted to avoid the problem, was there any way, any loophole, to avoid a complete and direct challenge and rejection of what everyone else seemed to accept?

So, to begin with, the reactions of any scientist confronted with results or ideas that conflict with current dogma are either to reject them ('Something must have gone wrong with his methods, or with my methods') or to view them as an exception ('Well, that's interesting, but it's not mainstream'). Sometimes, however, these difficulties and 'exceptions' start to become too numerous or varied, or they begin to arise from so many different parts of science as to suggest that something must be wrong. The scientist may then realize that the only way around it is to start again, starting from a completely different set of assumptions, and to see where that leads. Such a course is not easy, for it involves the tearing-up of everything that one has previously assumed and completely reworking the data. And, of course, the older you get, the more difficult it is to do so, for you have spent a longer time using the older ideas and publishing research that explicitly or implicitly accepts them. That is why, all too often, older workers take the lead in rejecting new ideas, for they see them as attacking their own status as senior, respected figures. Sometimes these workers also refuse to accept and use new approaches long after these have been thoroughly validated and widely used by their younger colleagues (see attitudes towards plate tectonic theory in Chapter 5). Another problem is that the debate can become polarized, with the supporters of two contrasting ideas being concerned merely to try to prove that the opponents' ideas are false, badly constructed and untrue (see dispersal vs. vicariance, discussed later in this chapter, and punctuated vs. gradual evolution, discussed in Chapter 6). Neither side then stops to consider whether it is perhaps possible that both of the apparently conflicting ideas are true, and that the debate should

instead be about when, under what circumstances and to what extent one idea is valid, and when the other is instead the more important. Also, too often, scientists have rejected the suggestions of another worker, not because the suggestions were in themselves unacceptable, but because the scientists rejected *other* opinions of that same author (e.g. Cuvier vs. Lamarck on evolution; see further in this chapter).

All of this is particularly true of biogeography, for it provides the additional difficulty of being placed at the meeting point of two quite different parts of science – biological sciences and earth sciences. This has had two interesting results. The first is that, from time to time, lack of progress in one area has held back the other. For example, the assumption of stable, unchanging geography made it impossible to understand past patterns of distribution. Nonetheless, it was a reasonable assumption until the acceptance of **plate tectonics** (continental drift) provided a vista of past geographies that had gradually changed through time. But it is also interesting to note that this major change in the basic approaches of earth sciences came in two stages.

To begin with, the problem was clearly posed and a possible solution was given. This was in 1912, when the German meteorologist Alfred Wegener (see later in this chapter) pointed out that many patterns in both geological and biological phenomena did not conform to modern geography, but that these difficulties disappeared if it was assumed that the continents had once lain adjacent to one another and had gradually separated by a process that he called **continental drift**. This explanation did not convince the majority of workers in either field, largely because of the lack of any known mechanism that could cause continents to move horizontally or to fragment. The fact that Wegener himself was not a geologist but an atmosphere physicist did not help him to persuade others of the plausibility of his views, for it was only too easy for geologists (who, of course, 'knew best') to dismiss him as a meddling amateur. Most biologists, faced with the uncertainties of the fossil record, did not care to take on the assembled geologists.

The second stage came only in the 1960s, when geological data from the structure of the seafloor and from the magnetized particles found in rocks (see Chapter 5) not only provided unequivocal evidence for continental movements, but also suggested a mechanism for them. Only then did geologists accept this new view of world history (known as plate tectonics; see Chapter 5), and only then could biogeographers confidently use the resulting coherent and consistent series of palaeogeographical maps to explain the changing patterns of life on the moving continents. Such a theory, based on a great variety of independent lines of evidence, is known as a **paradigm**, and the theory of plate tectonics is the central paradigm of the earth sciences.

The moral of this story is, perhaps, that it is both understandable and reasonable for workers in one field (here, biologists) to wait until specialists in another field (here, geology) have been convinced by new ideas before they feel confident in using them to solve their own problems. This, in turn, leads to the second topic that results from the position of biogeography between biology and geology. That is the temptation for workers in one field, frustrated by lack of progress in some aspect of their own work, to accept, uncritically and without proper understanding, new ideas in the other field that seem to provide a solution [1]. One must be particularly wary of new theories that are directed at explaining merely one difficulty in the currently accepted interpretations. This is because such suggestions sometimes simultaneously destroy the rest of the framework, without satisfactorily explaining the vast majority of the phenomena that were covered by that framework. For example, in the second half of the 20th century, some geologists suggested that the Earth had expanded, or that there had once been a separate 'Pacifica' continent between Asia and North America. Some biological biogeographers welcomed these ideas as the solution to some detailed problems of the distribution of terrestrial vertebrates, even though they were not supported by geological data and had not been accepted by geologists.

All of this has important lessons for us today, for it would be naive to believe that the assumptions and methods used in biogeography today are in some way the final and 'correct' ones that will never be rejected or modified. Similarly, every student should realize that those who teach science today have, of course, been trained to accept the current picture of the subject and may find it difficult to accept changes in its methodology. The price that we pay for gaining experience with age

is an increasing conviction of the correctness of our own methods and assumptions! (On the other hand, it is interesting to note that whereas in the physical sciences major new discoveries are usually made by intuitive leaps early in the scientist's career, those in the biological sciences are more often made only later, after the accumulation of data and knowledge.) It is also worth noting that erroneous assumptions are far more dangerous than false reasoning because the assumptions are usually unstated, and therefore far more difficult to identify and correct. So, the past with its false assumptions and erroneous theories is merely a distant mirror of today, warning us in our turn not to be too sure of our current ideas. Sometimes the limitations and problems of a new technique only become apparent gradually, some time after it has been introduced.

But, of course, those of us who carry out research and publish our ideas in books such as this also have a responsibility to use their experience and judgment in trying to choose between conflicting ideas, showing which we prefer and why. For example, in this book the author who wrote this chapter (Barry Cox) has criticized the methodology of a school of (mainly) New Zealand panbiogeographers (see later in this chapter). But, of course, he could be wrong, and interested students should read around the subject and come to their own conclusions. After all, the purpose of learning a subject at this level is for students to develop their own critical faculties, not merely to acquire attitudes and opinions. Even over the past 50 years, we have seen attitudes to a new idea, the Theory of Island Biogeography, change quite considerably (see later in this chapter, and Chapter 7). How many of the explanations and assumptions in this book will still seem valid in 50 years' time? But that is also one of the pleasures of being part of science, and of having to try continually to adapt to new ideas, rather than merely being part of some ancient monolith of long-accepted 'truths'.

Ecological versus Historical Biogeography, and Plants versus Animals

The most fundamental split in biogeography is that between the ecological and historical aspects of the subject. **Ecological biogeography** is concerned with the following types of questions. Why is a species confined to its present range in space? What enables it to live where it does, and what prevents it from expanding into other areas? What roles do soil, climate, latitude, topography and interactions with other organisms play in limiting its distribution? How do we account for the replacement of one species by another as one moves up a mountain or seashore, or from one environment to another? Why are there more species in the tropics than in cooler environments? Why are there more endemic species in environment X than in environment Y? What controls the diversity of organisms that is found in any particular region? Ecological biogeography is, therefore, concerned with short-term periods of time, at a smaller scale; with local, within-habitat or intracontinental questions; and primarily with species or subspecies of living animals or plants. (Subspecies, species, genus (plural: genera), family, order and phylum (plural: phyla) are progressively larger units of biological classification. Each is known as a **taxon** (plural: taxa).)

Historical biogeography, on the other hand, is concerned with different questions. How did the taxon come to be confined to its present range in space? When did that pattern of distribution come to have its present boundaries, and how have geological or climatic events shaped that distribution? What are the species' closest relatives, and where are they found? What is the history of the group, and where did earlier members of the group live? Why are the animals and plants of large, isolated regions, such as Australia or Madagascar, so distinctive? Why are some closely related species confined to the same region, but in other cases they are widely separated? Historical biogeography is, therefore, concerned with long-term, evolutionary periods of time; with larger, often sometimes global areas; and often with taxa above the level of the species and with taxa that may now be extinct.

Because of the different nature of plants and of animals, the ways in which their ecological and historical biogeography have been investigated and understood have differed in the two groups. Plants are static, and their form and growth are therefore much more closely conditioned by their environmental, ecological conditions than are those of animals. It is also far easier to collect and preserve plants than animals, and to note the conditions of soil and climate in which they live. But the fossil remains of plants are less common than those of animals, and they are also far more difficult to interpret, for several reasons. There are many more flowering plants than there are mammals – some 450 living families and 17 000 genera of plant, compared with 150 living families and 1250 genera of mammal. Furthermore, although the leaves, wood, seeds, fruit and pollen grains of flowering plants may be preserved, they are rarely found so closely associated that one can be sure which leaf belongs with which type of pollen grain, and so on. Finally, the taxonomy of flowering plants is based on the characteristics of their flowers, which are only rarely preserved. In contrast, the fossil bones of mammals are often associated as complete skeletons, which are easy to allocate to their correct family, and which provide a detailed record of the evolution and dispersal of these families within and between the continents through geological time.

For all these reasons, the biogeography of the more distant past was, until recently, largely the preserve of zoologists, whereas plant scientists were far more concerned with ecological biogeography – although studies of fossil pollen from the Ice Ages and postglacial times, which are easy to allocate to existing species, have been fundamental in interpreting the history and ecology of this most recent past (see Chapter 12).

In following the history of biogeography, it would be easy merely to follow a path through time, recounting who discovered what and when. But it is more instructive instead to take each thread of the components of biogeography in turn, to follow the different contributions to its understanding, and on the way to note the lessons to be learned from how the scientists reacted to the problems and ideas of their time.

Biogeography and Creation

Biogeography, as a part of Western science, began in the mid-18th century. At that time, most people accepted the statements in the Bible as the literal truth, that the Earth and all living things that we see today had been created in a single series of events. It was also thought that these events had taken place only a few thousands of years before, and it was believed that God's actions had always been perfect. It followed that the animals and plants that had been created were perfect, and had not changed (evolved) or become extinct, and that the world itself had always been as we see it today. The history of biogeography between then and the middle of the 20th century is the story of how that limited vision was gradually replaced by the realization that both the living world and the planet that it inhabits are continually changing, driven by two great processes – the biological process of evolution and the geological process of plate tectonics.

So, when the Swedish naturalist Linnaeus in 1735 started to name and describe the animals and plants of the world, he assumed that each belonged to an unchanging species, which had been created by God. But he soon found that there were species whose characteristics were not as constant and unchanging as he had expected. That might puzzle him, but he could only accept it. But there was a further problem, for, according to the Bible, the whole world had once been covered by the waters of the Great Flood. All the animals and plants that we see today must therefore have spread over the world from the point where Noah's Ark had landed, thought to be Mount Ararat in eastern Turkey. Linnaeus ingeniously suggested that the different environments to be found at different altitudes, from tundra to desert, had been colonized in turn by different animals and plants from the Ark as the floodwaters receded, progressively uncovering lower and lower levels of land. Linnaeus recorded in what type of environment each species was found, and so began what we now call ecological biogeography. He also recorded whereabouts in the world each species is found, but he did not synthesize these observations into any account of faunal or floral assemblages of the different continents or regions.

The first person to realize that similar environments, found in different regions of the world, contained different groupings of organisms was the French naturalist Georges Buffon; this important insight has come to be known as **Buffon's Law**. In various editions of his multivolume *Histoire Naturelle* [2], published from 1761 onward, he identified a number of features of world biogeography and suggested possible explanations. He noted that many of the mammals of North America, such as bears, deer, squirrels, hedgehogs and moles, were found also in Eurasia, and he pointed out that they could only have travelled between the two continents, via Alaska, when climates were much warmer than today. He accepted that some animals, such as the mammoths, had become extinct. Buffon also realized that most of the mammals of South America are quite different from those of Africa, even though they live in similar tropical environments. Accepting that all were originally created in the Old World, he suggested that the two continents were at one time adjacent and that the different mammals then sought out whichever area they found most congenial. Only later did the ocean separate the two continents and the two now-different faunas, whereas some other differences might have been due to the action of the climate. Buffon also used the fossil record to reconstruct a history of life that clearly had extended over at least tens of thousands of years. Only the last part had witnessed the presence of human beings, and included earlier periods within which tropical life had covered areas that are now temperate or even subarctic.

Buffon strongly felt that one had to be guided by study of the facts, and this conviction drove him to accept that geography, climate and even the nature of the species were not fixed, but changeable, and to suggest that continents might move laterally and seas encroach upon them. That was a truly courageous and visionary deduction to make in the late 18th century. So Buffon recognized, commented upon and attempted to explain many phenomena that other, later workers either ignored or merely recorded without comment. His observations on the differences between the mammals of the two regions were soon extended to land birds, reptiles, insects and plants.

The Distribution of Life Today

As 18th-century explorers and naturalists revealed more and more of the world, they also extended the horizons of biogeography itself, discovering a greater diversity of organisms. For example, in his second voyage around the world in 1772–1775, the British navigator Captain James Cook took the British botanist Joseph Banks and the German Johann Reinhold Forster, together with his son

Georg Forster, who collected thousands of species of plants, many of them new to science. Forster found that Buffon's Law applied to plants as well as to animals, and also applied to any region of the world that was separated from others by barriers of geography or climate [3]. He also realized that there are what we now call gradients of diversity (see Chapter 4), there being more plant species closer to the equator and progressively fewer as one moves towards the poles, and he made the first observations of island biogeography.

The concepts of ecological biogeography, botanical regions and island biogeography had, then, all been recognized by the end of the 18th century. But it was still generally accepted that there could be little or no change in the nature of each species, or in the pattern of the geography of the world. The early naturalists therefore still struggled to explain how all these different floras had come into existence, widely scattered over the Earth's surface. Perhaps the most plausible explanation was that of the German botanist Karl Willdenow, who in 1792 suggested that, although there had been only one act of creation, it had taken place simultaneously in many places. In each area, the local flora had been able to survive the Flood by retreating to the mountains, from which it was able later to spread downward to recolonize its own part of the world as the floodwaters receded. His book also included a chapter on the history of plants, and he noted that plants' growth habits were related to the conditions of their environment.

Despite the work of these two earlier botanists, the German Alexander von Humboldt is usually recognized as the founder of plant geography, perhaps because he was a far wealthier and more flamboyant figure. But Forster and Willdenow not only preceded Humboldt but also greatly influenced his life. It was Georg Forster who inspired Humboldt to become an explorer, and the slightly older Willdenow introduced him to botany and became his lifelong friend. Humboldt became famous for his 1799–1804 expedition to South America, during which he climbed to over 5800 m (19 000 feet) on the volcano Chimborazo – a world height record that he held for 30 years. He noticed that the plant life on the mountain showed a zonation according to altitude, much like the latitudinal variation that Forster had described. Plants at lower levels are of the tropical type, those of intermediate levels are

of the temperate type, and finally arctic types of plant are found at the highest levels. (Humboldt used the term *association* to describe the assemblages of plants that characterized each of these life zones; today, they are more commonly referred to as *formations* or *biomes*; see Chapter 3.) Humboldt believed that the world was divided into a number of natural regions, each with its own distinctive assemblage of animals and plants. He was also the first to insist that biological observations had to include detailed, accurate and precisely recorded data. He published a thorough account of his botanical observations in 1805, as part of a 30-volume series recording his findings in the New World [4].

Another early plant biogeographer was Augustin de Candolle of Geneva who, in 1805 together with Lamarck, published a map showing France divided into five floristic regions with different ecological conditions. Candolle later went on to study the dispersal of plants by water, wind or the actions of animals, pointing out that this would lead to the plants spreading until they encountered barriers of sea, desert or mountains. He was also the first to realize that another limiting factor was the presence of other plants that competed with them. The result of these processes would be the appearance of regions that, even though they might contain a variety of climatic zones and ecological environments, were distinct from one another because they contained plants that were restricted to that area, for which he coined the word **endemic** (see Chapter 2). The distinctions between these regions were thus partly dependent on their histories. Candolle went on to define 20 such regions, of which 18 were continents or parts of continents, and two were island groups [5]. He also noted that some plants had apparently worldwide distributions, that species pairs are to be found in Europe and North America and that some taxa are found in both the north and the south temperate regions (what we now call **bipolar distributions**). Finally, he realized that other plants have strangely 'disjunct' distributions (see Chapter 2) in locations that are widely separated from one another, such as the Proteas of southern Africa and Australia/ Tasmania. Candolle also commented on Forster's contributions to island biogeography.

All in all, Candolle made a massive and varied intellectual contribution to the botany of the early

19th century. However, he did not provide any world maps to illustrate these concepts, and most of the maps that botanists published in the later 19th century, and even in the 20th century, continued to be primarily 'vegetation maps' – maps of the relationships of vegetation to temperature and climate. So, even though the Danish botanist Joakim Schouw was the first to classify the world's flora and show the results on maps [6], these were mainly the distribution maps of particular groups of plants, rather than maps of regional floras. Grisebach's more detailed, coloured map of 1866 was similarly a vegetation map. So all of these maps were primarily concerned with ecological biogeography, rather than with systematic studies of the distribution of organisms, which would have demanded an historical explanation. But then, it was only after biologists had become convinced of the reality of evolution that they could start to integrate into their thinking the consequences of a fourth dimension – time.

Evolution – a Flawed and Dangerous Idea!

During the late 18th century, much of the leading work on biological and geological subjects had been carried out in areas of Europe that we now call Germany, but the French Revolution of 1789 led to a flowering of French science. To some extent, this was because the power of the Church, with its conservative influence on the generation and acceptance of new ideas, had been decisively broken. But the new government also carried out a complete reorganization of French science, liberally supported by the state and centred on the new National Museum of Natural History, which became a powerhouse of ideas and debate in Europe. One of those employed in this new museum was Jean-Baptiste Lamarck. As an older worker, he had been brought up to believe that there was some underlying pattern and structure to every aspect of the physical and biological world – a mind-set common among many 18th-century inquirers into the phenomena of nature. It should therefore be possible to recognize a 'scale of beings' in which different groups of organisms could be allocated to 'lower' or 'higher' places according to the level of 'perfection' of their organization – with, of course,

human beings at the apex of the resulting structure! In 1802, Lamarck suggested that the 'lower' organisms might also be found earlier in time and that they might gradually change into the 'higher' forms, due to an 'inherent tendency of life to improve itself' [7]. So there was no need to suggest that fossil organisms were in reality extinct, for it was possible that they had evolved into different and perhaps still-living descendants.

All of this was strenuously opposed by one of the new, young appointees in the museum, the great Georges Cuvier, who founded the science of comparative anatomy. Cuvier used this new branch of science to prove that such great fossil mammals as the mammoths of Europe and North America and the giant ground sloth of South America, as well as many others, belonged to quite different species from those of today and were extinct [8]. But, he believed, his detailed anatomical studies showed that even these creatures had been thoroughly and stably adapted to their environment. Their extinction must therefore have been due to a sudden catastrophic change in their environment. So, to Cuvier, Lamarck's theory of continual transformation was deeply unacceptable, for its suggestion that organisms were flexible and changeable challenged his own conviction that they were, on the contrary, irrevocably adapted to their existing environment. Cuvier was therefore opposed to Lamarck's views because they cast doubt on his belief in extinction (which was, perhaps, understandable). But this unfortunately also led him to reject the whole idea of evolution that Lamarck had championed – so throwing out the baby of evolution with the bathwater of extinction.

It is always very convenient if, in an argument, your opponent's views are championed by someone else of lesser ability. Lamarck's ideas were supported by another worker in the museum, Geoffroy St Hilaire. Over the years 1818–1828, Geoffroy suggested evolutionary homologies and links between such widely different animals as fish and cephalopods (octopus, squid, etc.) [9], but his ideas were ridiculed by other zoologists. Similarly, his supposed evolutionary sequences of fossils placed them in an order that was contradicted by the sequence of the rocks in which they were found. So it was easy for Cuvier to make a devastating attack on Geoffroy, and this had the effect of also discrediting Lamarck and the whole idea of evolution. In England, the case for evolution was further damaged in 1844, when the Scottish journalist Robert Chambers published a book, *Vestiges of the Natural History of Creation*, which contained astonishingly ignorant ideas. Chambers suggested, for example, that the bony armour of early fossil fishes was comparable to the external skeleton of arthropods (lobsters, crabs, insects, etc.) and that the fish might therefore have evolved from them. The progressively more detailed fossil record that was by then being revealed also gave no hint or indication that the major groups of organisms, traced back in time, converged towards a common, ancient ancestor. The fact that such people as Geoffroy and Chambers supported the idea of evolution unfortunately gave the impression that it was associated with the lunatic fringe of science. And, by now, Lamarck's explanation of evolution as due to an 'inherent tendency' seemed dreadfully old-fashioned.

When the geologist Robert Jameson translated Cuvier's ideas into English in 1813, he added notes suggesting that the most recent of Cuvier's continent-wide catastrophes could be interpreted as the biblical Flood. But Cuvier himself, and other scientists working in post-revolutionary France, accepted that science and religion should not interfere in each other's affairs. Matters were very different in England. There, the Church of England had become closely integrated into the power structure of a still-hierarchical Establishment, and entry to the universities (and so to the professions) was barred to non-Protestants. So both the authorities of the state (monarchy, aristocracy and wealthy landowners) and those of the Church (bishops and comfortable clergy) felt themselves threatened by the new-model social order of France, which they saw as encouraging a rising tide of atheism, republicanism and revolution. For, in the first half of the 19th century, English society was undergoing fundamental changes, fuelled by unemployment resulting from the end of the Napoleonic Wars and by the Industrial Revolution, which was driving people from the land and into overcrowded cities. In this conflict, the new ideas of evolution became a weapon that the rising middle class used in their attempts to gain entry to the universities and access to the professions and financial security. In response, to defend their own positions, the establishment portrayed evolution as atheistic, or even heretical.

Enter Darwin – and Wallace

So, in the early 19th century, evolution was seen as a slightly disreputable idea that also had links with a dangerously anarchic approach to the structure of society. It is therefore not surprising that the young Charles Darwin was cautious, secretive and reluctant to publish his ideas when he began to suspect that the problems he found in trying to interpret the patterns of life could only be explained by invoking evolution. He was the son of a fairly wealthy country doctor, whose father had been an atheist who believed in evolution – so the family was not exactly mainstream. As a student at Cambridge, Darwin had become interested in geology and natural history, and in 1831 he was invited to join the crew of a government ship, *HMS Beagle*, to act as a companion to the captain and also as a naturalist for what became a 6-year voyage to survey the coasts of South America [10]. Several experiences during this long voyage led him to wonder whether the idea of evolution might not, after all, contain some truth.

On the Galápagos Islands in the Pacific Ocean, isolated from South America by 960 km of sea, Darwin noticed that the mockingbirds on three islands were different from one another, suggesting that they had independently become different varieties on each island. He was also told that the giant tortoises of the different islands had differently shaped shells. Darwin also noticed great flocks of finches, with a variety of sizes of beaks, but they all fed together, and he couldn't make up his mind whether there were any different varieties. (Only later, when Darwin's collections were studied back in England by the ornithologist John Gould, was it realized that there were 13 different species of finch in the islands.) All of this suggested that species were not, perhaps, quite as unchanging as was then assumed. Equally disturbing were the fossils that Darwin had found in South America. The sloth, armadillo and guanaco (the wild ancestor of the domesticated llama) were represented by fossils that were larger than the living forms, but were clearly very similar to them. Again, the idea that the living species were descended from the fossil species that had existed in the same part of the world was a straightforward explanation, but one that contradicted the view that each species was a fixed, unchanging product of creation and had no blood relationship with any other species.

As explained, Darwin was not the first to suggest that organisms were related to one another by evolution; the British worker Alfred Russel Wallace was thinking along exactly the same lines. (In fact, Wallace was the first to realize and publish the significant fact that closely related species were often also found close to one another geographically, with the clear implication that the two were linked by an evolutionary process.) In the end, it was the receipt of a letter from Wallace, then working in the East Indies, that stimulated Darwin to finalize and publish his ideas after many years of agonizing over its possible hostile reception by the vociferously antievolutionary sections of British society. (It is interesting to note that, in the case of both workers, it was observation of the patterns of distribution of individual species of animals, i.e. their biogeography, that led them to consider the possibility of evolution.) Their great discovery was to deduce the driving mechanism of evolution – natural selection.

Any pair of animals or plants produces far more offspring than would be needed simply to replace that pair. There must, therefore, be competition for survival among the offspring. Furthermore, these offspring are not identical to one another, but vary slightly in their characteristics. Inevitably, some of these variations will prove to be better suited to the mode of life of the organism than others. The offspring that have these favourable characteristics will then have a natural advantage in the competition of life and will tend to survive at the expense of their less fortunate relatives. By their survival, and eventual mating, this process of natural selection will lead to the persistence of these favourable characteristics into the next generation. (More detail on how this takes place is given in Chapter 6.)

The idea of natural selection was announced by short papers from both Darwin and Wallace, read at a meeting of the Linnean Society of London on 30 June 1858; and Darwin quickly went on to publish his great book the next year [11]. There can be no doubt that Darwin has to share with Wallace the credit for identifying natural selection as the mechanism of evolution and identifying the patterns of biogeography as evidence for evolution. However, the lion's share of the credit for the almost immediate acceptance of the reality of evolution has to be given to Darwin and his book *On the Origin of Species*. For Darwin had spent the 40 years after his return from the voyage of the

Beagle in detailed research on many other areas of biology that provided evidence for evolution (see Box 6.3), and published this research in 19 books and hundreds of scientific papers. The essentials of this work were given in his great book (which sold out immediately on publication and had to be reprinted twice in its first year) and were far more convincing in their variety and detail than the short papers read to the Linnean Society.

Darwin's theory of natural selection was extremely logical and persuasive. His studies on the ways in which animal breeders had been able to modify the anatomical and behavioural characteristics of dogs and pigeons provided a neat parallel to what he believed had happened in nature over long periods of time, and were even more convincing. But, said his critics, all these different breeds of dog or pigeon were still able to breed with one another, which did not support Darwin's suggestion that this was the way in which new species could appear. Nor could Darwin provide any explanation of precisely how the different characteristics were controlled and passed from generation to generation. In fact, the outlines of the ways in which all this took place had been discovered by the Austrian monk Gregor Mendel in 1866, but his work remained unnoticed until the beginning of the next century. So, our modern science of genetics was still a closed book. Also, Darwin did not understand the nature of species. It was generally assumed at that time that each species was innately stable and resisted innovation – which would have impeded the action of natural selection in trying to alter its characteristics. In fact, we now know that the continual appearance of changed characters or 'mutations' (see Chapter 6) would quickly alter the nature of any species, and it is only the continual action of natural selection in weeding out most of these that gives the species the appearance of unchanging stability.

Another problem for Darwin was that most people believed that the Earth was only a few thousand years old. This was partly because some theologians considered that passages in the Bible could be interpreted as indicating that it had only been created some 8000 years ago – and, perhaps more fundamentally, also because few people could even imagine the enormous periods of time that were in fact required for evolution to take place. However, the British geologist Charles Lyell argued that many lines of evidence suggested that the Earth must be many millions of years old [12]. These included the evidence that sea levels had changed greatly over time, the presence of marine fossils at high levels in the mountains, the presence of tropical deposits such as coals or desert sandstones in what are now temperate regions and, even more dramatically, the time required to raise such great mountain chains as the Himalayas, Rockies or Andes. But this argument was weakened by the work of the physicist J.J. Thompson who, basing his work on calculations on estimates of the rate of cooling of the Earth from an original molten state, eventually concluded that it was less than 10 000 years old. He was unaware, of course, of the fact that much of the Earth's continuing warmth comes from radioactivity, for this was only discovered in the 20th century, leading to the eventual realization that the Earth is several billion years old. So, like any scientist, Darwin was a child of his time, unaware of future discoveries that might have explained his difficulties.

Despite these difficulties, the concept of evolution, and of natural selection as its mechanism, was very quickly accepted and is now a part of the basic philosophy of biological science. Just as the theory of plate tectonics is the central paradigm of the earth sciences (see Chapter 5), so the theory of evolution by natural selection is the central paradigm of the biological sciences. Biogeography provides a striking example of the concordance of the implications of these two paradigms. For example, the dates that plate tectonics theory indicates for the different islands in the Hawaiian chain are similar to those that evolutionary studies indicate for their animals and plants. The way in which biogeography provides interlocking support for these two paradigms is overwhelming evidence for the correctness of each and gives it a unique position in the natural sciences.

World Maps: Biogeographical Regions of Plants and Animals

Thanks to Darwin and Wallace, then, the process that was responsible for the living world's reactions to changes in the physical world was at last understood and accepted. Its mechanism (genetics) was yet to be identified, and it would take another century before the mechanisms of the geological process responsible for those changes were discovered. Nevertheless, it was now clear that some of the differences between the floras and faunas of the separate continents might have resulted from their having had separate evolutionary histories. The German botanist Adolf Engler (1879) was the first to make a world map (Figure 1.1) showing the limits

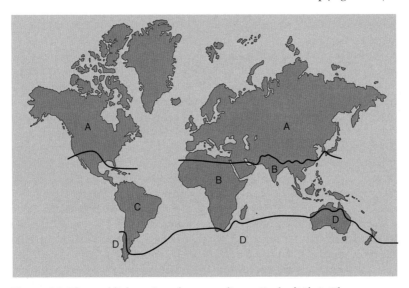

Figure 1.1 The world's botanic realms according to Engler [13]. A, The Northern Extratropical Realm; B, the Palaeotropical Realm, stretching from Africa to the East Indies; C, the South American Realm; and D, the Old Ocean Realm, stretching from coastal Chile via southernmost Africa, and the islands of the South Atlantic and Indian oceans, to Australia and part of New Zealand.

of distribution of distinct regional floras [13] – although his map also shows the different types of vegetation in each of his major areas. He identified four major floral regions, or 'realms', in the world, and attempted to trace the history of each of these back into what we now call the Miocene Epoch of the Tertiary Period, about 25 million years ago (see Figure 5.5). He also noted some of the plant families or genera that are characteristic or dominant in each realm. He had also read the work of the British botanist Joseph Hooker (discussed further in this chapter), who had found many similarities between the floras of the continents and islands of the Southern Hemisphere, and had suggested that these might be explained partly by the dispersal of floating seeds. This led Engler to distinguish what he called an *Old Ocean Realm*. Apart from comparatively minor modifications [14–16], the system of plant regions accepted today (Figure 1.2a) is very similar to that of Engler – although no one has yet

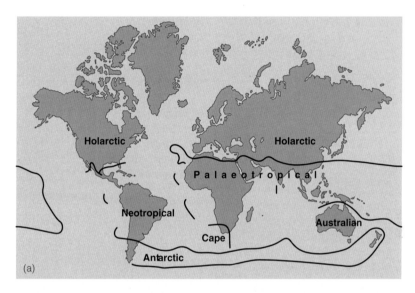

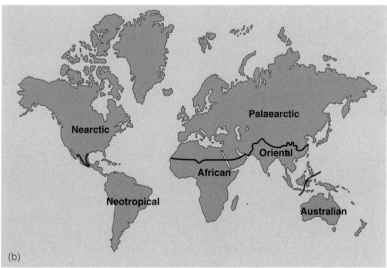

Figure 1.2 (a) Floral kingdoms, according to Good [15] and Takhtajan [16]. (b) Zoogeographical regions, according to Sclater [20] and Wallace [22]. From Cox [17]. (Reproduced with permission of John Wiley & Sons.)

provided any systematic comparison and contrast of the composition of the floras of these different realms [17]. He was also surprisingly perceptive in realizing that, scattered over the islands and lands of the southernmost part of the world, lay the remains of a single flora, which he called the *Ancient Ocean Flora*. (It was over 80 years before acceptance of the movement and splitting of continents at last explained this very surprising pattern of distribution.)

Zoogeography, too, had been developing from the early 19th century onward, but with a different emphasis. Because such dominant groups as the birds and mammals are warm-blooded, they are largely insulated from the surrounding environmental conditions and are often found in a great variety of environments. So, unlike the plants, they do not show a close correlation to local ecology. Even such early zoogeographers as Prichard in 1826 [18] and Swainson in 1835 [19] were therefore free to concern themselves with distribution at the world level, and they recognized six regions that corresponded with the continents. This was first formalized in 1858 by the British ornithologist Philip Sclater [20], who based his system on the distribution of the most successful group of birds, the passerines or 'songbirds', because he thought that they were less adept than other birds at spreading from place to place. He believed that all species had been created within the area in which they are found today, so that comparison of the different local bird faunas might identify where the centres of creation might have been. (He even thought that these might reveal where the different races of human being had been created.) As was normal in those days, he gave classical names to the six continental areas that he identified but, even though he listed or described the areas included in each region, he gave no maps to illustrate his views.

Alfred Wallace made his living by collecting bird skins, butterflies and beetles in the East Indies, and selling them to naturalists. (He had already made extensive collections in the Amazon rainforest.) These travels and collections had led him to become, like Darwin, interested in their patterns of distribution. He immediately accepted Sclater's scheme, including his names for the regions, and expanded it to include the distribution of mammals and other vertebrates (Figure 1.2b). Because of the pattern of barriers of ocean, desert and mountain between the zoogeographical regions, the only area where there is a significant overlap between the faunas of adjacent regions is precisely where Wallace was working: in the East Indies chain of islands between Asia and Australia. Wallace became fascinated by the unexpectedly abrupt north–south demarcation line that separated the more western islands, which had an overwhelmingly Oriental fauna, from those to the east that were, equally overwhelmingly, Australian. His map and the 'Line' that has been named after him have been largely accepted by zoogeographers ever since (cf. Figure 11.9).

Although he should always be remembered as the joint discoverer of natural selection, in many ways Wallace's greater claim to fame is as a profound thinker and contributor to the fundamentals of zoogeography. His books *The Malay Archipelago*, *The Geographical Distribution of Animals* and *Island Life* [21–23] were read by many people and were very influential. Wallace identified or commented on many aspects of biogeography that still occupy us today. These include the effects of climate (especially the most recent changes), extinctions, dispersal, competition, predation and adaptive radiation; the need to be knowledgeable about past faunas, fossils and stratigraphy, as well as about those of today; many aspects of island biogeography (see further in this chapter); and the possibility that the distributions of organisms might indicate past migrations over still-existing or even now-vanished land connections. He and Buffon were truly the giants in the development of zoogeography.

Getting around the World

The final acceptance of evolution gave a new importance to biogeography and posed new problems, which persisted over the century that elapsed before the mechanics of its geological counterpart, continental drift or plate tectonics, were revealed. If Darwin (and Wallace) were correct, new species arose in a particular place and dispersed from there over the pattern of geography that we see today, except where this had become modified by comparatively minor changes in climate or sea level. This concept of **dispersalism** therefore assumed that, where a taxon or two related taxa are found on either side of a barrier to their spread, this is because they had been able to cross that barrier after it formed.

But this was inadequate to explain many of the facts of world biogeography, especially some that were revealed by the rapidly expanding knowledge of patterns of distribution in the past. One might be able to invoke floating islands of vegetation, mud on the feet of birds or violent winds to explain dispersal between islands or otherwise isolated locations today. But even Darwin's old friend, the botanist Joseph Hooker, who had travelled and collected widely in the Southern Hemisphere continents and islands, found these explanations quite unconvincing. Hooker became one of a group who instead believed that the many similarities between the plants and animals of the separate southern continents, and of India, could only be explained by their having once been connected. This could have been by narrow land bridges, or by wider tracts of dry land across the present South Atlantic and Indian Oceans, that had later become submerged. But even by the end of the 19th century, this had been dismissed as a fanciful explanation for which there was no geological evidence.

The past, too, was providing more and more examples of puzzling patterns of distribution. For example, 300 million years ago, the plant *Glossopteris* existed in Africa, Australia, Antarctica, southern South America and, most surprisingly of all, India (Figure 1.3). A linkage between all these areas at that time was also suggested by the fact that all of them contained deposits of coal and traces of a major glaciation, contemporary in all those continents. Such facts, together with similarities in the outline of the Atlantic coasts of the Americas, Europe and Africa, and comparison of the detailed stratigraphy of the rocks along these coastlines, were what led the German meteorologist Alfred Wegener to present his theory of continental drift in 1912 [24]. Wegener suggested that all of today's continents had originally been part of a single supercontinent, **Pangaea** (Figure 1.4). But, as noted at the beginning of this chapter, in the absence of any known mechanism that might split and move whole continents, his suggestions were not accepted by either geologists or biologists. Biogeographers were instead driven back on progressively more desperate defences of dispersal as the only possible explanation of the patterns of distribution.

This was particularly true of what the botanist Leon Croizat called the New York School of Zoogeographers, a group of vertebrate zoologists founded by Walter Matthew. In his 1924 paper 'Climate and Evolution' [25], Matthew suggested

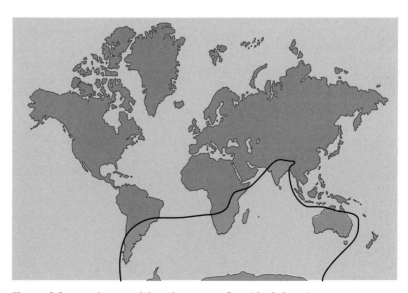

Figure 1.3 Distribution of the *Glossopteris* flora (shaded area).

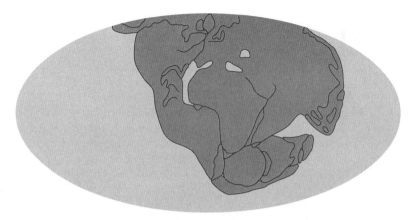

Figure 1.4 How today's landmasses were originally linked together to form a single supercontinent, Pangaea, according to Wegener. (Compare this with Figure 10.1 to see the modern, plate tectonic reconstruction of Pangaea.)

that all the patterns of mammal distribution could be explained if the different groups had originated in the challenging environments of the Northern Hemisphere. From there, they had dispersed across the intermittently open Bering land bridge between Asia and North America, and southward to the various continents of the Southern Hemisphere. Probably the most influential later member of the New York School was George Simpson, who not only wrote many papers on mammalian palaeontology and biogeography [26], but also several important books on evolutionary theory. He had no doubt that the patterns of distribution of mammals could be explained perfectly well without invoking continental drift. (This was largely true, for the radiations of the families of living mammals took place well after the fragmentation of Pangaea; only the presence of marsupials, but not placentals, in Australia provided an obvious problem.) Together with such other workers as the herpetologist Karl Schmidt, George Myers (who worked on freshwater fishes) and the zoogeographer Philip Darlington (who in 1957 wrote a major and influential textbook on zoogeography [27]), they provided a powerful and united body of opinion that was wholly opposed to the idea of continental drift and equally fervently supportive of the idea of dispersal.

Some idea of the lengths to which these workers were driven in trying to explain the facts of distribution is shown by Darlington's statement, in discussing the distribution of *Glossopteris*: 'The plants may have been dispersed partly by wind,

and, since they were frequently associated with glaciation, they may have been carried by floating ice, too. I do not pretend to know how they really did disperse, but their distribution is not good evidence of continuity of land' [28, p. 193]. Surely, one might think, this distribution, scattered across continents separated by thousands of miles of ocean (Figure 1.3), *was* evidence of continuity of land, but Darlington gave no reason why he thought that it was not *good* evidence.

It is not surprising that such attitudes provoked opposition, and this surfaced most strongly in the person of Leon Croizat. Born in Italy in 1894, Croizat was overwhelmed by the effects of fascism, World War I (1914–1918) and the Great Depression. After spending periods as an artist in New York and Paris, he became a botanist, at first in New York and eventually in Venezuela, where he lived from 1947 until his death in 1982. Croizat rightly felt that the dispersalists were going to extremes in their refusal to countenance any other explanation for the patterns of distribution that one could see in the world today, such as the widely disjunct distributions of many taxa, especially in the Pacific and Indian oceans. He amassed a vast array of distributional data, representing each biogeographical pattern as a line, or **track**, connecting its known areas of distribution. He found that the tracks of many taxa, belonging to a wide variety of organisms, could be combined to form a **generalized track** that connected different regions of the world. These generalized tracks (Figure 1.5) did not conform to

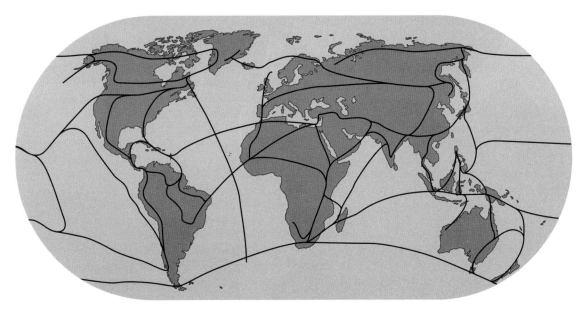

Figure 1.5 Croizat studied the distribution patterns of many unrelated taxa, and for each he drew lines or 'tracks' on the map linking the areas in which they are found. In many cases, these lines were similar enough in position to be combined as 'generalized tracks', shown here.

what might have been expected if these organisms had evolved in a limited area and had dispersed from there over the modern pattern of geography, as other biologists then believed. Croizat felt that it would be surprising if any single taxon had managed by chance to cross the intervening gaps, and incredible that a considerable variety, with different ecologies and methods of distribution, should have been able to do so. His method, which he called **panbiogeography**, argued that all of the areas connected by one of these tracks had originally formed a single, continuous area that was inhabited by the groups concerned. This theory therefore rejected both the concept of origin in a limited area and that of dispersal between subsidiary locations within that area. However, having rejected the facile use of dispersal as an explanation of each and every example of a trans-barrier pattern of distribution, Croizat then went to the other extreme and completely rejected dispersal in any shape or form – although, confusingly, he used the word 'dispersal' in a different sense, as describing the pattern of distribution of a taxon.

Instead, Croizat believed that the organisms had *always* occupied the areas where we now see them, together with the intervening areas, and that they had colonized all these areas by slow spread over continuous land. So, the flora of such isolated island chains as the Hawaiian Islands, or the scattered patterns of distribution of plants along the Pacific margins of North and South America, had arisen because land had once linked all these areas, or because islands containing the plants had moved to fuse with the mainland. Croizat believed that any barriers, such as mountains or oceans, that exist today within the pattern of distribution of the taxa had appeared *after* that pattern had come into existence, so that these taxa had never needed to cross them – a concept that came to be known as **vicariance**. To this extent, Croizat's theorizing anticipated the way in which plate tectonics would provide a geological contribution to the spread of organisms.

Croizat published his ideas in the 1950s and 1960s, his major presentation being his 1958 book *Panbiogeography* [29] – but little attention was paid to his work. This was partly because of the dominance of the New York School, with its acceptance of dispersalism, but also because of several weaknesses in Croizat's own work. He concentrated on the patterns of distribution of living organisms, was scornful of the significance of the fossil record and paid little attention to the effects of changes in geography or climate. In addition, because the idea of the stability of modern geography seemed to have successfully weathered Wegener's heresies, Croizat's theories of the movement of islands or of massive extensions of land into the Pacific and Atlantic cast him into that same mould – of passionate amateurs. And, even after the theory of plate tectonics had become well documented and widely accepted, Croizat refused to accept it and never integrated it into his methodology. He also became increasingly embittered by the way in which his work was largely ignored.

Ironically, the recognition of some of Croizat's perceptions and methods began in New York, where there arose a new generation of biogeographers who had not been brought up under the influence of the old New York School. For Croizat was correct, and ahead of his time, in believing that in many cases speciation had taken place *after* a barrier had emerged within an existing area of distribution of a taxon. But, unfortunately, the pendulum now swung to the opposite extreme – instead of 'Dispersal explains everything', their attitude was 'Vicariance explains everything', and dispersal is merely random noise in the system. Even more unfortunately, Croizat's supporters also inherited his confrontational approach, and the argument between the supporters of dispersal and the supporters of vicariance became increasingly bitter. (One problem that underlay this whole argument may have been that the available evidence was, in the majority of cases, inadequate for anyone to be able to prove whether dispersal or vicariance had been the cause. Although biogeographers were only too aware of this, they were nevertheless desperate to find some method, whether or not it was perfect, to explain the patterns of life that so intrigued them. Quite often, those who shout the loudest are those who are least secure of their case, and are trying to silence their own doubts as well as those of their opponents!)

Perhaps the most enthusiastic of Croizat's supporters was a group of biogeographers, most of whom worked in New Zealand, where the origins of the fauna and flora provide particularly difficult problems. These panbiogeographers accepted his generalized tracks running across the ocean basins, referring to them as **ocean baselines** (Figure 1.6), and viewed them as more useful and important than the conventional system of continental zoogeographical and plant geographical regions. Their methodology also considered the area where a taxon is most diverse in numbers, genotypes or morphology as the centre from which the track for that particular taxon had radiated – a dangerous assumption. The author of this chapter (Barry Cox) has reviewed the history and development of the New Zealand school of panbiogeographers [30], one of whom (John Grehan) responded to these criticisms [31]. More recently, the Mexican biogeographer Juan Morrone has written a defence of the concept of track analysis [32].

The long and bitter argument about dispersal versus vicariance only ended with the appearance of new molecular techniques of establishing the patterns of relationship of organisms and the time that has elapsed since the origin of each lineage. This now allows us to compare the timing of biological events and of the geological or climatic events that might have been associated with them. The result has been, rather ironically, to show the prevalence of dispersal to an extent far greater than the most optimistic dreams of the dispersalists!

The Origins of Modern Historical Biogeography

A century after Darwin had published his theory, acceptance of his ideas had revolutionized approaches to nearly every aspect of the biological sciences. These ideas had implicitly suggested that the contents of each biogeographical unit might have changed and diversified through time, and discoveries of the fossil record had in many cases documented these changes. But as long as the Earth's geography was assumed to have been stable, problems remained in the explanation of at least some of the patterns of disjunct distribution. Some of these could be explained by patterns of extinction. For example, the presence of fossil camelids and tapirs in North America and Asia showed that the disjunct distribution of these groups today, in South America and South-East Asia, did not have to be explained by some theory of the rafting of early members of these groups across the Pacific. However, the patterns of distribution shown by the ancient *Glossopteris* flora, or by the Antarctic Floral Kingdom today, still provided a major puzzle. How

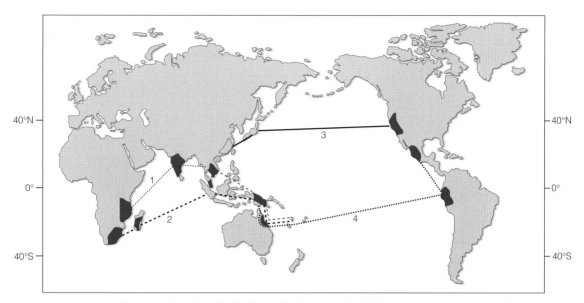

Figure 1.6 Craw's panbiogeographical method. The tracks link areas (dark blue) where related taxa are found. Tracks 1 and 2 are examples of an Indian Ocean baseline, and tracks 3 and 4 similarly are examples of a Pacific Ocean baseline. Adapted from Craw [58].

could organisms have dispersed across oceans to reach these scattered locations? As already mentioned, Wegener's theory of continental drift had provided an explanation to this conundrum early in the 20th century, but he had been unable to suggest any convincing mechanism that might have caused the movement and splitting of huge masses of land. As a result, his theory had been rejected by most geologists, and most biogeographers had reluctantly felt obliged to follow their lead. It was only in the 1960s that strong new evidence of the mechanism for Wegener's theory, now renamed *plate tectonics*, led to the acceptance of the reality of this phenomenon (see Chapter 5). It was only now that geologists were able to provide a series of palaeogeographic maps that showed, from the Silurian Period onward, the changing patterns of association of the various tectonic plates [33].

Until now, biogeographers had tried to analyse the biogeography of the past according to the different geological periods – the life of the Carboniferous, Permian and so on. But, as the new maps showed, there were major changes in the patterns of land and ocean within these periods of geological time. There would therefore have been corresponding changes in the likely biogeographical patterns, dooming to failure any attempt to detect a single pattern of biogeography for the time in question. However, the new maps also made it possible to identify stretches of time (*not* corresponding to the geological periods) within which the geographical patterns had remained constant. As the British biogeographer Barry Cox realized [34], these maps therefore provided the potential basis for appropriate biogeographical analysis, if to them one added the patterns of the shallow 'epicontinental' seas that lie on the edges of the continental plates – for these, too, are biological barriers. All that a palaeobiogeographer then had to do was to summate the faunas and floras from every locality within each of the resulting palaeocontinents. For the first time, the results made perfect sense, with elements of these faunas and floras showing clear evidence of endemicity (see Chapter 10) (Figure 1.7).

The theory of plate tectonics was soon accepted by nearly all biogeographers, but, perhaps unsurprisingly, some of the older ones held out against it. For example, Philip Darlington [28] rejected the idea of a general union of southern continents into a single supercontinent. He felt that such a geography would not have provided enough adjacent water for the development of the ice sheets that appeared to have covered it about 300 million years ago.

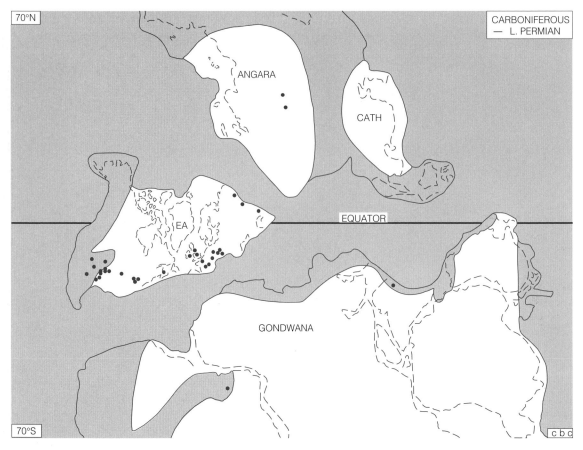

Figure 1.7 Palaeogeographical map of the Carboniferous–Lower Permian period of time, as reconstructed in 1973. The seas and oceans are tinted blue. The small dark blue circles show the positions of all the localities containing early terrestrial vertebrates. The one indicated in northern South America was later shown to belong to a later period of time, whereas those in northern India and in Siberia are doubtful fragments. The map thus strongly suggests that the earliest land vertebrates evolved in Euramerica. The four different floras recognized by palaeobotanists (the Angaran, Cathaysian (CATH), Euramerican (EA) and Gondwana floras) are also found to have lived on different palaeocontinents. This explains the previously puzzling fact that the *Glossopteris* flora, found in Gondwana, is found scattered over five of today's continents. From Cox [34]. (Reproduced with permission of John Wiley & Sons.)

At long last, biogeographers could build up a coherent, increasingly detailed set of pictures of the geography of the world over many millions of years. Now they could start to analyse the changing patterns of distribution of living organisms over that period of time and discover the historical roots of the patterns of biogeography that are seen in the world today. The results of this are reviewed in Chapters 10 and 11. But some biogeographers were particularly interested in the more recent past – partly because it encapsulated the origin and

spread of our own species. However, this period of time demanded quite different techniques of investigation, for it included the Ice Ages, with their major, oscillating effects on climate and sea level. In the end, it was found that the most reliable evidence of the general patterns of change in climate could be deduced from oxygen isotope studies of the fossil skeletons of plant microfossils in cores of the sediments in the deep ocean floors. The American Cesare Emiliani was the first, in 1958, to provide reliable temperature curves for the past

700 000 years. But in order to relate these general changes to more local climatic changes on land, biologists had to turn to fossil pollen, a technique pioneered by the Swedish worker Gunnar Erdtmann and the British worker Harry Godwin in the 1930s [35]. The pollen of different species of plant is often easily recognizable and is well preserved in sediments found in peats and lake deposits, so that study of these shows clearly how the vegetation of the area has gradually changed. The results of all of these studies, and their implications for the origins of our own species and of civilization, are dealt with in Chapters 12 and 13.

Biogeographers now had the tools with which, they thought, it should be possible to construct satisfying correlations between the patterns of geography and climate, on the one hand, and of life, on the other hand. But at first this was still disappointingly difficult to achieve, because different taxonomists had different opinions as to the taxonomy (and, therefore, the pattern of evolution) of the organisms involved. Two innovations have transformed this biological problem. The first, known as cladistics (see Chapter 8), provided a rigorous methodology for analysing the patterns of evolutionary relationship between the different members of a group. Still, as long as the characters used in this evaluation were morphological ones, the problems remained, for such characteristics can show convergent or parallel evolution, or may be dependent on one another for functional or developmental reasons. This problem has now been reduced by the development of molecular systematics, which uses more abstract and fundamental characteristics of the organisms, which lie in the detailed molecular make-up of their DNA and proteins (see Chapter 6). This not only provides more confidence in the accuracy of our reconstructions of the patterns of evolutionary divergence of the group under study, but also indicates the times at which the different branching events took place. This in turn allows us to make an informed decision as to whether a particular event was due to vicariance or dispersal (in those cases where the two explanations involve different periods of time).

These two advances have permitted major improvements in our methods of establishing biological relationships and are revolutionizing our understanding of biogeography at all levels (see Chapter 8).

The Development of Ecological Biogeography

As we have seen, ecological biogeography began with the simple observations of men such as Linnaeus, who recorded in what type of environment each plant was found, whereas Forster recognized latitudinal gradients of diversity, later matched by Humboldt's altitudinal gradients, and Candolle pointed out the importance of competition in limiting the distribution of plants. But the full development of this field of inquiry came much later, mainly in the 20th century, for it depended on the rise of modern science with its techniques of experimental, physiological studies. Unlike that of historical biogeography, its history was not complicated by the need to counter the attitudes of antagonistic philosophies or religion, nor by having to wait until data from another field, such as the earth sciences, could be understood. The development of ecological biogeography was, however, strongly dependent on the increasing application of chemical and physical concepts and techniques to the understanding of plant and animal function, and hence distribution. The birth of the science of genetics in the 20th century, leading ultimately to the development of molecular genetics, also expanded the horizons of ecological biogeography.

It was, of course, obvious to the earliest botanists that the distribution of plants was closely linked to climate. In trying to structure the results of this relationship, they could focus either on the demands of the environment on the physiology of the plants, or on the type of vegetation that resulted. Candolle, in 1855, was the first to contribute to this field of inquiry, recognizing three physiologically different types of plant that resulted from their adaptations to different levels of heat and moisture. He called these **megatherms**, **mesotherms** and **microtherms**, which, respectively, required high, moderate and low levels of heat and moisture, and **hekistotherms**, which live in the polar regions. Later, he added **xerophytes**, which can tolerate low levels of moisture.

Botanists soon also started to analyse the effects of the geology of the area in which the plants lived, and the interacting role of climate and of the plants themselves in breaking down the native rocks and converting them into soils of different characteristics.

The American botanist E.W. Hilgard showed, in 1860, how climate and plant life combined to gradually break down the native rock into smaller fragments and provide an increasing component of soil as a product of the biological activity; and the Russian V.V. Dokuchaev analysed the mineralogical and physical attributes of the soils that resulted from the breakdown of different types of rock.

The alternative focus, on the type of vegetation that resulted from the action of the climate, had begun with Engler's map (see Figure 1.1), which had shown the limits of various types of vegetation; but he had used a confusing system of classification. The first clear and simple system of categorizing the different types of vegetation was produced by the German botanists Hermann Wagner and Emil von Sydow in 1888 [36]. Their system was amazingly advanced for its time, for it recognized nine of the 10 categories that are still commonly used, such as tundra, desert, grassland, conifer forest and rainforest; only the Mediterranean type of scrubland was not identified. The many maps and systems produced by various workers since then have contributed different details and variations in emphasis, but have added little to Wagner and Sydow's basic system, although various terms have been coined to describe its elements. The earliest of these, introduced by Clements and Shelford in 1916, were **plant formation** or, with the addition of its animals, a **biome**. Tansley in 1935 [37] added the climatic and soil aspects of the complex, calling it an **ecosystem**, which became the basic unit of ecology. *Biome* has remained the usual term for classification at the macroscale level, but it is used in a variety of ways. If the main emphasis is on vegetation structure, ecophysiology and climate, then biomes can be seen to be the reactions of the living world to these conditions, and the 'same' biome can be found in different continents. If, instead, the emphasis is on the taxonomic or phylogenetic aspect of its plant components, then the biomes become regional, as in Takhtajan's 'floristic regions' [16]. On the whole, the word *biome* is best used in the former, nontaxonomic sense.

Living Together

The rise of ecology as a scientific discipline during the early part of the 20th century led to new approaches to biogeographical studies. Ecophysiology, the study of the ecological implications of plant and animal physiology, played an important role in these developments. The German botanist and plant physiologist Julius von Sachs had injected a strong physiological approach into the debates concerning adaptation to the environment that were prominent at the end of the 19th century [38]. Environmental stresses were seen as limiting factors in plant distribution patterns, and the morphology, anatomy and physiology of plants often reflected their capacity to cope with these stresses. Known as **plant form**, these features were recognized as a more effective way of defining the formations and biomes than any taxonomic or evolutionary system of classification. It was from this line of thinking that the Danish botanist Christen Raunkiaer developed his proposal of **life forms** of plants, based on their means of survival from one growing season to the next (see Chapter 3). A plant's growing points, he argued, are the most sensitive to environmental stress during an unfavourable period (be it cold or dry), and the position in which those growing points are held therefore provides an indication of the degree of stress to which it is exposed. He classified plants according to the height of their growing points above ground (or below ground). Plants growing in the unstressed conditions of the wet tropics could develop forms with their buds high above ground, whereas those in the polar regions or in deserts survived only if their buds were close to the ground or, in the case of the dry lands, below the surface. Annuals were a special case, as they survived an unfavourable period as dormant seeds.

The life form concept has been highly influential in plant geographical studies and generally fits well with observed facts. Plant formations, or biomes, are indeed characterized by the proportions of the different life forms of plants present – what Raunkiaer described in 1934 as the 'biological spectrum' of the vegetation. There are other important adaptations, however, apart from those associated with growing points of plants, or their means of surviving from one year to the next. Evergreen and **deciduous** leaf characters (deciduous plants shed their leaves during a cold or dry season of the year), rooting characteristics, drought and flooding physiology and symbiotic nitrogen fixation are all important aspects of coping with environmental stresses that are unrelated to bud positions.

In the latter part of the 20th century, the concept of **plant functional types** emerged, incorporating and moving beyond that of life forms. It is an approach that actually can be traced back over 2000 years to the work of the Greek botanist Theophrastus around 300 BC, but its use in recent times has drawn strongly upon the idea of **guilds**, a concept borrowed from animal ecology [39]. A guild is a group of animals, not necessarily related taxonomically, which all make use of the same resource or overlap significantly in their environmental requirements. It is a concept that has been used in a rather varied manner, sometimes being applied to organisms that respond in the same way when disturbed or have a particular management system applied. In one respect, all green plants are part of a single guild in that they all obtain energy directly from the sun, but with respect to other resources, such as water, nutrient elements, pollinators, seed dispersal vectors and so on, plants have different ways of coping with their environments. They can thus be classified as different functional types. The concept owes much to the work of Philip Grime, who developed the idea that plants have a range of survival strategies available to them [40]. It is an approach that is proving useful in studies such as those examining the nature of stability and resilience in communities, and is also being used in predicting the outcome of global change on vegetation.

The use of the word **community** has itself generated much debate in ecological biogeography. (Communities and ecosystems are examined in more detail in Chapter 4.) We observe organisms mixed together in groups, or assemblages, whose relative stability suggests that the different species are in an equilibrium, tolerating or perhaps even encouraged by the presence of others – perhaps because the different species may coevolve and adapt to the presence of those others. The American plant ecologist Frederic Clements was the first to suggest, in the early 20th century, that such integrated communities resemble individual organisms in their degree of internal organization, and may similarly behave as units in their patterns of distribution. The community concept was very convenient for biogeographers, as it facilitated the precise classification of vegetation, which is needed for it to be mapped effectively. But the voices of many ecologists were raised against it.

Henry Gleason formally set out the alternative approach in his Individualistic Hypothesis, stating that each species was distributed according to its own ecological requirements, and what we regard as a community is really little more than a chance assemblage of species with compatible ecological tolerances.

Emphasis on the concept of community led to the development of a distinct branch of plant geography, **phytosociology**, in which the plant communities are classified and may be arranged in a hierarchy – which is undoubtedly convenient, but may be unrealistic. Highly detailed systems of plant community classification have been established by using the techniques of phytosociology, pioneered by the botanist J. Braun-Blanquet [41]. The classification of vegetation, rather like the classification of organisms, is based on the idea that relatively sharp lines can be drawn around each defined unit. Field ecologists, however, soon recognized that in the case of vegetation there is usually gradual change from one type to another, leading to gradients along a continuum. Only where there are abrupt changes in the environment does one find sharp boundaries in vegetation. Recent developments in classification have therefore been based on the idea of defined reference points between which there may be a whole range of intermediates. Classification is necessary for the purpose of mapping, but in a situation of continuous variation, any system has to be considered relatively fluid.

Vegetation varies not only in space but also in time, adding to the complexity involved in classification. Increasing amounts of data from fossil pollen grains in lake and peat sediments have shown quite clearly that the distribution patterns of different plant species change quite independently of one another during periods of climatic change. What we currently regard as a community will change in its composition as the environment changes, and assemblages of the past will never be fully repeated. The community, therefore, is a convenient but artificial concept. Changes in assemblages of plants and animals are constantly taking place, and these sometimes follow a predictable pattern. The American botanist Henry Cowles, working in the Chicago region, showed that vegetation develops over the course of time, passing through several different assemblages of plants to finally reach what came to be known as the **climax vegetation** of the region, governed mainly by climate. This climax he regarded as both predictable and stable [42].

The linked concepts of **succession** and **climax**, first developed by Henry Cowles, also have been questioned over the last 100 years. Ecosystems certainly develop over time, and we can make some generalizations about this (see Chapter 4). But it is difficult to show that this somehow involves a predictable process, ending in a predetermined climax that is governed by climatic factors. The climax itself is never static but is in a constant state of change, so the idea of equilibrium has to be more dynamic than Cowles' original concept. An alternative approach is that of *chaos theory*, a concept which assumes that the outcome of a process is highly dependent on the initial conditions. If that is true, the development and outcome of successions may be determined by relatively minor differences in such original conditions as the availability of organisms and soil and weather conditions. So, although climate may in very general terms determine the end point (i.e. the biome) of succession, its detailed composition and nature will be affected by many other factors, including chance.

The ecosystem concept was one of the most influential ideas to emerge from ecological studies in the 20th century, and it has proved extremely useful in biogeographical studies. One of its most valuable features is that it can be applied at any scale, from a rock pool on the shore to the entire Earth. The concept owes much to the work of Raymond Lindemann, who in 1942 put forward a formal account of energy flow in nature. The idea was expanded by the work of American ecologists Howard and Eugene Odum, and named by the British botanist Arthur Tansley. It allows any selected portion of nature to be viewed as an entity, within which energy flows and elements cycle. The concept has recently proved especially valuable when applied on a large scale, where the global circulation of elements can be studied, and the relationships between human and natural processes can be identified. In the early 1960s, the first landscape-scale ecosystem was subjected to monitoring and manipulative management at Hubbard Brook, a forested mountainside in New Hampshire [43]. The budgets of chemical elements were examined in the undisturbed ecosystem, and again following deforestation, thus establishing an experimental approach to the study of large-scale ecosystems.

Ecophysiology, which examines how plants and animals vary in their physiological processes in response to the environment, also developed in new directions in the 20th century. Subtle differences between plants in their photosynthetic systems may provide some species with the capacity to survive in stressful environments. Similarly, animals vary in their capacities to cope with abiotic stresses, such as cold or high altitude, and in their tolerance to human-produced toxins. Thus, the explanation for the presence of a particular species in a given locality (one of the main questions underlying biogeography) may relate closely to the physiological capacity of the species to cope with local environmental stress. This area of research is now entering a new phase as it seeks to understand physiological processes at a molecular level. Molecular biology holds clues to many biogeographical problems and will undoubtedly increasingly be used to advance biogeographical science. Its value in determining taxonomic relationships is casting a new light on many controversial areas of historical biogeography, and its applications in physiological ecology will allow us to increase our understanding of the current distribution patterns of species and of their environmental limitations.

Advances in physiological research, together with ecological and behavioural studies, will help biogeographers to understand more fully the environmental requirements and the niches of organisms within ecosystems. The concept of the niche is complex, broadly being the role played by an organism in its particular setting. A very large number of variables contributes to the niche, including physical factors, chemical factors, food requirements, predation and parasitism, and competition from similar organisms. The concept of the niche was first devised by G.E. Hutchinson in the 1950s and has established itself as a valuable contribution to ecology and biogeography. Perhaps it is best viewed as a kind of conceptual envelope that has many dimensions relating to each requirement of an organism. An organism cannot survive outside these limits, so a full knowledge of those limits could be used to predict its theoretical geographical range [44]. Such knowledge, however, demands the accumulation of very large databases and very complex analyses, and both are becoming increasingly available to researchers as a result of the development of fast and powerful computers.

The application of niche theory in ecological biogeography places emphasis on the environmental factors that control the survival of a species in an area, but it does not really take into account the availability of a species and its dispersal capacity. An alternative approach has been developed called the **neutral theory of biodiversity**, which is based on the idea that the assemblage of species in a site is entirely a matter of chance [45]. The neutral theory claims that the arrival of a species is a stochastic process and that the best predictive models are based on this concept of chance dispersal. Certainly, the part played by chance needs to be taken into account when trying to explain the composition of communities.

Computers were first applied to problems in ecology and biogeography in the 1960s, and their use has expanded to the point where almost all such studies make use of them. Complex statistics, such as multivariate analyses, as used in niche research and community analysis, are vital analytical tools and can be performed rapidly and routinely on computers small enough to be carried in the field. Global positioning systems, using satellites to establish the precise location of an observer on the ground, have also revolutionized the mapping of distribution patterns in remote areas. Technological advances in the last half century must, therefore, be regarded as major steps forward in the history of ecological biogeography.

All of the above avenues of inquiry, using increasingly sophisticated methods of experimentation and analysis, are now used in modern research on ecological biogeography, as explained in Chapters 2, 3 and 4. They are also now used in trying to cope with the problems and questions that arise from humanity's use, and abuse, of an increasingly crowded planet, as explained in Chapter 14.

Marine Biogeography

As explained at the beginning of Chapter 9, the biogeography of the oceans is similar to that of the continents because it is concerned with the biota of vast areas of the surface of the globe. But it is also very different because of the nature of the environment and of the organisms that it contains. We ourselves are terrestrial and air breathing, so the oceans are a far more challenging environment for us to study and census, and they also contain little in the way of obvious demarcations between biogeographical regions or zones. As a result, marine biogeography has been relatively slow to develop, and we still have a great deal to learn about it.

Although earlier naturalists had published limited studies on the faunas of particular regions, the first worldwide survey, based on the distribution of corals and crustaceans, was that of the American scientist James Dana, who later became an eminent geologist. His brief paper, published in 1853, divided the surface waters of the globe into several different zones based on their mean minimum temperature. Three years later, the British zoologist Edward Forbes [46] published the first comprehensive work, recognizing five depth zones and 25 faunal provinces along the coasts of the continents. He was the first to recognize the enormous Indo-Pacific faunal region; stated that the coastal faunas varied according to the nature of the coast, seabed, local currents and depth; and placed the 25 faunal provinces in nine latitudinal belts. Forbes also later published a little volume on the natural history of European seas which made important contributions to marine zoogeography and ecology.

In 1880, the British zoologist Albert Günther published a book on fishes in which he recognized 10 different regions in the distribution of shore fishes, and the German Arnold Ortmann published a similar work based on the distribution of crustaceans such as crabs and lobsters. However, the great landmark in early studies of marine zoogeography was the 1911 *Atlas of Zoogeography* [47] assembled by three British zoologists (John Bartholomew, William Clark and Pery Grimshaw). Their 30 maps of the distributions of fishes were based on the patterns of distribution of 27 families. An influential review and synthesis of all the relevant literature was carried out by the Swedish worker Sven Ekman; it was published initially in German in 1935, followed by an English translation in 1953 [48]. This divided the faunas of the shallow seafloors into seven (mainly climatic) areas, and included the recognition of the unity of the faunas of the Indian and West Pacific oceans, as well as the unity of the faunas of the East Pacific and Atlantic oceans. Ekman suggested that the Panama barrier must formerly have been absent, and that the island-free East Pacific acted as a barrier to the dispersal of organisms; and he commented on the phenomenon of 'bipolarity', where a species is found on either side of the equatorial regions, but not within them.

Ekman's work was extended by the American marine zoologist Jack Briggs in 1974. In his book, *Marine Zoogeography* [49], Briggs used the patterns of endemicity of coastal faunas to identify locations where there appears to be a zone of unusually rapid faunal change, and then used this to distinguish 23 zoogeographical regions. Out in the oceans themselves, our knowledge of the distribution of plankton was greatly increased thanks to the work of the Dutch oceanographer Siebrecht van der Spoel and his co-workers. Their *Comparative Atlas of Zooplankton* [50] included over 130 maps of examples of different types of distribution, categorizing these and the different types of seawaters, the physical properties of the waters and diagrams of the relationships between the faunas of the different oceans.

The greatest of the more recent advances in our knowledge of marine biogeography have come partly from our increasing ability to explore the depths of the sea but also, surprisingly, from our ability to establish sensing and recording satellites in space. Our now-possible journeys into the deepest part of the oceans led to the discovery in 1977 of what is probably the last of the ecosystems of the world to be found, as well as perhaps the weirdest – the strange hydrothermal vent faunas. But, far more importantly, space satellites such as Nimbus have enabled scientists to monitor and record the changing patterns of planktonic life in the oceans continually and comprehensively. This has allowed the British marine biologist Alan Longhurst to propose a system of biomes and provinces within the oceans [51]. These provide for the first time a framework for their regional ecology that integrates their physical features with our increasing knowledge of the annual periodicity in the life, movements and reproduction of the plankton. We shall have great need of such studies in our efforts to comprehend and manage the life of the oceans, which we are increasingly affecting, and we also need it increasingly to feed the rapidly growing population of our planet.

Island Biogeography

As mentioned in this chapter, Georg Forster was the first biologist to remark on some of the particular features of island biogeography; he noted that island floras contain fewer species than the mainland, but that the number of species varies according to the size and ecological diversity of the island. Another early contributor was Candolle, who pointed out that the age, climate and degree of isolation of an island, and whether or not it was volcanic, would also affect the diversity of its flora. Nevertheless, the sheer variety and volume of the works on island biogeography published by Alfred Wallace mark him as the real founder of studies on this subject. His travels around the islands of the East Indies led him to make many profound observations on the reasons for their differing faunas and floras. He realized that the origins of the islands would affect the nature of their biota (i.e. their faunas and floras). Those of islands that had once been a part of a neighbouring continent were likely to contain most of the elements of the fauna and flora that they had inherited from the mainland. In contrast, islands that had arisen independently, as volcanic or coral-atoll islands, would only have organisms that had been able to cross the intervening stretch of sea. Wallace also pointed out that their distance from the mainland, or from one another, would affect the diversity of their biota. Finally, he realized that the diversity of islands made them good natural experiments, in each of which the processes of colonization, extinction and evolution had taken place independently, and so provided abundant material for comparative studies. These fundamental perceptions, as well as the sheer number of his books and research papers, leave no doubt that Alfred Wallace was the father of island biogeography.

But in Wallace's day, and for nearly a century afterward, island biogeography remained the preserve of the naturalist. There were so many islands whose biota needed to be described, for they are fertile breeding grounds for evolutionary innovation. Hundreds of papers were published on the plants of this group of islands, on the animals of that group of islands, or on the distribution of animals or plants over the islands of this or that part of the world. But each group of organisms or plants was treated as unique, with its own special history. Relatively few studies contained any attempts to be analytical and to identify underlying phenomena or processes that might explain some of this myriad diversity. An exception was Philip Darlington's observation in 1943 that larger islands contain a greater number of individuals, and a greater diversity of species, than smaller islands, the species diversity increasing by a factor of 10 for every doubling of island area.

Although science always tries to provide a unifying theory that can integrate a mass of data, it can only produce such an analysis once it has developed the tools to do so. It may be significant that such an integrated, synthetic approach to island biogeography only appeared after sophisticated mathematical techniques had been used to analyse biological phenomena in the new field of population genetics. The ground-breaking little paperback book, The Theory of Island Biogeography [52], published in 1967, was written by two American biologists: the mathematical ecologist Robert MacArthur and the taxonomist–biogeographer Edward Wilson. Other workers, such as the Swedish worker Olof Arrhenius in 1921, and the Americans Eugene Munroe in 1948 and Frank Preston in 1962, had noted the relationship between the area of an island and the number of species that it contains. But MacArthur and Wilson's book was on a quite different level, for it was a sustained (181 pages of text) exploration not only of the basic concepts but also of the ecological evidence and implications of the theory. The book put forward two main suggestions: that the changing, and interrelated, rates of colonization and immigration would eventually lead to an equilibrium between these two processes, and that there is a strong nonlinear correlation between the area of the island and the number of species it contains. The arguments for these ideas were mathematical, with detailed equations and graphs, and the results were very persuasive. Here, at last, it seemed as though biologists would be able to move beyond the raw data to understand the relationships between the simple biological processes. Even more importantly, in a world increasingly worried about the effects of human activity, the concept of an equilibrium of numbers promised to allow predictions as to what would happen under given circumstances, and so to optimize designs for conservation areas.

Over the years that followed the publication of The Theory of Island Biogeography, many papers were written that interpreted individual biota in terms of the theory. These papers were in turn taken as providing such a wide measure of support for the theory that it became almost uncritically accepted as a basic truth. In turn, therefore, results that did not conform to expectations based on the theory were re-examined in search of procedural or logical faults, or for unusual phenomena that might explain the 'anomalous' result. Sometimes they were simply ignored, rather than being seen to cast doubts on the applicability or universality of the theory. Unfortunately, this is far from unique as an example of the way in which new theories can come to so dominate the scientific field that critical evaluation, and even the concept that it may hold some of the truth but not necessarily all of it, is forgotten. This can happen especially either when the field in question has been seen as extremely difficult to interpret, as in the case of this theory, or when the field has been dominated previously by another, equally dominant and intolerant concept, as in the case of the confrontation between the dispersalist and vicariance schools of biogeography.

The story of the rise of the theory and of the later mounting wave of criticism has been told in the fascinating book The Song of the Dodo – Island Biogeography in an Age of Extinction, by the American science writer David Quammen (see the Further Reading list at the end of this chapter). It now seems clear that the theory cannot predict equilibrium levels for the biota of any island and that it is valid only in relating island area to biotic diversity. But MacArthur and Wilson nevertheless revolutionized the study of island biogeography, for they led the way in introducing mathematical techniques, and in providing a standard format for analysis and comparison. As we shall see in this volume, the ecology of island faunas and floras is far more fragile than that of the continents. We therefore greatly need to understand them, for, as a result of their number, diversity and role as natural laboratories for evolutionary change, they contain a high proportion of the biotic diversity that we now desperately need to conserve. For example, although New Guinea contributes only 3% of the world's land area, it contains some 10% of its species of terrestrial organism.

Biogeography Today

As explained in this chapter, the first aspect of biogeography to be recognized by scientists, during the 18th century, was its ecological component. Inevitably, its historical component could become recognized as a field of research only after the scientific community accepted the reality of evolution itself in the middle of the 19th century. Until quite recently, these two approaches to biogeography remained largely independent of one another. Ecologists began with the study of living species or subspecies, and with the factors that control, or alter, their patterns of distribution today. But if they attempted to extend their conclusions into the past, they soon ran into difficulties. This was because they were working at a scale of detail, both in geographical terms and in taxonomic terms, that could not be perceived in the historical record. Only in the study of the comparatively recent past, such as the Ice Ages, could the biogeographer be confident of the ecological preferences of the organisms under study, because they were closely related to those alive today. Only for that period of time was the fossil record sufficiently detailed for the palaeontologist to be confident of the nature and taxonomic level of the changes that were taking place. And only for that period of time were the records of changes in the environment sufficiently detailed, in both time and space, for it to be possible to make plausible correlations between the environmental changes and any biogeographical changes. For the more distant past, it was not possible to establish precisely when any evolutionary changes had taken place, and therefore it was impossible to correlate these to any ecological changes that might have occurred at that time.

The lack of integration between historical biogeography and ecological biogeography continued until the 1990s, when it was rapidly transformed by developments in two areas of study. The development of techniques of analysis of the details of the molecular structure of their genes provided an enormous quantity of data on the molecular characteristics of the organisms (see Chapter 6), showing precisely how they differed from one another. At the same time, as it became easier and cheaper to obtain this data, the number of organisms whose molecular characteristics had been analysed rapidly increased. So great was the quantity of data

that it would have been impossible to make any sense of it, had it not been for the parallel development of techniques of computer analysis. This, together with the use of cladistics, made it possible to work out the patterns of relationship between the different members of a group. But, even more importantly for biogeographers, these techniques made it possible to show precisely when two different lineages had diverged from one another. Now, for the first time, biogeographers could start to correlate the patterns of evolutionary divergence of the organisms and the patterns of change in the environment, over the timescales with which historical biogeographers worked. These advances have also made it possible to discover when related groups that live in different biomes diverged from one another. This in turn allows us to start to work out the history of the assemblage of the different components of the biomes – again, permitting an important linkage between historical and ecological biogeography (see Chapter 8). It now seems likely that the combination of cladistics and molecular analysis will allow us to solve many of the current problems in biogeography. So, today, the old distinction between the two approaches has largely disappeared. At last, it seems that biogeographical research is revealing, with increasing scope and detail, a single, consistent story of the history of the biogeography of the world today.

Ecological biogeography has also raised its level of research from the mainly local to larger scales of analysis, and is developing rapidly both in its establishment of a firm theoretical base and in its practical application to current global problems. In 1995, James H. Brown of the University of New Mexico proposed a new type of research programme, which he termed **macroecology** [53], dealing with ecological questions that demanded large-scale analysis. Range changes in response to climate change, patterns of diversity and analysis of ecological complexity all lend themselves to statistical and mathematical analysis on a larger scale than normally used by experimental ecologists. This is not a new discipline, but a fresh approach to old problems, and one that is increasingly appropriate in days of rapid global change.

During the latter part of the 20th century, it was progressively recognized that the human impact on the landscape was virtually ubiquitous. Throughout the world, landscapes have been so modified that they effectively can be considered *cultural landscapes*, a term that became increasingly used from the 1940s onward [54]. An entirely new discipline of **landscape ecology** appeared, pioneered by Richard Forman of Harvard University [55]. One of the main emphases of the study on the ecology of cultural landscapes was the predominance of fragmentation, as reflected in the title of Forman's classic book, *Land Mosaics*. Landscape ecology needed to examine the ecological consequences of habitat fragmentation on animal and plant populations (Figure 1.8), and so this discipline began to develop in a new direction, leading to the concept of **metapopulations**. A metapopulation consists of a series of separated subpopulations between which genetic exchange may be limited. Clearly, this is an important area of research in the study of gene flow in populations, and hence in the process of evolution. Not just populations but whole communities are fragmented as a result of human agricultural and industrial activities, so one can conceive of metacommunities of organisms

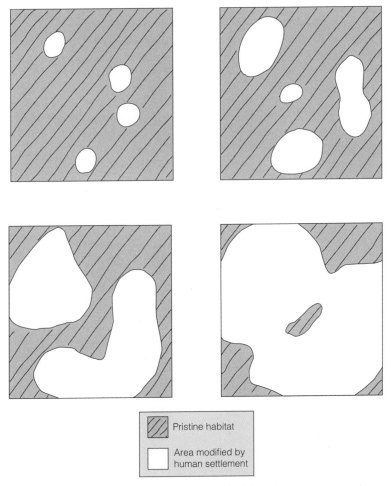

Pristine habitat

Area modified by human settlement

Figure 1.8 The impact of human settlement and disturbance on a natural habitat is progressive, leading to an increasing degree of fragmentation of the original habitat into isolated units. For some species, especially those animals that are of limited mobility and those plants that have limited seed dispersal, this can result in reduced gene flow. Genetic impoverishment can lead to an increased risk of local extinction, and the loss of such a species is not always compensated for by reinvasion.

that can be highly complex in their spatial dynamics [56]. This is precisely the type of problem one can assign to the area of macroecology.

One of the problems presented by habitat fragmentation and the development of metapopulations is the increased danger of genetic isolation and impoverishment, leading to possible extinction. Biogeographical studies thus come into contact with the discipline of wildlife conservation [57]. Many aspects of biogeographical research have a direct bearing on conservation, from the study of biogeochemical cycles and the monitoring of changing ranges of species in response to climate change, to the recording of the spread of invasive organisms and their impact on native populations. Thus, a growing body of work can be classified under the heading of **conservation biogeography** (see Chapter 14).

Biogeography today can be divided into three major areas. The first, and perhaps best understood, is that of the great continental areas, whose varied biota are continually changing as they evolve, compete and spread to new areas or become extinct.

This knowledge is helping us to confront, understand and cope with our need to conserve these biota. Secondly although there is an equally pressing need to conserve the faunas of the oceans, we are still in the process of even making an inventory of these faunas. We are now increasing our understanding of the oceans' basic patterns of biogeography and, especially, the nature of the environmental stimuli in an aquatic environment to which its animal life responds by evolutionary change. Thirdly, the enormous diversity of islands, each with a unique biota and history, provides a huge series of natural laboratories for our efforts to understand the processes of evolutionary change and the interactions between organisms in a developing ecosystem.

Biogeography today is thus developing both in its theoretical aspects and in its practical application to modern environmental problems. The remaining chapters of this book review our knowledge and techniques of analysis in biogeography today and identify those areas in which important new developments seem likely to take place.

Summary

1 Examining the history of biogeography helps us to understand the nature of the subject today and how biogeographers carry out their work within the current framework of the theories and assumptions of science and society.

2 The early biogeographers were inevitably preoccupied with the immense task of documenting the distributions of animals and plants on the surface of the planet, and trying to establish how these vary according to latitude, altitude and climate.

3 Increasing knowledge of the fossil record showed how the world's faunas and floras had undergone great changes, which could only have taken place over long periods of time. It was difficult to reconcile this with the doctrines of the Church that life on Earth was a comparatively recent creation and that species were unchanging. By providing a plausible explanation of how and why these changes might have taken place, Darwin's idea of evolution by natural selection was a major step in getting the general public to accept this very different view of the world's history.

4 However, as long as it was assumed that landmasses had always been stable in their positions, it was still very difficult to understand the patterns of life in the past, and biologists were driven to sometimes bizarre theories to explain these. It was only in the 1960s that the discovery of plate tectonics provided the key to understanding how the Earth's geography, as well as its living cargo, had varied through time.

5 Finally, two advances have transformed the whole field of research into the history of organisms and of their patterns of distribution. The first was the conception and acceptance of cladistic taxonomy. This gave biologists a rigorous system for establishing patterns of relationship that could then be used as a framework onto which patterns of distribution could be applied. Secondly, the use of molecular methods has provided biologists, for the first time, with reliable procedures for the analysis of relationships and the dating of divergences between lineages.

6 Meanwhile, ecological biogeographers were establishing a framework for the description of the varied types of vegetation, and were progressively coming to understand how climate affects the form of plants and how, together with the local geology and soil, it also affects the development and succession of plant communities.

7 Because of the alien nature of its environment, the study of marine biogeography is far more difficult

than that of the land. The general outlines of the distribution of shallow-sea marine faunas were documented in the 18th and early 19th centuries, along with the recognition of faunal zones controlled by latitude and depth. But the huge extent of the open oceans made it difficult to understand the dynamics of the annual changes in their faunas and floras until the recent introduction of satellite-based mapping and modern techniques of marine exploration. Even today, we have much to learn about the organisms of the oceans and about the processes that underlie their biogeography.

8 Islands, too, posed problems for the biogeographer because each is a unique natural 'experiment' in the evolution of floras and faunas. The radical concepts of *The Theory of Island Biogeography*, published by MacArthur and Wilson in 1963, introduced a major attempt to provide a framework for understanding this bewildering mass of data. The subsequent history of attitudes towards the theory, from initial almost uncritical acceptance through subsequent criticism and evaluation, provides a fascinating study of science at work today.

9 The introduction of molecular methods of analysis of the genetic basis of the taxonomy of living organisms, and their application to a large and increasing number of species, together with the development of powerful methods of computer analysis of the resulting mass of data have allowed us to extend our application and understanding of ecological biogeography into the past, blurring the old distinction between ecological and historical biogeography.

10 Biogeography today can be divided into three major areas of research which differ fundamentally in the nature of their environment and of the problems under investigation. These three are continental biogeography, marine biogeography and island biogeography.

Further Reading

Lomolino MV, Sax DF, Brown JH (eds.). *Foundations of Biogeography. Classic Papers with Commentaries*. Sunderland, MA: Sinauer Associates, 2005. (This gives detailed references to and translations of many of the 18th- and 19th-century works referred to in this chapter, as well as reprints and commentaries on later works.)

Quammen D. *The Song of the Dodo – Island Biogeography in an Age of Extinction*. London: Pimlico/Random House, 1996.

References

1. Cox CB. New geological theories and old biogeographical problems. *Journal of Biogeography* 1990; 17: 117–130.

2. Buffon G. *Histoire Naturelle Générale et Particulière*. Paris: 1867.

3. Forster JR. *Observations Made during a Voyage round the World, on Physical Geography, Natural History, and Ethnic Philosophy*. London: G. Robinson, 1778.

4. Humboldt A. de, Bonpland A. *Voyage de Humboldt et Bonpland aux régions équinoxiales du Nouveau Continent*, 30 vols. 1805–1834.

5. Candolle A. de. Essai elementaire de géographie botanique. In: *Dictionnaire des Sciences Naturelles*, vol. 18. Paris: 1820.

6. Schouw JF. *Grundzüge der einer allgemeinen Pflanzengeographie*. Berlin: 1823.

7. Newth DR. Lamarck in 1800: a lecture on the invertebrate animals, and a note on fossils. *Annals of Science* 1952; 8: 290–354. (An English translation of the original 1801 publication in French.)

8. Cuvier G. Extrait d'un ouvrage sur les espèces de quadrupèdes dont on a trouvé les ossemens dans l'intérieur de la terre. *Journal de Physique, de Chimie et d'Histoire naturelle* 1801; 52: 253–267.

9. Geoffroy St H. *Philosophie anatomique*, 2 vols. Paris: 1818–1822.

10. Darwin C. *Journal of the Researches into the Geology and Natural History of Various Countries Visited by H.M.S. Beagle, under the Command of Captain Fitzroy, R.N. from 1832 to 1836*. London: Henry Colburn, 1839.

11. Darwin C. *On the Origin of Species by Natural Selection*. London: John Murray, 1859.

12. Lyell C. *Principles of Geology*, 3 vols. London: 1830–1833.

13. Engler A. *Versuch einer Entwicklungsgeschichte der Pflanzenwelt*, vols 1 and 2. Leipzig: Engelmann, 1879, 1882.

14. Diels L. *Pflanzengeographie*. Leipzig: 1908.

15. Good R. *The Geography of the Flowering Plants*. London: Longman, 1947.

16. Takhtajan A. *Floristic Regions of the World*. Berkeley: University of California Press, 1986. (An English translation of the original 1978 book in Russian.)

17. Cox CB. The biogeographic regions reconsidered. *Journal of Biogeography* 2001; 28: 511–523.

18. Prichard JC. *Researches into the Physical History of Mankind*. London: Sherwood, Gilbert & Piper, 1826.

19. Swainson W. Geographical considerations in relation to the distribution of man and animals. In: Murray H (ed.), *An Encyclopaedia of Geography*. London: Longman, 1836: 247–268.

20. Sclater PL. On the general geographical distribution of the members of the Class Aves. *Journal of the Proceedings of the Linnean Society, Zoology* 1858; 2: 130–145.

21. Wallace AR. *The Malay Archipelago*. New York: Harper, 1869.

22. Wallace AR. *The Geographical Distribution of Animals*. London: Macmillan, 1876.

23. Wallace AR. *Island Life*. London: Macmillan, 1880.

24. Wegener A. *Die Entstehung der Kontinente und Ozeane*. Braunschweig: Vieweg, 1915.

25. Matthew WD. Climate and evolution. *Annals of the New York Academy of Sciences* 1915; 24: 171–218.

26. Simpson GG. *Evolution and Geography*. Eugene: Oregon State System of Higher Education, 1953.

27. Darlington PJ. *Zoogeography*. New York: Wiley, 1957.

28. Darlington PJ. *Biogeography of the Southern End of the World*. Cambridge, MA: Harvard University Press, 1965.

29. Croizat L. *Panbiogeography*. Caracas: Author, 1958.

30. Cox CB. From generalized tracks to ocean basins – how useful is panbiogeography? *Journal of Biogeography* 1998; 25: 813–828.

31. Grehan JH. Panbiogeography from tracks to ocean basins: evolving perspectives. *Journal of Biogeography* 2001; 28: 413–429.

32. Morrone JJ. Track analysis beyond panbiogeography. *Journal of Biogeography* 2015; 42: 413–425.

33. Smith AG, Briden JC, Drewry GE. Phanerozoic world maps. In: Hughes NF (ed.), *Organisms and Continents through Time. Special Paper in Palaeontology* 1973; 12: 1–47.

34. Cox CB. Vertebrate palaeodistributional patterns and continental drift. *Journal of Biogeography* 1974; 1: 75–94.

35. Godwin H. *Fenland: Its Ancient Past and Uncertain Future*. Cambridge: Cambridge University Press, 1978.

36. Wagner H, Sydow E von. *Sydow-Wagners Methodischer Schul Atlas*. Gotha: 1988.

37. Tansley AG. The use and abuse of vegetational concepts and terms. *Ecology* 1935; 16: 284–307.

38. Sachs J von. *Lectures on the Physiology of Plants*. Oxford: Oxford University Press, 1887.

39. Smith TM, Shugart HH, Woodward FI (eds.). *Plant Functional Types*. Cambridge: Cambridge University Press, 1997.

40. Grime JP. *Plant Strategies, Vegetation Processes, and Ecosystem Properties*. Chichester: Wiley, 2001.

41. Braun-Blanquet J. *Plant Sociology: The Study of Plant Communities*. New York: McGraw-Hill, 1932.

42. Moore PD. A never-ending story. *Nature* 2001; 409: 565.

43. Likens GE, Bormann FH, Pierce RS, Eaton JS, Noye MJ. *Biogeochemistry of a Forested Ecosystem*. New York: Springer-Verlag, 1977.

44. Hirzel AH, Le Lay G. Habitat suitability modelling and niche theory. *Journal of Applied Ecology* 2008; 45: 1372–1381.

45. Hubbell SP. *The Unified Neutral Theory of Biodiversity and Biogeography*. Princeton: Princeton University Press, 2001.

46. Forbes E. Map of the distribution of marine life. In: Johnston W, Johnston Ak (eds.), *The Physical Atlas of Natural Phenomena*. Edinburgh: W Johnston & AK Johnston, 1856: plate 31.

47. Bartholomew JG, Clark WE, Grimshaw PH. *Bartholomew's Atlas of Zoogeography*, vol. 5. Edinburgh: Bartholomew, 1911.

48. Ekman S. *Zoogeography of the Sea*. London: Sidgwick & Jackson, 1935.

49. Briggs JC. *Marine Zoogeography*. New York: McGraw-Hill, 1974.

50. van der Spoel S, Heyman RP. *A Comparative Atlas of Zooplankton*. Berlin: Springer, 1983.

51. Longhurst A. *Ecological Geography of the Sea*. New York: Academic Press, 1998.

52. MacArthur RH, Wilson EO. *The Theory of Island Biogeography*. Princeton: Princeton University Press, 1967.

53. Brown JH. *Macroecology*. Chicago: University of Chicago Press, 1995.

54. Birks HH, Birks HJB, Kaland PE, Moe D (eds.). *The Cultural Landscape: Past Present and Future*. Cambridge: Cambridge University Press, 1988.

55. Forman RTT. *Land Mosaics: The Ecology of Landscapes and Regions*. Cambridge: Cambridge University Press, 1995.

56. Holyoak M, Leibold MA, Holt RD. *Metacommunities: Spatial Dynamics and Ecological Communities*. Chicago: University of Chicago Press, 2005.

57. McCullough DR. *Metapopulations and Wildlife Conservation*. Washington, DC: Island Press, 1996.

58. Craw R. Panbiogeography: method and synthesis in biogeography. In: Myers AA, Giller PS (eds.), *Analytical Biogeography*. London: Chapman & Hall, 1988: 405–435.

The Challenge of Existing

Section

1

Patterns of Distribution: Finding a Home

Chapter 2

Biologists operate with the basic unit of the species, and most organisms can be assigned to a particular species. But since evolution is constantly taking place, some species may show further subdivisions or may hybridize with other species. The biogeographer is therefore faced with some problems when studying the ranges of different organisms and explaining them in climatic, geological and historical terms. Physical factors often limit the distribution patterns of species and subspecies, but this is not always the case. No species lives in isolation from other species, so sometimes the causes of range limitation may be due to biological factors, such as competition for food or space, predation or parasitism. The factors that influence the limits of a species can also interact in a complex pattern. Understanding how a species reacts to these factors, however, will prove increasingly important in predicting the biogeographical outcome of global environmental changes in the future.

The living world consists of many organisms, most of which can be arranged into groups that have many features in common, called **species**. Most species are reasonably clearly defined by their appearance, their structure, their physiology and their behaviour, but there is also a great deal of variation within species. Size, colour, feeding preferences and mate choice may all vary between individuals. Species normally restrict their breeding to individuals of the same species, but this is not always the case, and one cannot define a species on this basis. From geese to horses, we are very familiar with hybrid organisms that have intermediate characteristics. The science of classifying organisms is termed **taxonomy**, and since the naming of animals and plants has been going on since the invention of language, taxonomy could be regarded as the most ancient of biological disciplines. Early classification depended entirely on the form of organisms, and, on the whole, this has worked reasonably well as an approach to constructing a taxonomic order. But in the last half century, as we have come to understand more of the operation of genetics, particularly the way in which the structure of DNA determines features of form and physiology, it has become apparent that structural features are not always reliable as an indication of what comprises a species or how species are related to one another. Species can be grouped into higher units, genera, families and orders, in a hierarchical system, which is intended to be based on true evolutionary relationships, forming a natural systematic order. But structural features alone have proved misleading, and taxonomists rely increasingly on genetic studies to understand relationships in the natural world. Molecular genetics [1] have thus opened a new chapter in the development of taxonomic biology and consequently have had considerable impact in the field of biogeography.

One of the aims of biogeography is to understand the causes that underlie the patterns of distribution of organisms on our planet, but such studies must be based on an appreciation of the nature of the species with which one is dealing. Some species, for example, exist in a number of different forms that are sufficiently stable to be termed **subspecies**, and these often have different distribution patterns. A species that exists as a series of

Biogeography: An Ecological and Evolutionary Approach, Ninth Edition. Edited by C. Barry Cox, Peter D. Moore, Richard J. Ladle.
© 2016 John Wiley & Sons, Ltd. Published 2016 by John Wiley & Sons, Ltd.

subspecific forms is termed a **polytypic species**, as opposed to a less variable species that exists in just one form – called a **monotypic species**. As biologists apply genetic analyses more widely, the complex relationships within species are becoming increasingly evident, and such complexity is reflected in distribution patterns.

An example of a polytypic species whose taxonomy has been considerably revised in recent years is the herring gull (*Larus argentatus*). Until just a few years ago, this 'species' was regarded as polytypic with about a dozen subspecies spread around the entire Northern Hemisphere. Such a distribution is termed **circumboreal**. But genetic studies have revealed that the relationships between these 'subspecies' are more complex than had been realized. For example, the European herring gull, which was given the name *Larus argentatus argentatus* (the final name referring to its rank as a subspecies), is almost indistinguishable in its plumage from the American herring gull (*Larus argentatus smithsonianus*). But molecular studies on their DNA indicated that they were not as closely related as had been supposed. Another form that occurs around the coasts of the Mediterranean Sea had been given the name *Larus argentatus michahellis*, and this bird had the advantage of one very obvious structural difference – its legs were yellow rather than pink. Again, the DNA of this yellow-legged gull suggested a greater separation than was indicated by its subspecific rank. In Figure 2.1, the breeding distributions of the various herring gull **taxa** are shown [2,3]. (A **taxon**, plural taxa, is an undefined unit of classification, and it is used when an author is uncommitted regarding its precise taxonomic rank.) Some of the forms that were previously regarded as subspecies are now considered worthy of being species, whereas the jury is still out on some of the remainder, which are thus still regarded as subspecies, though not necessarily of the original European herring gull.

The spatial distribution of these gull taxa is not quite as clearly defined as the map in Figure 2.1 would suggest. In the border regions between taxa, interbreeding is common, even where the two forms are regarded as separate species. Hybrids

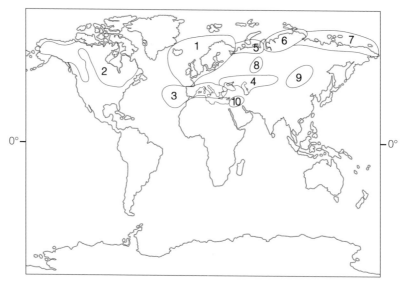

Figure 2.1 Approximate breeding distributions of various taxa within the herring gull complex [2,3]: (1) European herring gull (*Larus argentatus*); (2) American herring gull (*L. smithsonianus*); (3) yellow-legged gull (*L. michahellis*); (4) Caspian gull (*L. cachinnans*); (5) Heuglin's gull (*L. heuglini*); (6) Taymyr gull (*L. taimyrensis*); (7) Vega gull (*L. vegae*); (8) steppe gull (*L. barabensis*); (9) Mongolian gull (*L. mongolicus*); (10) Armenian gull (*L. armenicus*). There is still considerable dispute among taxonomists regarding the precise status of these taxa. 5, 6 and 7 may be subspecies of 1; and 8 and 9 may be subspecies of 4.

are frequent, and this makes field identification of these gulls even more difficult. Patterns are also constantly changing. Yellow-legged gulls are spreading northward along the west coast of Europe, and the Caspian gull is spreading north from the Black Sea area in eastern Europe and meeting up with the European herring gull in Poland. There is also a complication when the herring gull complex comes into contact with the group of black-backed gulls, which can also hybridize with them [4]. What this example illustrates is that evolution is still occurring in this gull complex as an ancestral species is splitting into new and separate forms, and then sometimes merging once more. For the biogeographer, it means that mapping the distribution of organisms and explaining such patterns that emerge are far from simple tasks. In the case of the gulls of northern Russia, it may prove impossible to detect precise divisions between the different types. Where there is gradual change in genetics and form along a gradient, taxonomists refer to a **cline**, and the variation is said to be **clinal**.

Complex variation within species is not confined to gulls; it is even found within primates. The chimpanzee (*Pan troglodytes*), for example, has four extant subspecies, all found in central and west Africa. Their distribution patterns are shown in Figure 2.2. Gabon and Congo hold populations

of *Pan troglodytes troglodytes*. A subspecies which is given the same subspecific name as the specific name is termed the **nominate subspecies**. Located farther west are *Pan troglodytes verus* (which is isolated from the other taxa) and *Pan troglodytes vellerosus*. To the east, in the Democratic Republic of Congo, is found *Pan troglodytes schweinfurthii*. At one time it was thought that there existed a fifth subspecies, but this has now been given full specific status as the pygmy chimpanzee, or bonobo (*Pan paniscus*), which is found to the south of the other taxa, as shown in Figure 2.2. The evolution of the bonobo into a full species is a consequence of separate genetic development resulting from the presence of a major barrier to interbreeding in the form of the River Congo. All of the chimpanzee subspecies lie to the north or east of this river and have developed in isolation from the bonobo.

Sometimes a species may form a circle around such a barrier, as is the case with the greenish warbler (*Phylloscopus trochilloides*) in Asia. This small, insectivorous bird is found mainly in eastern and central Asia, but sometimes wanders to western Europe. In central and eastern Asia, there are two main subspecies that can be distinguished by the presence of either one or two white bars in their wings. Birds with just one white bar belong to the subspecies *Phylloscopus trochilloides viridanus*, while those with two white bars are given the name

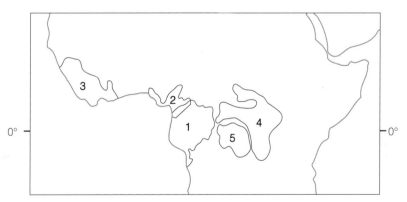

Figure 2.2 Distribution patterns of chimpanzee subspecies and bonobo in Central and West Africa: (1–4) the subspecies of chimpanzee, *Pan troglodytes*; (5) the bonobo or pygmy chimpanzee, *Pan paniscus*. More specifically: (1) the nominate *Pan troglodytes troglodytes*; (2) *Pan troglodytes vellerosus*; (3) *Pan troglodytes verus*; (4) *Pan troglodytes schweinfurthii*. The bonobo is found south of the River Congo and is hence separated from other chimpanzees by a formidable barrier to movement.

Phylloscopus trochilloides plumbeitarsus. The two-barred greenish warbler occupies a large area of eastern Asia (Figure 2.3), while the one-barred subspecies lies to the west, extending as far as southern Finland and the Baltic States. Where the two subspecies overlap in their range, they fail to interbreed, so subspeciation has reached a point where the two are regarded by some taxonomists as separate species. The greenish warbler's range also extends farther south, running west of the high Tibetan Plateau and then east along its southern edge, eventually turning north again into central China. Here, to the east of the Tibetan Plateau, is another subspecies of the greenish warbler, *Phylloscopus trochilloides obscuratus* [5,6]. As can be seen from Figure 2.3, the greenish warbler complex almost forms a complete ring, and it is indeed termed a **ring species**. It is questionable whether the ring could ever be completed. Much of the intervening land has been cleared of forest, and the greenish warbler is essentially a woodland bird, so human activity may have provided a final barrier to its completion of the ring. But genetic changes during the course of the development of the ring would almost certainly lead to incompatibility between populations even if they were to meet up. The processes by which these genetic changes take place will be discussed in greater detail in Chapter 6, but clearly such complications in the genetic diversity within species need to be taken into account when examining patterns of distribution.

Biogeographers, therefore, are dependent on taxonomists to define the units that they study, and when they examine the distributions of species they find that there are further complications in that no two species or subspecies are identical in their geographical ranges. Some correspond fairly closely, but others differ totally. When we use terms such as *distribution* and **range** (the latter being the area within which the species (or other taxon) is found), we must also be careful about the spatial scale we are considering. Two species may be widespread within a given area, such as the British Isles or the state of North Carolina, and yet occupy different types of habitats (such as woodland or grassland). Even within a habitat, species may occupy different **microhabitats**, such as forest canopies or forest floors. In a New Zealand forest, for example, one may find both the brown

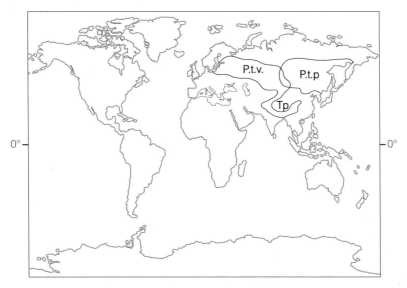

Figure 2.3 Breeding distributions of the greenish warbler (*Phylloscopus trochilloides*) in Asia. P.t.p. is the two-barred form of the greenish warbler, *Phylloscopus trochilloides plumbeitarsus*. P.t.v. is the single-barred form of the greenish warbler, *Phylloscopus trochilloides viridanus*. Between them, they form a ring around the harsh environment of the Tibetan Plateau (Tp). This is a rare example of an avian ring species. A further example of this process is shown in Plate 3.

kiwi (*Apteryx australis*) and the fantail (*Rhipidura fuliginosa*), a kind of flycatcher. But they occupy different microhabitats, for the kiwi is confined to the forest floor whereas the fantail nests in canopy branches. Therefore, scale, in both horizontal and vertical dimensions, is an important consideration when studying distribution patterns [7].

Limits of Distribution

Whatever the scale at which one studies the distribution of an organism, there are limits within which it is spatially confined. Beyond these limits, it is unable to sustain its population. These limits are determined by barriers, but the barriers may be of various kinds.

1. There are *physical barriers* that may prevent the spread of an organism. High mountain chains, expanses of water or areas of arid desert may confine a species to one particular region. But the conditions that prove insuperable to one species may not deter one that is more mobile. Birds, for example, may cross extensive areas of inhospitable terrain. The Himalayan mountain chain is a formidable barrier to most animals, but the barheaded goose (*Anser indicus*) is able to migrate over the Himalayas, flying at heights up to 10 175 m (33 382 feet). Their blood contains haemoglobin that is particularly efficient at oxygen absorption, so they can fly at altitudes that would prove impossible to most other birds. The River Congo is an impenetrable barrier to the bonobo, but animals that can fly or swim efficiently do not find it so.

2. *Climatic barriers* limit the distribution of many species. Frost can prove fatal to many tropical plants because the formation of ice within the cells of the plant, followed by melting, disrupts the cell membranes and results in death. Drought can similarly cause problems of desiccation in many plants and animals that have limited capacity for water conservation.

3. Geology, and its effect on soil chemistry and structure, is often limiting for plants and for soil-inhabiting invertebrate animals and microbes. Overcoming such *geological barriers* demands effective dispersal strategies, either by aerial transport in the case of some fruits, seeds and very small animals like spiders, or by hitching a ride with more mobile organisms, either in their gut as in the case of some digestion-resistant seeds, or on the surface, adhering to fur or feather.

4. At a lower level of scale, the *nature of the habitat* can impose limits to a species. A forest species may be deterred from crossing an area of grassland; or a marsh organism may fail to travel across dry habitats to reach the next area of wetland. Distribution patterns may thus be derived from habitat mosaics. This is particularly true in regions highly modified by human activity. At an even lower level of scale, organisms may occupy different microhabitats that are subjected to small-scale variations in physical conditions, or **microclimate**. *Microclimate* is a term that covers temperature, humidity and light variations on a very small scale. Animals may be restricted in their microhabitats because of limitations in their resistance to desiccation or temperature variation, but also in their dependence on food availability. These various factors may form barriers restricting species to their microhabitats. The insects that live in rotting logs, for instance, are adapted by their evolution to a microhabitat with a high water content and relatively constant temperatures. The logs provide the soft woody materials and the microorganisms that insects may need for food. Logs may also provide good protection from predators. Around the logs are areas with fewer or none of these desirable qualities, and, for many animals, attempts to leave their microhabitat would result in death by desiccation or starvation.

5. *Biological barriers* occur when an organism is subjected to increased predation, parasitism, disease or competition from more robust species if it were to move beyond specific area limits. The insect that leaves its forest-floor log, for instance, is exposed to a whole range of predators, such as beetles, shrews and insectivorous birds. The northern subspecies of the spotted owl (*Strix occidentalis caurina*) has become endangered because of the fragmentation of the forests of the Pacific northwest of North America. Its main problem is that of predation when crossing open areas from one forest fragment to another.

6. *Historical factors* may also create barriers that confine species to a limited area. Changes in the pattern of land masses over the surface of the Earth have resulted in the creation of physical barriers, sometimes between closely related taxa. Massive global climate changes, such as the expansion of ice masses in the past, have also contributed to the disruption of distribution patterns as the ice masses have arisen and then departed once more.

7. *Chance.* One additional complication is the role of sheer chance in the distribution of organisms. The arrival of a wind-borne insect or a seed at a particular point in space cannot be predicted with certainty, and the first arrival may well be at an advantage over those arriving later. Chance events are said to be **stochastic**, and these random elements within ecology and biogeography may be of great significance [8,9]. A consideration of the role of random factors in biogeography has led to the development of **neutral theory**, which will be considered when we examine the ways in which species become assembled together in communities (see Chapter 4).

Patterns of plants and animals over the surface of the Earth have thus been created by a range of different factors, many of them interacting with one another. Some plants and animals are confined in their distribution, sometimes (although not always) within the areas in which they evolved; these are said to be **endemic** to that region. Their confinement may be due to physical barriers to dispersal, as in the case of many island faunas and floras (termed **palaeoendemics**), or to the fact that they have only recently evolved and have not yet had time to spread from their centres of origin (**neoendemics**). These will be discussed in detail in this chapter.

In all these cases, the ultimate barriers are not necessarily the hostile factors of the environment but the species' own physiology, which has become adapted to a limited range of environmental conditions. In its distribution, therefore, a species is often the prisoner of its own evolutionary history.

The Niche

The demands that an organism places on its environment in terms of physical and chemical conditions, space and food supply help to define what ecologists call its **niche**. But the concept of the niche goes beyond the basic physics and chemistry of its habitat and covers all aspects of how the organism makes a living. It includes the food an animal requires, but also encompasses the way in which it acquires that food. The kestrel is a bird that hunts small mammals by day, whereas an owl performs a similar activity by night. Swallows catch aerial insects by day, and bats have the same feeding strategy at night. They overlap in their food requirements but obtain their food under quite different conditions. In the case of plants, they may have similar requirements for water and chemical elements from the soil, but may root at different depths, or flower at different times, and thus tap slightly different resources. In this way, they differ in their niche. The subdivision of resources in this way is termed **niche partitioning**.

On Lord Howe Island off the east coast of Australia, there are two closely related endemic species of palm [10]. One of these, *Howea forsteriana*, flowers approximately 7 weeks before the other species, *Howea belmoreana*. *H. belmoreana* also prefers more acidic soils than *H. forsteriana*. The two species thus differ in their niches, and these differences enable them to coexist on the island. Their niches can be seen to be multidimensional, in the sense that there are several requirements in which the two species vary, both in terms of the chemical environment and in the timing of their life cycles. One can think of these as separate axes of variation. Species may coincide in their requirements on one or more axes, but are unlikely to coincide on all axes. No two niches will be identical.

A more familiar example of the different niches of species is provided by the waterfowl that occupy shallow ponds in the temperate regions of North America and Europe. One very widespread duck in both continents is the gadwall (*Anas strepera*). It feeds mainly on submerged vegetation, which is most abundant in nutrient-rich (eutrophic) waters. It cannot dive, so it feeds by upending, and this generally restricts it to the shallow parts of a lake, where it may encounter some competition from other water birds, such as the mallard (*Anas platyrhynchos*). This species also upends, but it has a wider range of diet, including small invertebrates and seeds. So the specialization of the gadwall allows it to avoid much competition. It also avoids competition with wigeon (both the Eurasian wigeon, *Anas penelope*, and American wigeon, *Anas americana*) because these ducks spend much of their time on dry land eating terrestrial grasses and herbs. Teal (the Eurasian teal, *Anas crecca*, and green-winged teal, *Anas carolinensis* in North America) are also found in shallow waters, but they feed mainly by sieving the water for small invertebrates, so again they do not enter into direct competition with gadwall. But sometimes gadwall are found in deeper water, so how do they manage to obtain food in such situations? The answer is by theft. Coots (both Eurasian coot, *Fulica atra*, and the American coot, *Fulica americana*) dive for their food and bring vegetable matter to the surface from greater depths. The coots are messy eaters, and it is not difficult for gadwall to move in and collect some of the loot. This behaviour is called **kleptoparasitism**, and it is an effective way of widening the niche of the gadwall.

Ecologists have also developed two ways of looking at the niche. There is the theoretical or ideal type of niche, usually called the **fundamental niche**, which is the sum of all the niche requirements under ideal conditions when the species is given unimpeded access to resources. In the real world, such conditions are unlikely, usually because there are other species that compete for those resources (i.e. have overlapping niches) and may perform better in their efficiency of acquisition. The result is that the observed distribution of the organism is confined by species interactions, and the outcome is the **realized niche**, where the species is found over a smaller range than would have been predicted. These concepts are important in biogeography, especially when attempts are made to model potential niches as an aid to predicting distribution patterns.

Overcoming the Barriers

There are, therefore, many dimensions to the niche which restrict the habitats within which a species is found. Habitats, such as the rotting log mentioned in the 'Limits of Distribution' section, are often scattered or spatially fragmented, leading to individual organisms becoming confined in their distributions. If an organism is to spread, it needs to overcome spatial and physical barriers to gain access to new locations where its niche requirements can be satisfied. This may prove difficult, but a few inhabitants of rotting logs do occasionally make the dangerous journey from one log to another, for few environmental factors are absolute barriers to the dispersal of organisms and these factors vary greatly in their effectiveness.

Most habitats and microhabitats have only limited resources, and the organisms living in them must have mechanisms enabling them to find new habitats and new resources when the old ones become exhausted. These mechanisms often take the form of seeds, resistant stages or (as in the case of the insects of the rotting-log microhabitat) flying adults with a fairly high resistance to desiccation.

There is plenty of evidence suggesting that many spatial barriers are not completely effective. Organisms may extend their distribution by taking advantage of temporary, seasonal or permanent changes in weather conditions, overall climate or the distribution of habitats that allow them to cross barriers normally closed to them. The British Isles, for instance, lie within the geographical range of about 220 species of birds, but a further 50 or 60 species visit the region as casual vagrants. These birds do not breed in Britain, but one or two individuals are seen by ornithologists every few years. They come for a variety of reasons: some are blown off course by winds during migration, and others are forced in certain years to leave their normal ranges when numbers are especially high and food is scarce. Many of these accidental arrivals in Britain have their true home in North America, such as the ring-necked duck (*Aythya collaris*), a few of which are seen every year, but some come from eastern Asia, such as the olive-backed pipit (*Anthus hodgsoni*), or even from the South Atlantic, such as the black-browed albatross (*Diomedea melanophris*).

It is possible, although not very likely, that a few of these chance travellers may in time establish themselves permanently in Europe, as did the collared dove (*Streptopelia decaocto*) which since about 1930 has spread from Asia Minor and southern Asia across central Europe and into the British Isles and Scandinavia, perhaps the most dramatic natural change in distribution recorded for any vertebrate in recent times. This species is now common around the edges of towns and settlements in western Europe, and it seems to depend for food largely on the seeds of weed species common in farms and gardens, together with the bread that humans often put out for garden birds. Several factors may have interacted to permit this extension of range of the collared dove. Increased human activity during the last century, involving extensive changes in the environment, has produced new habitats and food resources, and it is possible, too, that changes in climate may have significantly favoured this species. It is, however, considered unlikely that the collared dove would have been able to take advantage of these changes without a change in its own genetic make-up, perhaps a physiological one permitting the species to tolerate a wider range of climatic conditions or to utilize a wider range of food substances. Its behaviour patterns have also changed, from nesting largely on buildings to nesting in trees, which may have favoured it in temperate Europe [11]. Since its introduction to the Bahamas in 1974, the collared dove has also spread rapidly through North America [12], as will be discussed in the 'Invasion' section, so this is one organism that has proved remarkably successful once dispersal barriers have been overcome.

Biogeographers commonly recognize three different types of pathway by which organisms may spread between one area and another. The first, easiest pathway is called a **corridor**; such a pathway may include a wide variety of interconnecting habitats, so that the majority of organisms found at either end of the corridor would find little difficulty in traversing it. The two ends would therefore come to be almost identical in their **biota** (the fauna plus the flora); for example, the great continent of Eurasia that links western Europe to China has acted as a corridor for the dispersal of animals and plants.

In the second type of dispersal pathway, the interconnecting region may contain a more limited variety of habitats, so that only those organisms that can exist in these habitats will be able to disperse through it. Such a dispersal route is known as a **filter**; the exclusively tropical lowlands of Central America provide a good example. Not all types of animal and plant are able to traverse this type of terrain.

Finally, some areas are completely surrounded by totally different environments, so that it is extremely difficult for any organism to reach them. The most obvious example is the isolation of islands by wide stretches of ocean, but the specially adapted biota of a high mountain peak, of a cave or of a large, deep lake is also extremely isolated from the nearest similar habitat from which

colonists might originate. The chances of such a dispersal are therefore extremely low, and largely due to chance combinations of favourable circumstances, such as high winds or floating rafts of vegetation. Such a dispersal route is therefore known as a **sweepstakes** route. It differs from a filter in kind, not merely in degree, for the organisms that successfully use a sweepstakes route are not normally able to spend their whole life histories *en route*. Such organisms are alike only in their adaptations to traversing the route, such as those aerial adaptations of spores, light seeds or flight in the case of insects and birds that enable them to disperse from island to island. Such a biota is therefore not a representative sample of the ecologically integrated, balanced biotas of a normal mainland area, and is said to be **disharmonic**.

A discussion of some patterns of distribution shown by particular species of animals and plants will show how varied and complex these may be, and will help to emphasize the various scales or levels on which such patterns may be considered. In fact, the number of examples that we can choose is quite limited, because the distribution of only a very small number of species has been investigated in sufficient detail. Even amongst well-known species, chance finds in unusual places are constantly modifying known distribution patterns, thereby demanding changes in the explanations that biogeographers give to explain these patterns.

Some existing patterns are continuous, the area occupied by the group consisting of a single region or of a number of regions which are closely adjacent to one another. These patterns can usually be explained by the distribution of present-day climatic and biological factors. Other existing patterns are discontinuous or **disjunct**, the areas occupied being widely separated and scattered over a particular continent, or over the whole world. The organisms which show such a pattern may, like the magnolias (see the 'Magnolias: Evolutionary Relics' section), be evolutionary relics, the scattered survivors of a once-dominant and widespread group, now unable to compete with newer forms. Others, the climatic relics or habitat relics, appear to have been greatly affected by past changes in climate or sea level. Finally, as will be shown in Chapters 10 and 11, the disjunct patterns of some extant groups, and of many extinct groups, have resulted from the physical splitting of a once-continuous area of distribution by the process of continental drift (see Chapter 5).

Climatic Limits: The Palms

An example of a family of plants confined to one particular climatic regime is the palms (family Arecaceae). Figure 2.4 shows the global distribution of this plant family, and it can be seen that members (about 2780 species) are found in all areas of the tropics and in many subtropical regions too. Such a distribution can be termed **pan-tropical**. In the temperate areas, however, such as Europe, there are very few species of palm that can be regarded as native. Indeed, there are only two truly native palms in Europe. One of these, *Chamaerops humilis*, is a very small species which grows in sandy soils in southern Spain and Portugal, eastwards to Malta (Figure 2.5). The second species, *Phoenix theophrasti*, is found on certain Mediterranean islands, mainly on Crete.

The United States is also relatively poor in palms, having just four native species. Florida and the southeastern states have the palmetto palms (*Sabal palmetto* and *Serenoa repens*), Texas has the Rio Grande palmetto (*Sabal mexicana*) and California has an impressive fan palm, *Washingtonia filiferia*. None of them is able to sustain a natural population farther north than the southernmost parts of North America, where the average annual minimum temperature ranges between –1°C and 7°C [13].

So, a family which is extremely successful and widespread in the tropics has failed to achieve similar success in the temperate regions. The real problem with the palms is the way they grow: they have only a single growing point at the apex of their upright stems, and if this is damaged by frost then the whole stem perishes. This weakness has even limited the use of palms as domesticated crop plants, for species such as the date palm (*Phoenix dactylifera*) cannot be grown in areas with frequent frosts. Even in the deserts of northern Iran, where the summers are hot and dry, the date palm is a rare sight because of the intense cold in the high-altitude deserts of Iran during winter (Figure 2.6). Perhaps the most successful palm in the temperate regions is *S. repens*, which is often cultivated an sometimes naturalized, and can reach 30°N in the United States; and also the *Trachycarpus* species, particularly *Trachycarpus martianus*, which grows to an altitude of 2400 m in Nepal [14]. But the family as a whole is limited geographically by its frost sensitivity.

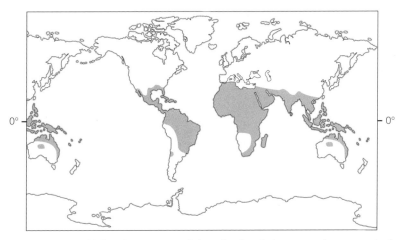

Figure 2.4 World distribution map of the palm family (Arecaceae), a pantropical family of plants.

Figure 2.5 The dwarf palm, *Chamaerops humilis*, one of the two native palms found in Europe.

Figure 2.6 The date palm, *Phoenix dactylifera*, at its most northerly site in the Great Kavir Desert of Iran.

A Successful Family: The Daisies (Asteraceae)

The palms are a successful family in that they have been able to spread through the entire tropics and subtropics, but they have failed to expand beyond these climatic limits. The daisy family (Asteraceae), on the other hand, seems to have overcome all barriers and occupied the entire globe (Figure 2.7), with the exceptions of the Antarctic, the Greenland ice cap and the most northerly of the Canadian Arctic islands. The daisy family provides an example of the way in which we need to invoke different explanations for distribution patterns at different spatial and taxonomic scales. The daisy family is extremely large (having about 25 000 species) and extremely successful, if you measure biogeographical success by the areal extent of distribution. It is a **cosmopolitan** family, which means that it is found throughout the world. In fact, the term *cosmopolitan* when used of the flowering plants is usually a slight exaggeration, since very few species of flowering plants have managed to establish themselves in Antarctica; even the Asteraceae have not achieved that, but they are present on all other continents. With the exception of the isolated cold regions of Antarctica and some parts of the northern Arctic, there has

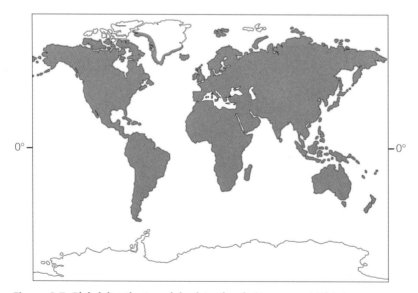

Figure 2.7 Global distribution of the daisy family (Asteraceae). This is a cosmopolitan family of plants that has successfully established itself in all parts of the world apart from Antarctica and the far north of the Arctic.

clearly been no insuperable climatic or other barrier to the geographical spread of the family during its evolutionary history.

When we look at those areas of the world where members of the Asteraceae are most abundant and diverse, we find that the mountainous regions of the tropics and subtropics, together with some of the semi-arid regions of the world, and those regions with **Mediterranean climates** (hot, dry summers and mild, wet winters) are the richest in members of this family. The equatorial rainforests are actually rather poor in daisy family species. Often, biogeographers use such information to try to reconstruct the evolutionary origins of a group. A great deal of generalization is involved, but it does seem as though this family has been most successful away from the competition of tall trees, in the more drought-prone habitats where their general adaptability and very diverse fruit-dispersal systems have given them many advantages.

Taking just one genus from within the family, the groundsel genus *Senecio*, we find that it reflects the whole family in many ways, being large (about 1250 species) and widely dispersed (cosmopolitan apart from Antarctica). Many members are efficient

weeds, being short-lived, having efficiently dispersed airborne fruits and having wide ecological tolerances of climate and soils (broad niches). Some taxonomists prefer to split this very large genus into subgenera, and one of these, the subgenus *Dendrosenecio*, is remarkable both for its form (Figure 2.8) and its restricted distribution pattern. This subgenus consists of just 11 species, often referred to as giant tree-groundsels. They are stocky, woody plants up to 6 m in height often with branched upper sections bearing terminal clusters of tough, leathery leaves. Botanists refer to thick-stemmed plants of this type as **pachycaul**. In distribution, this subgenus is restricted to East Africa and, examining the distribution pattern on a more detailed scale, only on the high mountains of East Africa (Figure 2.9) above the forest limits of bamboos and tree-heathers [15]. If we focus from the taxonomic level of subgenus to that of species, we find that each of the major mountains of East Africa has its own group of endemic species of *Dendrosenecio*, with never more than three species on any particular mountain (see Figure 2.9). Detailed analysis of the genetic material (the DNA) of the tree-groundsel species by Eric Knox

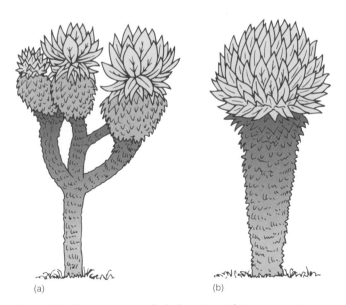

(a) (b)

Figure 2.8 Giant tree-groundsels from East African mountains. Family Asteraceae, genus *Senecio*, subgenus *Dendrosenecio*. (a) Branched form; (b) unbranched form.

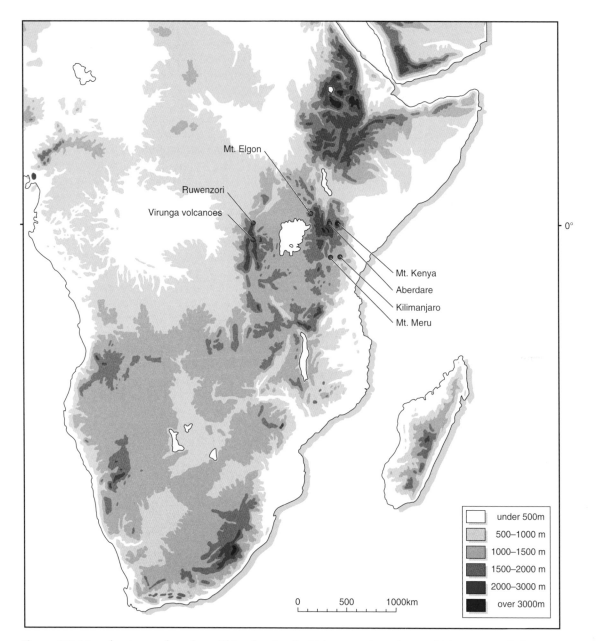

Figure 2.9 Map of eastern and southern Africa showing the high mountain peaks on which tree-groundsels are found. Mt Meru is an exception, having no tree-groundsels.

at Kew in London and Jeffrey Palmer at Indiana University [16] has shown that each species is more closely related to its neighbouring species on its own mountain than to the species of other mountains, despite the fact that in form, such as branching pattern (as shown in Figure 2.8), it may more closely resemble the giant groundsels from other mountains. It seems that chance has led to the colonization of each mountain peak (with the exception of one, Mount Meru, which is devoid of giant groundsels), and that the chance invader has in the course of time evolved into two or three separate species. (The time involved, incidentally, cannot be very long since Mount Kilimanjaro is

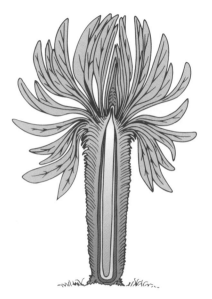

Figure 2.10 Cross-section of a tree-groundsel showing its thick central pith, surrounded by wood and cortex, together with the outer layer of dead leaves and leaf bases, forming an insulating sheath that protects the living tissues from frost.

only a million years old, which is quite young by geological standards.)

If we take the spatial scale of analysis one step lower and look at separate species on just one of the mountains, then additional factors come into play in the interpretation of their distribution patterns. On Mount Elgon (4300 m), situated on the border between Uganda and Kenya, north of Lake Victoria (see Figure 2.9), two species of the tree-groundsels are found, *Senecio elgonensis* and *Senecio*

barbatipes. In the open, alpine zone where these trees are found, *S. elgonensis* predominates below 3900 m and *S. barbatipes* above this level, so there is an altitudinal differentiation in their ranges on the mountain. Precisely what features of the morphology or physiology of the two species lead to these climatic preferences is not known, and there are no detailed meteorological measurements for the mountain, but temperature differences with altitude and, in particular, the frequency of frost during the night are likely to be the most important factors affecting the distribution of the two species. The giant senecios are more frost-tolerant than most tropical plants, being insulated by thick layers of leaves and leaf bases (Figure 2.10). When the night air temperature drops to −4°C, the temperature within the insulating layer of leaves only falls to 2°C. The insulation is made even more effective because the leaves alter their position during the night, closing together and trapping additional layers of air around the stem [17]. The vital, temperature-sensitive dividing cells of the main trunk system are thus protected from frost. Different insulating efficiencies or temperature sensitivities of the individual species may affect altitudinal limits, perhaps via seed production or germination. The two species may also be in competition with one another for space or some other resource, as will be illustrated by further examples later in this chapter.

Taking a final and even more detailed look at the distribution of *S. elgonensis* in a small valley within the lower part of the alpine zone of Mount Elgon, we find that the population is most dense around the valley floor (Figure 2.11), where a damp

Figure 2.11 Diagram of a cross-section of a small valley on Mt Elgon, Uganda, showing the higher density of the tree-groundsels in the valley bottom.

area fed by water seepage exists. The species is evidently affected at this habitat scale by the availability of deeper, moist soils, preferring these to the shallow, free-draining soils of the alpine slopes and ridges, where drought is likely during the hot conditions of the tropical alpine day.

This analysis of distribution patterns at increasingly detailed levels of scale and taxonomy within the Asteraceae demonstrates the way in which we must invoke different factors to account for the distribution patterns of organisms depending upon both the taxonomic and the geographical scale we are using.

Patterns among Plovers

Plovers (Charadriidae) are another cosmopolitan family, this time of shorebirds and waders. Within the family, which numbers about 67 species, the genus *Charadrius* itself is cosmopolitan and has representative species on all the continents of the world, with the usual exception of Antarctica. But the different species of *Charadrius* vary considerably in their geographical ranges and in their ecology. In North America, the most familiar member of the genus is the killdeer (*Charadrius vociferus*) which, as its Latin name implies, is an extremely vocal bird. It is also an extremely widespread one within the limits of North America, as the map in Figure 2.12 demonstrates. Its breeding range runs from northern Mexico and Baja California

to southern Alaska, and from Florida to Newfoundland. Only the most northern areas of Canada and Alaska lack any breeding populations of killdeer. One of the reasons for its success is its very wide tolerance of different habitats. It may be encountered on ocean shorelines, freshwater margins, marshes and other wetlands, but also on dry grasslands, roadsides, waste places and farmland, even airports and domestic lawns. Like all members of the genus, it nests on the ground; it prefers a shingle or gravel surface in which its eggs are highly camouflaged, but it may find suitable sites in a wide range of habitats, including those heavily disturbed by human influence. A species with a wide range of tolerance to ecological conditions is said to be **eurytopic**. This characteristic is valuable in enabling it to become widespread, as its distribution map in Figure 2.12 demonstrates.

Another species of the same genus in North America is the mountain plover (*Charadrius montanus*). In contrast to the killdeer, its breeding distribution is much more limited, as shown in Figure 2.13. It is confined to the United States, being found in a line running north from New Mexico, through Colorado and Wyoming, to Manitoba. This line runs along the eastern side of the Rocky Mountains and on the western edge of the Great Plains. Although its name suggests that it is a mountain bird, it is in fact a species of the prairie. The mountain plover's preferred nesting habitat is short-grass prairie, cropped close by grazing animals. These birds are particularly fond of nesting in the vicinity of prairie dogs, which are colonial,

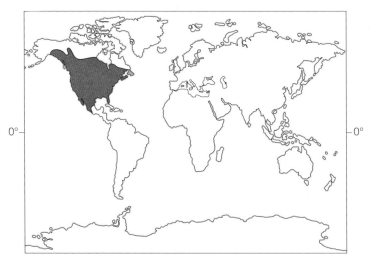

Figure 2.12 Breeding distribution of the killdeer, a widespread North American plover.

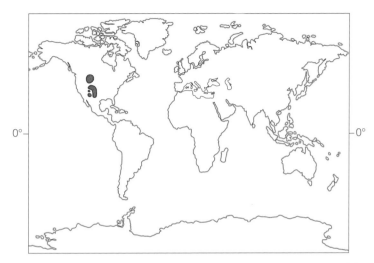

Figure 2.13 Breeding distribution of the mountain plover, a species of the western prairies with a limited range.

herbivorous mammals that crop the grass very short around their burrows, probably so that they can detect any approaching predator [18,19]. The combination of short grass and mammalian companions ever on the watch for danger makes such habitats ideal for their breeding. But such conditions are limited by the restricted range of the prairie dogs as well as the tendency for grassland in these regions to be converted into arable use by farmers. The mountain plover is thus a fastidious bird, and organisms with such limited habitat requirements are termed **stenotopic**. On the whole, organisms with very specific habitat requirements are restricted in their distribution patterns.

In Europe and Asia, the genus *Charadrius* is also widespread, as is another genus of the plover family, *Vanellus*. One of the most widespread members of the plover family is the lapwing (*Vanellus vanellus*). This breeds from the west of Europe, eastward through Central Asia to the Pacific coast. Its northern boundary runs from Fennoscandia and northern Russia to eastern China, and its southern limit lies in Turkey, Iran and Mongolia (Figure 2.14). Its habitat requirements are wide,

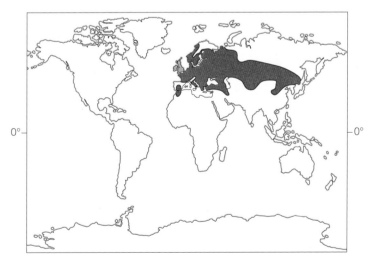

Figure 2.14 Breeding distribution of the lapwing, a widespread species of Old World plover.

preferring lowlands that are frost free during the breeding season, especially grasslands and arable fields. It is able to survive in steppe environments but prefers moist locations, from lake margins and marshes to dune slacks among sand dunes. Its tolerance of habitats modified by human activity, such as grazing and arable lands, has made it particularly successful. In western Europe, it has declined in recent years as a result of changes in land use and the widespread application of pesticides which kill its invertebrate prey species.

A close relative of the lapwing is the sociable plover (*Vanellus gregarius*), but unlike the lapwing it is far more restricted in its distribution, as shown in Figure 2.15. Its breeding range is restricted to Kazakhstan and neighbouring parts of southern Russia. Rather like the mountain plover of North America, the preferred breeding habitat of the sociable plover is short-grass prairie. Often, this lies in salty steppes that become very dry in summer. It does not tolerate the taller vegetation of the neighbouring semi-desert (see Plates 1 and 2), or the forest steppe of the moister soils. Like the mountain plover, therefore, this bird is extremely fastidious in its requirements; it is stenotopic. These two examples illustrate the relationship between habitat tolerance and the extent of the breeding range of a species.

A similar relationship is found in many species of animals and plants, including invertebrates, such as amphipods. Kevin Gaston and John Spicer [20] have examined this in their studies of various species of the crustacean genus *Gammarus*. They considered comparable pairs of species. For example, *Gammarus zaddachi* is an estuarine species with the capacity to tolerate a limited range of salinities. A similar species, *Gammarus duebeni*, is even more tolerant of salinity variation, occurring even in rock pools, where the salinity can become very low after rainfall but becomes high when exposed to long periods of sunshine. When we look at their global distribution patterns, it is the more tolerant, eurytopic species, *G. duebeni*, that has the greatest geographical range, being found on both sides of the Atlantic, whereas *G. zaddachi*, which is more stenotopic, is confined to northern Europe and Iceland (Figure 2.16). Similarly, when we look at *Gammarus locusta*, a strictly marine species, and compare it with another marine species, *Gammarus oceanicus*, we find that the more tolerant *G. oceanicus* also has the widest geographical range, as shown in the diagram. So, if we measure the success of an organism by its geographical distribution range, then the broadly tolerant species seem to have the advantage, at least as far as *Gammarus* species and the plovers are concerned.

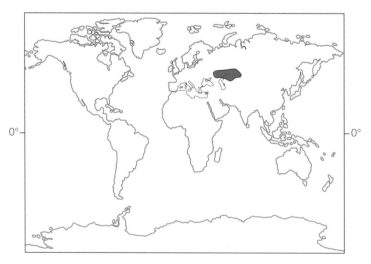

Figure 2.15 Breeding distribution of the sociable plover, which is almost entirely restricted to the steppes of Kazakhstan.

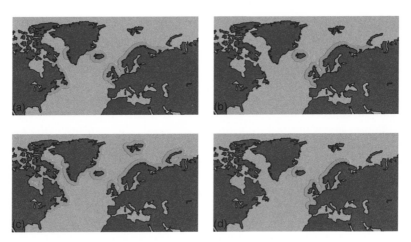

Figure 2.16 Global distribution maps of four species of the amphipod genus *Gammarus*: (a) *G. duebeni* (eurytopic) has a wide distribution pattern, (b) *G. zaddachi* (less tolerant) is more confined, (c) *G. oceanicus* (eurytopic) is widespread, and (d) *G. locusta* (stenotopic) is more confined. From Gaston and Spicer [20]. (Reproduced with permission of John Wiley & Sons.)

Magnolias: Evolutionary Relicts

Having looked at the various patterns of distribution that animals and plants display, we are now in a position to examine in greater detail the possible causes of those patterns. Some owe their origin to ancient evolutionary and geological developments, and among these is the flowering plant family, the magnolias.

The magnolias (family Magnoliaceae, genus *Magnolia*) have a very interesting modern distribution, as shown in Figure 2.17. Of the 80 or so species of the genus *Magnolia*, the majority are found in South-East Asia and the remainder, about 26 species, in the Americas, ranging from Ontario in the north, through Mexico, down into the northern regions of South America [21]. Their distribution is clearly disjunct, being separated into two main centres in this case. Unlike the palms, we cannot explain their distribution pattern simply in terms of the climatic sensitivities of the plants concerned, for the magnolias are reasonably hardy; they can be cultivated far into the north of the temperate area. Nor would climatic constraints explain why they

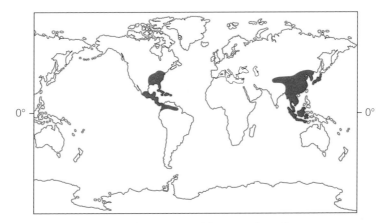

Figure 2.17 World distribution map of the magnolias, illustrating a disjunct distribution.

are not found in intermediate tropical, subtropical and temperate regions, as are the palms.

To understand the distribution of the magnolias, we need to look at their evolutionary history. Fossils of magnolia-like leaves, flowers and pollen grains are known from Mesozoic times, the age of the dinosaurs. Indeed, the magnolia family is regarded by botanists as one of the most primitive families of flowering plant groups. Its showy flowers were attractive to the rapidly evolving insects, and together they coevolved into a most successful team in which the insect visited the flowers for food and, in doing so, ensured the passage of pollen from one plant to another, thus taking the chance and the waste out of the highly risky wind-pollination process. As a result of their success, the magnolias spread and must at one time have formed a fairly continuous belt around the tropical, subtropical and temperate parts of the world, for their fossil remains have been found through Europe and North America, and even in Greenland. For perhaps as long as 70 million years, the magnolias remained widespread, right up to the last two million years, during which they have been lost from areas such as Europe that linked their current isolated centres of distribution in the Americas and East Asia.

Being small, slow-growing shrubs and trees, they were not able to compete well with the more robust and fast-growing tree species; and, when the climatic fluctuations of the last two million years began to disturb their stable woodland environment, they succumbed to the competitive pressures imposed by more competitive trees, and thus became extinct across much of their former range. Only in two parts of the world have they managed to escape and survive, as evolutionary **relicts**. The word *relict* was originally applied to a widow and implies being left behind, which is precisely what has happened to the magnolias.

It is interesting that another genus of the magnolia family, the tulip trees (genus *Liriodendron*), has a very similar distribution to that of the *Magnolia* genus and in all probability shares a similar fossil history. But, in the case of the tulip trees, only two species have survived: *Liriodendron tulipifera* is a successful component of the deciduous, temperate forests of eastern North America, and *Liriodendron chinense* survives only in a restricted area of South-East Asia (Figure 2.18).

The Strange Case of the Testate Amoeba

Very small organisms, especially microbes, tend to have very wide global geographical distributions; many are cosmopolitan [22]. The reason for this is their effective aerial dispersal, suspended in air currents. The rust and smut fungi, for example,

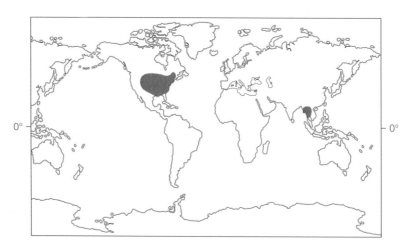

Figure 2.18 World distribution of the tulip trees (*Liriodendron* species). Only two species now survive, in widely separated localities, although it was once a widespread genus.

can travel many thousands of kilometres carried along in the atmosphere, and their abundance ensures that some will land in locations where conditions will be suitable for their survival and population growth. In the case of the rusts and smuts, that generally occurs on the leaves of the plant species that they parasitize. But the tendency to develop a cosmopolitan distribution does not apply to all microscopic organisms, as has been demonstrated for the testate amoeba *Nebela vas*, the mainly Southern Hemisphere distribution of which is shown in Figure 2.19.

Testate amoebae are tiny protozoans that live in wet habitats, often in the spongy mosses of bogs and marshes. They differ from other amoebae in having a permanent tough shell, which enables them to survive periods of drought. Other amoebae are capable of producing a cyst when subjected to adverse conditions, but the testate species, rather like snails, constantly carry a cyst around with them just in case disaster strikes. Due to their small size, the problems of finding them and the difficulties involved in identifying testate

amoebae, information on their distribution patterns is naturally less abundant than for birds and flowering plants. But studies by Humphrey Smith and David Wilkinson [23], collating records from around the world, have revealed that some species of testate amoebae have surprising geographical distributions, including *N. vas*, as shown in the map in Figure 2.10.

Sometimes the distribution pattern of species that are not easily recognized simply reflects the locations of experts in the field, coupled with the intensity of field survey. But the restriction of this species to the tropics and Southern Hemisphere is not a consequence of more thorough searching in those regions; indeed, the Northern Hemisphere has probably seen more extensive survey work than regions farther south [24]. Nor is this a case of stenotopic ecology (being ecologically intolerant as opposed to eurytopic or ecologically tolerant species) on the part of the protozoan because it has been recorded in a wide range of habitats, including bog mosses, forest floors and even high-altitude forests of bamboo and rhododendron. *N. vas*

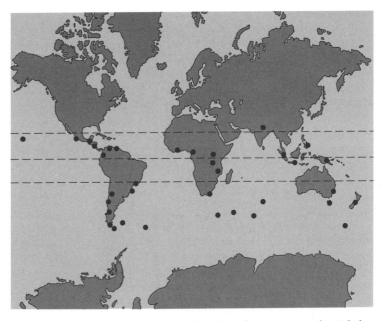

Figure 2.19 Locations around the world where the testate amoeba *Nebela vas* has been recorded. Note its predominantly Southern Hemisphere and tropical distribution. This may be explained by former linkage of the land masses in the supercontinent of Gondwana. From Smith and Wilkinson [23]. (Reproduced with permission of John Wiley & Sons.)

is even found over a wide range of pH conditions, from 3.8 to 6.5, thus ranging from very acidic to neutral environments. Its climatic requirements are also broad, encompassing temperate lowlands and high-altitude tropical sites, and extending into regions of the sub-Antarctic. So this is not a species that has highly specific requirements for its habitat.

The geographical range of *N. vas* is also wide, covering the whole of southern America from Costa Rica to Tierra del Fuego, Africa south of the Sahara as well as Australasia. Its island records range from tropical Hawaii and Java south to the sub-Antarctic island of South Georgia. But this tiny amoeba has not been recorded north of the Tropic of Cancer apart from one location in the Himalayas of Nepal. What can possibly account for such an odd distribution pattern?

The one feature that links all the regions occupied by *N. vas* is the fact that they were once part of a huge supercontinent called Gondwana. It was not until the late Cretaceous, around 90–100 million years ago, that this great southern continent finally broke up into the pattern of continents with which we are familiar today (see Chapter 7). The strange distribution pattern of this protozoan may well be due to its evolution and spread within Gondwana prior to its fragmentation. Even so, it is surprising that it has failed to occupy the new land masses now available to it.

The examples of the magnolias and the testate amoebae illustrate the possibility that some organisms owe their distributions to events from the deep geological past. Others can be explained by more recent geological changes.

Climatic Relics

Many species of animals and plants which in the past were widely distributed have been affected by the climatic changes of the past two million years or so, which in geological terms are relatively recent. Some of these organisms now survive only in a few 'islands' of favourable climate. Such species are called **climatic relicts**. They are not necessarily species with long evolutionary histories, because many major climatic changes have occurred quite recently, even the past 20 000 years since the maximum extent of the last glacia-

tion of the Ice Age. The Northern Hemisphere has an interesting group of **glacial relict** species whose distributions have been modified by the northward retreat of the great ice sheets that extended as far south as the Great Lakes in North America, and to Germany in Europe, during the Pleistocene Ice Ages (the last glaciers finally retreated from these temperate areas only about 10 000 years ago). Many species that were adapted to cold conditions at that time had distributions to the south of the ice sheets almost as far as the Mediterranean in Europe. Now that these areas are much warmer, such species survive there only in the coldest places, usually at high altitudes in mountain ranges, and the greater part of their distribution lies far to the north in Scandinavia, Scotland or Iceland. In some cases, species even appear to have become extinct in northern regions and are represented now only by relict populations at high altitude in the south, such as in the Alpine ranges. The places where relicts have managed to survive through a time of stress are called **refugia**.

An example of a climatic relict is the springtail *Tetracanthella arctica* (Insecta, Collembola). This dark-blue insect, only about 1.5 mm long, lives in the surface layers of the soil and in clumps of moss and lichens, where it feeds on dead plant tissues and fungi. It is quite common in the soils of Iceland and Svalbard, and it has also been found further west in Greenland and a few places in Arctic Canada. Outside these truly Arctic regions it is known to occur in only two locations – in the Pyrenean Mountains between France and Spain, and in the Tatra Mountains on the borders of Poland and Slovenia (with isolated finds in the nearby Carpathian Mountains) (Figure 2.20). In these mountain ranges, the species is found at altitudes of around 2000 m in Arctic and sub-Arctic conditions. It is hard to imagine that the species can have colonized these two areas from its main centre farther north, because it has very poor powers of dispersal (it is quickly killed by low humidity or high temperatures) over land and is not likely to have been transported there accidentally by humans. Springtails are capable of survival in the surface layers of ocean waters and could be transported around the Arctic in this way, but that would be of no help in reaching the land-locked mountains of Europe. The likely explanation for the existence of the two southern populations is

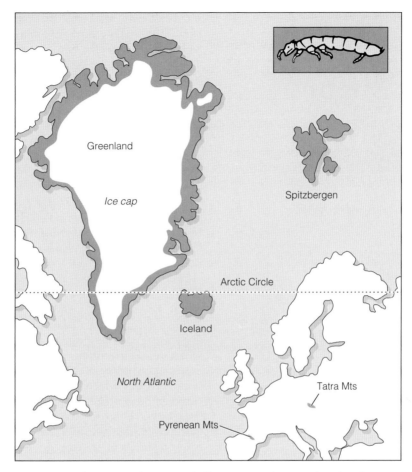

Figure 2.20 The springtail *Tetracanthella arctica*, and a map of its distribution. It is found mostly in northern regions, but populations exist in the Pyrenees and in mountains in central Europe. These populations were isolated at these cold, high altitudes when the ice sheet retreated northward at the end of the Ice Age.

that they are remnants of a much wider distribution in Europe in the Ice Ages. But it is surprising that *T. arctica* has not been found at high altitudes in the Alps, despite careful searching by entomologists. Perhaps it has simply not yet been noticed, or perhaps it used to occur there but has since died out. One interesting feature of this species is that whereas representatives from the Arctic and the Tatras have eight small eyelets (ocelli) on either side of the head, specimens from the Pyrenees have only six. This suggests that the Pyrenean forms have undergone some evolutionary changes since the end of the Ice Ages while they have been

isolated from other populations of the species, and perhaps they should be classified as a separate subspecies.

A plant example of a glacial relict (Figure 2.21) is the Norwegian mugwort (*Artemisia norvegica*), a small alpine plant now restricted to Norway, the Ural Mountains and two isolated localities in Scotland. During the last glaciation and immediately following it, the plant was widespread, but it became restricted in distribution as forests spread. Its pollen grains have been found as far south as Wales, dating from the closing stages of the last glaciation.

(a)

(b)

Figure 2.21 The Norwegian mugwort, *Artemisia norvegica*: (a) the plant; (b) a distribution map showing its restricted range in only two mountainous areas of Europe.

It is very difficult to document the movements of organisms in the past and to test various possibilities of migration and disjunction. Recently, molecular studies have offered a new approach to the problem, sometimes with surprising results. Take the glacial buttercup (*Ranunculus glacialis*), for example. This plant has a distinctly disjunct distribution, being found in the mountains of northern Europe, in the Alps and Pyrenees and in the region known as Beringia around the Bering Straits, including western Alaska and the eastern extremity of Russia. In Europe, the glacial buttercup can be termed an *arctic-alpine species*, meaning that it is found both in the high latitudes of the Arctic and in the high-altitude mountains of lower latitudes. Studies on DNA from various populations of the plant revealed some unexpected results. One might suppose that the European populations would show differences from the far-removed Beringian populations, but this was not the case. The populations from the eastern Alps showed close similarities to the northern European populations, as expected, but plants from the western Alps and the Pyrenees showed greater similarity to the Beringian plants [25]. It must be concluded that the Alps was colonized in two waves, presumably at the end of the last glaciation. One of these came from surviving populations to the north, and the other arrived from the far west [26]. The idea that disjunct populations from Arctic and Alpine locations can be explained simply by reference to the idea of glacial relicts must be called into question. Further molecular studies of disjunct distributions of plants and animals may well result in substantial changes in our ideas.

There are probably several hundred species of both animals and plants in Eurasia that are glacial relicts of this sort, and they include many species that, in contrast to the springtail, have quite good powers of dispersal. One such species is the mountain or varying hare, *Lepus timidus*, a seasonally variable species (its fur is white in the winter and bluish for the rest of the year) which is closely related to the more common brown hare, *Lepus capensis*. The varying hare has a circumboreal distribution, including Scandinavia, Siberia, northern Japan and Alaska, being replaced in Canada by the snowshoe hare (*Lepus americanus*), a closely related species with similar seasonal variations in its pelt. The southernmost part of the main European distribution is in Ireland and the southern Pennine mountains of central England, but there is a glacial relict population living in the Alps that differs in no important features from those in the more northerly regions. There is, however, an interesting complication. *L. timidus* is found throughout Ireland, thriving in a climate that is no colder than that of many parts of continental western Europe. The brown hare is absent from Ireland, presumably because the sea between Ireland and Britain has proved too great a barrier, so the varying hare has no competition in that geographically isolated region. There seems to be no climatic reason why this hare should not have a wider distribution in many parts of the world, but it is probably excluded from many areas by its inability to compete with its close relatives, the brown hare (*L. capensis*) in Europe and various other species of hare (*Lepus* spp.) and rabbit (*Sylvilagus* spp.)

Displaced dung beetle

One very remarkable example of a glacial relict is the dung beetle species *Aphodius holdereri* (Figure 2.22). This beetle is now restricted to the high Tibetan plateau (3000–5000 m), and its southern limit is the northern slopes of the Himalayas. In 1973, G. Russell Coope, of London University, found the fossil remains of at least 150 individuals of this species in a peaty deposit from a gravel pit in southern England [28]. The deposit dated from the middle of the last glaciation. Subsequently 14 sites have yielded remains of this species in Britain, all dated between 25 000 and 40 000 years ago. Evidently, *A. holdereri* was then a geographically widespread species, possibly ranging right through Europe and Asia, but climatic changes, especially the warmer conditions of the last 10 000 years, have severely restricted the availability of suitable habitats for its survival. Only the remote Tibetan mountains now provide *A. holdereri* with the extreme climatic conditions within which it is able to survive, free from the competition of more temperate species of dung beetle.

Figure 2.22 *Aphodius holdereri*, a dung beetle now found only in the high plateau of Tibet but which has been found fossilized as far west as Great Britain.

in North America, for food resources and breeding sites. Relict populations of the varying hare survive in the Alps because, of the two species, it is the better adapted to cold and snowy conditions [27].

The varying hare is an example of a species which has clearly distinct fundamental and realized niches. It is capable of a much wider ecological and geographical spread, but fails to achieve this as a consequence of its relatively weak competitive interaction with closely related and ecologically similar species.

Box 2.1 discuses another relict species, the dung beetle *Aphodius holdereri*.

Climatic relicts are not confined to the temperate regions. The last two million years of geological history, when what is now the temperate zone was being subjected to glaciation, have seen considerable changes in the vegetation of the tropics. Many areas now occupied by rainforests were modified as a result of climate changes. Some authors maintain that the rainforest was partially replaced by drier vegetation of a tropical woodland or grassland type, the savanna. Fragments of rainforest undoubtedly remained in the most favourable of locations, possibly being widespread [29], and partial forest fragmentation may account for the disjunct distribution of certain rainforest species right up to

the present time. Termites (Isoptera) are important arthropods in tropical forests that make a living by attacking dead wood, leading to its decomposition. Many of the termite species found in South-East Asia are stenotopic and are particularly sensitive to environmental disturbance, such as the removal of the forest canopy. Recovery from disturbance is slow because termites are poor dispersers and find it difficult to reinvade regions where populations have been eliminated. The distribution patterns of termites could therefore provide an indication of which areas of forest have experienced long-term stability. Such areas would have provided refugia for termites and possibly other species also. Using this approach, Freddy Gathorne-Hardy and coworkers have identified regions of Sumatra, Brunei, northern Sarawak and eastern Kalimantan that served as rainforest refugia during the main glacial advances in the higher latitudes [30]. Evidence from other sources, including geological and botanical data, helps to confirm these conclusions, so analysis of termite assemblages in the area provides a clue to the existence of a climatic refugium.

The strawberry tree (*Arbutus unedo*) in Europe is a good example of what may be termed a **postglacial relict** (Figure 2.23), for its current distribution is a reflection of climatic changes that have

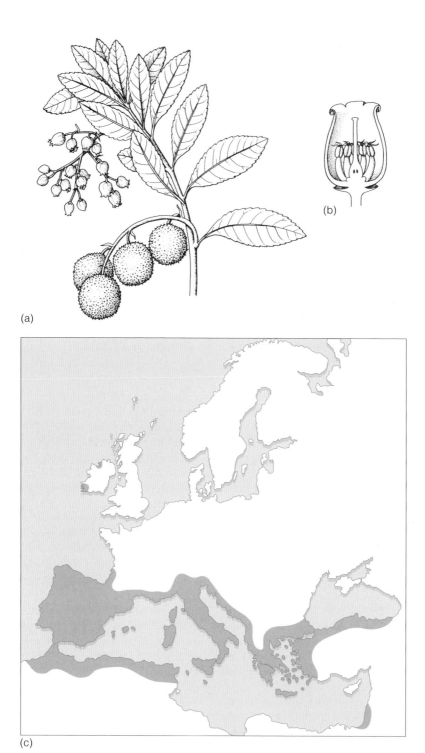

Figure 2.23 The strawberry tree, *Arbutus unedo*: (a) plant showing leathery leaves, and swollen fruit which are red in colour; (b) cross-section of a flower; (c) map of European distribution, showing a relict population in Ireland.

taken place since the last glaciation ended. *A. unedo* is disjunct, having its main centre of distribution in the Mediterranean region but with outliers in western France and western Ireland. The Irish population is particularly surprising because it lies far north of the tree's limits on the mainland of Europe. The Ice Age closed with a sudden warming of the climate, and the glaciers retreated northward; behind them came the plant and animal species that had been driven south during glacial times. Warmth-loving animals, particularly insects, were able to move northward rapidly, but plants were slower in their response because their rate of spread is slower. Seeds were carried northward, germinated and grew, and the mature plants finally flowered and sent out more seeds to populate the bare northlands. As this spread of vegetation continued, melting glaciers produced vast quantities of water that poured into the seas, and the ocean levels began to rise. During the height of the glaciation, so much water was locked up in the ice that sea levels fell by about 100 m, so many areas that are now covered by oceans were then exposed. Some of the early plant and animal colonizers reached new areas by land connections that were later severed by rising sea levels.

The maritime fringe of western Europe must have provided a particularly favourable migration route for southern species during the period following the retreat of the glaciers. Many warmth-loving plants and animals from the Mediterranean region, such as the strawberry tree, moved northward along this coast and penetrated at least as far as the southwest of Ireland, before the English Channel and the Irish Sea had risen to form physical barriers to such movement. The nearness of the sea, together with the influence of the warm Gulf Stream, gives western Ireland a climate that is wet, mild and frost-free, and this has allowed the survival of certain Mediterranean plants that are scarce or absent in the rest of the British Isles. Perhaps this explanation also accounts for the presence of the cold-tolerant varying hare in Ireland as well as the absence of the warmth-demanding brown hare, which arrived after any land bridges were severed by the rising sea.

Like many Mediterranean trees and shrubs, the strawberry tree is **sclerophyllous**, which means

it has hard, leathery leaves (Figure 2.23). This is a plant adaptation often associated with arid climates and seems out of place in the west of Ireland. Flowering in many plant species is triggered by a response to a particular day length, a process called **photoperiodism**. *A. unedo* flowers in late autumn, as the length of night is increasing, and this is an adaptation which is again associated with Mediterranean conditions, since at this season the summer drought gives way to a warm, damp period. The flowers, which are cream-coloured, conspicuous and bell-shaped, have nectaries that attract insects, and in Mediterranean areas they are pollinated by long-tongued insects such as bees, which are plentiful in late autumn. In Ireland, however, insects become increasingly scarce in the autumn and pollination is therefore much less certain. Thus, the strawberry tree reached Ireland soon after the retreat of the glaciers and has since been isolated there as a result of rising oceans. Although the climate has steadily grown colder since its first colonization, *A. unedo* has so far managed to hold its own and survive in this outpost of its range, despite having features in its structure and life history that seem ill-adapted to western Ireland. The warm waters of the Gulf Stream Drift arriving from the Caribbean regions have undoubtedly contributed to the survival of the strawberry tree by reducing the incidence of prolonged winter frosts.

Several plants and animals, in addition to the strawberry tree, have this disjunct distribution pattern between Spain and Portugal and the west of Ireland, and they are termed **Lusitanian species**. Lusitania was a province of the Roman Empire in the Iberian Peninsula. Among Lusitanian animals, perhaps the most remarkable is the Kerry slug (*Geomalacus maculosus*) [31]. Its distribution pattern is shown in Figure 2.24. It is highly unlikely that a relatively immobile animal, such as a slug, could have crossed the waters of the eastern Atlantic and found its way to Ireland, so its movement must have taken place while sea levels were considerably lower. Any intermediate populations have evidently been lost. In 2010, the hazel dormouse was found for the first time in Ireland. It is present in southern Britain and western Europe, but was not known in Ireland. Genetic analysis has shown that it is more closely related to the French

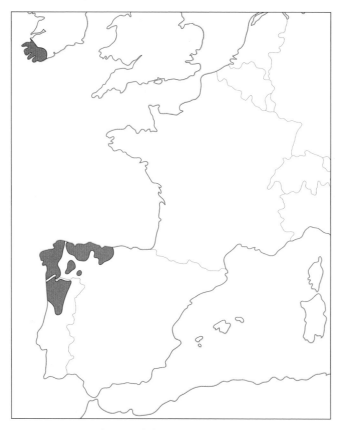

Figure 2.24 Distribution of the Kerry slug in western Europe. Like the strawberry tree, it is a Lusitanian species that is likely to have spread north following the retreat of the glaciers some 10 000 years ago, assisted by much lower sea levels at that time. Rising sea levels have now obliterated any intervening populations. Adapted from Beebee [31].

populations than to those of Britain, again suggesting the Irish animals arrived by a land route directly from France, avoiding Britain.

A further example of a disjunction that has taken place in relatively recent times is the gorilla. The western gorilla (*Gorilla gorilla*) is found in an area of lowland tropical rainforest in the extreme west of tropical Africa. It is considered to be represented by two subspecies, *Gorilla gorilla gorilla* in the far west of its range, and *Gorilla gorilla diehli* to the eastern side of the range. The eastern gorilla (*Gorilla beringei*), as its name implies, inhabits regions farther to the east of Africa, but it is not limited to lowland forest, being also found in mountains (Figure 2.25). There are two populations of the eastern gorilla, which are regarded as separate subspecies, *Gorilla beringei beringei* in the mountains, and *Gorilla beringei graueri* in the eastern lowland forest. The two gorilla species and their constituent subspecies have diverged as a result of the changing patterns of vegetation in central Africa over the last two million years or so [32], during which the composition of the forest has passed through a number of changes in response to climatic changes.. This pattern of disjunction exhibited by the gorillas is reflected in the distributions of many African plants and animals [33].

Figure 2.25 Distribution map of the gorilla (*Gorilla* species), a mammalian genus with a disjunct distribution. The two populations are now regarded as distinct species: *Gorilla gorilla*, the western lowland gorilla, and *Gorilla beringei*, the eastern gorilla, which consists of two subspecies, the mountain gorilla (*G. b. beringei*) and the eastern lowland gorilla (*G. b. graueri*).

Topographical Limits and Endemism

Many of the examples of distribution patterns given so far are now confined to a specific range by topographical factors, such as expanses of ocean or, in the case of the giant groundsels of East Africa, areas of lowland forest between their mountain habitats. Such isolation may lead to the evolution of new forms or even new species in these localities (see Chapter 8). Other species may evolve in one region, spread to other locations and then become extinct in all but a restricted area where it survives, as in the case of the Kerry slug. Species restricted in this way are said to be endemic to that area. As time goes by, increasing numbers of organisms may evolve within an area, or become confined there, and the percentage of that region's biota which is endemic is therefore a

good guide to the length of time for which that area has been isolated.

As these endemic organisms continue to evolve, they will also become progressively different from their relatives in other areas. Taxonomists often recognize this by giving higher taxonomic rank to the organisms concerned. So, for example, after two million years the biota of an isolated area might contain only a few endemic species. After 10 million years, the descendants of these species might be so unlike their nearest relatives in other areas that they might be placed in one or more endemic genera. After 35 million years, these genera might appear to be sufficiently different from their nearest relatives as to be placed in a different family, and so on. (The absolute times involved would, of course, vary depending upon the rate of evolution of the group in question.) Therefore, the longer an area has been isolated, the higher the taxonomic rank of its endemic organisms is likely to be, and vice versa.

Figure 2.26 shows the percentage of the flora in various European mountain ranges that are endemic to their particular area. The more northerly of the mountain ranges shown have a lower proportion of their flora that is endemic, whereas the southern, Mediterranean mountains have higher proportions [34]. The mountains of southern Spain and Greece have more endemics than the Pyrenees and the Alps. This could be interpreted to mean that the southern mountains have been isolated for longer periods. But montane plants, such as the glacial relicts described in this chapter, are now limited in range because of the increasing warmth of the last 10 000 years. The northern mountains may be poorer in endemics simply because local glaciation there was more severe and some of the species that still survive further south consequently became extinct. On the other hand, the richness of the southern mountains could be explained by the fact that the geographical barriers between the northern mountain blocks are less severe (less distance, no sea barriers), and hence migration and sharing of mountain floras are more likely than in the south, where barriers are considerable. This illustrates that the interpretation of patterns of endemicity must be undertaken with care. Biogeographers are increasingly looking to genetic evidence to elucidate such problems.

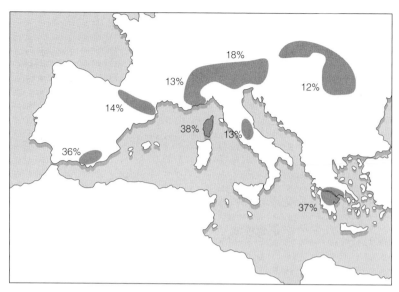

Figure 2.26 The percentage of endemic plants in the floras of the mountain ranges of southern Europe. From Favarger [34]. (Reproduced with permission of Elsevier.)

In general, there are two major factors influencing the degree of endemism in an area: isolation and stability. Thus, isolated islands and mountains are often rich in endemics. The island of Australia, for example, has long been isolated from outside influences, until the arrival of European people with their associated invasive organisms. Although not particularly stable in its climate, Australia covers a very considerable area, so on a simple area basis it should be expected to have an extensive range of endemics. Australia also contains few physical barriers to movement during times of change, so extinction by local isolation has not been an important factor. This would also contribute to high levels of endemism, and as a result of these factors, Australia is indeed rich in endemics, many of which have a long geological history. This kind of 'fossil endemism' is called **palaeoendemism**, in contrast to **neoendemism** resulting from recent surges in the evolutionary process and the generation of new species that have not yet had an opportunity to spread beyond their current limits.

California, for example, is rich in neoendemics, including such plant genera as *Aquilegia* and *Clarkia*, which are undergoing rapid evolution. Several bird species are also endemic to California, such as Nuttall's woodpecker (*Picoides nuttallii*) and the yellow-billed magpie (*Pica nuttallii*). California is isolated from much of the North American continent by the high mountains of the Sierra Nevada, and by the Mojave and Sonoran Deserts, so the evolution taking place there has not been able to disperse at all easily. The richness of the flora of California, however, is typical of many regions of the Earth with a 'Mediterranean' type of climate, including the Mediterranean basin itself, Chile, the southern tip of South Africa and the southwestern extremity of Australia. Much debate has surrounded the high floral richness of these regions, and it may well be that the long history of recurrent fires has created conditions under which small, isolated populations of plants have diversified, leading to a high density of species, many with restricted distributions [35].

Physical Limits

The range of a species is not always determined by the presence of topographical barriers preventing its further spread. Often, a species' distribution is limited by a particular factor in the environment that influences its ability to survive or reproduce adequately. These factors in the

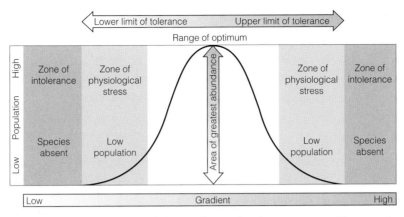

Figure 2.27 Graphic model of the population abundance maintained by a species of animal or plant along a gradient of a physical factor in its environment.

environment include physical factors such as temperature, light, wetness and dryness, as well as biotic factors such as competition, predation, parasitism or the presence or absence of suitable food. All of these factors contribute to the niche of the organism, as described in this chapter.

Taking a single environmental variable, such as temperature, humidity, soil pH and so on, any species will have certain limits along a gradient of that factor, as illustrated in Figure 2.27. It will also have an optimum for that factor, at which its populations will grow most effectively. Between its limits and its optimum, the species will suffer various degrees of physiological stress that will render it less efficient in competition with other species. The diagram is thus a representation of the fundamental niche of the species with respect to this particular factor. The realized niche may be more abbreviated as a result of competition and exclusion, especially as the species approaches its limits.

Of these factors, there is often one that is particularly important and that may be over-riding in determining survival and hence distribution. This is called the **limiting factor**. Anything that tends to make it more difficult for a species to live, grow or reproduce in its environment may prove to be a limiting factor for the species in that environment. To be limiting, such a factor need not be lethal for a species; it may simply make the operation of its physiology or behaviour less efficient, so that it

is less able to reproduce or to compete with other species for food or living space.

The grey hair grass (*Corynephorus canescens*) is widespread in central and southern Europe and reaches its northern limit in the British Isles and southern Scandinavia (Figure 2.28). Examination of the factors that may be responsible for maintaining its northern limit shows that both flowering and germination are affected by low temperature [36]. This grass has a short life span (about 2–6 years), so it relies upon seed production to maintain its population. Any factor interfering with flowering or with germination could therefore limit its success in competitive situations. At its northern limit, low summer temperature delays its flowering with the result that the season is already well advanced when the seeds are shed. Seed germination is slowed down at temperatures below 15 °C, and seeds sown experimentally after October have a very poor survival rate. This may explain why its northern limit in western Europe so closely matches the 15 °C July mean isotherm. Temperature can thus be regarded as its limiting factor on the northern edge of its range. Other factors, however, must be in operation to prevent its spread in southern and central Britain and southern Ireland. Its eastern limit may also be determined by a separate factor, possibly the duration and the severity of winter conditions in northeastern Europe.

Box 2.2 illustrates some limiting factors of bird species.

Even mobile animals, like birds, may have their distributions closely linked to temperature, as in the case of the eastern phoebe (*Sayornis phoebe*), a migratory bird of eastern and central North America. Analysing data collected by ornithologists of the National Audubon Society, ecologist Terry Root has been able to check the winter distribution of this bird against climatic conditions [37]. She found that the wintering population of the eastern phoebe was confined to that part of the United States in which the mean minimum January temperature exceeded −4°C. The very close correspondence of the bird's winter range to this isotherm, shown in Figure 2.29, probably relates to the energy balance of the birds. Warm-blooded animals, such as birds, use up large quantities of energy to maintain their high blood temperature, and in cold conditions they can lose a great deal of energy in this way, which means they therefore have to eat more. Terry Root found that birds in general do not occupy regions where low temperature forces them to raise their resting metabolic rate (i.e. their energy consumption) by a factor of more than 2.5. In the case of the eastern phoebe, this critical point is reached when the temperature falls below −4°C, so the bird fails to occupy colder regions. Other birds have different temperature limits because they have different efficiencies in their heat generation and conservation, but they still seem to draw the line at raising their resting metabolism by a factor of more than 2.5.

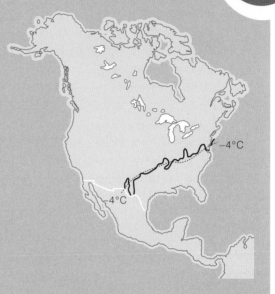

Figure 2.29 Northern boundary (solid line) of the distribution of the eastern phoebe (*Sayornis phoebe*) in North America in December and January, compared with the −4 °C January minimum isotherm (dashed line). From Root [37].

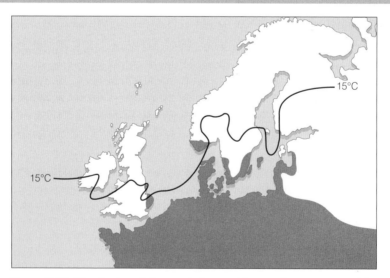

Figure 2.28 Distribution of the grey hair grass (*Corynephorus canescens*) in northern Europe (shaded) and its relationship to the 15°C July mean isotherm.

Many plants have their seeds adapted to a specific temperature for germination, and this often relates to conditions prevailing when germination is most appropriate for the species. P.A. Thompson of Kew Gardens, United Kingdom [38], examined the germination requirements of three members of the catchfly family, Caryophyllaceae. The catchfly *Silene secundiflora* is a Mediterranean species that is found in southern and eastern Spain, as well as the Balearic Islands. Its optimum germination temperature was relatively low, peaking at 17°C. The best time for a plant to germinate in the Mediterranean is the autumn, when the hot, dry summer is over and the cool, moist winter is about to begin, so a low germination temperature requirement is appropriate. The related ragged robin (*Lychnis floscuculi*) occurs throughout temperate Europe, and for this species the cold winter is the least favourable period for growth, hence there are advantages to be gained by germinating only with the onset of warmer conditions in the spring. Optimum germination occurs at about 27°C. The third species, the sticky catchfly (*Silene viscosa*), is an eastern European steppe species. The invasion of open steppe grassland is an opportunistic business; each chance that offers itself must be taken, so any temperature limitation is likely to be an unacceptable restriction on a plant in its struggle for space. Wide tolerance of temperature for germination is thus an advantage, and *S. viscosa* seeds germinate over the wide temperature range of 11–31°C. The temperature requirements for germination are thus tuned to their overall distribution and their respective ecologies.

Germination is not the only plant process affected by temperature, however. Most metabolic activities in plants and animals are assisted by the activity of enzymes, proteins that act as catalysts in biochemical interactions. All enzymes become deactivated at very high or very low temperatures, but different enzymes vary in their optimum temperature for operation. Photosynthesis, in which atmospheric carbon dioxide is reduced and fixed into organic materials, is critical to the function of green plants, and, like all processes mediated by enzymes, it is sensitive to temperature. Most plants use an enzyme called rubisco to fix atmospheric carbon dioxide and produce initially a sugar with three carbon atoms. For this reason, they are called C3 plants. In some green plant species,

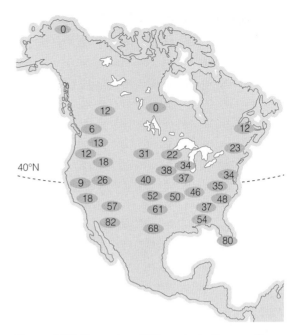

Figure 2.30 Proportion of C4 species in the grass flora of various parts of North America. From Teeri and Stowe [40]. (Reproduced with permission of Springer Science + Business Media.)

however, there is a supplementary mechanism at work in which carbon dioxide is temporarily fixed into a four-carbon compound, which is later fed into the conventional C3 fixation process. These are called C4 plants, and they use a different enzyme represented by the abbreviation PEP carboxylase. The grass family (Poaceae) has both C3 and C4 species within it, and examination of the distribution of these two photosynthetic types in North America reveals that C4 species are more abundant in the south, and C3 species in the north (Figure 2.30) [40].

In general, C4 plants prove more efficient when light intensities and temperatures are high, and where drought is a problem. Theoretical considerations based on the temperature requirements of the enzymes involved and the relative advantages conferred by the C4 system in water conservation suggest that the latitude at which the balance shifts is 45° north, as shown in Figure 2.31. This corresponds quite well to the observed biogeography of the grass species. The C4 photosynthetic system in plants is evidently a mechanism for coping with high temperature and high illumination.

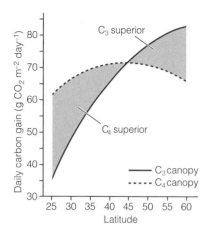

Figure 2.31 Predicted levels of photosynthesis for C3 and C4 species over a range of latitudes in the Great Plains during July. The C4 advantage is lost in latitudes higher than 45°N. From Ehleringer [39]. (Reproduced with permission of Springer Science + Business Media.)

Light in its daily and seasonal fluctuation also regulates the activities of many animals. The concentrations of oxygen and carbon dioxide in the water or air surrounding organisms are also important. Oxygen is essential to most animals and plants for the release of energy from food by respiration, and carbon dioxide is vital because it is used as the raw material in the photosynthesis of carbohydrates by plants. Many other chemical factors of the environment are of importance, particularly soil chemistry where plants are concerned. Pressure is important to aquatic organisms; deep-sea animals are specially adapted to live at high pressures, but the tissues of species living in more shallow waters would be easily damaged by such pressures.

In marine environments, variation in the salinity of the water affects many organisms, because many marine organisms have body fluids with much the same salt concentration as seawater (about 35 parts per thousand), in which their body tissues are adapted to function efficiently. If they become immersed in a less saline medium (e.g. in estuaries), water moves into their tissues due to the physical process called osmosis, by which water passes across a membrane from a dilute solution of a salt to a concentrated one. If the organisms cannot control the passage of water into their bodies, the body fluids are flooded and their tissues can no longer function. This problem of salinity

is an important factor in preventing marine organisms from invading rivers, or freshwater ones from invading the sea and spreading across oceans to other continents.

In a coastal rock pool, the salinity can change very quickly. Once isolated from the main body of the sea, such a pool can become increasingly saline as a result of evaporation. But if there is rain, then the salinity can be rapidly lowered, placing any organisms present under great osmotic stress. An estuary is rather more predictable because the salinity varies regularly both in space and in time. The distance from the sea influences salinity as the input of seawater becomes less, but salinity at any given location will vary with time because of the impact of tidal flows. The crustacean genus *Gammarus* is found in estuaries but is represented by different species according to the nature of the salinity conditions (Figure 2.32). Each species has its optimum set of conditions for salinity, but also has its distribution limits which result from a combination of its reduced tolerance and also competition from other species which may perform more efficiently under the new conditions [41], as illustrated in the diagram in Figure 2.27.

Any regular change in physical or chemical conditions through space thus creates a sequence of replacement of one species by another, among both animals and plants. This is known as **zonation** and is common where habitats gradually merge from one type to another. Such conditions are familiar in many locations where vegetation changes from one form to another as conditions vary in a linear manner. Margins of lakes and pools, for example, often demonstrate zonation patterns as floating aquatic plants are replaced by emergent aquatics and reeds as the water becomes shallower. Finally, species requiring better soil oxygenation, often represented by shrubs and trees, take over in conditions where the water table is at or below the soil surface. Each species has its optimum location and its limits within such a sequence. Sometimes, as in the case of the pool margin, the zonation pattern can also be related to a temporal sequence. Growth of plants leads to increased sedimentation and hence to changing conditions, leading to a change in vegetation. But other types of zonation, such as the estuarine example given above, or the pattern of organisms along seashores, regulated by

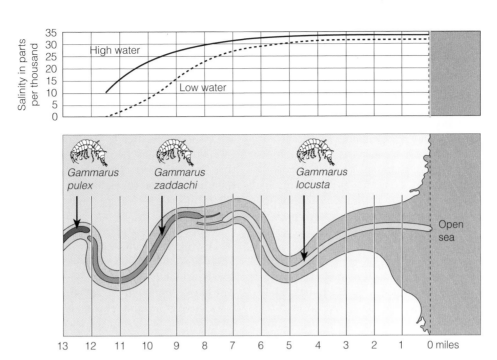

Figure 2.32 Distribution along a river of three closely related species of amphipod (Crustacea), relative to the concentration of salt in the water. *Gammarus locusta* is an estuarine species and is found in regions where the salt concentration does not fall below about 25 parts per thousand (ppt). *Gammarus zaddachi* is a species with a moderate tolerance of saltwater and is found along a stretch of water between 11 and 19 km (8–12 miles) from the river mouth, where salt concentrations average 10–20 ppt. *Gammarus pulex* is a true freshwater species and does not occur at all in parts of the river showing any influence of the tide or saltwater. From Spooner [41].

periods of emersion and immersion in seawater, do not vary over the course of time.

Ecological factors do not always act independently of one another. The environment of any species consists of an extremely complicated series of interacting gradients of all the factors, biotic as well as physical, and these influence its distribution and abundance. Populations of a species can live only in those areas where favourable parts of the environmental gradients that affect it overlap. Factors that fall outside this favourable region are limiting ones for the species in that environment. The species must also, of course, be available to invade the area when given the opportunity.

Some of the interactions between the various factors in an organism's environment may be very complex and difficult for the ecologist to interpret, or for the experimentalist to investigate. This is because a series of interacting factors may have

more extreme effects on the behaviour and physiology of a species than any single factor in isolation. To take a simple example: temperature and water interact strongly on organisms, because both high and low temperatures reduce the amount of water available to an organism in an environment. High temperatures cause evaporation and low ones cause freezing, but it may be very hard to discover if an organism is being affected by the direct effects of heat or cold, or by lack of water. Similarly, light energy in the form of sunlight exerts a great influence on organisms because of its importance in photosynthesis and in vision, but it also has a heating effect on the atmosphere and on surfaces, and therefore raises temperatures. A shade-preferring organism may be seeking low light intensities, or may simply be avoiding high temperatures or the low humidity associated with high temperatures. In natural situations, it is often almost impossible to tell which of many possible factors is mainly responsible for limiting the distribution of a particular species.

Defining the parameters of a species' niche in this way can also lead to important applications in conservation. The leopard (*Panthera pardus*), for example, is a scarce and threatened species in western Asia, and it is valuable to be able to map the areas where the species would be able to survive. With a large mammal such as the leopard, one cannot set up experiments to measure its preferences with respect to various environmental factors: data on the environmental preferences of the organism have to be obtained by field observation. Ecologists from Russia and Georgia [42] have located populations of wild leopards in western and central Asia and have documented various features of the habitats where they are found. They noted various aspects of climate, terrain features such as vegetation and tree cover, and the proximity to, or distance from, human activity. Pooling their observations, they were able to construct a model that described the conditions tolerated by the leopard. The cats were found to avoid deserts, urban developments and regions with prolonged snow cover. The researchers could then produce maps in which areas according with the requirements of the leopard were all met, and thus they could highlight sites appropriate for survey work to enumerate leopard populations and sites suitable for leopard conservation and possible reintroduction. They could also examine the likelihood of leopard movements between these regions, which is important for the maintenance of genetic flow.

This study did not examine in detail certain other factors that could influence the presence and survival of leopards, such as the availability of prey or the intensity of hunting and poisoning by people. Potential distributions of species cannot be fully understood without reference to the influence of other organisms and their respective distribution and ecological requirements.

Species Interaction: A Case of the Blues

Physical factors clearly play an important part in determining the distribution limits of many plants and animals, but organisms also interact with one another, and this can place constraints on geographical ranges. One species may depend strictly on another for food, as in the case of some butterflies which may be limited to a single food plant, or a parasite may be limited to a specific host. Some species may be unable to colonize an area because of the existence of certain efficient predators or parasites in that area, or because some other species is already established there and can compete more efficiently for a particular resource that is in demand. These are biotic factors, and they are often responsible for limiting the geographical extent of a species within its potential physical range.

Several examples of such limitations are found within the family of blue butterflies (Lycaenidae). There are about 5000 species in this cosmopolitan family, which in North America is given the poetic title of *gossamer wings*. The blue butterflies form a distinctive group and are found in both the Old and New Worlds. Several of the species of blue butterfly have complex relationships with other organisms, three of which will be described here.

The Adonis blue (*Lysandra bellargus*), as its name suggests, is a spectacularly beautiful insect, especially the male, which is coloured an electric blue. It is found throughout central and southern Europe, from England to Spain and from France eastward to Iran and Iraq. The caterpillars feed upon only one species of plant, the horseshoe vetch (*Hippocrepis comosa*), which grows only on chalk or limestone, so the distribution of this butterfly is limited by its feeding requirements and, consequently, by geology. The caterpillars are distinctive in possessing glands on their bodies that secrete a kind of sweet honeydew, which is very attractive to ants (Figure 2.33). Wherever you find the caterpillars, they are sure to be attended by ants that constantly lick the glands. Two small tentacle glands can be expanded when required, and these release volatile chemicals to attract ants when they are no longer in attendance. The ants obviously benefit from this arrangement in that they obtain easy food, but it turn they protect the caterpillar from predators. They guard it all day while it is feeding on the vetch, and at night they often hide it by burying it, sometimes collecting several caterpillars together and constructing an underground cell for them. The ants also look after the chrysalis, and the final emergence of the butterfly is often accompanied by frenzied excitement among

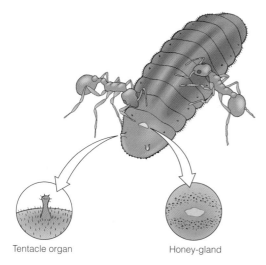

Tentacle organ   Honey-gland

Figure 2.33 Ants attending the caterpillar of an Adonis blue butterfly. Also shown are the two types of gland, the tentacle organ that secretes volatile attractants, and the honey gland that secretes honeydew. After Thomas and Lewington [43]. (Reproduced with permission of Bloomsbury Publishing plc.)

the ants. The relationship is evidently mutually beneficial, but it does mean that the butterfly has one more possible restriction on its distribution, requiring the presence of ants. Fortunately for the Adonis blue, it is associated with several species of ants, so this is not a severe limitation.

In the case of the large blue (*Maculinea arion*), the relationship with ants has taken a different turn. The large blue extends further north than the Adonis blue, reaching into Scandinavia, but is restricted in Spain to the northeast of the country. It extends east through Italy and Greece, and north through Russia and Siberia, Mongolia, China and Japan. The larvae feed on various species of wild thyme (*Thymus* spp.), which between them cover a very wide range of geology and habitat, from acid to alkaline. This may account for the greater range of this butterfly. Like the Adonis blue, its caterpillars have a honey gland that secretes honeydew and this is attractive to ants, but in this case it is just one genus of red ant, *Myrmica*, that takes charge of them. After feeding for several hours, the ants seem to adopt the caterpillar and carry it off to their nest, treating it as though it were one of their own grubs. Perhaps it secretes chemical stimuli that fool the ants into believing this. But once in the nest, the

caterpillar turns predator and begins to eat the ant grubs, soon becoming a hundred times bigger as a consequence. It may spend up to two years in the ant nest, during which time it is thought to consume up to 1200 grubs [43]. Although the caterpillars may be adopted by several species of *Myrmica* ants, their success is far greater when the species *Myrmica sabuleti* is involved as host. When in the nests of other *Myrmica* species the grub is often killed, suggesting that its mimetic scents so closely resemble *M. sabuleti* that other species become suspicious and destroy it.

This degree of specialization on the part of the large blue may account for the fact that, although widespread, it is a relatively rare species through its range. Its success is closely tied to that of its host ant, *M. sabuleti*, and that ant has its own ecological requirements and limits. In particular, this ant species demands very closely cropped vegetation, which usually means heavy grazing. Figure 2.34 described an experiment in which scrub vegetation was burned and grazed. At first it was another ant species, *Myrmica scabrinodis*, that rose to prominence, only to be replaced by *M. sabuleti* as grazing continued. When grazing was relaxed, the turf became tall and neither ant species performed well, but the return of grazing allowed the recovery of the ants. The complexity of the relationships became clear when in the 1950s the virulent disease myxomatosis spread through the rabbit population of Europe, resulting in the decline in grazing pressure on grasslands, the growth of taller turf, the decline in *Myrmica* ants and the collapse in large blue populations.

A third illustration of species interdependence among the blue butterflies is the case of the holly blue (*Celastrina argiolus*). The life cycle of this species is simpler than that of the other two blues considered here, and its food preferences are much wider, which may account for its more widespread global distribution. It is found throughout Europe, North Africa, Asia (east to Japan) and North America (from Alaska east through Canada, and throughout the United States south to Mexico and Panama). Some taxonomists, however, regard the New World taxon as a separate but closely related species, *Celastrina ladon*, with the common name spring azure. Food plants are many and various, belonging to such families as Rosaceae (roses), Cornaceae (dogwoods), Fabaceae (peas), Ericaceae

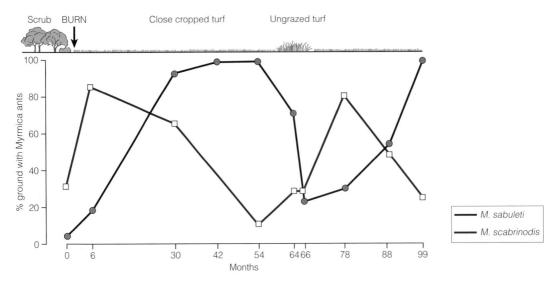

Figure 2.34 Changes in the abundance of two species of red ant in the course of time as the management regime of an area of grassland is modified. The host species of the large blue butterfly, *Myrmica sabuleti*, requires sustained intense grazing to produce a short turf. From Thomas [44].

(heathers) as well as many more. Many of its food plants are relatively tall, even trees, so although the caterpillar is equipped with ant-attracting glands, only ants that are prepared to climb are likely to encounter them. Perhaps they need more ant protection than they receive because their most formidable enemies are the parasitic wasps, *Cotesia inducta* (which attacks young larvae) and *Listrodomus nycthemerus* (which concentrates on older larvae). The intensity of parasitism is such that the population of holly blues can be virtually extinguished on a local level. When the prey becomes scarce, however, the wasp population declines, and the loss of wasps means that the caterpillars thrive during the next year or two. This cyclic process is shown in Figure 2.35.

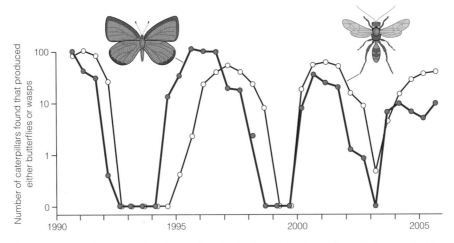

Figure 2.35 Cyclic population fluctuations in the holly blue butterfly and the parasitoid wasp *Listrodomus nycthemerus*. The butterfly population tends to peak every 2–3 years while the wasp population is low. From Thomas and Lewington [43]. (Reproduced with permission of Bloomsbury Publishing plc.)

All three of these examples from the Lycaenidae butterflies show how the dependence of one species on another, in this case ant species, can be beneficial, but it also exposes the species to additional degrees of limitation to its population numbers and consequently distribution pattern.

Competition

When two species are attempting to tap a resource in the same way, and when that resource is in short supply, the two are said to be in competition. The outcome of competition will depend on the relative efficiency with which each species manages to tap the resource, and if one species is significantly more efficient than the other, then the weaker species may be excluded from access to the resource entirely. In spatial terms, this may mean that the presence of one species precludes the presence of its competitor.

When a species is prevented from occupying an area by the presence of another species in this way, it is termed **competitive exclusion**. It is not always easy to observe this ousting of one species by another in nature, but an example of its occurrence is found in the barnacle species that occupy the rocky seashores of western Europe and northeastern North America. Adult barnacles are firmly attached to rocks and feed by filtering plankton from the water when the tide is in. Two common species are *Chthamalus stellatus*, which is found within an upper zone of the shore just below the high-tide mark, and *Balanus balanoides*, which occupies a much wider zone below that of *C. stellatus*, down to the low-water mark. The distribution of the two species does not overlap by more than a few centimetres of elevation. This situation was analysed by the ecologist J.H. Connell [45], who found that when the larvae of *C. stellatus* ended their free-swimming existence in the sea and settled down for life, they did so only over the upper part of the shore above mean tide level. The larvae of *B. balanoides*, on the other hand, settled over the whole zone between high and low water, including the area already occupied by the adults of *C. stellatus*. Despite overlapping patterns of distribution of the larvae, different distributions of the adults of the two species prevail as a result of two separate processes. One process acts on the

zone at the top of the shore. The young *B. balanoides* are eliminated from this region because they cannot survive the long period of desiccation and the extremes of temperature to which they are exposed at low tide. *C. stellatus* larvae are more resistant to desiccation and survive. Lower down the shore, the *B. balanoides* larvae persist because they are not exposed for so long, and here the larvae of *C. stellatus* are eliminated by direct competition from the young *B. balanoides*. These grow much faster and simply smother the *Chthamalus* larvae or even prise them off the rocks. Connell also performed experiments on these species and found that, if adult *B. balanoides* were removed from a strip of rock and young ones prevented from settling, the *C. stellatus* were able to colonize the full length of the strip right down to low-tide level. This showed that the competition with *B. balanoides* was the main factor limiting the distribution of *C. stellatus* to the upper part of the shore.

This example provides an illustration of the difference between the fundamental niche and the realized niche of an organism. The fundamental niche of *C. stellatus* is much wider than is apparent in the field. Once competition from *B. balanoides* is taken into account, then its niche is much narrower; this is its realized niche. From this, we learn that one cannot necessarily determine the physical limits of a species simply by observing its pattern of distribution in the field. The modifying factor of competition must be taken into account, and this can be difficult to discern without further experimentation.

The example of the barnacles is a relatively simple one because only two species are involved. In most communities of animals and plants, many species interact, and this makes it even more difficult to sort out the full picture of the relationships between species. One experimental approach to the problem is to remove one species from the community and to observe the reaction on the part of the others. This method has been tried in salt-marsh communities in North Carolina by J.A. Silander and J. Antonovics [46], who removed selected plant species and recorded which of the other plants present in the community expanded into the spaces left behind (Figure 2.36). They found a great range of responses. The removal of one grass species, *Muhlenbergia capillaris*, resulted in an equal expansion on the part of five

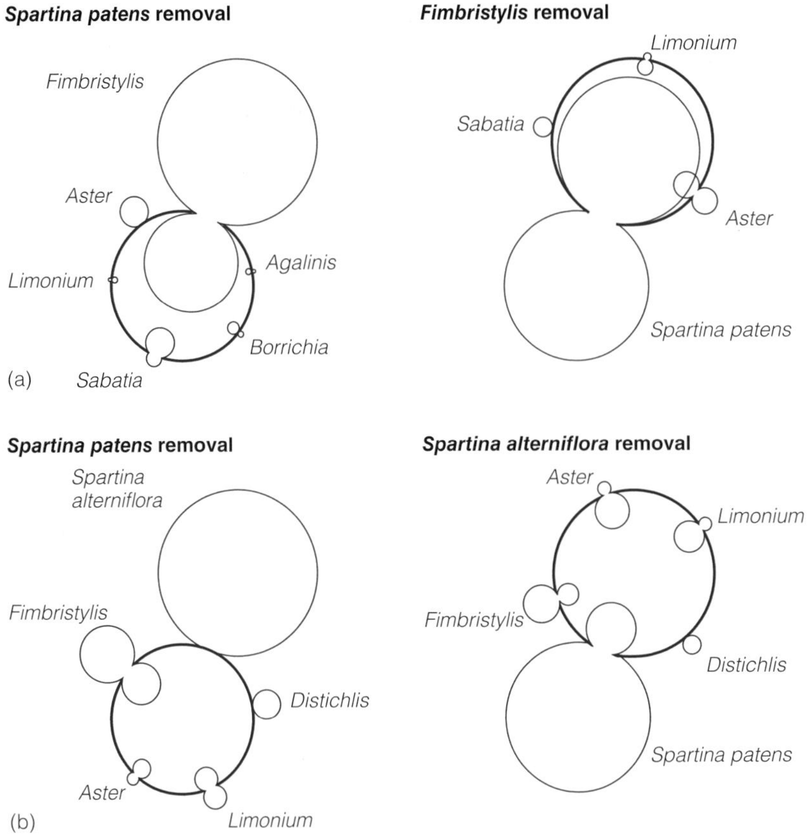

Figure 2.36 Graphic illustration of the effect of removing a single plant species from a salt marsh community. The sizes of the circles represent the abundance of the plant species concerned, and the heavy circle refers to the species that has been removed. Circles intruding into the heavy circle denote the responses of different species to the perturbation of removal. (a) High marsh site, where the removal of *Spartina patens* mainly results in the expansion of *Fimbristylis*, and the removal of *Fimbristylis* results in the expansion of *Spartina*. The two species seem to be in competition, and the effects of removal are roughly reciprocated. (b) Low marsh site where two species of *Spartina* predominate. The removal of *S. patens* results in no response by the other *Spartina* species, whereas the removal of *Spartina alterniflora* does permit some expansion of *S. patens*. Competition here is thus not reciprocal. From Silander and Antonovics [46].

other plants, suggesting that this grass was in competition with many other species. In the case of the sedge *Fimbristylis spadiceae*, however, removal led to the expansion of only one other plant, the chord grass species *Spartina patens*. The reciprocal experiment in which *Spartina* was removed similarly led to *Fimbristylis* taking full advantage of the new opportunity. In this case, we seem to

have only two species that are competing for this particular niche.

Selective removal of species in this way, however, is somewhat artificial and can result in disturbance to the physical environment that alters the very nature of the habitat, so it can only provide a preliminary guide to the relationships of species in the community.

Reducing Competition

An organism may find considerable advantage in avoiding competition, whether with other species or other members of its own species. Many different ways of reducing competition between organisms have evolved. Sometimes, species with similar food or space requirements exploit the same resources at different seasons of the year, or even at different times of day. A common system amongst predatory mammals and birds is for one species (or a group of them) to have evolved specialized night-time activity whilst another species or group of species are daytime predators in the same habitats. Many species of owl hunt at night, judging the location of their prey mostly by ear, whereas the hawks and falcons are daytime hunters with extremely keen eyesight, especially adapted for judging distances accurately. Thus, both groups of predators can coexist in the same stretch of country and prey on the same limited range of small mammals. Many bats are night-active insectivores, avoiding competition for prey with insectivorous birds during the day, and also avoiding the predatory attention of day-active hawks and falcons. Cases of this sort are described as **temporal separation** of species, and this is an effective method of tapping food resources amongst several species. Among plants, this process can be seen operating in deciduous forest habitats, where many woodland floor herbs flower and complete the bulk of their annual growth before the leaf canopy emerges on the trees. In this way, the light resources of the environment are used most efficiently. It also means that two or more species can tap the same resource in different ways and thus avoid being in direct competition.

A different type of temporal separation is found in the complex grazing communities of the East African savanna [47]. During the wet season, all the five most numerous grazing ungulates (buffalo, zebra, wildebeest, topi and Thomson's gazelle) are able to feed together on the rich forage provided by the short grasses on the higher ground. At the beginning of the dry season, plant growth ceases there. The herbivores then descend to the lower, wetter ground in a highly organized sequence. First are the buffalo, which feed on the leaves of very large riverine grasses, which are little used by the other species. The zebra, which are highly efficient at digesting the low-protein grass stems, move

down next. By trampling the plants and eating the grass stems, they make the herb layer available for the next arrivals, topi and wildebeest. These two are found in slightly different areas. The jaws and teeth of the topi are adapted to the short, mat-forming grasses common in the north-western part of the Serengeti. Those of wildebeest are instead adapted to eating the leaves of the upright grasses commoner in the southeastern Serengeti. These two species reduce the amount of grass, facilitating the grazing of the last species, Thomson's gazelle, which prefers the broader-leaved dicotyledonous plants to the narrow-leaved monocotyledonous grasses. The whole community therefore interacts in a complex manner, utilizing the pasture in a highly organized and efficient fashion and avoiding competition. This sharing of resources is an example of niche partitioning.

Probably much more common than niche partitioning by temporal allocation of resources, however, are cases where the resources of a habitat are divided up by the restriction of each species to only part of the available area, to specialized microhabitats. This is called **spatial separation** of species; it means that each species must be adapted to live within the fixed set of physical conditions of its particular microhabitat. It also means that such a species is not as well adapted to live in other microhabitats, and may find it difficult to invade them even if they were for some reason vacant and their food resources untapped. An example of spatial separation has been described in the extensive marshlands of the Camargue in southern France, where various wading bird species have different preferences for the available feeding areas. The greater flamingo (*Phoenicopterus ruber*) has very long legs and is thus able to wade into deep water, where it can sieve planktonic organisms with its highly specialized bill. In shallower water, the avocet (*Recurvirostra avosetta*) and the shelduck (*Tadorna tadorna*) feed in a similar way, by sweeping, side-to-side actions of their necks. On the water's edge is the Kentish plover (*Charadrius alexandrinus*), which feeds on mobile invertebrates and is restricted to these regions by its shorter legs.

The spatial patterns of distribution and feeding of predatory birds, such as these waders, sometimes reflect the patterns of their preferred food species. For example, the oystercatcher (*Haematopus ostralegus*) has a strong predilection for the bivalve mollusc *Cerastoderma edulis*, the common cockle, and this is found mainly on sandy and muddy shores just below the mean high-water mark of neap tides [48]. Therefore, this is the favourite feeding zone of the oystercatcher. Similarly, the mud-dwelling crustacean *Corophium volutator* is a favoured food species for the redshank (*Tringa totanus*), and, because it thrives best in the upper regions of mudflats, usually above the mean high-water mark of neap tides, this is often where large numbers of feeding redshanks can be found.

A high level of niche specialization may be found even within a closely related group of species, such as the tanagers of Central America (Figure 2.37). Three closely related species of tanager, the speckled tanager (*Tangara guttata*), the bay-headed tanager (*Tangara gyrola*) and the turquoise tanager (*Tangara mexicana*), may be found coexisting and feeding alongside one another without any apparent competitive interaction. The reason for this harmony is that each feeds in a slightly different location in the forest canopy. The speckled tanager takes insects from the underside of leaves, the turquoise tanager from fine twigs and the bay-headed tanager from main branches. Each occupies its own niche, and there is little overlap between them. Niche specialization of this sort helps to explain how speciation can take place even in the absence of geographical barriers preventing interbreeding and gene flow.

Even where overlap of niches does occur, animals may often coexist because they have their own distinctive location or way of life which they share with no other. This can be seen in Figure 2.38, which displays diagrammatically the niches of various primate species in the tropical forest of Ghana [49]. Each species has its own

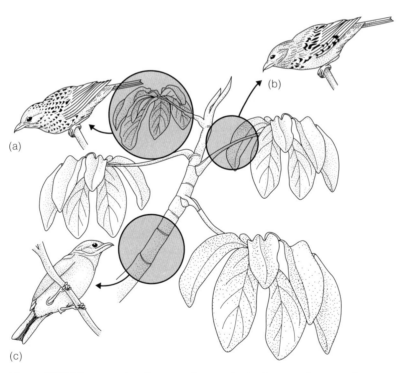

Figure 2.37 Three species of tanager that coexist in the same forest on the island of Trinidad in the West Indies. All feed on insects, but they exploit different microhabitats within the canopy and thus avoid direct competition. (a) The speckled tanager takes insects from the underside of leaves; (b) the turquoise tanager obtains its insects from fine twigs and leaf petioles; and (c) the bay-headed tanager preys upon insects on the main branches.

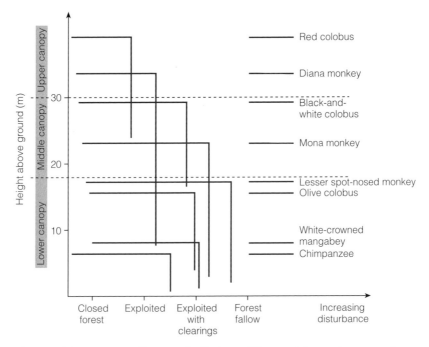

Figure 2.38 Diagram to illustrate the limits to different niche requirements of a range of primate species in the tropical forest of West Africa in relation to canopy height and degree of human disturbance. Each species indicated occupies the space below and to the left of the lines. Although the demands of the various species overlap, each has a particular height in the canopy or a type of site where it is most efficient as a competitor, and therefore more successful. From Martin [49].

preferred position in the forest canopy, some preferring undisturbed forest and others coping with exploited and cleared areas. There is considerable overlap in their tolerances, but each has its own individual specialist location where it can hold its own in the face of competition for resources from other species.

Predators and Prey, Parasites and Hosts

Predators may be another biological factor influencing the distribution of species, as may the presence and abundance of parasites in a habitat, but their effects have been much less studied than those of competition. The simplest influences that predators might have is to eliminate species by eating them or, alternatively, to prevent the entry of new ones into a habitat. There is very little evidence

that either of these processes is common in nature. One or two experimental studies have shown that predators sometimes eat all the representatives of a species in their environment, particularly when the species is already rare. However, all such studies have been made in rather artificial situations in which a predator is introduced into a community of species that has reached some sort of balance with their environment in the absence of any predator; such communities are not at all like natural communities which already include predators. In general, it is not in the interests of predatory species to eliminate a prey species, because if they do this they destroy a potential source of food. Probably most natural communities have evolved so that there is a great number of potential prey species available to each predatory species. Thus, no species is preyed upon too heavily, and the predators can always turn to alternative food species if the numbers of their usual prey should be reduced by climatic or other influences. This is termed **prey switching**.

Prey switching of this type has been described on the island of Newfoundland, where the grey wolf (*Canis lupus*) and the lynx (*Lynx lynx*) were major predators in the 19th century, but where the wolf is now extinct as a result of human persecution. The lynx was a rare animal until a new potential prey animal was introduced to the island in 1864, namely, the snowshoe hare (*Lepus americanus*). The hares multiplied rapidly, and so did the lynxes in response to the newly available food source. But the snowshoe hare population crashed to low levels in 1915 and the lynx, faced with starvation, switched its attentions to caribou calves, which had once been a major food source for the wolf. The snowshoe hare has now developed a 10-year cycle of high and low population levels, and the lynx has continued to switch between hare and caribou depending upon whether the former is in a peak or a trough [50]. This pattern of prey switching allows the lynx to maintain a fairly stable population level, and, as a consequence, it also permits the recovery of snowshoe hare populations.

This type of behaviour on the part of a predator may thus serve to prevent the extinction of its prey. Sometimes, the relationship is even more complex; the predator may prevent the invasion of more efficient or voracious predators which could reduce the prey population yet further. A good example of this is the Australian bell miner (*Marlorina melarlophrys*) [51]. This is a communal and highly territorial bird which occupies the canopy of eucalypt forests and which feeds largely on the nymphs, secretions and scaly covers of psyllid plant bugs. Yet the psyllid bugs survive well under this predation and, what is even more remarkable, they seem to require the attentions of the bell miners, for when these birds are removed, the populations of bugs crash and the eucalypt trees become healthier. It seems that the aggressive behaviour of the miners towards other birds prevents their entering bell miners' territories and eating all the psyllids. While the bell miners stay the psyllids are safe, but the trees suffer!

Parasites generally reduce the rate of growth of populations because of their negative influence on general fitness, survival and fecundity. Like predators, however, they rarely cause the complete extinction of the host species; this would hardly be in their interests. This was seen in the case of the holly blue butterfly and its parasitoid wasp (Figure 2.35). Although parasitism causes a collapse in the butterfly population, this is only temporary and recovery takes place after the collapse of the parasite population, creating a cycle. Detailed studies concerning the influence of parasites on the range and geographical spread of species are often based on introduced species. It has been found that introduced species often have lower parasite loads than native species. One study showed that there were 40% fewer parasites associated with introduced species of plants and animals than were found infesting the native species [52]. In addition, the parasites that the invaders carry are often the least virulent because these tend to be more prevalent among hosts simply because they cause less harm. If the invader is also less susceptible to the local parasites than the native species, then its success potential is further enhanced. So an invasive organism may be at an advantage with respect to parasite limitation, and this may be a clue to the success of many invasive organisms.

As mentioned in this chapter, competition may prevent two species from living together in a habitat and may modify the distribution of species, because the resources of the habitat are inadequate to support both of them. Probably the most important effect of predators and of parasites and disease on the distribution of species is that, by feeding on the individuals of more than one species, they reduce the pressures of competition between them. Thus, by reducing pressures on the resources of the habitat, predators may allow more species to survive than would be the case if the predators were not there. This possibility was first demonstrated experimentally by classical studies on flour beetles and their susceptibility to a sporozoan parasite [53]. When two flour beetles, *Tribolium castaneum* and *Tribolium confusum*, were kept together in a single container, *T. castaneum* inevitably drove *T. confusum* to extinction because it proved a more effective competitor for the food resources. On the introduction of the parasite *Adelina tribolii* to the system, however, the competitive *T. castaneum* proved more susceptible to the parasite and hence its performance was impeded, with the result that the two species were able to coexist. Thus, the presence of a parasite in this situation enabled the weaker competitor to survive and thereby increased the species diversity of the system. The experiment was repeated more recently using a different parasite, *Hymenolepis diminuta*, and the result was very different [54]. The weaker competitor, *T. confusum*, proved more susceptible to this parasite than its dominant companion, so the presence of the parasite drove it to extinction even quicker. So, one cannot generalize about the influence of parasitism on coexistence as it depends on the differential impact of the parasite on the host species involved. A parasite may increase diversity, but it may also reduce it.

A general conclusion, then, is that the presence of predators in a well-balanced community is likely to increase rather than reduce the numbers of species present, so that, overall, predators broaden the distribution of species. Only a few experiments similar to Paine's (described in Box 2.3) have been performed, and so one must be cautious about applying this conclusion to all communities. As in the case of parasites, much depends upon the susceptibility of the various component species in the community to the predator. There is some independent evidence, however, that herbivores, which can loosely be regarded as predators or parasites of plants, may similarly increase the number of plant species that can live in a habitat. In the last century, Charles Darwin noticed that in southern England, meadowland grazed by sheep often contained as many as 20 species of plants, whereas neglected, ungrazed land contained only about 11 species. He suggested that fast-growing, tall grasses were controlled by sheep grazing in the meadow, but that in ungrazed land these species grew tall so that they shaded the small slow-growing plants from the sun and eliminated them. A large-scale experiment occurred in the chalk grassland areas of Britain, when the disease myxomatosis caused the death of large numbers of rabbits in the 1950s; the resulting reduction in grazing allowed considerable invasion by coarse grasses and scrub. As a result, many of these areas are much less rich in species than they were under heavy 'predation'.

On the Washington coast in North America, Robert Paine, who had investigated the effects of predation on animal communities (Box 2.3), performed another series of experiments in which he removed the herbivorous sea urchin *Strongylocentrotus purpuratus*, which grazes on algae [56]. Initially, there was an increase in the number of species of algae present; the six or so new species were probably ones that were normally grazed too heavily by the sea urchin to survive in the habitat. But over 2 or 3 years, the picture changed as the community of algae gradually became dominated by two species, *Hedophyllum sessile* on exposed parts of the shore, and *Laminaria groenlandica* in the more sheltered regions below the low-water mark. These two species were tall and probably shaded out the smaller species, as did the tall grasses studied by Darwin. The total number of species present was in the end greatly reduced after the removal of the herbivores.

Predatory starfish

Concept Box 2.3

Many studies of natural communities have confirmed the hypothesis that predators may increase the number of different species that can live in a habitat. The American ecologist Robert T. Paine made an especially fine study on the animal community of a rocky shore on the Pacific coast of North America [55]. The community included 15 species, comprising acorn barnacles, limpets, chitons, mussels, dog whelks and one major predator, the starfish *Pisaster ochraceus*, a generalist which fed on all the other species. Paine carried out an experiment on a small area of the shore in which he removed all the starfish and prevented any others from entering. Within a few months, 60–80% of the available space in the experimental area was occupied by newly settled barnacles, which began to grow over other species and eliminate them. After a year or so, however, the barnacles themselves began to be crowded out by large numbers of small but rapidly growing mussels, and when the study ended these completely dominated the community, which now consisted of only eight species. The removal of predators thus resulted in the halving of the number of species, and there was additional evidence that the number of plant species of the community (mainly rock-encrusting algae) was also reduced, because of competition from the barnacles and mussels for the available space.

In the floristically rich grasslands of the European Alps, it has been found that the presence of a semi-parasitic plant, the yellow rattle (*Rhinanthus minor*), is associated with increased plant diversity, and conservationists have taken to including this species in seed mixtures when rehabilitating areas of damaged grassland as a means of maintaining plant biodiversity [57]. The parasite taps the roots of the more competitive grasses for part of its nutrition, thus reducing their productivity and giving smaller and less competitive plant species an opportunity to survive.

The activities of carnivorous predators in a community can also have an effect on the plants since, by limiting to some extent the number of their herbivorous prey, they prevent overgrazing, and thus reduce the risk of rare species of plants being eliminated. Such interactions may be quite complex, however, as in the case of the Hawaiian damselfish, which is a predator of herbivorous fish in coral reef habitats. In an experimental study of the influence of this fish [58], plates were constructed which were suitable for algal colonization, and these were placed in three types of location: (i) within cages which excluded all herbivorous fishes; (ii) uncaged, but within the territories of the carnivorous damselfish; and (iii) uncaged and placed outside damselfish territories. The diversity of the colonizing algae was highest on the uncaged plates inside damselfish territories and least in the uncaged samples outside the territories. In other words, where there was no grazing at all the algal diversity was higher than when there was intense grazing, but aggressive algae became dominant and excluded some smaller plant species. The highest diversity was found in sites where grazing was controlled to an extent by the predation of the damselfish upon the grazers. The presence of some light grazing suppressed the more robust algae and thereby allowed colonization of more delicate species.

The complicated sets of interactions between predator, grazer and plant can lead to the development of a finely balanced and diverse community, as shown by this coral reef example. In all of these experiments, based on the manipulation of communities of organisms, it has been found that any one species exerts an influence over many other components of the community, not just its prey organism. The removal of this one species could create effects far in excess of what may originally have been expected. Influential species of this kind are known as **keystone species**. Identifying the keystone species in an ecosystem is clearly a very important task, especially if biodiversity is to be maintained. The loss of a keystone species can cause an avalanche of local extinctions.

Migration

Environmental conditions alter with seasons, especially in the higher latitudes, and some animals alter their distribution patterns in concert with the seasons. This is called **migration**. It should not be confused with range expansion, or spread, of a species, because it consists of the temporary occupation (usually seasonal) of a region while conditions are suitable and then mass movement to an alternative region when the seasonal conditions demand it. Only motile organisms can partake in migration, but even microscopic plankton are capable of changing the depth at which they live, depending on conditions, and this can be regarded as a form of vertical migration. Such migration on the part of plankton is often diurnal rather than seasonal.

Migration often takes the form of latitudinal movements of animals in order to take advantage of long summer days and high productivity in the high latitudes, and then to retreat to lower latitudes to avoid the stresses of the winter season. The movements of caribou (*Rangifer tarandus*) in North America illustrate such migratory patterns [59]. The females give birth to calves in the early summer, and the various distinct herds of the North American population migrate to the north as the snows melt to find the most appropriate location where calving takes place. Each herd has a traditional location (Figure 2.39), usually reflecting the high quality of vegetation productivity that will best ensure survival of the young calves. Consequently, the spring sees a northward migration of caribou herds heading for the breeding grounds. In the fall, as the tundra becomes colder, productivity falls, and the snow accumulation begins, the herds move south again. Migration is energetically expensive and also exposes animals to the risks of predation, in the case of caribou by wolves that follow the herds. But the benefits of migration in terms of food availability and quality must outweigh these costs.

Birds are among the most mobile of animals, and many species resort to migration in order to maxi-

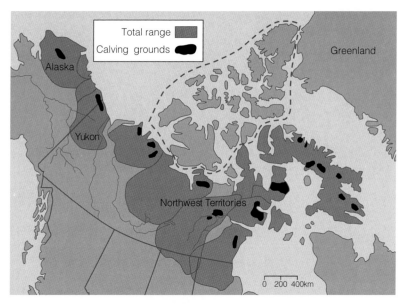

Figure 2.39 Ranges of caribou herds in North America, also showing their calving grounds to which they migrate each spring. Caribou are also located on the islands enclosed within the dashed line. From Sage [59].

mize food supply, especially during the breeding season. The white-fronted goose (*Anser albifrons*) has a circumpolar distribution pattern, breeding in the long summer days of the Arctic and sub-Arctic in North America, west Greenland and Siberia.

The geese spend their winter in southern parts of North America and Central America, in Europe and the Persian Gulf, and in Japan and eastern China, depending upon their breeding locations (Figure 2.40). Perhaps the most extraordinary of all

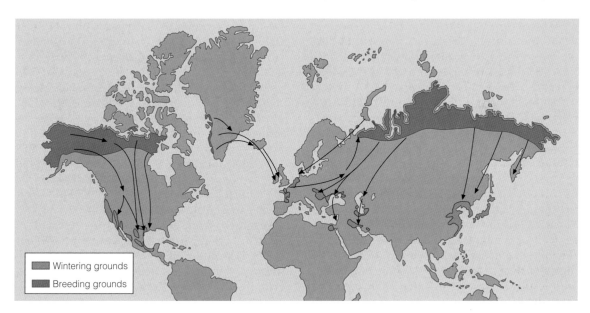

Figure 2.40 Breeding grounds, migration routes and wintering grounds of the white-fronted goose (*Anser albifrons*). From Mead [60].

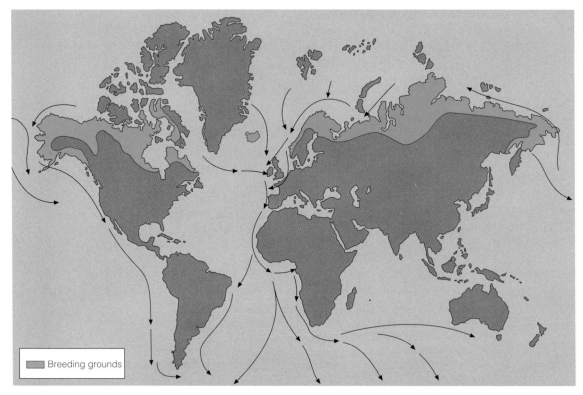

Figure 2.41 Breeding grounds, migration routes and wintering area of the Arctic tern (*Sterna paradisaea*). From Mead [60].

migrant birds, however, is the Arctic tern (*Sterna paradisaea*) which, as its name suggests, nests in the Arctic and yet travels to the Antarctic during the northern hemisphere winter. (Figure 2.41). This bird must enjoy more daylight in the course of its life than any other organism.

Even small songbirds, such as thrushes and warblers, undertake seasonal migrations. Swainson's thrush (*Catharus ustulatus*), for example, migrates between Canada and the Pacific Northwest to Central and South America each fall, returning in the spring. This journey costs a lot of energy. It has proved possible to trap and mark birds along the route to determine just how much energy is expended [61]. It takes a Swainson's thrush about 42 days to fly from Panama to Canada, but during this time the actual travel involved consists of 18 nights of flying. The rest of the time is taken up with resting at stopover locations along the route. Over the 4800 km journey, 4450 kJ of energy is expended, so the cost is just a little less than 1 kJ

for each kilometre. What is surprising is that only 29% of the energy lost is expended on the actual flight; the remainder is lost during the stopover rests, mainly because so much time is spent recuperating and seeking food at these locations. The fact that resting periods are so energetically costly underlines the importance of choosing the right weather conditions for the migration. If conditions during the stopovers are cold and energetically costly, it could result in the failure of the bird to survive the migration. But once established in the northern breeding locations, the long days provide ample time for food acquisition, so the bird is better able to feed its young.

Migration, then, provides a means of changing the distribution pattern of a species with the season. It also means that the organism belongs to different communities and ecosystems at different times of its life. A salmon, for example, spends much of its life as part of the oceanic biome, but then moves upriver and into tributary streams for its breeding. There it spends the last days of its life, as a temporary member of a freshwater community. When it dies, the nutrients that it contains then become a part of this inland ecosystem and can contribute significant quantities of certain nutrients, such as nitrogen, to the other organisms that share the ecosystem [62]. The implications of migratory behaviour, therefore, extend beyond mere distribution patterns.

Invasion

The capacity to spread is important to all organisms. The success of a species can, in part, be measured by its geographical distribution, and the ability to move into new areas is one of the attributes required in order to achieve this. Habitats may alter or be lost, so a species needs to be capable of moving to more appropriate sites. Climate may change, and species will then need to alter their ranges to cope with the new conditions. Often this involves overcoming physical barriers that may seem insuperable. Mosses and ferns, for example, may seem to have little hope of long-distance dispersal, because most are relatively small and very static. Both demonstrate an alternation of generations in which spores are produced, however; these spores are of dust-like dimensions, usually less than 30 μm in diameter, and therefore have the capacity to be carried thousands of miles in the atmosphere. Some ferns, and many mosses, also produce gemmae, organs of vegetative propagation consisting of just a few cells that can be dispersed in the same way as spores. Species with this capacity have proved particularly effective in long-distance movements, such as between islands in the South Pacific [63].

Among animals, the small flightless springtails, just one or two millimetres in length, may also seem unlikely candidates for long-distance movements, as discussed earlier in this chapter. But these small organisms have hydrophic, unwettable surfaces, and, despite being terrestrial soil inhabitants, they can walk on the surface of water. Indeed, experiments have shown that they can survive 16 days on the surface of agitated seawater. Given this capacity, they are able to travel many hundreds of miles in the oceans. They can also cope with freezing and can live for as long as 4 years at a temperature of –22°C, so they could be incorporated into sea ice and be carried over considerable distances in the polar regions [64]. The dispersal capacity of organisms, therefore, can be much greater than might be expected.

All organisms need to disperse to ensure survival. Many of the animals that have failed to survive and have become extinct have failed in this respect. The flightless birds, such as the great auk [65] and the dodo, are examples. Environmental conditions, including climate, are constantly changing so species need to be able to move to new areas if local conditions deteriorate for them. But range expansion for organisms with little or no mobility of their own presents problems, and these have been overcome in a variety of ways. Plants often use wind or water [66]. A survey of the plants of the South-East Asian island of Rakata showed that 49% had arrived by aerial dispersal, 17% by flotation on the sea and the remaining 34% by animal (usually bird) transport, either attached to the external surface or carried in their guts [67]. This will be discussed in greater detail in Chapter 8.

Overcoming barriers to dispersal, such as mountain ranges and oceans, has been greatly assisted by humans in the last few centuries. *Ecological imperialism* is a term sometimes used to describe the wave of biological invasions that has occurred in the wake of human invasions, by either accidental transport or deliberate introduction. A typical example is the white sage, or tickberry (*Lantana camara*) [68]. This is a shrub native to Central and South America that has very attractive red and yellow flowers, initially endearing it to gardeners from Europe. It did not prove a pest in northern Europe, but when it was carried to warmer temperate regions and to other tropical and subtropical countries, it became an invasive plant especially of disturbed soils. The map in Figure 2.42 shows the history of its spread with human aid during the 19th century, when imperialism was at its height. The consequences of this dispersal are now being felt acutely in areas such as South Africa, India and the Galapagos Islands, where native species of plant are threatened by the vigorous and competitive growth of *Lantana*. Its remarkable success in so many parts of the world results from many attributes, all of which contribute to its invasive

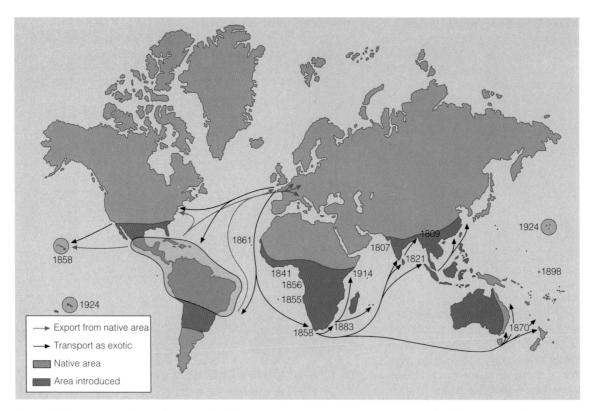

Figure 2.42 The spread around the world of the weed species *Lantana camara* as a result of human introductions. Its transport from its native range in South and Central America was due to its horticultural attractiveness. But it has subsequently become a serious invasive pest in many tropical and subtropical parts of the world. From Cronk and Fuller [68].

capacity. It is spread by birds that disperse its fruit (especially the mynah birds of India); it flowers profusely (hence its appeal to gardeners); it grows rapidly; it fragments easily, and its vegetative parts take root and grow; it is toxic to many grazing animals, including both mammals and insects; its chemical content can even poison other plants, making it a strong competitor; and it has broad ecological tolerance to environmental factors. These characters are typical of many invasive plant species.

An aggressive interloper may prove a threat to native organisms, and there are many examples of the displacement of native species by an invader. The European starling (*Sturnus vulgaris*), for instance, was introduced into Central Park, New York, in 1891. Since then, it has spread widely and is now present throughout the United States (Figure 2.43). It is mostly found in urban areas, and in the east it has partially displaced the bluebird (*Sialia sialis*) and the yellow-shafted subspecies of the northern flicker (*Colaptes auratus*). These birds nest in tree holes or in man-made holes, and starlings can occupy and defend many of the limited supply of these nest sites. In the towns, then, the starling successfully competes with the native species for living space. But when flocks of starlings invade the countryside, they compete for food, insects and seeds with the meadowlarks (*Sturnella* spp.), and the latter also have declined in some areas.

Even more rapid in its colonization of North America has been the Eurasian collared dove

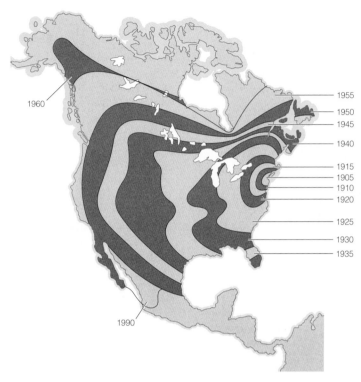

Figure 2.43 Map of North America showing the range extension of the European starling (*Sturnus vulgaris*) following its introduction to the continent late in the nineteenth century. Adapted from Baughman [12].

(*S. decaocto*). This bird was introduced deliberately into the Bahamas in 1974 and had spread to the mainland of North America by 1986. Since then, its progress has been remarkably rapid (as shown by the map in Figure 2.44), moving north into the Carolinas and up the Mississippi river system into Montana and beyond. The secret of the collared dove's success is its adaptability and, like the starling, its willingness to avail itself of human settlements and gardens. Its arrival in North America is a consequence of human introduction, but the species had already shown itself a capable invader even without such aid. In 1900 it was restricted to Asia, especially the subtropics, but then began its westward spread, moving from the Middle East into Egypt and Turkey [69]. It continued westward through central Europe, reaching Britain in the 1950s and becoming an abundant inhabitant of suburban habitats. It is difficult to establish what

caused its sudden expansion, but it appears to have changed its behaviour pattern, becoming more closely associated with people, and also its tolerance to more temperate climates. There are concerns that the expansion of this dove may have a detrimental effect on similar native species, such as the turtle dove (*Streptopelia turtur*) in Europe and the mourning dove (*Zenaida macroura*) in North America. In Europe the turtle dove has certainly declined recently, but this could be due to other factors, such as drought and land use changes in its wintering grounds in Africa and a high intensity of hunting in the Mediterranean region. The mourning dove may well prove more susceptible because its niche in North America is more closely linked with human settlements than the turtle dove.

The invasion of North America by European species has not been a one-way process. Some North American species have been successful as invaders

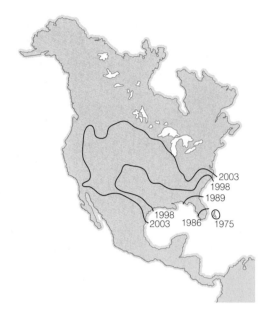

Figure 2.44 Map of North America showing the range extension of the Eurasian collared dove (*Streptopelia decaocto*) since its introduction to the Bahamas in the 1970s. Its spread in North America follows a similarly rapid extension of range in Europe over the last century.

in new regions, such as Europe and Australia. An example is the American grey squirrel (*Sciurus carolinensis*), which was introduced into the British Isles in the nineteenth century. Between 1920 and 1925 the native red squirrel (*Sciurus vulgaris*) suffered a dramatic decline in numbers in Britain, largely due to disease. The spread of the grey squirrel has been accompanied by the disappearance of the red squirrel from many areas, where it has probably acted as a disease carrier. Where the grey squirrel has replaced the native red, it probably has done so also by virtue of its superior adaptability to the niche of herbivore at the canopy level in deciduous woodland. In the few locations where the grey squirrel has not succeeded in invading, such as the Isle of Wight, an island off the south coast of Britain, the red squirrel still thrives. It also seems to maintain a competitive advantage in the conifer forests of upland Britain.

Dispersal to a new area does not ensure that an invader will persist there. When an organism arrives it must be able to establish itself, possibly by outcompeting and displacing native species at its point of arrival. The resident community may present a degree of **biotic resistance** to any invasive species. The invader must also be able to survive the pressures of predation and parasitism in its new environment. In the case of plants, one of the most important factors is its ability to cope with the many pathogens that exist in the soil and that often threaten its survival soon after arrival [70]. Some invasive organisms may prove successful simply because the parasites and predators that have evolved with them in their native regions are not found in the new location. Tamarisk shrubs (*Tamarix* spp.) are native to Europe and western Asia but have become widespread in the western United States, especially along riverbanks, since their introduction to that country [71]. As they spread they displace native willows and cottonwoods, but have the advantage of an absence of predators. It is hoped that the introduction of Chinese leaf beetles (*Diorhabda elongata*) will solve the problem, but the use of such **biological control** methods carries dangers that the introduced predator will find new sources of food among native species. Invasive species can thus become 'pest' or 'weed' species, but these terms can be applied only in relation to human attitudes towards species causing problems with agriculture, horticulture, transport, industry or nature conservation. Invaders are likely to be successful, and pest problems are thus likely to arise only if the biotic resistance of the invaded region is inadequate to prevent the establishment of the alien.

An invasive species is thus faced with both the problem of coping with the physical environment of its new home, plus the resistance presented by the native species as they compete for resources. This biotic resistance is not accounted for in the climate modelling described for the potential weed, garlic mustard (Box 2.4); it is possible that the native flora in some of the areas that the invader could occupy will not succumb to its competition and will resist its invasion. In other words,

It is not easy to predict whether an introduction or a new arrival is likely to prove invasive and whether it may become a pest species. We can make some predictions on the basis of the physical environment, including climatic conditions, and this may help in determining what areas of an invaded land may be at risk. Take the garlic mustard (*Alliaria petiolata*), for example [72]. This is a European herb that has reputed medicinal properties and was probably brought into North America by the early settlers for this reason. In Europe it grows in dense shade beneath a forest canopy, and hence it found the woods of New England very much to its liking, so it soon spread through the region. It now has a range that extends from Ontario to Tennessee and continues to expand into the Midwest. In order to determine which areas are at risk of invasion, ecologists have studied its distribution pattern in its native Europe and have established what climatic factors limit its range. These have been used in the construction of a computer model that can predict the likely outcome of its expansion in North America, based upon climatic matching (Figure 2.45). This type of work can provide advance warning of future problems; in this case, there are evidently areas of the western United States that are still liable to be invaded by this aggressive plant. This example illustrates the value of computer-based models of species' niches in order to predict potential ranges, as in the case of the leopard mentioned earlier in this chapter.

it may not fulfil its fundamental niche, of which the climate pattern is a partial representation. Its realized niche may prove smaller. An analysis of data concerning the introduced birds of the world, however, has revealed that the success of an invader is more often determined by the physical, abiotic aspects of the region invaded than by resistance from the native species [73].

There is much discussion among biogeographers and ecologists regarding what makes a community resistant to invasion. In his classic book on the ecology of invasions, first published in 1958 [74], Charles Elton proposed that more complex communities (i.e. those with a greater diversity of species) would be more resistant to invaders than simple communities with few species. But this assumes that the community is in a state of equilibrium and is saturated with species, which is probably rarely the case. Perhaps an invader is successful because it fills a vacant niche in the community, but if this were the case then no other species would be displaced by its arrival.

When an invasive species takes hold in an area, the consequences may extend beyond its immediate competitors. In North America, various alien species of honeysuckle (*Lonicera* spp.) and buckthorn (*Rhamnus* spp.) have become established and are now widespread, especially in the east. *Lonicera maackii*, for example, is now well established in 25 states east of the Rocky Mountains. These shrubs are highly favoured by certain songbirds, especially the American robin (*Turdus migratorius*), as nesting locations because they produce a suitable branch structure and have a flush of leaves providing cover early in the spring. But compared with the native shrubs, such as the hawthorns (*Crataegus* spp.), they lack spines and they encourage nesting close to the ground. The result is that robin nests in the invading shrubs have a much higher rate of predation than is the case with those in the native vegetation [75]. The impact of an invader can thus extend far into the web of ecological interactions within the ecosystem.

As we have seen, invasion and displacement of species in recent times are often consequences of human transport of organisms from one part of the world to another. Human colonists of new lands have often carried with them the familiar plants and animals from the old country, and, where the climate of the new lands has proved appropriate for their survival, these species have often gained a permanent foothold and thus extended their

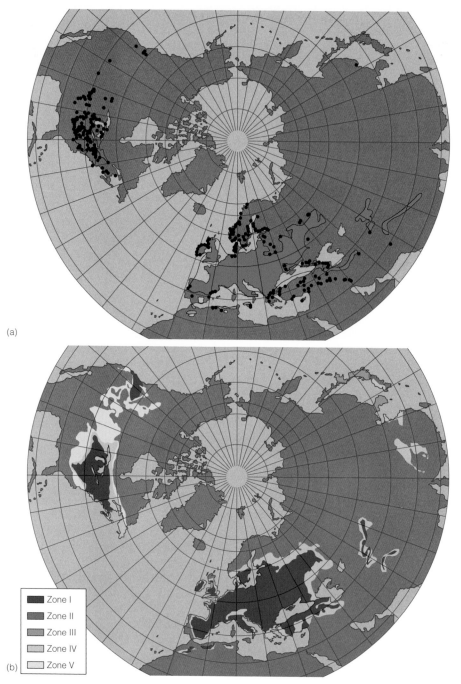

(a)

(b)

■	Zone I
■	Zone II
■	Zone III
■	Zone IV
☐	Zone V

Figure 2.45 (a) Distribution of the garlic mustard (*Alliaria petiolata*) in Europe and Asia, where it is native, and in North America, where it has been introduced and is proving invasive. (b) The modelled potential range of the species based upon its climatic requirements. There are clearly extensive areas of North America that are at risk of further invasion. The zone grades refer to the proportion of the year that is supportive of the species, where zone I = 100%, II = >96%, III = >92% IV = >88%, and V = <88%. From Welk et al. [72].

Table 2.1 The proportion of introduced plant species in the floras of various regions of the world; note the high proportion of alien species on islands. Data from Lovei [76].

Region	Native species	Alien species	Percentage of aliens
Hawaii	1143	891	44
British Isles	1255	945	43
New Zealand	2449	1623	40
Australia	15 638	1952	11
United States	17 300	2100	11
Continental Europe	11 820	721	6
Southern Africa	20 573	824	4

geographical range. Table 2.1 shows the numbers of alien plant species that have established themselves in different parts of the world, compared with the number of species that were already present. It can be seen that the localities with the highest proportion of newcomers are islands; these seem to be particularly sensitive to invasion by aliens [76].

New sets of environmental conditions can lead to shifts in the balance of communities and give rise to waves of invasion. One concern about the current rate of atmospheric change and its climatic consequences is that some species will inevitably suffer and others benefit in any given area. In the Mojave Desert of California, for example, elevated levels of atmospheric carbon dioxide are likely to favour non-native annual grasses, such as cheatgrass (*Bromus tectorum*), at the expense of native species [77].

Despite these dramatic examples of invasion and competitive displacement, it is most likely that, in natural situations, species that compete for food or other resources have evolved means of reducing the pressures of competition and of dividing up the resources between them. This is mutually advantageous since it reduces the risk of either species being eliminated and made extinct by competition with the others. This is an advantage not only to the species directly involved but also to the whole community of species in the habitat, since it results in more species being able to cohabit an area, depending on as many different sources of food as possible. Over evolutionary time scales, therefore, one might expect the species richness of an area to increase as this process of division of the resources gradually takes place. This assumes, however, that the environment is relatively stable, allowing organisms time to evolve and equilibrate. In such communities, competition would occur between many different species, each with its own specialized adaptations, so that no single species could become so numerous as to totally displace others. This equilibration could result in a greater degree of stability for the community, and stable communities are likely to be strongly resistant to the invasion of any new species that might disrupt the highly evolved pattern of competition within them. There is certainly evidence that disturbed, unstable habitats are more susceptible to invasion by non-native species than stable, undisturbed habitats [78], but this idealized scheme, together with the functional consequences of high species diversity, is hotly debated. Evidence suggests that only a small proportion of non-native species arriving at a site ever manages to establish themselves, an even smaller proportion manages to persist over the course of time and only a fraction of these can be regarded as a threat to native species. In the light of these facts, there are ecologists who consider the dangers associated with invasive species to be over-rated [79].

Although one can study the requirements of individual species of plants and animals in isolation from one another, this chapter has demonstrated that the interaction between different species is often critical in determining the extent to which the potential range of an organism is actually achieved. Species have both negative and positive effects upon one another, so it is important for biogeographers to consider whole assemblages of plants and animals as well as look at the individual components. This will be the subject of the next chapter.

Summary

1 Organisms are distributed over the face of the planet in a variety of patterns, and there are many possible factors underlying these patterns.

2 The causes of patterns vary according to the taxonomic level that we are dealing with. Many species are polytypic, consisting of several different subspecies, each with its own genetic makeup and environmental preferences.

3 The causes of patterns also vary with the spatial scale at which we are considering the organism, whether global, regional or local – the habitat scale.

4 Factors that need to be considered in the explanation of patterns include geological history, climate and microclimate, availability of food, chemistry of the environment, competition, predation and parasitism.

5 Interactions between species, such as competition and predation, create delicate balances in species assemblages that affect both individual species ranges and the biodiversity of communities.

6 The spatial and temporal separation of organisms, as well as the specialization of groups within populations, can lead to the formation of new species and an increase in biodiversity.

7 Migration constitutes a special kind of dynamic pattern that is developed in response to diurnal or, more frequently, seasonal changes in food supply.

8 Human beings have modified several biogeographical patterns as a result of species introductions, often leading to unforeseen consequences for native species.

9 It is possible to study the physical requirements of potentially invasive organisms and thus create predictive models of their likely future success and their threat to human interests as pest species.

Further Reading

Beeby A, Brennan A-M. *First Ecology: Ecological Principles and Environmental Issues.* 2nd ed. Oxford: Oxford University Press, 2004.

Bullock JM, Kenward RE, Hails RS (eds.). *Dispersal Ecology.* Oxford: Blackwell Publishing, 2002.

Fuller RJ. *Birds and Habitat: Relationships in Changing Landscapes.* Cambridge: Cambridge University Press, 2012.

Lomolino, MV, Riddle BR, Brown JH. *Biogeography.* 3rd ed. Sunderland, MA: Sinauer Associates, 2006.

References

1. Beebee T, Rowe G. *An Introduction to Molecular Ecology.* Oxford: Oxford University Press, 2004.

2. Olsen KM, Larsson H. *Gulls of Europe, Asia and North America.* London: Christopher Helm, 2003.

3. Howell SNG, Dunn J. *Gulls of the Americas.* Boston: Houghton Mifflin, 2007.

4. Collinson JM, Parkin DT, Knox AG, Sangster G, Svensson L. Species boundaries in the herring and lesser black-backed gull complex. *British Birds* 2008; 101: 340–363.

5. Aye R, Schweizer M, Roth T. *Birds of Central Asia.* London: Christopher Helm, 2012.

6. Brazil M. *Birds of East Asia.* London: Christopher Helm, 2009.

7. Rosen BR. Biogeographic patterns: a perceptual overview. In: Myers AA, Giller PS (eds.), *Analytical Biogeography.* London: Chapman & Hall, 1988: 23–55.

8. Hubbell SP. *The Unified Neutral Theory of Biodiversity and Biogeography.* Princeton: Princeton University Press, 2001.

9. Leibold MA. Return of the niche. *Nature* 2008; 454: 39–40.

10. Savolainen V, Anstett M-C, Lexer C, *et al.* Sympatric speciation in palms on an oceanic island. *Nature* 2006; 441: 210–213.

11. Cramp S (ed.). *Handbook of the Birds of Europe, the Middle East and North Africa IV.* Oxford: Oxford University Press, 1985.

12. Baughman M (ed.). *Reference Atlas to the Birds of North America.* Washington, DC: National Geographic, 2003.

13. Sibley DA. *The Sibley Guide to Trees.* New York: Alfred A. Knopf, 2009.

14. Gibbons M. *A Pocket Guide to Palms.* London: Salamander, 2003.

15. Hedberg O. Features of Afroalpine plant ecology. *Acta Phytogeographica Suecica* 1995; 49: 1–144.

16. Knox EB, Palmer JD. Chloroplast DNA variation and the recent radiation of the giant senecios (Asteraceae) on the tall mountains of eastern Africa. *Proceedings of the National Academy of Sciences USA* 1995; 92: 10349–10353.

17. Crawford RMM. *Plants at the Margin: Ecological Limits and Climate Change.* Cambridge: Cambridge University Press, 2008.

18. Kaufman K. *Lives of North American Birds*. Boston: Houghton Mifflin, 1996.

19. Sibley DA. *The Sibley Guide to Birds*. 2nd ed. New York: Alfred A. Knopf, 2014.

20. Gaston KJ, Spicer JI. The relationship between range size and niche breadth: a test using five species of *Gammarus* (Amphipoda). *Global Ecology and Biogeography* 2001; 10: 179–188.

21. Dandy JE. Magnolias. In: Horai B (ed.), *The Oxford Encyclopedia of Trees of the World*. Oxford: Oxford University Press, 1981: 112–114.

22. de Wit R, Bouvier T. '*Everything is everywhere*, but, *the environment selects*'; what did Baas Becking and Beijerinck really say? *Environmental Microbiology* 2006; 8: 755–758.

23. Smith HG, Wilkinson DM. Not all free-living microorganisms have cosmopolitan distributions – the case of *Nebela (Apodera) vas* Certes (Protozoa: Amoebozoa: Arcellinida). *Journal of Biogeography* 2007; 34: 1822–1831.

24. Charman D. *Peatlands and Environmental Change*. Chichester: John Wiley & Sons, 2002.

25. Ronikier M, Schneeweiss GM, Schoenswetter P. The extreme disjunction between Beringia and Europe in *Ranunculus glacialis* s.l. (Ranunculaceae) does not coincide with the deepest genetic split – a story of the importance of temperate mountain ranges in arctic-alpine phylogeography. *Molecular Ecology* 2012; 21: 5561–5578.

26. Crawford RMM. Gaps in maps: disjunctions in European plant distributions. *New Journal of Botany* 2014; 4: 64–75.

27. Harris S, Yalden DW. *Mammals of the British Isles*. 4th ed. London: The Mammal Society, 2008.

28. Coope GR. Tibetan species of dung beetle from Late Pleistocene deposits in England. *Nature* 1973; 245: 335–336.

29. Colinvaux P. *Amazon Expeditions: My Quest for the Ice-Age Equator*. New Haven, CT: Yale University Press, 2007.

30. Gathorne-Hardy FJ, Syaukani, Davies RG, Eggleton P, Jones DT. Quaternary rainforest refugia in south-east Asia: using termites (Isoptera) as indicators. *Biological Journal of the Linnean Society of London* 2002; 75: 453–466.

31. Beebee T. Ireland's Lusitanian wildlife: unravelling a mystery. *British Wildlife* 2014; 25: 229–235.

32. Jolly D, Taylor D, Marchant R, Hamilton A, Bonnefille R, Buchet G, Riollet G. Vegetation dynamics in central Africa since 18,000 yr BP. Pollen records from the interlacustrine highlands of Burundi, Rwanda and western Uganda. *Journal of Biogeography* 1997; 24: 495–512.

33. Hamilton AC. *Environmental History of East Africa*. London: Academic Press, 1982.

34. Favarger C. Endemism in the montane floras of Europe. In: Valentine DH (ed.), *Taxonomy, Phytogeography and Evolution*. London: Academic Press, 1972: 191–204.

35. Cowling RM, Rundel PW, Lamont BB, Arroyo MK, Arianoutsou M. Plant diversity in Mediterranean-climate regions. *Trends in Ecology and Evolution* 1996; 11: 362–366.

36. Marshall JK. Factors limiting the survival of *Corynephorus canescens* (L.) Beauv. in Great Britain at the northern edge of its distribution. *Oikos* 1978; 19: 206–216.

37. Root T. Energy constraints on avian distributions. *Ecology* 1988; 69: 330–339.

38. Thompson PA. Germination of species of Caryophyllaceae in relation to their geographical distribution in Europe. *Annals of Botany* 1978; 34: 427–449.

39. Ehleringer JR. Implications of quantum yield differences on the distribution of C3 and C4 grasses. *Oecologia* 1978; 31: 255–267.

40. Teeri JA, Stowe LG. Climatic patterns and the distribution of C4 grasses in North America. *Oecologia* 1976; 23: 1–12.

41. Spooner GM. The distribution of *Gammarus* species in estuaries. *Journal of the Marine Biological Association* 1974; 27: 1–52.

42. Gavashelishvili A, Lukarevskiy V. Modelling the habitat requirements of leopard *Panthera pardus* in west and central Asia. *J. Applied Ecology* 2008; 45: 579–588.

43. Thomas J, Lewington R. *The Butterflies of Britain and Ireland*. Gillingham: British Wildlife Publishing, 2010.

44. Thomas J. The return of the large blue butterfly. *British Wildlife* 1989; 1: 2–13.

45. Connell J. The influence of interspecific competition and other factors on the distribution of the barnacle *Chthamalus stellatus*. *Ecology* 1961; 42: 710–723.

46. Silander JA, Antonovics J. Analysis of interspecific interactions in a coastal plant community – a perturbation approach. *Nature* 1982; 298: 557–560.

47. Bell RHV. The use of the herb layer by grazing ungulates in the Serengeti. In: Watson A (ed.), *Animal Populations in Relation to Their Food Resources*. Oxford: Blackwell Scientific Publications, 1970: 111–127.

48. Hale WG. *Waders*. London: Collins, 1980.

49. Martin C. *The Rainforests of West Africa*. Basel: Birkhäuser Verlag, 1991.

50. Bergerud AT. Prey switching in a simple ecosystem. *Scientific American* 1983; 249 (6): 116–124.

51. Lyon RH, Runnalls RG, Forward GY, Tyers J. Territorial bell miners and other birds affecting populations of insect prey. *Science* 1983; 221: 1411–1413.

52. Thomas F, Renaud F, Guegan J-F. *Parasitism and Ecosystems*. Oxford: Oxford University Press, 2005.

53. Park T. Experimental studies of interspecies competition. I. Competition between populations of the flour beetles, *Tribolium confusum* Duval and *Tribolium castaneum* Herbst. *Ecological Monographs* 1948; 18: 265–308.

54. Yan G, Stevens L, Goodnight CJ, Schall JJ. Effects of a tapeworm parasite on the competition of *Tribolium* beetles. *Ecology* 1998; 79: 1093–1103.

55. Paine RT. Food web complexity and species diversity. *American Naturalist* 1966; 100: 65–75.

56. Paine RT, Vadas RL. The effect of grazing in the sea urchin *Strongylocentrotus* on benthic algal populations. *Limnology and Oceanography* 1969; 14: 710–719.

57. Moore PD. Parasite rattles diversity's cage. *Nature* 2005; 433: 119.

58. Hixon MA, Brostoff WN. Damselfish as keystone species in reverse: intermediate disturbance and diversity of reef algae. *Science* 1983; 220: 511–513.

59. Sage B. *The Arctic and its Wildlife*. London: Croom Helm, 1986.

60. Mead C. *Bird Migration*. Feltham: Country Life, 1983.

61. Wikelski M, Tarlow EM, Raim A, Diehl RH, Larkin RP, Visser GH. Costs of migration in free-flying songbirds. *Nature* 2003; 423: 704.

62. Ben-David M, Hanley TA, Schell DM. Fertilization of terrestrial vegetation by spawning Pacific salmon: the role of flooding and predator activity. *Oikos* 1998; 83: 47–55.

63. Dassler CL, Farrar DR. Significance of gametophyte form in long-distance colonization by tropical epiphytic ferns. *Brittonia* 2001; 53: 352–369.

64. Coulson SJ, Hodkinson ID, Webb NR, Harrison JA. Survival of terrestrial soil-dwelling arthropods on and in seawater: implications for trans-oceanic dispersal. *Functional Ecology* 2002; 16: 353–356.

65. Gaskell J. *Who Killed the Great Auk?* Oxford: Oxford University Press, 2000.

66. Cousens R, Dytham C, Law R. *Dispersal in Plants: A Population Perspective.* Oxford: Oxford University Press, 2008.

67. Thornton I. *Island Colonization: The Origin and Development of Island Communities.* Cambridge: Cambridge University Press, 2007.

68. Cronk QCB, Fuller JL. *Plant Invaders: The Threat to Natural Ecosystems.* London: Earthscan, 2001.

69. Blackburn TM, Duncan RP. Determinants of establishment success in introduced birds. *Nature* 2001; 414: 195–197.

70. Klironomos JN. Feedback with soil biota contributes to plant rarity and invasiveness in communities. *Nature* 2002; 417: 67–70.

71. Knight J. Alien versus predator. *Nature* 2001; 412: 115–116.

72. Welk E, Schubert K, Hoffmann MH. Present and potential distribution of invasive garlic mustard (*Alliaria petiolata*) in North America. *Diversity and Distributions* 2002; 8: 119–133.

73. Blackburn TM, Duncan RP. Determinants of establishment success in introduced birds. *Nature* 2001; 414: 195–197.

74. Elton C. *The Ecology of Invasions by Animals and Plants.* Chicago: University of Chicago Press, 2000.

75. Schmidt KA, Whelan CJ. Effects of exotic *Lonicera* and *Rhamnus* on songbird nest predation. *Conservation Biology* 1999; 13: 1502–1506.

76. Lovei GL. Global change through invasion. *Nature* 1997; 388: 627–628.

77. Smith SD, Huxman TE, Zitzer SF, *et al.* Elevated CO_2 increases productivity and invasive species success in an arid ecosystem. *Nature* 2000; 408: 79–82.

78. Chytry M, Maskell LC, Pino J, Pysek P, Vila M, Font X, Smart SM. Habitat invasions by alien plants: a quantitative comparison among Mediterranean, sub continental and oceanic regions of Europe. *Journal of Applied Ecology* 2008; 45: 448–458.

79. Davis MA. *Invasion Biology.* Oxford: Oxford University Press, 2009.

Communities and Ecosystems: Living Together

Chapter 3

No organism lives in total isolation from all others. Different organisms interact with one another in both the long and the short term, competing for resources and sometimes excluding one another from certain areas. Over evolutionary time, this can lead to populations specializing in certain ways, perhaps in the way they obtain food, or the type of food they eat or the type of microclimate in which they perform best. One animal may feed exclusively upon a specific source of food, so that the consumer is associated with its food species in its distribution. Species can thus become dependent on one another: predator on prey, parasite on host and so on. Alternatively, species may simply have similar environmental requirements and histories and hence tend to be found together in the same area. The outcome is an assemblage of organisms that appears bound together in a community. This chapter will examine the concepts underlying the community and will also consider the interactions between the living, biotic community and the non-living environment – a combination which has come to be called the ecosystem. Such communities and ecosystems themselves have global patterns of distribution, related to climate and other environmental factors.

The Community

Species rarely occur as single populations; usually they occur in a mixture of different species, and such an assemblage is termed a **community**. The concept of community can be applied at different spatial scales and may sometimes be limited to specific groups of organisms that are distinguished by their shared taxonomy or their role in the total assemblage. Thus we can speak of the bird communities of cliffs, or the microbial community in a soil or the plant community in a meadow. So, although the term *community* can refer to the total assemblage of living species found in a site, interacting in a whole range of different ways and forming a complex grouping of plant, animal and microbial components [1], it can also be used in a more limited sense. When it is used of animals that play a particular role in the community, such as plant-sucking insects or insectivorous birds, then the term **guild** is applied rather than community. Some animal species, such as those herbivores with very specific food requirements, may form close, dependent unions with certain plants, but others may have requirements for certain spatial architectural conditions that are best supplied by particular assemblages of plants. Again, the outcome of such associations is the existence of communities of plants and animals in nature which, within a geographical area, may be repeated in similar topographical and environmentally comparable sites, and which may be very predictable in their species composition.

Some plants and animals occur together regularly and are associated with specific habitats, such as wet meadow, sand dune, pine forest and so on. As a consequence, ecologists and biogeographers have come to use the term *community* as though such assemblages were clearly defined units with distinct boundaries and predictable component parts. Such a concept is neat and pleasing, especially if the biogeographer wishes to divide the natural world into

Biogeography: An Ecological and Evolutionary Approach, Ninth Edition. Edited by C. Barry Cox, Peter D. Moore, Richard J. Ladle.
© 2016 John Wiley & Sons, Ltd. Published 2016 by John Wiley & Sons, Ltd.

discrete units for the sake of mapping. But are such units real? This has long been a bone of contention, and several different approaches have emerged. It can be argued that species adapt during the course of their evolution not only to cope with the physical demands of their environment, but also to cope with the presence of other species. The process of **co-evolution** thus results in species not only tolerating one another, but sometimes becoming dependent on one another. At the same time, those species that are in competition for a particular, limited resource (which may be simply space) adjust their demands in such a way as to partition the resource between them (see Chapter 2). Coevolution and resource partitioning may thus result in an assemblage of different species being able to occupy the same habitat in a sustainable way. In this way, the community comes into being and should lend itself to recognition and definition by the scientist.

Perhaps the concept of community is most clearly expressed if one confines attention to a single group of organisms, such as plants. This is appropriate because it is often vegetation that forms the basis for biological mapping and habitat definitions. Plants are almost all doing precisely the same job, namely, fixing the energy of sunlight and taking up water and mineral elements from the soil. Interactions between plants and the environment can therefore be described in relatively simple terms in which different species have their optima and limits along any given environmental gradient. This was illustrated in Figure 2.27, where a species forms a bell-shaped curve with respect to any given factor, such as soil wetness, soil acidity, light intensity, the concentration of any given element and so

on. This simple model can be extended by adding other species and observing how the curves relate to one another. In Figure 3.1a, as might be expected, each curve is different from all others, but the different curves tend to form groups. Such groups can be called communities, each of which consists of predictable collections of species. If this model was applied in nature, it would be ideal for the construction of vegetation maps because the assemblages are so clearly defined and could undoubtedly be recognized in the field.

The idea of discrete plant communities that can be described, and even named and classified, is very attractive to the inherent neatness of the human mind and would certainly be valuable in the process of mapping areas and also for assessing their value for conservation and in determining plans for management. Whether plant communities have such an objective reality as discrete entities has been energetically debated throughout the twentieth century, when a distinct discipline of vegetation science, or **phytosociology**, came into being. The debate was summarized by the views of the two American ecologists, Frederic Clements and Henry Gleason, who began the discussion in the 1910s and 1920s. Essentially, **Clements** regarded the plant community as an organic entity in which the positive interactions and interdependencies between plant species led to their being found in distinct associations that were frequently repeated in nature. He felt that the community assembled itself in a manner comparable to the embryology of an organism, and could thus be conceived as an integrated entity. The view, which is essentially that illustrated in Figure 3.1a, proved both

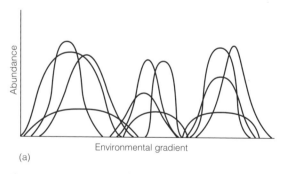
(a)

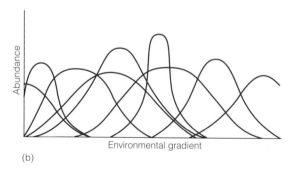
(b)

Figure 3.1 Diagrammatic representation of two models of vegetation. (a) The model of Clements in which species' requirements coincide, leading to the separation of distinct 'communities'. (b) The 'individualistic' model of Gleason in which each species is distributed independently and no clear 'communities' are apparent.

attractive and pragmatically useful, forming the basis for early attempts at describing and classifying vegetation by such ecologists as Braun Blanquet in France and Arthur Tansley in Britain.

But is the model on which such an approach is based really the case in nature? Do plant species fall into groups along environmental gradients, or are they more independent of one another? **Gleason's argument** emphasized the individual ecological requirements of plant species, pointing out that no two species have quite the same needs. Very rarely do the distributional or ecological ranges of any two species coincide precisely. Even the degree of association between ground flora and tree canopy in a woodland is often weaker than one might assume from casual observation. The application of statistical techniques to the problem soon demonstrated that, although species often overlap in frequently occurring assemblages, the composition of these groups varies geographically as the physical limits of species are encountered. This model, sometimes referred to as the **individualistic concept**, can be illustrated by Figure 3.1b in which the bell-shaped curves of different species do not cluster into groups. If this is the case in nature, then discrete communities do not really exist. But humans find it helpful to divide the continuum of variation because it is required if we are to describe nature. In such circumstances, the best we can do is to make relatively arbitrary definitions and boundaries, defining artificial communities on the basis of convenient reference points where certain combinations of species tend to recur.

Studies of the past history of plant and animal species over the last 10 000 years or so have also demonstrated that assemblages of species change in the course of time. Species come into contact at certain times in their history, but have also been periodically separated as the climate changed. Often the associations we now observe are of relatively recent origin and should be regarded as transitory, a moment in history when certain species happen to have coincided in their distribution patterns. The concept of the community, according to this school of thought, must be looked upon as useful but somewhat artificial; vegetation is actually a continuum both in space and in time.

The continuum model seems to work particularly well in situations where vegetation has experienced little modification by human activity, and where there are no sharp changes in landscape or geology. When humans have modified vegetation by clearing forest, burning grassland or cultivating the land, then sharp boundaries are created and there are often clearly defined changes from one vegetation type to another. Many modern systems of vegetation classification, such as that devised by John Rodwell to describe the vegetation types of Britain [2], are based on the idea of frequently repeated combinations being selected and described which can then be used as reference points. This type of scheme allows the possibility of a wide range of intermediate types, which is what one would expect if Gleason's ideas are valid.

In areas such as eastern North America, where forest remains a major element of the vegetation, the simplest approach to vegetation classification is to use the tree canopy dominants as a basis for recognizing units. Hazel and Paul Delcourt [3], for example, in their studies of the deciduous forests of eastern North America, find that there are just six major tree dominant species, and the subdivision of forest vegetation they propose is based on the proportions of these species, plus the relative abundance of minor tree and shrub components. In general, the main tree species will have broader ecological tolerances than the species present at ground level, but they still have optima and limits that place them in specific positions along an environmental gradient. Typical observed combinations are then used to define vegetation units, but intermediate types occur between these units.

The existence of discrete communities depends upon positive interactions between species, and this is not always easy to account for, especially in the case of plants that are basically all seeking the same resources of sun, space, water and soil nutrients. It is much easier to visualize competitive, negative interactions between species than mutually beneficial ones. The Darwinian approach to ecology has also led us to be suspicious of any hint of altruism in a hypothesis involving the interaction of species. But positive relationships can occur [4]. Often we find, for example, that tree species survive best as seedlings when under the cover of specific shrub plants, as in the case of the coast live oak (*Quercus agrifolia*) from California [5]. It has been found that 80% of the seedlings of this tree are located beneath the cover of just two species of shrub. This 'nurse' effect of one species upon another is not uncommon in nature and is an example of **facilitation**, which is an important element in the process of succession, as described in Chapter 4. In all vegetation studies, the possibility of facilitation interactions that lead to positive associations between species must be balanced against the competitive interactions that lead to negative associations.

The importance of scale in this argument is also very evident. As we increase the size of the area under study, increasing numbers of species are recorded together and thus appear positively associated, while reducing the sampling area must inevitably lead to more negative associations. In addition, vegetation itself can often be visualized as a mosaic [6]. Patches of vegetation may be in different stages of recovery from disturbance or regional catastrophe, or from the death of an old tree or human clearance, or the vegetation may simply reflect underlying environmental patterns of geology, soils or hydrology. All of these can affect the pattern of species distributions and patterns of associations, in both space and time, and each type of disturbance leads to a different spatial scale of vegetation pattern (Figure 3.2). If the boundaries between the different elements in

the mosaic are sharp and well defined, then the distinction between 'communities' will also be more clear-cut. Perhaps this is why the community concept, especially in vegetation studies, has proved more useful in countries like France and Switzerland which have very old cultural landscapes that have evolved over millennia of intensive human fragmentation and cultivation. In countries where such intensive agricultural activity is a relatively recent event, such as in North America and Australia, landscapes still persist that are occupied by a continuum of vegetation rather than a clear patchwork mosaic.

The Ecosystem

The concept of community encompasses only living organisms and takes no account of the physical and chemical environment in which those organisms live, even though these factors may define the composition of those communities. If we include environmental features, including the underlying rock and soil, the water moving through the habitat, and the atmosphere permeating the soil and surrounding the vegetation, then we have an even more complex, interactive system that is called the **ecosystem**. Whereas the idea of community concentrates on the different species found in association with one another, the concept of the ecosystem is largely concerned with the processes that link different organisms to one another. The concept of the ecosystem can be applied at various levels. The entire Earth can be viewed as an enclosed ecosystem, but so can a lake, a forest, a stream or even a clump of grass.

There are two fundamental ideas that underlie the ecosystem concept; these are **energy flow** and **nutrient cycling**. Energy, initially fixed from solar radiation into a chemical form by green plants, moves into herbivores as a result of their feeding upon plants, and then moves on into carnivores as the herbivores are themselves consumed. Since herbivores rarely consume all the available plant material, and since carnivores do not eat every individual of their prey organisms, some living tissues are allowed to die naturally, which itself may involve parasites and pathogens [7], before being eaten. These dead tissues constitute an energy resource which can be exploited by scavengers,

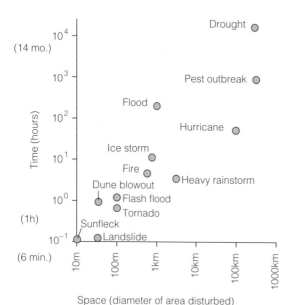

Figure 3.2 Disruption of habitats by disturbance creates a mosaic of patches in different stages of recovery. This leads to an uneven landscape where separate 'communities' of organisms may be perceived. The spatial scale of patches varies with the type of disruption as shown here. Note the log/log scale on the axes. From Forman [5].

detritus feeders and decomposers. Thus, we can classify all the organisms in any community in terms of their feeding relationships. In practice, a series of complex feeding webs are usually formed, relating each species to many others, whether as feeder or food.

Food webs can be very complicated, especially in ecosystems with a high species diversity. Figure 3.3 shows the food web of a tropical rainforest at El Verde, Puerto Rico [8], and the complexity of interactions even within this simplified diagram is immediately apparent. Here the major groups of organisms are not split into their individual species, because to do so would make the system too complicated to be able to visualize. But the real ecosystem has an abundance of species within each of the boxes shown, each one with its own particular way of life, or niche, and each occupying one or more specific locations in the interactive net of feeding relationships. The arrows shown in a diagram of this sort indicate the linkages between consumers

and those they consume: grazer and plant, parasite and host, predator and prey. But they can also be regarded as routes along which energy flows. The base level in the diagram consists of the plants and their products that have captured the energy of sunlight in the process of photosynthesis, and have stored this energy in organic materials by taking carbon dioxide gas from the atmosphere and building large molecules with it. These compounds include the sugars of nectar, the lignins of wood, the cellulose from leaves and so on, and they are available to grazing animals or to detritus feeders and decomposers. The carbon derived from the atmosphere is also combined with elements obtained from the soil; nitrogen is taken up from the soil as nitrates and is combined to form amino acids and hence proteins; phosphorus, also from the soil, is used in constructing phospholipids that form an essential component of living cell membranes. All of these, and many more, form a food resource for consumers and decomposers. From the

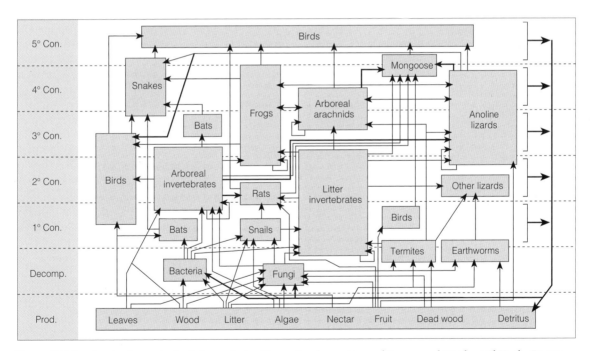

Figure 3.3 Food web of a tropical rainforest derived from the observations of Reagan and Waide at El Verde, Puerto Rico [8]. Organisms are grouped first taxonomically (into boxes) and then arranged in a series of layers (trophic levels) according to their feeding positions in the system. Several taxonomic groups are represented in more than one trophic layer (e.g. birds) because of the varied feeding habits of their component species. Detritus from dead organisms in this diagram is brought back into the base layer with the primary producers (green plants). Together, these form the basic sources of organic energy for the food web.

primary consumers, energy is passed through the feeding web, sometimes being stored for a short while in the bodies of the living and dead organisms, often being lost in the process of respiration as the animals conduct themselves in an energetic manner, and eventually all being dissipated in the respiratory heat output of the ecosystem. In Figure 3.3, the dead residues of animals and plants are shown returning to the detritus that is available to decomposers and that leads back into the food web.

It is convenient to conceive the ecosystem as a series of layers, as represented here, since organisms obtain their energy in a kind of stratified sequence as it passes from one animal to another. But it is an oversimplification to place whole groups of animals within particular layers; hence, birds are represented three times on the diagram according to whether they eat fruit (primary consumers), eat herbivorous and carnivorous invertebrates (secondary and tertiary consumers) or feed as predators at the top of the food web. For these top predators, the energy they eventually receive has passed through many organisms and many feeding levels, termed **trophic levels**. Since energy is lost at each transfer from one trophic level to the next, the energy reaching the top carnivores is more limited than that available at the base of the food web. This is one reason why top carnivores will generally be fewer in number than animals lower down the system.

Ecosystems have inputs and outputs of energy, but generally these are relatively simple. Solar energy is the principle source of energy for most ecosystems, and respiration and consequent energy dissipation as heat comprise the main loss. But some ecosystems are exceptional, such as the deep-sea vents where bacteria capture the energy released from chemical reactions involving inorganic materials such as iron and sulphur compounds and from the methane that belches out of the vents. Some ecosystems, such as streams running through forests, and mudflats in estuaries, may receive most of their energy second-hand from other ecosystems in the form of plant detritus. In this way, ecosystems can exchange energy with one another.

Whereas energy is eventually dissipated as low-grade heat, nutrient elements, such as calcium, nitrogen, potassium and phosphorus, are not irretrievably lost. They are cycled within ecosystems along the same paths as the energy, but most are eventually returned to the soil from which they can be reused by plants. Carbon is an exception because this is released to the atmosphere as carbon dioxide as the plants, animals and microbes respire. It is recycled, however, as plants photosynthesize once more. Some other elements, such as nitrogen and sulphur, may also find their way into the atmosphere if the materials containing them are subjected to fire. Movements of elements between ecosystems occur, and in terrestrial ecosystems the **hydrological cycle** (the movement of water from oceans through atmospheric water vapour, through precipitation and back to the oceans via streams and rivers) plays a major role in both delivering and removing elements to and from ecosystems. Apart from the nutrients arriving in rainfall, the other major source for most ecosystems is the gradual degradation of underlying rocks (weathering), which replenishes elements removed by plants, and by water percolating through the soil. The water movement through the soil also leaches away unbound nutrients and takes them from one ecosystem to another, often from a terrestrial ecosystem to an aquatic one. A knowledge of the quantities and flow rates of ecosystems helps conservationists manage them efficiently [9]. If we wish to crop the ecosystem at any given trophic level (e.g. hay at the first trophic level, or sheep at the second trophic level), then we must ensure that the rate of removal of nutrients can be compensated for by natural inputs from rainfall and weathering. If this is not so, then we must either reduce the level of exploitation or add those elements in deficit as fertilizers. An element in short supply may limit the rate of ecosystem productivity.

The ecosystem concept therefore involves all the complexity of species interactions within the system, but views these in relation to the processes in which the ecosystem is engaged, including productivity, energy flow, nutrient cycling and so on. This approach has proved a very useful one, not only in the assistance it provides in understanding the relationships between organisms and the interactions with the physical environment, but also because it gives us a basis for the rational use of natural resources for the support of human populations.

Ecosystems and Species Diversity

How many species are really necessary to keep an ecosystem functioning? Is there a basic minimum number of species that is needed for any given ecosystem to operate, above which other species are simply excess to requirements? In other words, are some species redundant? If this were the case, according to the *redundant species hypothesis*, the removal of certain species from an ecosystem would have little or no effect on the functioning of that ecosystem. Here, the function of an ecosystem could refer to its overall productivity, its rate of nutrient cycling or its general self-sustainability. Following the removal of these redundant species, the ecosystem would only begin to suffer as a result of further losses. This is illustrated in Figure 3.4a, where initial removal of species has little impact on the functioning of the ecosystem, but as further species are removed the reduction in ecosystem function becomes more acute. A possible alternative model is based on the supposition that all species are equally important to ecosystem function, in which case the loss of each species renders it a little less efficient, as shown diagrammatically in Figure 3.4b. This is a *linear relationship model*. The removal of any species from this ecosystem would render it less efficient in functional terms, but would not necessarily cause its collapse. A third option, shown in Figure 3.4c, is that certain species play key roles in the ecosystem, and when they are lost there is a sudden drop in the capacity of the ecosystem to function. This is sometimes referred to as the *rivet hypothesis*. Some species could be lost from such an ecosystem, and it would result in very little, if any, change in the ecosystem function. But if one of the key species is removed, then the entire structure is weakened, just as would happen if certain vital rivets were removed from the panelling of a ship's hull. There is also the possibility, of course, that the entire structure could collapse if a final and vital rivet were removed. This concept links with that of *keystone species* (see Chapter 2), which are species that play a particularly significant role in the function of an ecosystem, and upon whose presence and activities many other species depend. Such a species may be large, like a beaver in a pond, or very small, such as the nitrogen-fixing lichens on an alpine scree slope. Keystone species

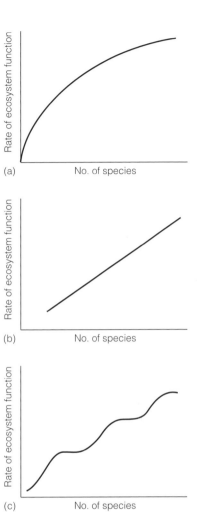

Figure 3.4 The possible relationships between the number of species in an ecosystem and its rate of function (e.g. primary productivity, decomposition, nutrient turnover etc.). (a) Redundant species hypothesis, where at high species density some species can be lost without affecting ecosystem function. (b) Linear model, in which every species is equal in functional importance and the loss of any species reduces the efficiency of ecosystem function accordingly. (c) Rivet hypothesis, in which certain species are critical in supporting ecosystem function ('rivets'). Loss of these species has a disproportionate effect on ecosystem function.

would act as vital rivets in the model of ecosystem stability. There is a final option, sometimes called the *idiosyncratic model*, in which the stability and functioning of the ecosystem are entirely independent of how many species are present [10].

Many experiments have been conducted using controlled artificial ecosystems consisting of a limited number of species in order to test these possible outcomes. In general, increasing the number of species in such systems does lead to an increase in productivity, but the practical problems involved in experimentation are considerable and the number of species relatively few (generally below 30), so one cannot eliminate the possibility that the system would eventually become saturated and any more species added would simply be redundant. Evidence has also emerged from field experiments that not all species are equal in importance. For example, plant members of the legume family (Fabaceae) often have the capacity to fix atmospheric nitrogen as a result of bacterial colonies in their root nodules. The presence of a legume in a grassland can thus have a considerable impact on ecosystem productivity. Similarly, loss of such a species would have the reverse effect. The roles that species play within an ecosystem can therefore vary in importance to general ecosystem function, and it is possible to divide species into **functional types** [11] according to their physiological and ecological capacities. The efficiency of an ecosystem may depend not simply on the number of species present but also on the appropriate array of functional types.

Small-scale laboratory experiments have largely focussed on plants and plant functional types. A field study of a natural ecosystem in the savanna grasslands of the Serengeti National Park in East Africa by Sam McNaughton and his colleagues from Syracuse University uses a larger scale approach and has provided an example of how more animal species could result in better ecosystem function [12]. Comparing sites that have been grazed by the herds of large mammalian herbivores of the tropical grasslands with other sites where these animals had been excluded by the erection of fences, they demonstrated that several nutrients, such as nitrogen and sodium, were cycled more efficiently in the grazed ecosystems. Therefore, grazing animals actually enrich the nutrient availability in the ecosystems they occupy. Again, it is probable that some species are more effective than others in this role, and their loss would be particularly serious for ecosystem function.

A related question, but perhaps even more difficult to answer, is whether diversity has an influence on the stability of an ecosystem. It was the British ecologist Charles Elton who first proposed (in the 1950s) that a more complex and rich ecosystem should also be more stable, meaning that it was less prone to violent fluctuations such as those caused by epidemic disease or pest outbreaks [13]. It seemed a reasonable proposal, since experience showed that such instability, as defined in these terms, was often a feature of simple ecosystems, as in the case of agricultural monocultures. It was also argued that a complex food web could provide a buffer against any perturbation in which certain species might become scarce. With a complex web there are more opportunities for prey-switching, as with the lynx and hare relationship described in Chapter 2.

But the development of mathematical modelling as an approach to the understanding of populations did not provide the expected results. Such models have generally shown that a species in a diverse ecosystem is no less subject to fluctuations caused by unfortunate events, such as drought or disease, than is a species in a simple ecosystem. It remains possible, however, that although individual species may still fluctuate, the function of the entire ecosystem could be less vulnerable to such chance events if the system is diverse and complex than if it is species-poor and simple. Thus, individual species may rise and fall in abundance, but the ecosystem survives because many other contenders are available to replace the unfortunate sufferer in the event of catastrophe. The American ecologist David Tilman and his co-workers [14] have illustrated this possibility in their field experiments with natural vegetation plots. Some of these suffered an unplanned natural disturbance in the form of drought, and the plots with higher species numbers suffered lower declines in biomass than species-poor plots. Therefore, although species richness in an ecosystem does not guarantee the success, or even survival, of individual species, it does provide that ecosystem with a greater capacity to cope with disaster. There are clear lessons here for both conservationists and agriculturalists. On the conservation side, the loss of global biodiversity that we are currently experiencing may well be affecting the functioning of the entire biosphere, as will be discussed in Chapter 4. In the field of agriculture, the use of multicropping systems rather than single-species stands can provide advantages in terms of both productivity and stability of the system, which is a particular concern in marginal areas, such as arid regions.

Some confusion can arise because of the different ways in which the term *diversity* is used. Very often, it is simply used as an alternative to the number of species present within an ecosystem, the **species richness**, but an ecosystem could be regarded as more diverse if the representation of the various species is reasonably equal, rather than dominated by one or two species, as shown in Figure 3.5. In Figure 3.5a, a hypothetical community of 10 species with a total of 100 individuals is represented, and one species dominates the system, accounting for 55 of the individuals. The other nine species thus have just five individuals each. In Figure 3.5b, the 100 individuals are equally divided among the 10 species, resulting in a perfectly even distribution. It can be argued that Figure 3.5b shows a more diverse community than 3.5a despite the fact that both have the same species richness.

Evenness can thus contribute to diversity. Various formulae have been constructed that incor-porate the concept of evenness into a species diversity index. But most studies considering the relationship between diversity and stability have concentrated on richness rather than any measure of diversity that involves evenness. However, experiments with microbial microcosms, in which both the richness and evenness of the original community were varied, showed that the functioning of the system (such as the process of dentrification) was sustained more effectively when placed under stress (increasing salinity) if the original composition was even [15]. This could be an important consideration when managing ecosystems. For example, by selectively harvesting one particular species from an ecosystem, the evenness and therefore the stability of that system might be put at risk.

One final point in this debate needs to be clarified, and that is: what precisely do we mean by **stability**? Is a stable ecosystem one which is difficult to deflect from its current composition or function? This approach defines stability in terms of **inertia**, or **resistance** to change. Alternatively, a stable ecosystem could be defined as one which rapidly returns to its original state following disturbance. This uses the concept of **resilience** as a basis for defining stability. Both ideas, of course, are inherent in the concept that most people have of stability. The two properties are illustrated in Figure 3.6, where the more resistant ecosystem is deflected to a lesser extent by a perturbation, but is slower in its recovery, whereas a more resilient ecosystem may be more easily and more drastically disturbed by a perturbation, but returns to its original state more rapidly. Resilience is a less satisfactory measure of stability in that simpler ecosystems may prove more resilient by definition. For example, a neglected arable field covered with weeds may be disturbed and yet return to its original state very rapidly. But it would be misleading to regard such an ecosystem as inherently stable as a consequence of this resilience. Perhaps the most effective approach would be to combine the two ideas by employing **predictability** as a measure of stability [17]. A stable ecosystem should behave in a predictable manner no matter what fate may cast in its path, and biodiversity does appear to render an ecosystem predictable by providing a kind of 'biological insurance' against the failure of certain sensitive species when exposed to particular stresses.

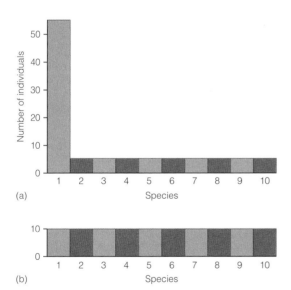

Figure 3.5 Hypothetical community of 10 species and 100 individuals. In (a), one species dominates; and in (b), all species have equal representation. It can be argued that (b) represents the more diverse of the two communities despite their having identical species richness.

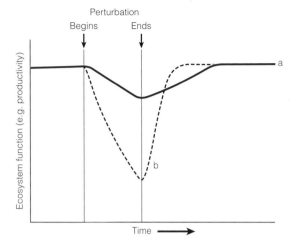

Figure 3.6 Two possible responses of an ecosystem to disturbance. Line a (solid) represents an ecosystem that is resistant to perturbation. Its response to disturbance is slower and less severe, but its return to its original state is slow. Line b (dashed) shows a resilient ecosystem that is more severely affected by the disturbance, but rapidly returns to its original state. Either could be regarded as an illustration of ecosystem stability. Adapted from Leps [16].

At present, there is no general consensus on the question of how many species are needed to maintain ecosystem functions, or precisely how diversity affects stability. Continuing research is revealing the sheer complexity of the interactions within an ecosystem, sometimes described as an **ecological network**, but how complexity relates to stability seems to depend upon what species are gained or lost. The species with the most complex interactions with other species may be the ones whose removal results in the greatest threat to the ecosystem [18]. The question of ecosystem function is equally vexed. The more ecosystem 'functions' that are considered, the more species are required to sustain them [19]. Any generalization is thus likely to prove an oversimplification.

These various debates are combining the concepts of community and ecosystem and are asking questions regarding the ways in which community composition affects ecosystem operation. These are very important questions in biogeography, because we need to be able to appreciate how the presence or absence of particular species in an area will affect the behaviour and survival of the remaining species. As we begin to realize the extent to which the world is changing, largely as a consequence of the activities of our own species, we need to be able to predict the outcome of biogeographical alterations in species distributions and assemblages.

Biotic Assemblages on a Global Scale

Assemblages of species of plant, animal and microbe can be viewed in different ways and at different scales. We may adopt a taxonomic approach, identifying all the organisms that are found in an area, analysing their associations and defining clusters of species that can be classified and named as communities. As has been pointed out, such definitions are likely to be somewhat artificial because they usually lack sharp boundaries between them, but they may be very useful, enabling biogeographers to map the biota within regions. It is also possible to view communities in a different way, based on how their living and non-living components interact. Using this ecosystem view, the taxonomic identity of species is of lesser interest than their respective roles within the system. Thus, each species falls within a particular functional type, initially determined by its feeding systems but also by other aspects of its mode of life, such as its nitrogen-fixing capacities and so on.

The application of the community concept on a global scale is complicated by the restriction of individual species to certain regions. The use of lists of species and the study of species inter-relationships are useful only within a limited area. Broader-based studies, covering continental-scale surveys, require a different approach. The vegetation of northern Iran in central Asia can serve as an example of the importance of scale in such studies (Figure 3.7). Much of the semi-desert region of central Iran is occupied by sparse scrub, as shown in the photograph Plate 1a. The vegetation over large areas of flat, sandy expanses is relatively uniform, as seen in the photograph, and is composed largely of the deciduous woody shrub, *Zygophyllum eurypterum*, interspersed with a smaller woody species of wormwood, *Artemisia herba-alba*. If a sample plot of about 50 × 20 m

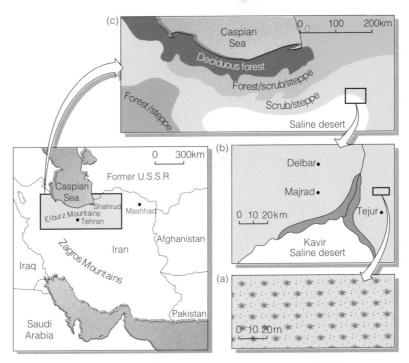

Figure 3.7 The effect of sampling scale on the appropriateness of different types of vegetation description, illustrated by the vegetation of northern Iran. (a) At a scale of tens of metres, individual shrubs are apparent and their specific identity can be determined. Bushes of the desert shrub *Zygophyllum eurypterum* tend to be evenly spaced, and smaller woody plants of *Artemisia herba-alba* (shown as dots) occur between them. (b) At a scale of kilometres, different groupings of species form distinct assemblages ('communities') having their own constituent species that can cope with various environmental stresses (drought, salinity, grazing, etc.). The saline desert (no shading) is almost devoid of vegetation except for a halophyte, *Salsola* sp. The moister alluvial margins and riverside (dark shading) bear a community of the salt bush *Haloxylon persicum* and shrubs of tamarisk (*Tamarix* sp.). Most of the land surface is covered with *Zygophyllum* scrub (grey shading), except where villages have caused vegetation loss by overgrazing and firewood gathering. (c) At a scale of hundreds of kilometres, vegetation is perceived in terms of physiognomic form (deciduous forest, forest-scrub-steppe, scrub-steppe, saline desert, etc.) rather than as species assemblages.

is studied in detail, as shown in Figure 3.7a, the individual plants of the two species can be mapped in relation to one another, and it is found that the spacing between individuals is extremely even, the outcome of intense competition for water among the plants [20]. When observing the same region at a scale of tens of square kilometres rather than square metres, a more complex vegetation pattern emerges. Now it is possible to make out landscape patterns related to water movements in the

alluvial plain, together with disturbances due to human settlements, resulting in variations in grazing intensity. The low-lying salt desert is practically devoid of vegetation, with only the extremely salt-tolerant plants surviving, such as the saltworts (*Salsola* spp.). The fringe of moister regions surrounding the salt flats, together with the valleys of temporary streams (shaded dark on the diagram), is occupied by other salt-tolerant shrubs, *Haloxylon persicum* and species of *Tamarix*. This vegetation type is shown in Plate 1b. Much of the area is covered with *Zygophyllum* and *Artemisia* scrub, but the surroundings of the rural villages Delbar and Majrad, although lying within the scrub belt, are often stripped bare of shrubs because of intensive sheep grazing and collection of fuel wood by the villagers. A few grazing-resistant species, including toxic species such as *Peganum harmala* and *Ephedra* spp., are able to survive. At this scale of observation, it is therefore possible to pick out 'communities' of species that share an ability to exist under certain types of environmental stress (salinity, drought, grazing etc.).

The communities described here are characterized by particular species, most of which are confined to this region of the world. The harsh desert environment has resulted in communities that are relatively species poor in their vegetation, but this simplifies the classification of plant assemblages. Similar desert scrub vegetation, however, is found in the Mojave Desert of California, which closely resembles the Iranian scrub in general appearance but has different species, such as creosote bush (*Larrea tridentata*), which belongs to the same plant family as *Zygophyllum*, together with white bur-sage (*Ambrosia dumosa*), which is in the same family as *Artemisia*. The Californian scrub is thus ecologically equivalent but consists of different species of plants. Its overall appearance and structure, however, are strikingly similar. Both vegetation types have evolved in response to very similar climatic and soil conditions.

When considering vegetation on a global scale, therefore, the use of species for vegetation classification is inappropriate. More useful is a classification by the **physiognomy** (general form and lifestyle) of the vegetation. This permits a comparison of the vegetation of similar climatic locations in different parts of the world. The need for a physiognomic approach to the classification and mapping of vegetation is evident if we move up a level of scale on the Iranian map, shown in Figure 3.5c, and observe the entire northern part of Iran, where the units fall into groups such as temperate deciduous forest, forest-scrub, scrub-steppe and saline desert. This level of classification is defined on the general nature of the vegetation, expressed in terms of its gross structure and appearance, and this is controlled by major factors such as climate, hydrology and type of soil. Similar patterns to that of Iran occur

in different parts of the world, so that a very similar type of vegetation may evolve in each of these areas, as in the case of the desert scrub of California. This comparability of vegetation forms may lead to the evolution of similar types of animals in each area. Both regions have tortoises, toads, small burrowing rodents such as kangaroo rats, jack rabbits and hares, quails and sandgrouse, buzzards and hawks, eagles and vultures, and coyotes and foxes. The species are all different, but they represent **ecological equivalents** in different parts of the world.

A general classification of terrestrial, large-scale ecosystems has gradually evolved, which includes the following major types: deserts; the cold tundra of high latitudes and high altitudes; northern coniferous forest, or taiga; temperate forest, including temperate rainforest; tropical rainforest; tropical seasonal forest; tropical grassland and open woodland, usually termed savanna; temperate grassland (known as prairies in North America, steppes in Eurasia, pampas in South America and veld in South Africa); and finally the chaparral of areas with a Mediterranean climate. This classification was first based simply on vegetation, and the resulting units were called **formations**, but in modern usage animal life is included in their descriptions and definitions and they are called **biomes**.

For reasons already outlined in this chapter, any system of classifying the world's biota into units must avoid the simple use of particular groups of species for their definition. Although the rainforests of Brazil are comparable in many respects to those of West Africa or South-East Asia, the actual assemblage of species is quite different. Their similarity is mainly due to the fact that they are structurally comparable, being dominated by tall trees arranged in a series of layers, most of which are evergreen and broad-leaved. There are other similarities, such as the presence of canopy-dwelling primates, and the fact that pollination is often brought about by birds, bats and so on. Essentially, we are comparing the communities in terms of the functional types of plants and animals present rather than their taxonomic affinities. Since vegetation forms the basic template for life within the biomes, it is natural that early ideas on global community classification came from botanists.

The concept of a **life form** among plants, for example, was first put forward by the Danish botanist Christen Raunkiaer in 1934. He observed that the most common or dominant types of plants in a climatic region had a form well suited to survive in the prevailing conditions. Thus, in arctic conditions the most common plants are dwarf shrubs and cushion-forming species that have their buds close to ground level. In this way, they survive the winter conditions when windborne ice particles have an abrasive effect on any elevated shoots. In warmer climates, buds are usually carried well above the ground and the tree is an efficient life form, but periodic cold or drought may necessitate the loss of foliage and the development of a dormant phase. This has resulted in the evolution of the deciduous habit. More prolonged drought results in a different type of vegetation, with shrubs that have smaller above-ground structures. Some plants of areas with seasonal drought survive the unfavourable period as underground organs (e.g. bulbs or corms), or as dormant seeds. Animals also show distinct life forms adapted to different climates, with cold-resistant, seasonal or hibernating forms in cold regions and forms with drought-resistant skins or cuticles in deserts. Nevertheless, animal life forms are usually far less easy to recognize than plants, and, consequently, biomes are primarily distinguished and defined by the plant life forms they contain. Raunkiaer analysed the flora of different parts of the world into component functional types and found that each region had its own distinctive **biological spectrum**.

Using this physiognomic approach to vegetation, it has proved possible to define and characterize units, initially referred to as plant formations since they were based on purely botanical criteria. Global maps were constructed by plant geographers that used these plant formations as their basic units, and these were found to have a broad correspondence with climatic zones. Indeed, in the early days the vegetation of understudied areas of the world was often taken as a basis for predicting climate, so it was inevitable that vegetation maps and climate maps had a general similarity to one another.

Figure 3.8 shows an idealized northern hemisphere continent with its typical pattern of vegetation types (based on Box and Fujiwara [21]). There is a basic latitudinal banding in these plant

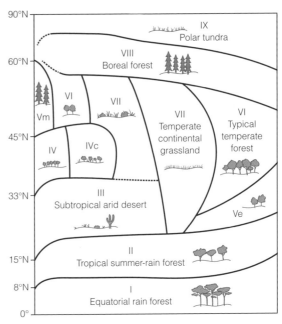

Figure 3.8 An idealized Northern Hemisphere continent showing the pattern of climate and biome types over its surface. I, Equatorial rainforest. II, Tropical summer rain deciduous forest and savanna forest. III, Subtropical arid desert and semi-desert. IV, Mediterranean forest/scrub (maritime and continental forms). Vm, (West coast) temperate rainforest. Ve, (East coast) warm temperate forest. VI, Temperate forest. VII, Dry temperate grassland (increasingly continental in the east). VIII, Boreal coniferous forest. IX, Polar tundra. Adapted from Box and Fujiwara [21].

formations, but the west and east coasts of a continent differ to a certain extent as a result of climatic differences resulting from wind patterns and oceanic currents. The tropical forest belts, for example, extend further north on the east of a continent than they do on the west, and the polar and boreal belts extend further south on the east. The mid-latitude regions differ because of the varying degrees of aridity in the centre and on either coast of the continent. Plate 2a shows the actual pattern of biomes over the surface of the Earth as derived from satellite imagery, and this can be compared with the idealized model in Figure 3.8.

The Raunkiaer life form approach to vegetation classification is limited simply because it depends solely on one aspect of the plant's response to environmental conditions, namely, the protection of its perennating organs during unfavourable seasons. There are many other aspects of a plant's form, structure and life history that can be taken into account, and the development of the idea of functional types has endeavoured to take such features into account. Evergreen or deciduous foliage, the association of nitrogen-fixing microbes and the mechanisms of pollination and fruit dispersal all contribute to a plant's capacity to cope with the stresses and strains of the environment and to establish a niche within a competitive community. There are also characteristics of plant life histories that can be regarded as **strategies** for survival, although the word does suggest a degree of volition on the part of the plant, which is, of course, not intended. Strategies are themselves the outcome of many generations of selection of individuals and genotypes, conserving those best fitted for prevailing conditions. Particularly important among such strategies are the reproductive processes (fast versus slow breeding), individual longevity and the allocation of resources to robust growth. Shade tolerance, fire tolerance and nutrient requirements are additional factors that contribute to the survival and competitive strategy of a plant. The concept of strategies was first developed by the British ecologist Philip Grime [22] when he was examining the ways in which communities are assembled, but it has subsequently been expanded and applied to communities on the biome scale [23].

There is no full agreement among biogeographers about the number of biomes in the world. This is because it is often difficult to tell whether a particular type of vegetation is really a distinct form or is merely an early stage of development of another, and also because many types of vegetation have been much modified by the activities of human beings. This is very apparent from the global vegetation map in Plate 2a [24]. Satellite imagery has increased the accuracy with which vegetation can be mapped on a global scale, but the very considerable impact of humans, especially in the temperate zone and around the edges of the arid regions, means that the expected close relationship between vegetation patterns and climate patterns is no longer clear.

Using satellite data, it has proved possible to elucidate the precise climatic requirements of particular biomes as defined by plant life forms and functional types. Each biome is said to fit within a certain **climatic envelope**, which is the sum of all the climatic variables that limit that biome. This approach has led to a refinement of conventional biome definitions, particularly in the case of forests and grassland. Ian Woodward of the University of Sheffield in England has proposed the following classification of these vegetation types [25].

1. Evergreen needleleaf forests – tall (over 2 m high), dense (over 60% cover) forests of evergreen trees with narrow leaves (e.g. boreal coniferous forests).

2. Evergreen broadleaf forests – tall, dense forests of evergreen trees with broad leaves (e.g. tropical rainforests).

3. Deciduous needleleaf forests – tall, dense forests of narrow-leaved trees that seasonally lose their leaves (e.g. larch).

4. Deciduous broadleaf forests – tall, dense forests of broad-leaved trees that seasonally lose their leaves (e.g. beech, maple, some oaks).

5. Mixed forests – tall, dense forests with an intermixture or a mosaic of deciduous and evergreen trees.

6. Woody savannas – trees exceed 2 m high, but cover only 30–60% of the land surface; intermixed with herbaceous vegetation.

7. Savanna – trees exceed 2 m high, but are widely scattered, covering only 10% to 30% of the surface, the remainder being dominated by herbaceous vegetation.

8. Grasslands – land with herbaceous cover and with less than 10% tree or shrub cover.

9. Closed shrublands – lands with woody vegetation less than 2 m tall and with a shrub cover (evergreen or deciduous) of more than 60%.

10. Open shrublands – lands with woody vegetation (evergreen or deciduous) less than 2 m tall and with a shrub cover of between 10% and 60%.

Extreme deserts have very little vegetation or none at all. This classification system has the objective advantage that it can be easily recognized from satellite imagery. Each of these types can then be assigned a particular climatic envelope.

The link between biomes and climate becomes even more apparent if biomes are mapped against climatic variables, such as precipitation and temperature [26], as has been done in Figure 3.9. Again, the divisions between biomes are not as sharp as those indicated here, but it is plain that each biome occupies a region where specific climatic requirements are met.

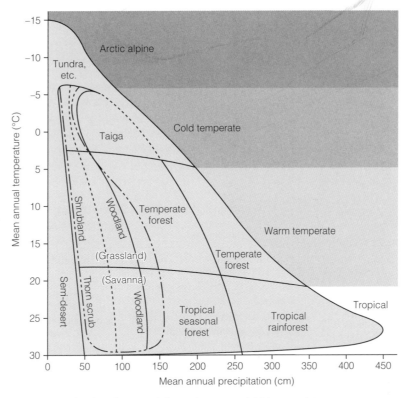

Figure 3.9 The distribution of the major terrestrial biomes with respect to mean annual precipitation and mean annual temperature. Within regions delimited by the dashed line, a number of factors including geographical location, seasonality of drought and human land use may affect the biome type which develops. From Whittaker [26].

Mountain Biomes

We have seen that there is an overall latitudinal zonation of biomes, which is apparent as we move from the rainforests of the equatorial regions to the dwarf-shrub tundra of the Arctic, and a similar trend is reflected with increasing altitude.

Air at the base of a mountain is heated by the sun. The component molecules of gas move faster, and as a consequence they expand and the gas becomes less dense. This packet of air is then displaced by denser air descending, and the warm air is forced upward. But as it rises, the pressure of the atmosphere above lessens, so it expands again, which involves the loss of energy. As a consequence, the air packet becomes cooler. The outcome is a fall in atmospheric temperature with increasing altitude,

the rate of which is termed the **lapse rate**. Average figures for lapse rates range generally between about 5.5°C and 9.8°C for every 1000 m rise in altitude, depending in part on the humidity of the air. But this change has a profound effect on the living organisms on a mountain.

Figure 3.10 shows the changes in biomes with altitude in the Himalayan Mountains. Ascending from the northern Indian savanna and thorn scrub, we move up through subtropical monsoon forest, largely occupied by drought-deciduous tree species. A zone of scrub follows, which owes its existence largely to the impact of human activity and that of domesticated animal grazing. Above this is temperate deciduous forest with oak and rhododendron, its general form and appearance being very similar to that found in western Europe or the eastern part of the United States. Coniferous forest

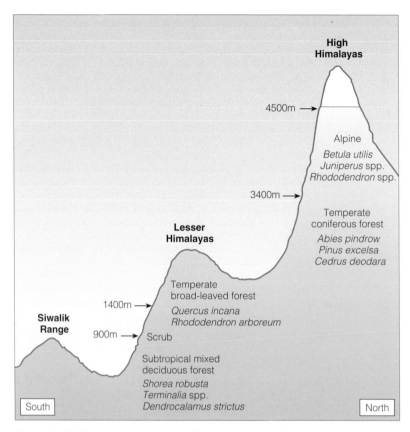

High
Himalayas

4500m →

Alpine

Betula utilis
Juniperus spp.
Rhododendron spp.

3400m →

Temperate
coniferous forest

Abies pindrow
Pinus excelsa
Cedrus deodara

Lesser
Himalayas

Temperate
broad-leaved forest

1400m →

Quercus incana
Rhododendron arboreum

Siwalik
Range

900m →

Scrub

Subtropical mixed
deciduous forest

Shorea robusta
Terminalia spp.
Dendrocalamus strictus

South

North

Figure 3.10 Diagrammatic section of the western Himalayas in northern India, showing the approximate altitudinal limits of the major vegetation types. With increasing altitude, one passes through vegetation belts similar to those found on passing from lower into higher latitudes. The scrub zone (900–1400 m) is strongly modified by human deforestation and subsequent grazing.

lies above the oak zone, dominated mainly by the deodar cedar, and above this are the alpine birch and juniper scrub that lead on up to the tundra and the permanent snows of the high mountains. The altitudinal sequence thus broadly mirrors the latitudinal zones but in a much shorter distance.

The **timberline** in the Himalayas (about 30°N), where forest vegetation gives way to alpine scrub, lies at about 3400 m. At Mount Kenya, practically on the equator, it lies at 3658 m. The Sierra Nevada of California (40°N) has its timberline at about 3000 m. In the Alps of central Europe (48°N) the timberline occurs at about 2300 m, while in Scotland (60°N) it is at about 800 m. The higher the latitude, the lower the timberline, and other biome boundaries would similarly be at lower altitudes.

Biome boundaries may also vary with the aspect of the mountain. One side of the mountain may receive more sunlight than another, usually the southern side for a Northern Hemisphere mountain and the northern side for one in the Southern Hemisphere. Prevailing winds may also bring more rain to one side than another. A good example is the Sierra Nevada Mountains of California, where the western slopes receive cool, moisture-bearing winds from the Pacific Ocean, whereas on their eastern side lie the hot, arid deserts of the continental interior [27]. Consequently, the boundaries of the different forest types are lower on the western slopes than on the east, as shown in Figure 3.11. An even greater disparity of biome boundaries with aspect is apparent in the Galápagos Islands, as shown in Figure 3.12.

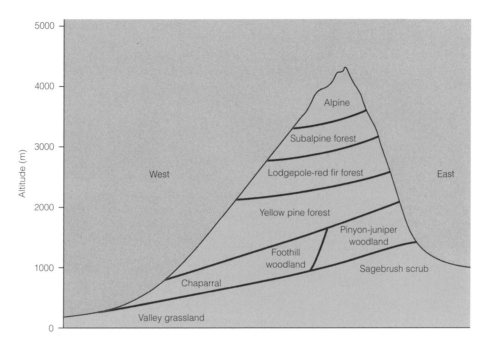

Figure 3.11 Cross section of the Sierra Nevada Mountains in California, showing the approximate boundaries of the different forest types. Warmer, drier conditions on the eastern side of the mountain chain permit the forest to survive at higher altitudes. From Schoenherr [27]. (Reproduced with permission of University of California Press.)

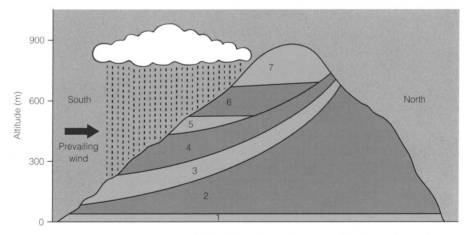

Figure 3.12 Cross section of a typical island from the Galápagos archipelago. The southern side receives substantially more rainfall that the northern side because of the prevailing southerly winds. Consequently, the vegetation is richer in the south [28]. The basal zone (1) is the littoral fringe, followed by (2) an arid zone with cacti; (3) a transitional zone, dominated by trees; (4) a *Scalesia* zone, a cloud forest consisting of various species of *Scalesia* trees up to 20 m; (5) a *Zanthophyllum* zone, consisting of spiny shrubs; (6) a *Miconia* zone with dense shrubs; and (7) a pampa, or fern-sedge, zone.

The northern (equatorial) side receives more sunlight and is therefore hotter, but even more important is the prevailing wind direction from the south, which bears the bulk of the precipitation to these arid islands. As a consequence, vegetation is richer on the southern side and extends to considerably higher altitudes [28].

Although temperature in general falls as one ascends mountains, other environmental conditions do not precisely mirror those found at higher latitudes. For example, the seasonal variations in day length that are typical of high-latitude tundra areas are not found in the 'alpine' regions of tropical mountains, where the variation in the duration of day and night is considerably less. Also, there is a high degree of insolation on tropical mountains resulting from the high angle of the sun, and this results in extreme diurnal fluctuations in temperature in tropical alpine locations that are not found in high-latitude tundra regions. Daytime temperatures can thus rise very steeply following very cold nights. It is not surprising, therefore, that the altitudinal zonation of plants and animals should not precisely reflect the global, latitudinal zonation. Also, where a species is found in both arctic and alpine tundra, the arctic and alpine races of the species often differ in their physiological make-up as a consequence of these climatic differences and the differing selective pressures that have been placed upon them during their evolution [29].

Climate thus has its impact on vegetation and biome patterns through altitude as well as latitude. It is important, therefore, for the biogeographer to understand climatic variables and their causes.

Global Patterns of Climate

The **climate** of an area consists of the whole range of weather conditions experienced within that area, including temperature, rainfall, evaporation, sunlight and wind through all the seasons of the year. Many factors are involved in the determination of the climate of an area, particularly latitude, altitude and location in relation to oceans and land masses. The climate in turn is a very important factor in determining the species of plants and animals, and even the life forms or functional types, that can live in an area.

Climate varies with latitude for two reasons. The first reason is that the spherical form of the Earth results in an uneven distribution of solar energy with respect to latitude. As the angle of incidence of the sun's rays approaches 90°, the area over which the energy is spread becomes smaller, so that there is an increased heating effect. In the high latitudes, energy is spread over a wide area; thus, polar climates are generally cold (Figure 3.13a). The precise latitude that receives sunlight at 90° at noon varies during the year; it is at the equator during March and September, at the Tropic of Cancer (23°28′N) during June and at the Tropic of Capricorn (23°28′S) during December. These two tropics mark the limits beyond which the sun is never directly overhead. The effect of this seasonal fluctuation is more profound in some regions than in others, with the greatest contrasts experienced at high latitudes.

Variations in climate also result from the pattern of movement of air masses (Figure 3.13b). Air is

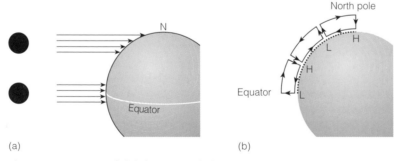

(a)   (b)

Figure 3.13 Causes of global patterns of climate. (a) Due to the spherical shape of the Earth, polar regions receive less solar energy per unit area than the equatorial regions. (b) The major patterns of circulating air masses (cells) in the Northern Hemisphere: H, high pressure; L, low pressure.

heated most strongly over the equator and therefore rises (causing a low-pressure area) and moves towards the pole. As it moves towards the pole, it gradually cools and increases in density until it descends, where it forms a subtropical region of high pressure, known as the Horse Latitudes. Air from this high-pressure area either moves toward the equator or else moves poleward. The poleward-heading air mass eventually meets cold air currents moving south from the polar regions where air has been cooled, and descends (causing a high-pressure area at the poles). Where these two air masses meet, a region of generally low pressure results in which the weather is very unstable and therefore variable.

This simplified and idealized picture is complicated by the Coriolis effect (named in honour of the French mathematician Gaspard Coriolis, who analysed it), which is a consequence of the west–east rotation of the Earth. This spinning force tends to deflect any freely moving object to the right of its course in the Northern Hemisphere and to the left in the Southern Hemisphere. As a result, the winds moving toward the equator in both hemispheres are deflected and consequently blow from east to west. These 'trade winds', found in both the Northern and Southern Hemispheres, meet in the region of the equator, and this is known as the **intertropical convergence zone (ITCZ)**. Where these easterly winds have passed over oceans they become moist, and this moisture is deposited as rain, generally over the eastern portions of the equatorial latitudes of the continents. Similarly, the winds that move poleward from the high-pressure Horse Latitudes are deflected by the Earth's spin and come to blow from a more westerly direction, and provide rain along the western regions of the higher latitudes of the continents. The Horse Latitudes themselves are regions in which dry air is descending, and arid belts form along these latitudes on the continental land masses, especially in the Northern Hemisphere where there are more continental masses that occupy these latitudes.

As the ITCZ shifts north or south of the equator at different seasons, this causes an alteration in global wind patterns, as shown in Figure 3.14. The strongest impact is in the Indian Ocean, where there is a complete reversal in prevailing wind direction. In the Northern Hemisphere summer, strong southerly winds are generated in the Indian

Ocean, carrying warm, moist air into India and East Africa. These winds bring the monsoon rains to these regions.

The global pattern of oceans and land masses modifies the climatic patterns yet further. Because heat is gained or released slower by oceans than by land, heat exchange is slower in maritime regions, while at the same time humidity is higher. In summer, therefore, continental areas tend to develop low-pressure systems as a result of the heating of land and the conduction of this heat to the overlying air masses. Conversely, in winter, the reverse situation occurs: continental areas become cold faster than the oceans, and high-pressure systems develop over them, as shown in Figure 3.14. One effect of this process is that the continental low-pressure systems draw in moist air from neighbouring seas. The winter season in these areas, on the other hand, is usually dry.

The global circulation patterns of water within the oceans are also of great importance in determining world climate patterns (Figure 3.15). Warm surface waters from the equatorial regions of the Atlantic Ocean are carried northeastward towards Iceland and Norway, warming this section of the Arctic Ocean, keeping it ice-free through the summer and bringing mild conditions to the whole of western Europe. As this water body cools in the North Atlantic, it becomes denser and sinks, reversing its flow pattern as it heads back to the southern Atlantic at greater depth. From here, the cold, dense water may flow up into the Indian Ocean where it receives new warmth, or it may continue eastward around the southern latitudes and eventually move northward into the Pacific Ocean and pick up heat there. Once heated, the less dense waters move westward along the ocean surface and back into the Atlantic. This **thermohaline oceanic circulation**, sometimes referred to as an oceanic conveyor belt, is responsible for dispersing much of the warmth received in tropical regions to higher latitudes, particularly along the Atlantic seaboard. Without this energy distribution, the high latitudes would be much colder. Indeed, one of the main features of the glacial episodes in the Earth's recent history has been the shutting down of the global oceanic heat conveyor, which has led to rapid cooling in the high latitudes. The influence of the thermohaline conveyor on biome distribution is apparent if one compares the northwest of

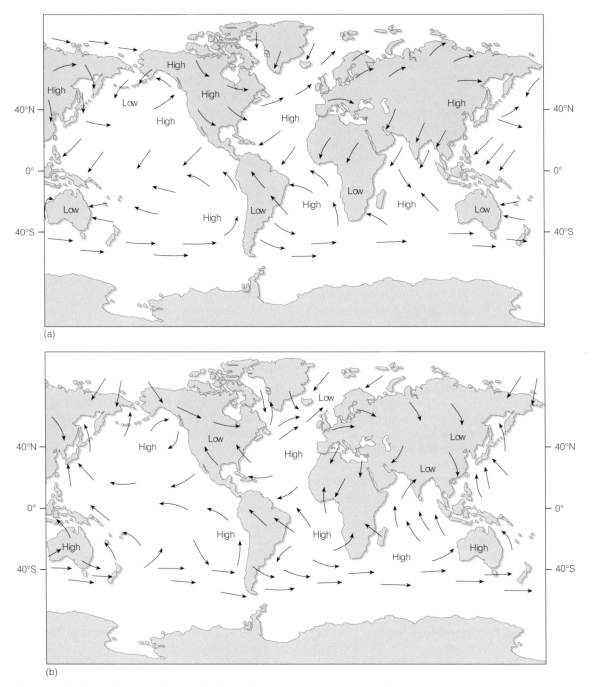

Figure 3.14 General pattern of areas of high and low pressure across the Earth's surface, and the consequent patterns of movement of air masses: (a) in January, and (b) in July. Note the reversal of wind directions in the Indian Ocean in July, taking monsoon rains into northern India and east Africa.

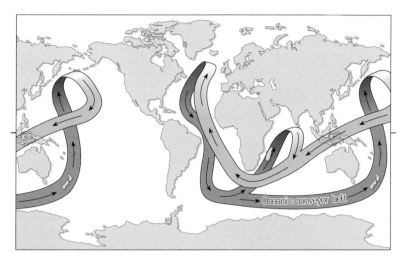

Figure 3.15 The oceanic conveyor belt carrying warm, low-salinity surface waters northwards into the North Atlantic and deep, higher salinity cold waters from west to east into the Indian Ocean and the Pacific. From here, water receives a renewal of its heat content, rises to the surface and returns westward into the Atlantic.

Europe with the northwest of North America. The boreal coniferous forest extends to much higher latitudes in Scandinavia than it does in Alaska, as shown in Plate 2a.

The topography of the land surface also affects patterns of climate. We have already seen how mountains create their own climates and thus affect the altitudinal zonation of biomes. Mountain barriers may interrupt the movement of air masses and can thereby modify climate, an example being the Himalayan Mountain chain that effectively blocks the moist air of the Indian monsoon from reaching the Tibetan Plateau and the Gobi Desert to the north. The sheltered plateau lies in the rain shadow of the Himalayas.

Climate Diagrams

As we have seen, individual species of plants and animals, and also general life forms, are affected by a whole range of physical factors in their environment, many of these being directly related to climate. Biogeographers, therefore, have long sought a means of portraying climates in a simple, condensed form that would give at a glance an indication of the main features that might be of critical importance to the survival of organisms in the area. Mean values of temperature and rainfall may be of some use, but one also needs to know something of seasonal variation and of extreme values if the full implications of a particular climatic regime are to be appreciated. It is with this aim in view that Heinrich Walter, of the University of Hohenheim in Germany, devised a form of climate diagram that is now widely used by biogeographers [30]. An explanation of the construction of these diagrams is given in Figure 3.16, and a selection of climate diagrams displaying the climates of some of the major biomes of the Earth is given in Figure 3.17.

The essential features of the climate diagram reveal changing patterns of conditions through the year at any given location. Mean monthly temperatures and precipitation are overlaid in a manner that reveals the time of year when water is likely to be scarce or abundant. This is achieved by careful selection of the scales at which temperature and precipitation are depicted. Above the baseline, periods of drought or of excess water availability are clearly displayed, while below the baseline an indication is given of periods of frost. Thus, in a single diagram, the main features that influence living organisms can be made immediately apparent.

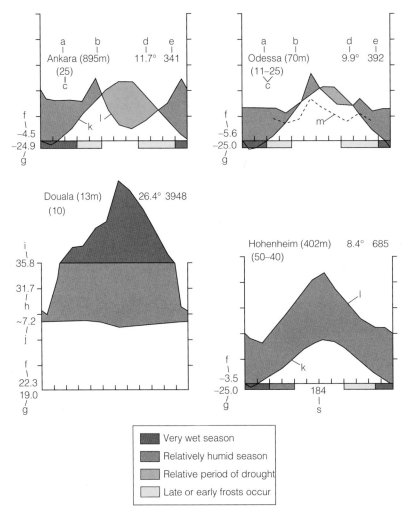

Figure 3.16 Climate diagrams convey many aspects of seasonal variation in climate at a site in a manner that is easily assimilated visually. The diagrams concentrate upon those climatic features of greatest relevance to the determination of vegetation in the area: temperature and precipitation. Key to the climate diagrams: abscissa: months (Northern Hemisphere, January–December; Southern Hemisphere, July–June); ordinate: one division = 10°C or 20 mm rainfall. (a) Station; (b) height above sea level; (c) duration of observations in years (if there are two figures, the first indicates temperature and the second precipitation); (d) mean annual temperature in °C; (e) mean annual precipitation in millimetres; (f) mean daily minimum temperature of the coldest month; (g) lowest temperature recorded; (h) mean daily maximum temperature of the warmest month; (i) highest temperature recorded; (j) mean daily temperature variations; (k) curve of mean monthly temperature; (l) curve of mean monthly precipitation; (m) reduced supplementary precipitation curve (10°C = 30 mm precipitation); (s) mean duration of frost-free period in days. Some values are missing from some diagrams, indicating that no data are available for the stations concerned. Adapted from Walter [30].

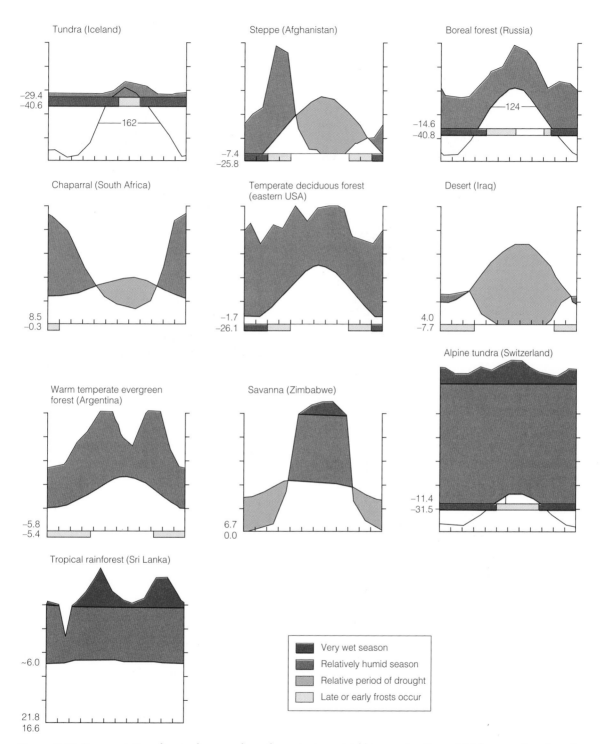

Figure 3.17 Representative climate diagrams from the major terrestrial biomes. See Figure 3.16 for an explanation of their construction.

Modelling Biomes and Climate

The general link between biomes, defined in terms of life forms or functional types, and climate is evident, but modern biogeographers require more robust methods in the study of connections between vegetation and climate, especially if they are to be in a position to predict future global responses to climate change. Much effort has been expended in improving the definitions of biological units, the biomes, and fitting them to specific climatic envelopes. We now have much more detailed information about the physiology of different plant types than was available to Raunkiaer, including their tolerance of cold or heat and their ability to cope with drought or flooding. When all of this information is put together, it is possible to define much more precisely the range of plant functional types that can be useful in classifying, mapping and understanding vegetation and its relation to climate. It is also possible to escape the notion that biomes are fixed in their make-up; instead, we can admit the possibility of regional variations and transitions between biomes. We need to accept that the composition of biomes may change with time and altering climate, not just the location of biome boundaries. Studies of vegetation history over the past 10 000 years or so have shown that the composition of biological assemblages is constantly changing, so we need to introduce a time variable into our concept of the biome, which gives it a dynamic rather than static status.

One of the most satisfactory attempts at describing the complex relationships between vegetation and climate has been that of E.O. Box [31], who compiled a list of 90 plant functional types. He was able to assess the climatic tolerances and requirements of his functional types, involving temperature, precipitation and the seasonal variation in these. Precipitation alone, however, can be misleading because other conditions, such as high temperature and intense solar radiation, can lead to the rapid evaporation of water; therefore, a moisture index expressing the ratio of precipitation to potential **evapotranspiration** (the combination of evaporation of water from surfaces and the upward movement and loss of water from plants) is required. This is similar to the approach used by Walter in his construction of climate diagrams. Once these data are available, it is possible

to inspect a set of climatic conditions and predict the plant functional types that would be found, together with their relative abundance and importance. The predictions from such a model [32] can then be checked against observations in the field to see how closely the model matches reality. It must be borne in mind, of course, that the predictions based on this model will show potential vegetation on the assumption that climate is the sole determining factor; it makes no allowance for human modification of the natural habitats, nor for geological or historical factors. It is essentially the study of ideal climatic envelopes.

A classic and widely used model relating global vegetation to climate is that constructed by Colin Prentice, working at the time of this work from the University of Lund, Sweden, together with a number of colleagues [24]. These researchers have sought to maintain a simple approach in their efforts to build a computer model of world vegetation, and they used just 13 functional types based not simply on morphology but also on physiology, especially the temperature tolerances of the plants. They regard minimum temperature tolerance as particularly important for this seems to determine woody plant distribution, as Raunkiaer correctly asserted. Most broad-leaved evergreen tropical trees, for example, are killed if exposed to frost. Temperate deciduous trees, such as many of the oaks, are damaged by temperatures below $-40°C$, and many of the boreal evergreen, needle-leaved conifers, such as spruce and fir, can cope with temperatures between -45 and $-60°C$. Some of the needle-leaved deciduous conifers, such as larch, are able to tolerate even lower temperatures.

On the other hand, many temperate trees need to be chilled during the winter if they are to bud effectively and produce flowers the following spring, so this requirement needs to be built into the model. It means that some of the needle-leaved evergreens will demand winters in which the coldest month has an average temperature of below $-2°C$. Warmth in the growing season, however, is also needed, and even cold-dwelling trees need about $5°C$ for adequate photosynthetic activity. Lower-growing plants, on the other hand, such as dwarf shrubs and low-growing cushion plants, may survive effectively where the average temperature of the warmest month is only $0°C$. To allow for the wetness of an area, Prentice and his colleagues used a moisture index of the type developed by Box, building in an allowance for seasonal variation in wetness.

The Prentice model, given the name BIOME 3, works by taking into account all of these, and several other, climatic variables and determining which of the functional types could survive, then assessing which is likely to be dominant. In this way, the potential biome for the area is deduced [33]. The outcome of the model is shown in Plate 2b, which depicts the potential biome distribution in both the Old and New Worlds. The model begins from the requirements of the plant functional types, and assembles groupings on the basis of what types could tolerate given climatic conditions in different parts of the world. There is no initial assumption about how many biomes exist or what comprises them. In fact, 17 terrestrial biomes emerge, as can be seen from the maps. The model predictions of the distribution of these biomes are in very reasonable agreement with the actual information available, the 'ground truth', as shown in Plate 2a. We have to bear in mind, however, that the information is incomplete (hence the gaps in the real-world maps), and often reflects the consequences of human cultural activities rather than potential vegetation under the existing climatic conditions.

This approach to the global classification and mapping of biomes provides us with a robust model on which we can project new circumstances and observe their possible outcome. As information accumulates, so such models can be improved. Climatic scenarios of different types can then be imposed on the computer model, and the consequences for biome distribution can be estimated. Since we know the relationships of certain types of agriculture to these biomes, we can also examine the impact of climate change on agricultural potential of different parts of the world. All such models, however, constantly need to be checked against the real world, the so-called 'ground truth', so that their accuracy can be determined.

One of the most important reasons why such predictions should be made is to protect the very varied assemblage of living organisms on our planet from becoming reduced in its biodiversity as a result of rising levels of extinctions. To achieve this, we need to understand the global patterns of biodiversity on Earth and locate areas that may be sensitive to climatic and other changes. This is the subject of the next chapter.

Summary

1 The idea of 'communities' of organisms that occur in discrete units, which are predictable in terms of species composition, is attractive and useful to biogeographers, but nature often exhibits a gradual and continuous change in species assemblages depending on the individual requirements of species.

2 In a landscape consisting of a fragmented mosaic of different habitats, communities are more likely to have distinct boundaries and therefore to be recognizable in nature.

3 The ecosystem is a useful way of considering biotic (animal, plant, and microbial) assemblages in relation to the non-living world. It is a concept based on the ideas of energy flow through a series of feeding (trophic) levels and the circulation of elements between living organisms and the non-living world.

4 The use of the ecosystem concept and the notion of functional types of organisms (producers, decomposers, nitrogen-fixers etc.) within the community provides a way of investigating the implications of biodiversity for natural systems. It allows us to ask the question: are all species really necessary for the maintenance of the stability of an ecosystem, or are some redundant? Current research suggests that some species can be lost without necessarily destabilizing an ecosystem. More critical is the maintenance of a balance of functional types.

5 Global ecosystems, often called biomes, are best defined in terms of vegetation physiognomy or of functional types, either morphological or physiological. Models relating biome distribution to climate can then be developed.

6 Climate-biome models provide a means of predicting the outcome of climate change on the Earth's biogeography and will have implications in both conservation and agriculture. But predictions of biome shifts are only as good as the climatic predictions that underlie them.

Further Reading

Archibold OW. *Ecology of World Vegetation*. London: Chapman & Hall, 1995.

Crawford RMM. *Plants at the Margin: Ecological Limits and Climate Change*. Cambridge: Cambridge University Press, 2008.

Grime JP. *Plant Strategies, Vegetation Processes, and Ecosystem Properties*. 2nd ed. Chichester: John Wiley, 2001.

Loreau M, Naeem S, Inchausti P. *Biodiversity and Ecosystem Functioning*. Oxford: Oxford University Press, 2002.

Ohgushi T, Craig TP, Price PW. *Ecological Communities*. Cambridge: Cambridge University Press, 2007.

Van der Maarel E, Franklin J. *Vegetation Ecology*. 2nd ed. Chichester: Wiley-Blackwell, 2013.

Woodward FI, Lomas MR. Vegetation dynamics: simulating responses to climatic change. *Biological Reviews of the Cambridge Philosophical Society* 2004; 79: 643–670.

References

1. Ohgushi T, Craig TP, Price PW (eds.). *Ecological Communities*. Cambridge: Cambridge University Press, 2007.

2. Rodwell JS. *British Plant Communities, vol. I. Woodlands and Scrub*. Cambridge: Cambridge University Press, 1991.

3. Delcourt H, Delcourt P. Eastern deciduous forests. In: Barbour MG, Billings WD (eds.), *North American Terrestrial Vegetation*. 2nd ed. New York: Cambridge University Press, 2000: 357–395.

4. Callaway RM. Positive interactions among plants. *Botanical Review* 1995; 61: 306–349.

5. Callaway RM, D'Antonio CM. Shrub facilitation of coast live oak establishment in central California. *Madroño* 1991; 38: 158–169.

6. Forman RTT. *Land Mosaics: The Ecology of Landscapes and Regions*. Cambridge: Cambridge University Press, 1995.

7. Thomas F, Renaud F, Guegan J-F. *Parasitism and Ecosystems*. Oxford: Oxford University Press, 2005.

8. Reagan DP, Waide RB. *The Food Web of a Tropical Rain Forest*. Chicago: University of Chicago Press, 1996.

9. Vitousek P. *Nutrient Cycling and Limitation*. Princeton: Princeton University Press, 2004.

10. Naeem S, Loreau M, Inchausti, P. Biodiversity and ecosystem functioning: the emergence of a synthetic ecological framework. In: Loreau M, Naeem S, Inchausti P (eds.), *Biodiversity and Ecosystem Functioning*. Oxford: Oxford University Press, 2002: 3–17.

11. Smith TM, Shugart HH, Woodward FI. *Plant Functional Types: Their Relevance to Ecosystem Properties and Global Change*. Cambridge: Cambridge University Press, 1997.

12. McNaughton SJ, Banyikwa FF, McNaughton MM. Promotion of the cycling of diet-enhancing nutrients by African grazers. *Science* 1997; 278: 1798–1800.

13. Elton CS. *The Ecology of Invasion by Animals and Plants*. Chicago: University of Chicago Press, 2000.

14. Chapin FS, Walker BH, Hobbs RJ, Hooper DU, Lawton JH, Sala OE, Tilman D. Biotic control over the functioning of ecosystems. *Science* 1997; 277: 500–504.

15. Wittebolle L, Marzorati M, Clement L, *et al.* Initial community evenness favours functionality under selective stress. *Nature* 2009; 458: 623–626.

16. Leps J. Diversity and ecosystem function. In Vander-Maarel E, Franklin J (eds.), *Vegetation Ecology*. Chichester: Wiley-Blackwell, 2013: 308–346.

17. Naeem S, Li S. Biodiversity enhances ecosystem reliability. *Nature* 1997; 390: 507–509.

18. Montoya JM, Pimm SL, Sole RV. Ecological networks and their fragility. *Nature* 2006; 442: 259–264.

19. Hector A, Bagchi R. Biodiversity and ecosystem multifunctionality. *Nature* 2007; 448: 188–190.

20. Moore PD, Bhadresa R. Population structure, biomass and pattern in a semi-desert shrub, *Zygophyllum eurypterum* Bois. and Buhse, in the Turan Biosphere Reserve of north-eastern Iran. *Journal of Applied Ecology* 1978; 15: 837–845.

21. Box EO, Fujiwara K. Vegetation types and their broad-scale distribution. In Vander Maarel E, Franklin J. (eds.), *Vegetation Ecology*. Chichester: Wiley-Blackwell, 2013: 455–485.

22. Grime JP. *Plant Strategies, Vegetation Processes, and Ecosystem Properties*. Chichester: John Wiley, 2001.

23. Shugart HH. Plant and ecosystem functional types. In: Smith TM, Shugart HH, Woodward FI (eds.), *Plant Functional Types*. Cambridge: Cambridge University Press, 1997: 20–43.

24. Prentice IC, Cramer W, Harrison SP, Leemans R, Monserud RA, Solomon AM. A global biome model based on plant physiology and dominance, soil properties and climate. *Journal of Biogeography* 1992; 19: 117–134.

25. Woodward FI, Lomas MR, Kelly CK. Global climate and the distribution of plant biomes. *Philosophical Transactions of the Royal Society of London B* 2004; 359: 1465–1476.

26. Whittaker RH. *Communities and Ecosystems*. 2nd ed. New York: Macmillan, 1975.

27. Schoenherr AA. *A Natural History of California*. Berkeley: University of California Press, 1992.

28. Fitter J, Fitter D, Hosking D. *Wildlife of Galapagos.* London: Collins, 2007.

29. Crawford RMM. *Plants at the Margin: Ecological Limits and Climate Change.* Cambridge: Cambridge University Press, 2008.

30. Walter H. *Vegetation of the Earth.* 2nd ed. Heidelberg: Springer-Verlag, 1979.

31. Box EO. *Macroclimate and Plant Forms: An Introduction to Predictive Modeling in Phytogeography.* The Hague: Dr W. Junk, 1981.

32. Cramer W, Leemans R. Assessing impacts of climate change on vegetation using climate classification systems. In: Solomon AM, Shugart HH (eds.), *Vegetation Dynamics and Global Change.* London: Chapman & Hall, 1992: 190–217.

33. Haxeltine A, Prentice IC. BIOME 3: an equilibrium terrestrial biosphere model based on ecophysiological constraints, resource availability, and competition among plant functional types. *Global Biogeochemical Cycles* 1996; 10: 693–709.

Patterns of Biodiversity

Biodiversity is a term that encompasses all of the living things that currently exist on Earth. It includes all animals and plants that zoologists and botanists have discovered and described, and also all those that remain undiscovered and still await a scientific description. In addition, biodiversity includes all fungi, bacteria, protozoa and viruses which, on the whole, are even less well known than the animals and plants. But biodiversity goes even beyond this and includes the genetic variation found within each species, and some would even extend the use of the term to cover the great range of habitats that exist on Earth and that support all living things. Habitat diversity is undoubtedly closely linked to biodiversity in its biological sense. In this chapter, we examine what is known of the Earth's biodiversity and whether one can discern patterns in biodiversity distribution.

Given the inconceivable immensity of the universe, it is unlikely that the Earth is the only body on which life exists. But as far as present knowledge is concerned, it remains unique in supporting a wide range of living organisms. As astronomical research continues, the discovery of evidence for life existing on some other distant planet in another solar system becomes statistically greater, but one thing is sure: life is an exceedingly rare commodity in the universe and should therefore be greatly valued. **Biodiversity** is an expression of the great variety of living things on our planet, but it is far more than a simple count of species.

As discussed in Chapter 2, when we take a species and analyse its composition, we find that it consists of a series of populations, sometimes adja-

cent to one another and sometimes fragmented and isolated. A fragmented population is termed a **metapopulation**. Some of these isolated populations are clearly genetically distinct and can be classified as subspecies, but even within single populations there is often great variation between individual organisms. Biodiversity includes the whole range of populations, together with all the genetic variations found within each species.

Just as we can regard a species as a collection of component populations, so we can interpret communities of organisms as assemblages of many populations of a variety of different species, all interacting together (see Chapter 3). When these communities are placed in the setting of their non-living environment, they make up ecosystems. Our concept of biodiversity should therefore include the rich variety of communities and ecosystems that occupy the Earth, many of which have an influential human component. If we wish to conserve biodiversity, and most ecologists believe that this is basically a sensible aim, indeed a responsibility, then the conservation of whole ecosystems and their habitats is the most appropriate starting point.

The subject of conservation will be examined in detail in Chapter 14, but it is helpful at this stage to consider why biodiversity should be regarded as important and why it is appropriate to retain it. Why do we need, or why do we want, to maintain the biotic richness of the Earth? Needing and wanting are very different experiences. We need something when it is useful to us, and one could argue that the other living organisms of the Earth are indeed useful. Some provide food, or materials

Biogeography: An Ecological and Evolutionary Approach, Ninth Edition. Edited by C. Barry Cox, Peter D. Moore, Richard J. Ladle.
© 2016 John Wiley & Sons, Ltd. Published 2016 by John Wiley & Sons, Ltd.

for building homes and making fabrics; others are a source of pharmaceuticals. Very many organisms are part of our general support system on Earth, being involved in the maintenance of the gaseous balance of the atmosphere, the healthy structure of the soil and even the modification of our climate. So there is a strong utilitarian argument for the maintenance of the Earth's biodiversity. Even those species that are not currently regarded as useful may one day be found to be so.

There is a further argument, however, that may be associated with 'wanting' rather than 'needing'. We may want to have birds, flowers and butterflies in the world simply because we enjoy having them there; this is a kind of aesthetic argument. We may regret that we shall never have the opportunity to see a passenger pigeon, a dodo or a great auk. Or we may take a view that is even less human-centred and claim that the organisms have legitimate rights of their own to exist, perhaps even as great a right as ours. This is an ethical argument that lies beyond the remit of science, but this does not deny its validity. It is our responsibility, therefore, as a species with a high impact on the Earth's ecosystems, to ensure that extinction is minimized. Conservation itself then becomes in part an ethical issue.

If we accept any or all of these arguments, then it leads us to a position where we need to know whether the Earth is actually losing species – and, if so, how fast? We also need to be concerned about whether human beings are contributing substantially to the rate of loss and whether there is anything that we can do about it. But to understand extinction rates and their causes, we have to go back further still and ask just how many species there are on Earth so that we can calculate how rapidly we may be losing them.

How Many Species are There?

No one likes to lose things, but the outcome of any loss can best be measured on the basis of what is left. The loss of a dollar will be felt more severely by a pauper than by a millionaire. The importance to humanity of the loss of species from the Earth can only be judged, therefore, if we can view it in proportional terms and view the loss from the perspective of what remains. We need to

appreciate how many species occupy the Earth in order to evaluate the importance of current extinction rates.

One of the most surprising things about science is how little we know. You might assume, for example, that biologists would have a reasonably good idea of how many species of living organisms exist on Earth, but in fact this is not so. The question is still hotly debated, and the estimates that biologists have made range between 3 and 500 million species! What they all agree upon is that only a very small proportion of the total is currently known to science and has been adequately described. Some parts of the world have been little studied, especially the tropical regions where the diversity of species is particularly high. Many very abundant species are extremely small, so they may have gone unnoticed in the past. Many species are difficult to identify, so very careful study is needed to distinguish them, and the numbers of expert scientists in the field of **taxonomy** (the study of plant and animal classification) are relatively few, so the task of counting species is much more difficult to accomplish than might at first appear.

The confusion about the number of species present on Earth may seem surprising to non-biologists, but the sheer wealth of species makes it difficult to be sure that all of those that have been described are valid and are not duplicates, or that those described as a single species do not, in fact, consist of a number of species that we have ignorantly lumped together. The purple swamphen (*Porphyrio porphyrio*), for example, is found throughout the Old World, from southern Spain, through Africa, India and South-East Asia, Indonesia, Australia, New Zealand and the islands of the South Pacific [1]. It has also been introduced into Florida, but in North America it is largely replaced by the purple gallinule (*Porphyrio martinica*) [2]. But recent work on the Old World species has shown that there are 14 subspecies, and these fall into six genetically related groups on separate evolutionary branches, or **clades** (see Chapter 6). The genetic separation between these groups has been examined and is now regarded as adequate for the recognition of six full species instead of one [3]. Thus one species has been translated into six, and global biodiversity has been increased at a stroke! There are likely to be many such examples in the future as analysis of DNA provides greatly improved ways of examining biodiversity. One of the problems here is actually defining a species. The convenient idea that two species cannot interbreed is not actually workable because it is often found that interbreeding is possible between several animal and plant species that are morphologically distinct. If populations of an organism are sufficiently distinct in their outward form, their physiology or their behaviour patterns, and if they are distinct in their molecular genetic patterns, then they are regarded as distinct species (see Chapter 6). But the definition is quite plastic and the lines drawn around species will vary from time to time as more information becomes available, so the task of determining how many species there are on Earth becomes even more of a problem.

Although there are undoubtedly many species yet to be discovered, there are still approximately 1.8 million species of organism that have been described. Many may have been described and named twice, or even more times, however! Among beetles, for example, 40% of the species described have only ever been recorded at the site of their first description. This is very unlikely to be a true reflection of their distributions, and it is quite likely that many of the species are in fact duplicates with different names in different localities. This so-called 'alias problem' will inflate the numbers of species described [4], but the likelihood is that there are far more species remaining to be found, so that the true number of species still living on the Earth must almost certainly greatly exceed the number currently described. Box 4.1 discusses counting species in further detail.

The diversity of microbes (bacteria, fungi, viruses etc.) is particularly difficult to estimate, because these groups are not nearly as well known as, say, the mammals or the flowering plants. Of the bacteria, for example, only about 4000 species have so far been described (Table 4.1), and this may well represent only about one-tenth of one per cent of the total, so much remains to be done in this area. The study of biodiversity among microbes is complicated by the range of genetic and biochemical variation found among wild populations [9,10]. It is also made more difficult by the fact that bacteria

Counting species

Concept Box 4.1

A conservative estimate for the possible number of species on Earth is 12.5 million [5], but the tropical ecologist Terry L. Erwin [6] has proposed that the total is far greater than this, perhaps as high as 30 million for tropical insects alone. He came to this conclusion as a result of a study of beetles on a single tree species, *Luehea seemannii* (a tropical tree related to the lime tree of temperate regions) in Panama, which he sampled by 'fogging'. This is an efficient technique for stunning the insects in a canopy by smoking them with an insecticide. The dazed insects fall from the tree and are collected in trays placed beneath the canopy. Erwin examined just 19 individual trees of *L. seemannii* in the Panamanian forests and managed to obtain 1200 species of beetles alone from this analysis. This large number is not entirely surprising, since beetles are extraordinarily successful insects and may comprise as much as 25% of the total number of species of living organisms. But this study does illustrate the remarkable richness of beetles in the tropical forest.

From these data, Erwin made a number of assumptions about the numbers of beetle found specifically on particular tree species, the numbers of tree species found and the proportions of different organisms in the forest in relation to one another. He extrapolated from the information gathered and came to the conclusion that, if this number of beetles is truly representative of the forest richness, then one might predict a total of 30 million species of insect on Earth. The uncertainty of many of his assumptions, however, should make us very cautious in accepting this figure uncritically. Other entomologists, such as Nigel Stork and Kevin Gaston [7], have checked Erwin's estimates using data from studies in the tropical forests of Borneo. Stork has generated estimates ranging from 10 million to 80 million for the arthropods (a group of invertebrate animals including the insects). Another independent estimate [8] supports the lower end of this scale, placing tropical arthropods at 6 to 9 million. The range of error in all estimates is still so wide that there is bound to be a great deal of discrepancy in the figures arrived at, but the world's wealth of species is likely to exceed 10 million.

Table 4.1 The numbers of described species in selected groups of organisms, together with the likely total numbers on Earth, and the percentage of the group that is currently known. Data from Groombridge [15].

Group	Number of described species	Likely total	%
Insects	950 000	8 000 000	12
Fungi	70 000	1 000 000	7
Arachnids	75 000	750 000	10
Viruses	5000	500 000	5
Nematodes	15 000	500 000	3
Bacteria	4000	400 000	1
Vascular plants	250 000	300 000	83
Protozoans	40 000	200 000	20
Algae	40 000	200 000	20
Molluscs	70 000	200 000	35
Crustaceans	40 000	150 000	27
Vertebrates	45 000	50 000	90

can survive deep in geological deposits, far beyond those surface layers of the Earth which were once supposed to represent the limits of the **biosphere**, that part of the Earth that is inhabitable by living organisms [11]. Their capacity to live in extreme environments and their great potential in the service of humankind make them particularly interesting and important to humanity, so it is in our interests to improve our knowledge of microbial diversity. But even the concept of the species has to be reconsidered when dealing with microbes [12].

An example of the immensity of the problem facing those studying microbial diversity is afforded by the Rio Tinto in southern Spain [13]. This river was known to the ancient Phoenicians as the 'River of Fire' because of its deep red colour, caused by the high concentration of iron and other metals dissolved in its highly acidic (pH 2) waters. These conditions are now often associated with highly polluted waters, usually as a result of human mining activities, and the Rio Tinto has indeed been affected in this way for over 5000 years. But even before the times of human pollution, this river was rich in metals because of the high metal content of the rocks over which it flows. Research into the biodiversity of the river has employed molecular techniques to examine the range of microorganisms present and has found a great wealth of previously undetected species. Over 60% of the biomass of the river consists of highly tolerant strains of microscopic algae that have evolved and adapted over millennia of exposure to these extreme conditions.

Even the most unlikely habitats, therefore, may contain a wealth of life.

Among the fungi, some 70 000 species have been described so far, but David Hawksworth, formerly head of the International Mycological Institute, England, believes that the true total could be around 1.6 million species [14]. In areas of the Earth where the fungi have been thoroughly investigated, each higher plant species supports about five or six fungal species. Therefore, if the total flowering plants (once all have been described) is assumed to be about 300,000, the total number of fungi must be in the region of one-and-a-half million or more.

From this example, it can be seen that the estimation of possible numbers of organisms can be derived by a process of extrapolation. If we have certain facts and make further assumptions about proportional representation of different groups, then we can begin to project from what we know into the misty realms of uncertainty. The outcome is not satisfactory, but it is the best we can do so far. Table 4.1 provides some idea of what we presently know about the richness of the Earth's diversity of species for a few groups of organisms and also approximately what proportion of each group is currently thought to have been described [15]. The vertebrates (animals with backbones) are reasonably well known, and relatively few new species may be expected in this area, and the same is true for flowering plants. In surveys, 4327 species of mammal were listed, and 9672 species of bird. It is unlikely that these totals will grow very substantially even with further survey work and research into their classification, although new species do continue to be described at a rate of about 100 species per decade in the case of mammals [16]. In fact, there have been over 40 new species of primates described since 1990, including two new monkey species from the Brazilian Amazon region in 2002. But the spiders and mites (arachnids), the algae and the nematode worms, among others, are still very poorly understood and many more species can be expected in these groups. There is the additional frustration, from the biologist's point of view, that smaller organisms tend to be both more abundant and more diverse than are larger ones, thus adding to the workload at the most difficult end of the spectrum [17].

The process of listing all the species present at a site, quite apart from the genetic and habitat variation that contribute to biodiversity, would be a long, costly, time-consuming and finally inaccurate process. An alternative approach, illustrated above by its use with fungi, is to assess the richness of certain groups of well-known organisms that are easily observed and identified (such as higher plants, mammals, birds or butterflies) and to assume that they have a consistent proportional relationship to the less easily observed and identified groups. This method should work well where species are closely dependent on one another (as host and parasite, or as predator and prey), as in the case of gall-forming organisms and plant species in the scrub grasslands of South Africa (Figure 4.1) [18]. Here, there is a strong linear relationship between the two groups of organisms. This is not always the case, however. The work of John Lawton and colleagues in the Mbalmayo Forest Reserve in Cameroon [19], covering birds, butterflies, beetles, ants, termites and nematodes, has shown little overall relationship between the abundance of one group and that of another. But other workers have analysed data from numerous biological surveys of different sites in the United Kingdom and Canada and have found that it is possible to ignore a proportion of the total species without significant loss of data [20]. Eliminating the 10% of species from a survey that are the most difficult and time-consuming to identify, for example, resulted in no significant change to the assessed pattern of biodiversity.

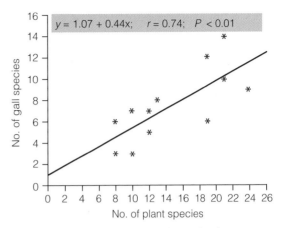

Figure 4.1 Graph showing the relationship between the richness of gall-forming species and the number of woody shrub species in the Aynsberg Nature Reserve, South Africa. Data from Wright and Samways [18].

Another attempt to find a short-cut to the estimation of the total number of living things has been the use of body size. It is quite obvious that there are fewer large organisms than small ones, but is there a simple relationship between species number and body size? Since we know the numbers of large-bodied species reasonably well, we could then extrapolate from what we know of them to estimate how many small creatures exist on Earth. If we consider animals of 5–10 m in length (e.g. elephants and whales), there are fewer than 10 species that come into that size range. In the range of 1–5 m (e.g. horses and deer), there are perhaps a few hundred. In the 0.5–1 m range (e.g. fox), there may be up to a thousand species, and rat-sized organisms (0.1–0.5 m) could run to 10 000. This body-size–abundance relationship continues to hold as we move down the size range, so that at 0.005–0.01 m (ant-sized), there could be around a million species. But for smaller creatures (below 0.005 m, such as fleas, mites and smaller), the number of known species declines from the straight-line logarithmic relationship that holds at higher levels of size. Is this lack of apparent diversity in the smaller organisms a consequence of our own ignorance of their taxonomy and their diversity, or are we wrong to assume that they should follow the same linear pattern as that of larger species? If we assume that the linear graph represents a real pattern of size–diversity relationship, then we can calculate an expected number of small organisms, and calculate that total biodiversity should reach around 10 million.

Researchers at the University of Minnesota have tested the assumption that small creatures must be more diverse than they appear by examining single ecosystems in great detail and working out the relationship between size and diversity for these specific sites. Figure 4.2 shows the results of analysing in detail the invertebrate species of a North American grassland ecosystem in relation to their body size (measured here as volume rather than length) [21]. It can be seen that in a wide range of insect groups, there are in fact fewer species at the small end of the spectrum than the linear model predicts, and in this case we can be assured that it is not because small species have been overlooked. We must therefore conclude that the linear model itself is at fault and that this is not an appropriate way to calculate total biodiversity. Perhaps studies of this kind will result in the establishment of a more reliable model than the linear one that can in future be used for extrapolation and biodiversity estimation. There is also a need to investigate the size–diversity relationship among plants. At present, there is evidence that there are considerably more species of small plants than there are large ones, possibly because they have narrower niches or higher fecundity [22], but no precise model of the relationship is currently available.

Because the estimation of how many species are present on Earth is so difficult and uncertain, it is clear that rates of extinction can only be vaguely estimated. Although many known extinctions in the recent past have resulted from human hunting or persecution, the organisms involved have been large, conspicuous and easily targeted. Nothing is known of the smaller creatures, especially the microbes. It is likely, however, that extinction of small organisms is mainly due to a loss of habitat, and some habitats are richer than others, so calculating possible extinction rates depends on what kind of habitats are being destroyed. Assessing rates of extinction is also made difficult because we can rarely be sure that a species is actually lost, that no isolated members remain. There is still a possibility that the ivory-billed woodpecker (*Campephilus principalis*) survives somewhere in the southern United States or in Cuba, even though it has been regarded as extinct [23]. In the case of plants, extinction is even more difficult to record. Some plants are able to survive for decades or even centuries as buried seeds, so they can unexpectedly reappear after long episodes of presumed extinction. Many of the plant species of the Atlantic forests of Brazil, now occupying only about 10% of their original cover, were last recorded back in the 1850s and have not been seen since. But it is difficult to be sure that they have finally gone and that not even dormant seeds survive. There are

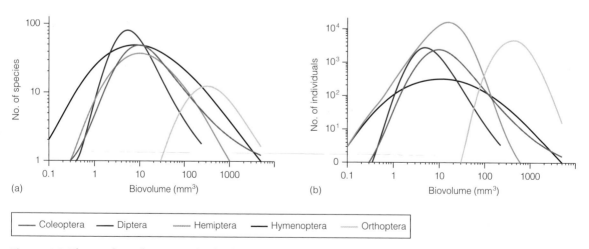

Figure 4.2 The numbers of species and individuals of various insect groups in relation to their body size in grassland in Minnesota. The groups shown are Orthoptera (grasshoppers), Hymenoptera (wasps and bees), Hemiptera (bugs), Coleoptera (beetles) and Diptera (flies). Adapted Siemann *et al.* [21].

occasions when plants are discovered that have formerly been known only in a fossil state. Even as recently as 1994 a new gymnosperm tree species was discovered in a deep gorge near Sydney, Australia [24], which has been given the name *Wollemia nobilis* and, like the 'living fossil' ginkgo, closely resembles a fossil plant of the Cretaceous that was presumed to be long extinct.

But, despite the problems in recording the process accurately, extinction is undoubtedly occurring all around us. The biologist Edward O. Wilson [5] has calculated that the loss of species from the tropical forest area alone could currently be as high as 6000 species per year. This amounts to 17 species each day, and the tropical forests cover only 6% of the land surface area of the Earth, so the global rate of extinction will be even higher if we include other biomes. Although extinction has always been an inevitable element in the evolutionary process,

it is calculated that recent extinction rates may be 100–1000 times greater than those before the emergence of our species. It is also feared that they might accelerate by a further 10 times in the next century [25]. There have already been at least five major extinction events in the history of the Earth, but this sixth extinction may prove greater and more rapid than all those preceding it.

Latitudinal Gradients of Diversity

The distribution of biodiversity over the land surface of the planet is far from even [26]. The tropics contain many more species, of both plants and animals, than an equivalent area of the higher latitudes. This seems to be true for many different groups of animals and plants, as can be seen from Figure 4.3, which illustrates the number of

Figure 4.3 Numbers of breeding birds and mammal species in different parts of Central and North America.

breeding birds and mammals found in various Central and North American countries and states. The tropical country Panama, only 800 km north of the equator and a close neighbour of Costa Rica, has 667 species of breeding birds, three times the number found in Alaska, despite the much greater area of Alaska.

A similar pattern is seen in the number of mammal species at different latitudes in North America (Figure 4.3). Considering just the forest areas from southern Alaska (65°N) in the north, through Michigan (42°N) into the tropical forest of Panama (9°N), these have 13, 35 and 70 mammalian species, respectively. Breaking the mammals down into their component groups by taxonomy, we find that the bats account for a large part of the difference between the three locations. Moving from the north, one-third of the increase in species from Alaska to Michigan is due to the larger number of bats, and so also is two-thirds of the increase between Michigan and Panama. Yet more information becomes available if we consider diet among the mammals. Much of the tropical diversity among mammals is due to the greater predominance of a fruit-eating way of life and to the greater number of insectivores, many of which eat insects that in turn feed on the fruits of the forest. This may also account for the diversity of frogs [27] in the lower latitudes (Figure 4.4). Therefore, diet is evidently an important aspect underlying the diversity gradient found among animals.

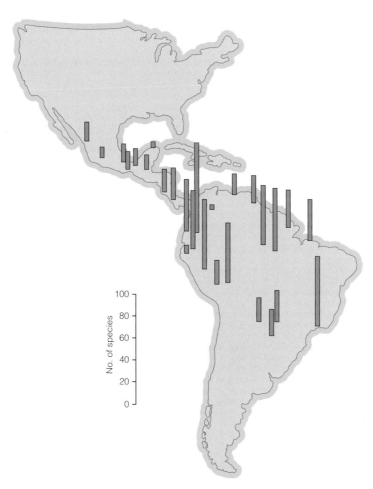

Figure 4.4 Numbers of species of frog in different parts of Central and South America. Data from Groombridge [27].

As has been discussed in this chapter, insect diversity is difficult to measure because the insect groups are generally not fully described. Butterflies are among the best recorded of the insects, and Figure 4.5 shows the latitudinal gradients of richness in just one group of butterflies, the swallowtails (Papilionidae) [28]. The high numbers of species found in the tropics is again apparent, and here it can be seen that this applies to all tropical areas of the world. One easily explained anomaly in the gradients can be seen in the African–European section, where there is a dip in species richness that coincides with the North African desert region.

Diet may be an important factor in determining animal diversity, but what of plants? They also show a general trend towards increasing diversity in the tropics, as shown in the map of tree diversity in North and South America (Figure 4.6), but they do not vary in their diet since they all need sunlight energy for their photosynthesis.

The relationship between plant species richness and latitude is not at all a simple one. David Currie and Viviane Paquin of the University of Ottawa constructed a map of the richness of tree species across North America [29], and this is shown in Figure 4.7. From this map, it can be seen that the contours of richness do not simply follow the lines of latitude, especially in the areas south of Canada. Patches of low diversity occur in the mid-west, and there is an exceptionally high diversity of trees in the southeast. When these workers examined the possible environmental factors that may be asso-

ciated with this pattern, the one which correlated most closely was the sum of evaporation (directly from the ground) and transpiration (from the surface of vegetation) combined to give a value for the loss of water from the land surface (**evapotranspiration**). Evidently, those regions with the highest evapotranspiration are able to support the highest diversity of tree species. But evapotranspiration itself also correlates closely with the potential **productivity** of a region (the amount of plant material that accumulates by photosynthesis in a given area in a given time), so perhaps plant diversity is essentially determined by how much photosynthesis can be carried out in a given site. Figure 4.8 shows this relationship between primary production and tree species richness, and it can be seen that there is indeed a good correlation between the two.

In general, the equatorial regions are the areas in which highest productivity is possible because of the prevailing climate, which is hot, wet and relatively free from seasonal variation. Figure 4.9 illustrates this by displaying the world distribution of mean annual primary productivity. From this map, it can be seen that very high productivity is concentrated in the equatorial belt and that this drops off as one moves towards higher latitudes. The picture is complicated by the arid belts in northern Africa and central Asia, of which more will be said in the 'Biodiversity Hotspots' section, but the general trend is decreasing productivity at higher latitudes. An examination of the North American part of this map shows a good correlation with the

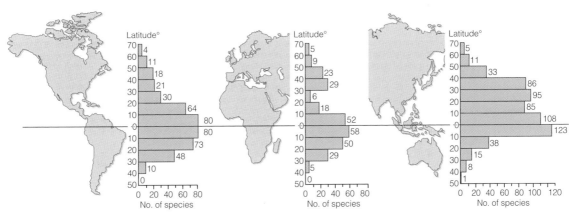

Figure 4.5 Latitudinal gradients of species richness for swallowtail butterflies in three different parts of the world. Data from Collins and Morris [28].

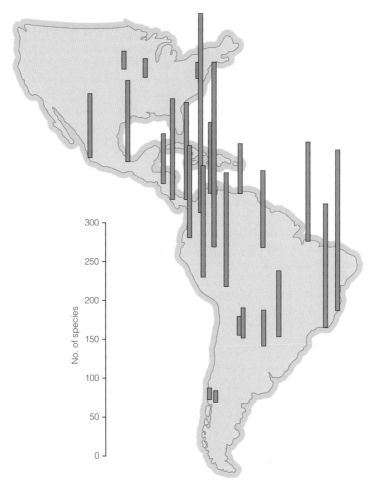

Figure 4.6 Latitudinal gradients of tree species richness in the Americas. Data from Groombridge [15].

tree-richness map (see Figure 4.7), especially with regard to the high diversity of tree species and high productivity in the southeastern United States. This approach to explaining the latitudinal gradient of species diversity suggests that the critical factor is how much energy is captured by the vegetation. This has come to be known as the **energy hypothesis**, and it is particularly well supported by plant-based data [30].

When developing the hypothesis to account for animal diversity gradients, several factors have to be considered. Higher plant productivity in general means greater energy stores available for consumers. Higher productivity often results in higher biomass and more complex vegetation architecture,

producing more microhabitats, which can result in greater opportunities for animals. And warm, moist climates in which high productivity occurs can also lead to greater metabolic rates in organisms. This latter approach has led to the development of what has been termed a **metabolic theory** to account for the latitudinal gradients in biodiversity [31]. Researchers from the University of Ottawa, Canada, have tested the metabolic theory by examining the relationship between temperature and species richness for a wide range of animal groups in North America, including amphibians, reptiles, tiger beetles, butterflies, blister beetles as well as trees [32]. They found that the metabolic theory failed to fully account for the observed

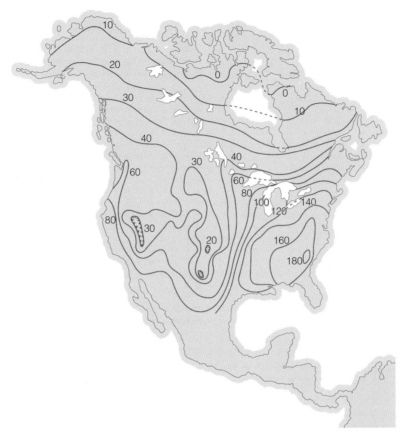

Figure 4.7 Number of tree species (i.e. any woody plant over 3 m in height) found in different parts of North America. The contours indicate areas where particular numbers of tree species were recorded within large-scale quadrats (mean area of 70 000 km²). Data from Currie and Paquin [29].

patterns, and they suggested that predictive models need to take water availability into account in addition to temperature. Using a water-energy model for predicting the richness of mammals, birds, amphibians, reptiles and plants in Europe, Robert Whittaker and colleagues from Oxford University [33] have shown good correlations, so the general climatic theory of biodiversity determination seems to hold up well in this case.

Productivity, therefore, seems to be involved, but perhaps its influence is indirect. Where conditions are most suitable for plant growth, that is, where temperatures are relatively high and uniform and where there is an ample supply of water, one usually finds large volumes of vegetation – in

other words, high biomass. This leads to a complex structure in the layers of plant material. In a tropical rainforest, for example, a very large quantity of plant material builds up above the surface of the ground. There is also a large mass of material developed below ground as root tissues, but this is less apparent and, in the case of the tropical rainforest, confined to the upper layers of soil. Careful analysis of the above-ground material reveals that it is arranged in a series of layers, the precise number of layers varying with the age and the nature of the forest. The arrangement of the biomass of the vegetation into layered forms is termed its **structure** (as opposed to its **composition**, which refers to the species of organisms forming the community).

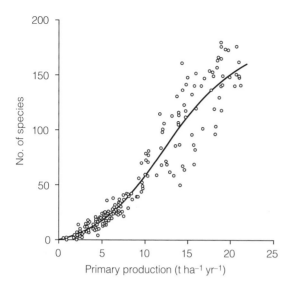

Figure 4.8 Number of tree species in North American sites (see Figure 4.7) plotted against the primary productivity of those sites. A distinct positive relationship can be observed. Data from Currie and Paquin [29].

Structure is essentially the architecture of vegetation and, as in the case of some tropical forests, can be extremely complicated. Figure 4.10 shows a profile of a mature flood plain tropical forest in Amazonia [34] expressed in terms of the percentage cover of leaves at different heights above the ground. There are three clear peaks in leaf cover at heights of approximately 3, 6 and 30 m above the ground, and the very highest layer, at 50 m, corresponds to the very tall, emergent trees that stand clear of the main canopy and form an open layer of their own. So this site contains essentially four layers of canopy.

Forests in temperate lands are simpler, often with just two canopy layers, so they have much less complex architecture. Structure, however, has a strong influence on the animal and plant life inhabiting a site. It forms the spatial environment within which an animal feeds, moves around, shelters, lives and breeds. It even affects the climate on a very local level (the **microclimate**) by influencing light intensity, humidity and both the range and extremes of temperature. Figure 4.11 shows a profile through an area of grassland vegetation that has a very simple structure, and it can be seen that the ground level has a very different microclimate from that

experienced in the upper canopy of the grasses. Wind speeds are lower, temperatures are lower during the day (but warmer at night) and the relative humidity is much greater near the ground. The complexity of microclimate is closely related to the complexity of structure in vegetation; and, generally speaking, the more complex the structure of vegetation, the more species of animal are able to make a living there per unit area of land surface. This is illustrated in Figure 4.12, which relates the number of bird species found in woodland habitats to the number of leaf canopy layers that can be detected [35]. The high plant biomass of the tropics leads to a greater spatial complexity in the environment, and this will lead to a higher potential for diversity in the living things that can occupy the region. The climates of the higher latitudes are generally less favourable for the accumulation of large quantities of biomass, hence the structure of vegetation is simpler and the animal diversity is consequently lower.

There is one important extension to this line of argument that is worth pursuing. Complexity, or the conception of complexity, depends upon the size of the observer. It was stated here that grassland has a relatively simple structure, but this is only the case if one views it from a human perspective. From an ant's point of view, on the other hand, a grassland environment may be highly complex. For this reason, an area of grassland offers a home to far more ants than it does to humans, cows or bison. As discussed in this chapter, the Earth as a whole can support far more small creatures than it can large ones. In part, this can be explained by the greater number of opportunities offered to very small organisms, even within habitats of relatively simple structure. In habitats of more complex structure, of course, small organisms find even more microhabitats and different ways of making a living. The latitudinal diversity of small organisms, therefore, tends to follow the same patterns as larger ones, being richest in the tropics.

There is one group of very small animals, however, that goes against the rules. The aphids (such as plant-feeding greenflies), of which about 4000 species are known, are less diverse in the tropics than they are in the temperate regions [36]. Most aphid species feed on only one type of plant and are rather poor at locating that plant from a distance, relying on the sheer chance of air-flow patterns to

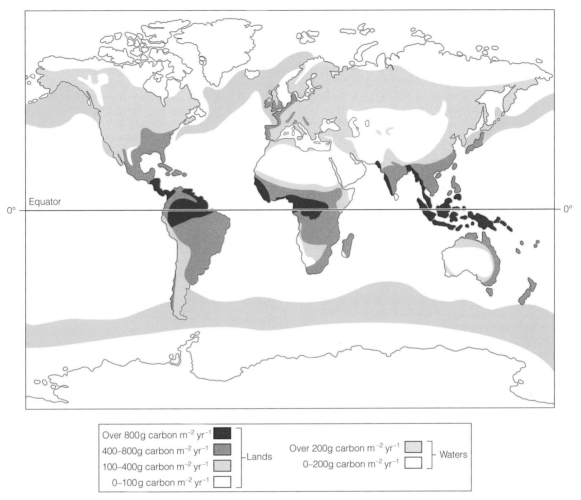

Over 800g carbon m^{-2} yr^{-1} ▮
400–800g carbon m^{-2} yr^{-1} ▮
100–400g carbon m^{-2} yr^{-1} ▮ ⎤ Lands
0–100g carbon m^{-2} yr^{-1} ▯

Over 200g carbon m^{-2} yr^{-1} ▯ ⎤ Waters
0–200g carbon m^{-2} yr^{-1} ▯

Figure 4.9 World distribution of plant productivity. The data displayed here are simply estimates of the amount of organic dry matter that accumulates during a single growing season. Full adjustments for the losses due to animal consumption and the gains due to root production have not been made. Map compiled by H. Leith.

carry them from one suitable host to another. They do best therefore where populations of particular plant species are dense, as is the case in agricultural crops. Unfortunately for the aphids, tropical plant assemblages consist of many species each of which is present only at low density, so aphids are not well suited to such conditions. Aphid diversity is therefore inversely related to plant diversity, so they transgress the general pattern of latitudinal gradients.

The richness of the tropics as far as animal life is concerned may thus be a consequence not simply of the high productivity of these latitudes, but also of their great structural complexity resulting from their high biomass, which can support many species of small animal.

In a sense, we may be asking the wrong question when we try to analyse the factors that contribute to the high species diversity of the tropics, for this is really the perspective of biologists viewing the problem from the temperate regions, which happens to be where most ecologists live and work. From a tropical angle, the question should be why the higher latitudes have lower diversities than the tropics [37]. Perhaps it is simply a matter of land area? The tropics contain a larger surface area of

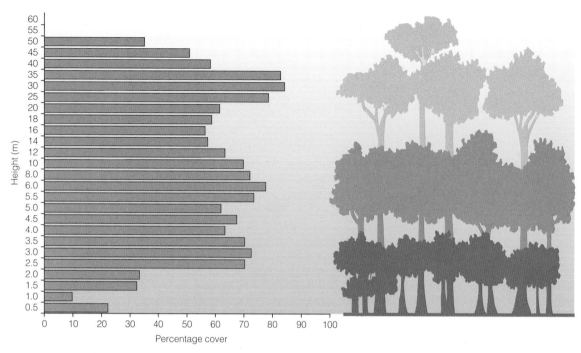

Figure 4.10 Profile of a tropical rainforest with the percentage leaf canopy cover recorded at different heights above the ground. Note the stratification of the leaf cover into distinct layers. From Terborgh and Petren [34].

land than higher latitudes (a fact that is not always evident when we examine commonly used projections of the Earth's curved surface since this tends to exaggerate the areas of land in the higher latitudes), and some biogeographers regard the latitudinal gradients of diversity as a reflection of this effect [38]. But an analysis of the data by Klaus Rohde [39] does not support this explanation.

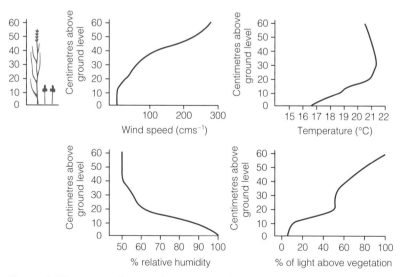

Figure 4.11 Diagram showing the structure of grassland vegetation and the effect this has upon the microclimate of the habitat.

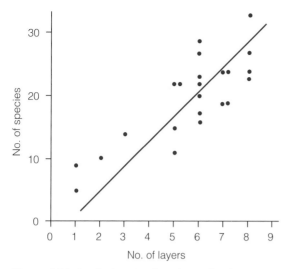

Figure 4.12 Graph showing the relationship between the number of bird species and the number of layers in the vegetation stratification. Data from Blondell [35].

Although area may contribute to biodiversity, it is certainly not the major story; otherwise, large land masses would always be richer.

Overall, the proposal that the latitudinal gradient in biodiversity, demonstrated by so many plant and animal groups, is largely determined by energy availability and hence climate, and is operating through the productivity and complexity of the structure of vegetation, is supported by many observations. There are, however, alternative or additional possible causes that may have an influence on global patterns.

Is Evolution Faster in the Tropics?

Evolution, the process by which new species are generated, is discussed in detail in Chapter 6. It essentially involves genetic variation and subsequent selection of the most suitable genetic combinations within a particular environment. The important question, as far as the explanation of latitudinal gradients of diversity is concerned, is whether evolution (either the generation of genetic variation or its selection) proceeds more rapidly under tropical conditions. If it does, then the richness of the tropics could simply be due to the continual evolution of new forms there [40].

It is possible to approach this question historically by asking whether the tropics have acted as a centre of evolution for groups of organisms in the past. Plants and their fossils provide a useful group in which to investigate this possibility. If we examine the modern distribution patterns of the various families of flowering plants, we find that they have a tendency to centre on the tropics. Very roughly, about 30% of flowering plant families are widespread in distribution, about 20% mainly temperate and about 50% mainly tropical. These figures have led to the suggestion that the tropics have been a centre for the evolution of many of the flowering plant (angiosperm) groups. This proposal can be tested by looking at the fossil record to see whether the tropics have always been richer than the temperate latitudes as far as flowering plants are concerned. This approach has been attempted by Peter Crane and Scott Lidgard of the Field Museum of Natural History in Chicago [41], and some of their results are represented in Figure 4.13. This shows an analysis of fossil plant material covering the period from 145 to 65 million years ago, and depicts the relative abundance of the angiosperms in different latitudes during this period of time. From this, it is clear that the flowering plants first rose to some prominence in the tropics, and that their predominance in the tropics was maintained throughout this period of time as they gained importance in the plant kingdom. The latitudinal gradient in diversity for flowering plants, therefore, goes back a very long way – right

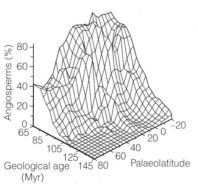

Figure 4.13 Estimated percentage representation of flowering plants (angiosperms) at different times in geological history and at different latitudes. Angiosperms have always been most abundant in the low-latitude (tropical) regions. From Crane and Lidgard [41].

back, in fact, to the evolutionary origins of the group. Tropical biodiversity is clearly an ancient phenomenon and may be related, in this case, to tropical origins and enhanced tropical rates of evolution and diversification.

This line of argument leads back to the metabolic theory. In an environment where energy is abundant and temperature is consistently high, such as the tropics, metabolic rates in organisms tend to be faster. Fecundity is greater, and generation times may be shorter. Consequently, genetic modification by mutation (see Chapter 6) is faster, so that species are constantly generating new variations. Ultraviolet radiation is generally greater in the tropics, and this can increase mutation rates. At the same time, there is a positive feedback in that the new varieties are in competition with one another, leading to an intense degree of selection for the most fit. Together, these two mechanisms would lead to a more rapid rate of evolution in the tropics.

Is it also possible that diversity itself generates further diversity? In other words, evolutionary diversification is itself promoted by high diversity. It is difficult to erect a theoretical model on which such a proposal can be based, but there is circumstantial evidence from a study of island endemism [42] that regions with high diversity exhibit a faster rate of diversification. It is possible that this argument is circular, however. If some external factor, such as conditions promoting productivity as suggested by the metabolic model, results in high diversity, it is also likely to create the required environment for enhanced diversification. Thus, there may be a link between diversity and diversification that is not directly causative. In biogeography, as in many areas of study, one should never assume that correlation implies causation.

Jonathan Davies and his colleagues from Imperial College London and from Kew Gardens have surveyed data relating to the rate of diversification of plant groups in the tropical regions [30]. Although they found that molecular evolution was indeed rapider in high-energy environments, they were not convinced that this is the driving force resulting in increased diversity. They found that the accumulation of species was indeed faster in regions of high biomass (and high energy), but this could be the result either of speciation being more rapid, or of extinction rates being lower. Either

way, it looks as though the energy–biomass model remains the most robust explanation currently available for latitudinal gradients in diversity.

There remain, however, additional factors that need to be considered.

The Legacy of Glaciation

Just as a fast rate of speciation in the tropics can be regarded as a factor leading to their increased biodiversity, a slower rate of extinction in the tropics could also be involved. One possible explanation for greater extinction rates in the higher latitudes is the instability of climatic conditions over the past 2 million years (see Chapter 12).

The Earth's climate has been constantly changing, and it has been considerably colder over the past 2 million years than was the case for the previous 300 million years. There has also been a high amplitude of variation between warm and cold conditions during that time. During cold episodes, the high latitudes have been disrupted by the development of glaciers over the land surface. The effects of these changes on the biogeographical patterns of plants and animals will be considered in Chapter 12, but it is evident that the most severe disruption, in the form of ice masses that have spread and destroyed all vegetation over major areas, has occurred largely in the high latitudes, and the tropics, apart from the high mountains, have been ice free and consequently subjected to less obvious climatic stress. This idea of a climatically stable tropical belt, if it is indeed true, could account for some of the diversity still found in the tropics as the plants and animals could be a relic accumulation of species from a former age. But has the tropical region actually been climatically more stable than the temperate region? The general conclusion that has emerged, particularly from the studies of Paul Colinvaux of the Smithsonian Institute, Panama, and his co-workers [43], is that the tropical lowlands of Amazonia have also been considerably colder (perhaps 5–6°C colder than at present) during recent times (the last million years or so).

One theory is that this climatic shift has meant that the tropical rainforest has periodically been restricted in its altitudinal range and has also been at least partially fragmented as a result of cold and drought during the glacial periods of higher latitudes.

It is then suggested that fragmentation and isolation of pockets of forest could have led to a degree of genetic isolation and hence to speciation and diversification. But Colinvaux comes to a very different conclusion based on his pollen studies in the Amazon [44]. He claims that the entire basin was clothed with evergreen lowland tropical forest throughout the 'glacial' episodes of the high latitudes. Forests that remained during the cooler episodes are likely to have changed in their specific composition, with a higher proportion of trees that are now associated with higher altitudes and hence cooler conditions. But the uninterrupted tract of forest remained. In such a case, the diversity of the tropics could be a consequence of their stability rather than their fragmentation. The equatorial forests have thus had to endure less disturbance than their temperate counterparts, and many areas have probably maintained themselves in a forested form throughout the period of stress, even though their species composition and architectural structure may well have changed.

An alternative approach to the question of maintained stability or fragmentation of tropical forests during the Pleistocene is available in the analysis of the genetic composition of forest organisms. It is possible to trace the history of the evolution of groups of species by checking their genetic diversification, and it is even possible to estimate the date at which different species separated from common ancestors (see Chapter 6). If the tropical forests have been stable and uninterrupted in their history in recent times (the last 2 million years or so), then speciation is likely to have taken place in the more distant past. If they have been fragmented in recent time, then the speciation should also be more recent. Evidence has been collected from molecular studies of some American songbirds by John Klicka and Robert M. Zink of the University of Minnesota [45]. They found that the time of evolutionary divergence could be estimated from the similarity or dissimilarity of their mitochondrial DNA, and most of the species examined seem to have diverged from common ancestors up to 5 million years ago, considerably further back in time than can be accounted for by the Ice Ages of the last 2 million years. Such genetic data support the uninterrupted forest hypothesis of Colinvaux.

It is reasonable to conclude, therefore, that many factors contribute to the species richness of the tropical regions and the lower richness of higher latitudes.

Latitude and Species Ranges

Apart from latitudinal gradients of diversity, it has also been observed that high-latitude organisms have broader geographical ranges than those from the low latitudes. This apparently general feature of biogeography was first pointed out by E.H. Rapoport in the 1970s, but rose to prominence as a result of the work of George C. Stevens [46], who coined the term **Rapoport's rule**. Much work has now been carried out to test the generalization, and the species of high latitudes do, on the whole, display wider geographical ranges, greater altitudinal ranges and broader ecological tolerances in comparison with tropical species. But there remains doubt as to whether this is a local effect that only makes itself felt in more northerly latitudes (above about 40–50°N), or whether it continues to be operative in the equatorial regions. Klaus Rohde [47], of Armidale, Australia, considers that Rapoport's rule is of local application only and cannot be applied in the tropical regions. The occurrence of broad-range species in the high latitudes could itself be a consequence of the impact of successive glaciations, leaving only the most adaptable species behind. It can also be argued that the greater seasonal fluctuations of the high latitudes will select for wide-tolerance organisms. Detailed testing of the 'rule' using a whole range of statistical techniques has failed to support a general global relationship between latitude or elevation and range among species [48].

An unusual extrapolation of Rapoport's rule has been illustrated by Katherine Smith and James Brown of the University of New Mexico, who have examined the diversity of fish as one proceeds deeper into the ocean [49]. Overall diversity peaks in the upper 200 m of water and then declines with depth, and the species of greater depth have wider tolerance, being able to tolerate shallow waters also. Narrow-range species were restricted to the upper layers of the ocean. Again, this suggests that species tolerating more extreme conditions tend to occupy wider ranges.

The use of Rapoport's rule (perhaps better termed 'the Rapoport effect' [50] since its general application is now questioned) as an explanation for the cause of latitudinal gradients of diversity is now virtually dismissed. Given lower species richness in the high latitudes, it is only to be expected that there will be less competition for resources and that ecological and geographical ranges of species will therefore be more extensive than in the species-dense tropics. The Rapoport effect is likely to be a consequence of latitudinal gradients in species richness rather than its cause.

Although the Rapoport effect has proved more restricted in its application than was originally anticipated, it provides an excellent example of the kind of question that biogeographers are now asking. Stepping beyond the framework of maps and patterns of distribution [51], many biogeographers are examining the mechanisms that underlie their observations. In many respects, they are asking ecological questions within a much larger scale framework of space and time. James Brown of the University of New Mexico has coined the term **macroecology** to cover this approach to biogeographical and ecological research [52,53].

Diversity and Altitude

As discussed in Chapter 2, patterns of vegetation and hence of biomes are generally related to latitude, and these are broadly repeated with respect to altitude, with higher altitudes often bearing biome types more typical of higher latitudes. One might expect, therefore, that the global patterns of biodiversity, declining from equator to poles, will be reflected with increasing altitude. Many studies have been conducted on different groups of organisms investigating the changes in species richness with altitude. Most of these have concentrated on tropical mountains, where the full range of climatic variation is found. But the outcome has not generally conformed with expectation. Many studies have demonstrated that the richness of species, particularly plants, increases with altitude, reaches a peak and then declines again at very high altitudes. Some of the results are summarized in Figure 4.14.

Vascular plants (those with a water-conducting system, including ferns, conifers and flowering plants) are among the least difficult to survey in tropical regions, so it is not surprising that more information is available for these than most other groups of organisms. Surveys from various parts of the tropical and subtropical world indicate an increase in overall richness with altitude to a peak

at about 1500–2000 m, depending on location, followed by a decline at higher altitudes. On Mount Kinabalu in Borneo (Figure 4.14a) there is a very clear and distinct peak [54], whereas in the Himalayan Mountains of Nepal (Figure 4.14b) the peak is rather more spread and diffuse [55]. There are some problems in interpreting these data, partly because the lower altitudes tend to be more extensive in area than higher ones (leading to inflated diversity), and the lower altitudes are also often subject to more intensive disturbance by human settlement and agriculture than higher ones, which could decrease or increase diversity, depending on the nature of the human impact. Data on the vegetation of the European Alps suggest that human management, including fire and pastoralism, has served to increase plant diversity in these montane habitats over the past 5500 years [56].

Tropical mountains create complex local climates, and these may in part explain the humped pattern of plant richness with altitude. Moderate altitudes often receive more rainfall than either low or very high altitudes in the tropics and subtropics. In the case of the Himalayas, for example, the lower elevations lie in the monsoon rain belt, so there is a distinct dry season alternating with a period of heavy rainfall. Cloud formation in the cooler air of the mid-altitudes leads to high humidity and more generally distributed rainfall, so that the forests are permanently moist. Under these conditions there is often an abundance of **epiphytes**, plants that use other vegetation for support, rooting upon trunks, branches and even the leaves of more robust species, but limiting their demands to support rather than any form of direct parasitism. Mosses and lichens often live as epiphytes in such forest, as do ferns and certain flowering plant families, such as orchids and bromeliads.

Studies concentrating on ferns and epiphytes in general, such as those shown in Figure 4.14c and 4.14d from Nepal [57] and Bolivia [58], respectively, demonstrate that these groups have very strong peaks in richness around the 2000 m altitude, where the mountains are clothed in a **cloud forest**, a highly humid forest type in which plants living entirely in the canopy with no roots reaching the ground are unlikely to experience desiccation. The mid-altitude peak in plant richness owes much to this abundance of epiphytes in the cloud forest zone [59]. At higher altitudes, conditions for growth become more difficult as average temperatures fall and frost becomes increasingly likely. The alpine zone of mountains, like the polar tundra regions, is a zone in which few plants can survive and the extinction rate of immigrant species is likely to be high.

The study of altitudinal gradients of diversity in animal groups has proved more challenging than similar studies of plants. This is because it is much easier to survey static plants rather than mobile, often inconspicuous animal species. Survey methods for animals are consequently more complex, often involving trapping. Birds and bats have proved the most amenable to survey, and some general outcomes are shown in Figure 4.14e and 4.14f. Studies of the birds of Taiwan, South-East Asia [60],

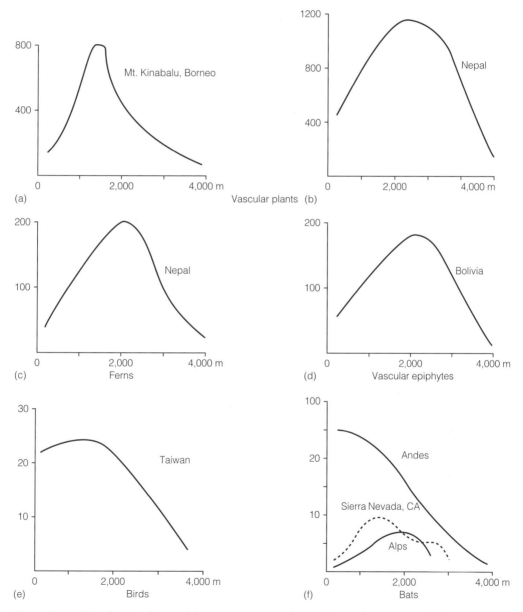

Figure 4.14 Altitudinal gradients of diversity in certain plant and animal groups. Simplified summaries from a range of sources.

showed that bird richness exhibited a slight tendency to a hump-shaped relationship with respect to altitude, and more detailed analysis of its causes demonstrated that there is a strong positive correlation between the productivity of vegetation and bird diversity. In the case of birds, therefore, it seems that the energy hypothesis, which has been widely employed as an explanation for latitudinal diversity, is also appropriate for the interpretation of variations in bird diversity with altitude. Any mid-altitudinal peak is a consequence of altitudinal variations in plant productivity as determined by local climate. Bird studies in the Andes of Columbia [61], however, show a more complex picture. The overall change in bird richness with altitude shows a steady decline from highest values at low altitude. Analysis of these data shows that the bulk of the birds at low altitude are widely distributed. Birds associated particularly with the Andes mountains peak in diversity at around 2000 m. In this region, the lowland occupies a large area and is continuous with the lowland forests of the interior, so the lower parts of the mountains share in the overall forest diversity. The mid-altitudes maintain relatively high diversity because isolation of mountains has resulted in a long-term process of speciation leading to a richness of endemics. The high altitudes, as usual, are associated with few bird species.

Analysis of the bird assemblages of Norwegian mountains has shown that there are two main types, specialists and generalists [62]. The specialists, like ptarmigan and snowy owl, are largely confined to alpine habitats all year round, whereas most of the generalists are migratory and breed in the tundra habitats but leave for lower areas in the winter. These include various wading birds, such as snipe and sandpipers that spend the winter in wetlands and on seashores, and various thrushes and warblers that migrate to milder regions in winter. But the overall bird richness is low.

Bats have been studied in a very wide range of mountain regions of the world, and the results have been surveyed in detail by Christy McCain of the University of California, Santa Barbara [63]. A few samples are shown simplified in Figure 4.14f. Some locations in the world show a general decline in species with altitude (such as Peru), whereas other locations (such as the European Alps and Yosemite, California) show a mid-altitude peak in richness. The general conclusion from these studies is that bat diversity, like plant diversity, is largely controlled by the availability of water and the associated temperature regime. Good conditions for primary production of vegetation are also good for bats, presumably because their insect food is more abundant under such circumstances, and the same is true for fruits.

The overall conclusion regarding patterns of diversity with altitude, therefore, is that the energy model used in latitudinal studies probably holds good for many types of organism in relation to altitude. The complications found in altitude patterns of diversity probably largely relate to the peculiarities of local climates in mountain regions (see Chapter 3). It is also undoubtedly true that many such patterns have been obscured by a long history of human modification of montane environments, especially in the temperate regions of the world [64].

Biodiversity Hotspots

Although species richness does generally follow a latitudinal gradient, its pattern often proves more complex when we examine the picture in detail. For example, the Amazon basin contains approximately 90 000 flowering plant species, whereas equivalent areas in Africa and in South-East Asia contain only about 40 000 each. There appear to be certain areas of the world that are exceptionally rich in species, termed **biodiversity hotspots** by the conservationist Norman Myers [65]. He originally proposed 10 hotspots, largely identified on the basis of plant diversity, for his argument was based on the idea that if vegetation is diverse all else will follow. But this is not always entirely true, as we have already seen in the case of aphids. Figure 4.15 shows the areas of high plant diversity in Africa compared with areas of high bird diversity, and it can be seen that there is relatively little overlap, so one cannot assume parallelism in trends of biodiversity between different groups of organisms.

The original work of Myers has been developed and expanded. Figure 4.16 shows the location of 25 of the hottest spots for biodiversity on Earth based on a consideration of many of the better known groups of organisms (plants, mammals, birds, reptiles, amphibians etc.). When we add all the land surface area of the hotspots together, they

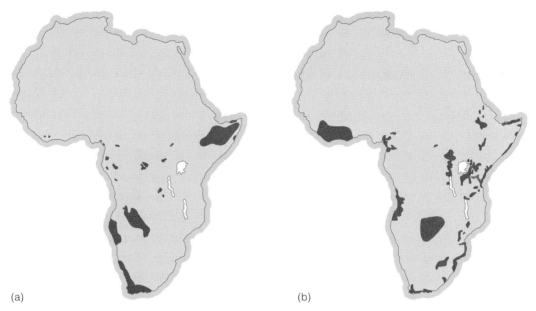

Figure 4.15 The areas of Africa that are particularly rich in (a) plant species, compared with those areas that are rich in (b) endemic birds. As can be seen, the two do not always correspond.

comprise only about 1.4% of the Earth's terrestrial total, yet they contain about 44% of the world's vascular plants and 35% of the vertebrates from the four main groups [66]. Stuart Pimm and Peter Raven [67], research conservationists from Columbia University and the Missouri Botanical Garden, respectively, have calculated that even if the 25 hotspots were given protection the likely extinction rate of species would be about 18%. If protection is delayed then the extinction rate could be as high as 40%. Obviously these areas must form a focus for global conservation activity, but it would be a mistake to confine attention to them because more than half of the world's species are located elsewhere. There are also regions of the world, such as the polar regions, where diversity is not high but the organisms present are very distinctive and often restricted in distribution. This topic will be covered in greater detail in Chapter 14.

The map displayed in Figure 4.16 also shows the pattern of human settlement density around the world, and it can be seen that many of the biodiversity hotspots lie in high human population density areas [68]. Demographic analyses have shown that about 20% of the world's population lives within the biodiversity hotspots. This must represent a serious

threat to habitat survival, especially because some of these areas also have rapid rates of human population expansion and the pressures on space and natural resources are intense. But biodiversity hotspots are not necessarily those areas that have suffered least from human impact. In the case of the Mediterranean region of Europe, for example, there is high biodiversity accompanied by a very long history of human activity that has often proved very destructive. The island of Crete is only 245 km long by 50 km wide and has been isolated as an island for about 5.5 million years. It has supported human populations since at least the arrival of Neolithic peoples about 8000 years ago. Since then, climatic changes have resulted in the development of very dry conditions in summer, and additional disturbance by earthquake and volcano activity has been experienced. The increase in human populations, their need for agricultural land and their intensive pastoralism have resulted in the stripping of much of the original vegetation [69]. Yet, despite all this, Crete has 1650 species of plant, 10% of which are endemic to the island. The fossils of Crete tell us that many species have become extinct during its recent history, yet it still remains a remarkably species-rich island. Indeed, the Mediterranean climate areas generally are very rich in plant

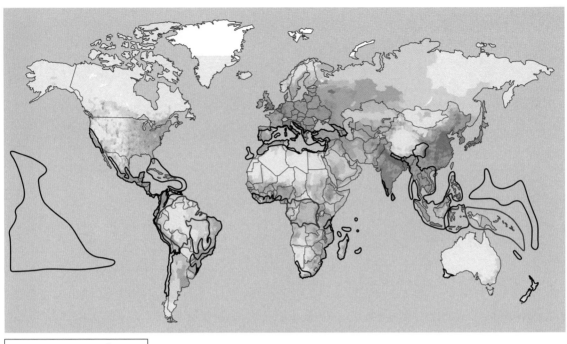

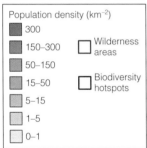

Population density (km^{-2})

- 300
- 150–300
- 50–150
- 15–50
- 5–15
- 1–5
- 0–1

Wilderness areas

Biodiversity hotspots

Figure 4.16 The location of areas of exceptionally high biological richness (biological hotspots) in relation to human population density. From Cincotta *et al.* [68].

species, perhaps as a result of the high intensity of habitat patterns and the severity of the local impacts of drought and fire [70]. It is possible that moderate human pressures can increase biodiversity as a result of diversifying habitats; the studies on Alpine vegetation referred to here illustrate the same point. There is certainly some evidence that biological diversity and human population density are positively related, at least over the less extreme human density range [71]. One must be careful before assuming that such correlation implies causation, however.

Explaining why hotspots are so rich is even more difficult than solving the latitudinal gradient question. There is, indeed, a general low-latitude concentration in the hotspots, so the problem

of latitudinal gradients is clearly confused with that of hotspots. But additional mechanisms are also at work. Perhaps these hotspots are centres of evolution, or perhaps they are relict fragments of former diverse communities. If these explanations are correct, then one might expect hotspots to be rich in endemic organisms, those confined to the region because of either their recent evolution or their extinction in surrounding areas. Several attempts have been made to test the correlation of hotspots with endemism, with some conflicting results. Analysis of the global richness pattern of bird species alone showed no significant correlation between overall richness and the richness of endemic species [72]. But a wider analysis using data from amphibians, reptiles, birds and mammals [73] showed some degree of correlation between diversity and endemicity. Endemics have long been the subject of special attention for conservation simply because of their rarity or their narrow geographical range. It is reassuring, therefore, to know that the establishment of protection for those areas where endemic vertebrates are abundant also provides a safeguard for a wide range of other species and is good overall policy for biodiversity conservation. Work on marine fish diversity in the South Pacific [74] also suggests that fish biodiversity hotspots coincide with centres of evolution and that diversity in surrounding areas is related to distance from such hotspots. This supplies hotspots with another good reason for conservation as they may prove to be the focal points for future species generation (see Chapter 14).

Diversity in Space and Time

No ecosystem is entirely static. The processes involved in the colonization of bare areas of coastal sand, volcanic debris or glacial detritus have been studied extensively by ecologists, and the growth and development of an ecosystem in such situations are called **succession**. Often the changes that take place over time are relatively predictable, especially in general terms [75].

A simple example of succession is the invasion of vegetation following the retreat of a glacier, as has been illustrated by studies in Glacier Bay, Alaska [76], shown in Figure 4.17. Warmer conditions have caused the melting of ice, and as the ice front gradually receded it left bare rock surfaces and crushed rock fragments in sheltered pockets and crevices. Such primitive soils may be rich in some of the elements needed for plant growth, such as potassium and calcium, but are poor in organic matter. These soils usually have a very limited capacity for water retention, poor microbial populations, little structure and low levels of nitrogen. A plant that can grow even under these stressed conditions is the Sitka alder (*Alnus sinuata*). This is a low-growing bushy tree that owes its success in part to its association with a bacterium that grows in association with its roots. This microbe forms colonies in swollen nodules on the alder's roots and is able to take nitrogen from the atmosphere and convert it

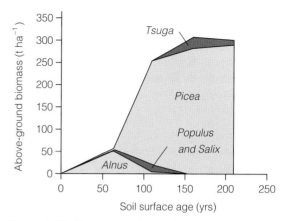

Figure 4.17 Change in major species composition (expressed in terms of biomass, i.e. above-ground dry weight) during the development of forest following ice retreat in Alaska. The dominance of one species, spruce (*Picea*), is established as the biomass increases with successional development of the plant community. From Bormann and Sidle [76].

to ammonium compounds that can subsequently be used (together with materials derived from alder photosynthesis) to build up proteins. Therefore, the alder manages perfectly well despite the low levels of nitrate in the soil. In fact, because of the gradual death of roots and the return of litter to the soil from the alder, the growth of the tree increases the amount of nitrate in the soil and thus fertilizes it. But this very process of modifying the soil environment eventually proves the downfall of the alder, because it permits the invasion of other, less highly adapted plants, among them the Sitka spruce (*Picea sitchensis*). After about 80 years, the Sitka spruce trees, which are more robust and faster growing than the alder shrubs, assume dominance in the vegetation and begin to shade out the pioneer alders. Thus, by their very existence at the site, the alders have effectively sealed their own fate and made the next step in the succession inevitable. This driving mechanism that underlies the successional process is termed **facilitation**, and it ensures a progressive and reasonably predictable development within the vegetation.

The course of succession also leads to an accumulation of biomass during the course of time. In Figure 4.17, the biomass of the major tree species can be seen increasing over the course of 200 years of the succession. Alder, poplar (*Populus*) and willow (*Salix*) are replaced by Sitka spruce and hemlock (*Tsuga*), and, whereas alder only achieved a maximum biomass of about 50 t/ha, the spruce–hemlock forest grows to a biomass of over 300 t/ha. Such an increase in biomass naturally involves the development of a more complex canopy structure, and, as we have seen, the diversity of animal species often follows the increase in structural complexity in vegetation.

Successions affect species diversity because the maturation of an ecosystem is usually accompanied by increasing numbers of species. We have already seen how higher biomass in the tropics is a possible cause of the high species diversity of low latitudes because it offers greater spatial heterogeneity and more opportunities for small organisms to find a home. Similarly in succession, increasing biomass and architectural complexity in the vegetation lead to higher species diversity. Climate and soils usually limit how much biomass can accumulate at any given location, so successions must come to an end

and form a stable, self-perpetuating system, called the **climax state**. But this final condition is far from static. In a forest ecosystem, for example, old trees die and open up clearings (Figure 4.18), and light penetration into such gaps allows the growth of plants, including some tree species, that are normally associated with an earlier stage in succession.

In the northeastern United States, a typical sequence of events in the hardwood forest, following the fall of a mature beech (*Fagus grandifolia*) tree, is a yellow birch (*Betula alleghaniensis*) invasion, followed by sugar maple (*Acer saccharum*) and eventually the regrowth of beech [77]. But since this cyclic process of forest healing is taking place wherever a gap has resulted, the climax forest actually consists of a mosaic of patches all in different stages of recovery, together with some patches of mature beech. Therefore, the climax vegetation is actually a collection of different-aged patches. This, in fact, adds to the diversity of the whole system, for the vegetation is not uniform but extremely heterogeneous, and many of the species that would

(a)

(b)

(c)

(d)

Figure 4.18 Gaps in forests, created by the death of old trees or minor catastrophes such as wind-blow or fire, become filled by the regrowth of young trees, often passing through a succession of different species. Habitat heterogeneity and canopy complexity are an outcome of this process.

be lost from an area, if successional development had effectively ceased, are still present in some of the forest openings.

Patchiness in a habitat may thus be caused by the constant turnover of vegetation within an ecosystem. It can also result from small variations in topography and hence microclimate. In the boreal forests of Canada, for example, the open black spruce and lichen forest of the more southerly regions contains a scatter of frost hollows dominated by cushion lichens and small shrubs of dwarf birch and ericaceous evergreens. These hollows act as accumulators of cold air, which sinks and fills them. The white surface of lichens reflects sunlight and contributes to the overall coldness of these so-called 'frost hollows' [78]. Thus, the natural topographical variability of a habitat can create new opportunities for the enhancement of biodiversity.

Returning to the question of why the tropics, and tropical forests in particular, are so rich in species, we have now established a further means by which high diversity can be maintained. The forest is constantly undergoing disturbance from storms, local fire and the meandering and flooding of rivers. All of these leave the forest in a state of turbulence and active regeneration that contributes to the diversity of the whole. Thus, time-related, cyclic processes in vegetation development allow even more species to be packed into a given area. Different species have what one may term a **temporal niche**, a preferred stage in a successional development when their attributes are most effective in competition and ensure the establishment of a sustainable population [79]. In the complexity of a mature ecosystem, such as a forest, spatial heterogeneity leads to a range of species with different temporal niches existing in close proximity. Spatial heterogeneity thus leads to higher biodiversity. Ecologists sometimes refer to this spatial element of biodiversity as **beta diversity**, which differentiates it from the more commonly used **alpha diversity**, referring to the species of a uniform area. Conservationists and habitat managers often attempt to encourage such heterogeneity in order to enhance diversity, as in the case of grassland butterflies where experimental management has demonstrated that the maintenance of patches of diverse management (mowing, grazing, ploughing, burning etc.) can lead to increased diversity

[80]. This theme will be covered more extensively in Chapter 14.

As discussed in Chapter 2, *diversity* is often used simply to describe species richness, or the number of species in a given area, whereas the degree of evenness of those species can also be regarded as an indication of true 'diversity' [81]. In the case of successions, both richness and true diversity tend to increase with time, especially for the animal component because the increasing abundance of small-bodied animals is closely related to architectural complexity. But this may not always be the case with plants. Later stages in succession, as in the case of the alder–spruce sequence, may become dominated by a few large-bodied species, which effectively reduces plant diversity. In the early stages of succession, however, increasing diversity seems to hold generally true for both plants and animals.

Intermediate Disturbance Hypothesis

It appears, therefore, that some degree of disturbance in an ecosystem often leads to an increase in the richness of species. A site that is totally static, if such a situation ever existed, would theoretically develop a stable and uniform biota. In practice, there is always some degree of change, at least on a local scale, including the death and collapse of individual trees, as we have discussed. Even such local changes create niches for species that require more open conditions and might be considered to belong to earlier stages in succession, and their arrival (in the case of plants, often from dormant seeds in the soil) increases diversity. Thus, even in a 'natural' habitat (i.e. one in which human activity is absent), there is a regular cycle of disturbance leading to the development of a patchwork of habitats.

Suburban human settlements, parks and gardens can provide a rich variety of habitats that promote diversity, which is why human density and biodiversity can be positively related (see Figure 4.16). Habitat management in the form of moderate disturbance, preferably in a mosaic pattern (e.g. felling clearings in forest and mowing sections of reedbeds), can lead to increased diversity, and this has become a widespread approach of conservationists, who are applying biogeographical principles.

A note of caution is necessary, however, before such management should be universally applied. Although so-called 'intermediate disturbance' may result in more species overall, it may also result in the elimination of sensitive species, those that require a complete lack of disturbance for their survival. Species that need extensive ranges of forest, reedbed or grassland for the maintenance of populations will not be favoured by disturbance and habitat fragmentation.

Dynamic Biodiversity and Neutral Theory

It is easy to assume that patterns of biodiversity are fixed and static, but this is not so. Although the tropics have always been richer in species than the temperate and polar regions, there is a constant flux of species over geological time. Species are always coming and going (evolving and becoming extinct) and are always on the move (changing their global distribution patterns), so the idea that the natural world has attained some kind of equilibrium state is never valid. Some biogeographers maintain that compatible species, each with its distinctive and individual niche, become sorted over the course of time to assemble themselves into stable, equilibrium communities. But this is imposing more order onto nature than is in fact present. Stephen Hubbell of the University of Georgia has adopted the other extreme in his proposal that species vary in their abundance in a random manner. He has set out a **neutral theory of biodiversity** that considers assemblages of species as a collection of randomly selected individuals [82]. The proposal can be used to form the basis for computer simulations and often provides a good match with what is actually found in nature, especially for plants, which are all doing very much the same job in the ecosystem, namely, fixing energy from the sun and carbon dioxide from the atmosphere. The fact that they all have similar functions suggests that it should not matter which species are present in the system. The neutral theory seems to work best when considering species at a selected trophic level. But at such a level,

chance, according to Hubbell, plays a great part in determining what species are present in an ecosystem, and perhaps such randomness plays a far greater part than has been appreciated in the past. According to this approach, communities assemble themselves over time as a result of chance arrivals and establishments rather than the shuffling of species with their different niche requirements, ultimately becoming sorted into a collection of co-existing components, each with its distinctive role.

On the basis of neutral theory, therefore, the species present in a community depend more on powers of dispersal and availability for immigration than on the compatibility of niches [83]. Despite the robust nature of the predictions resulting from neutral models of community assemblage, there have been questions and criticisms [84]. There is evidence for stabilizing forces at work in nature that may lead to particular end points (in terms of community composition) being achieved. An analysis of long-term vegetation change using stratified fossil pollen grains preserved in the lakes of North America, for example, shows a great deal of variation in assemblages over time, but there is a tendency for variability to decline with time as a general stability is attained. Such results, based on relatively long-term data (over a period of 10 000 years), indicate that random forces and neutral drift alone cannot account for the composition of communities and their biodiversity [85].

Chance can operate, however, in a number of different ways apart from the stochastic and relatively unpredictable nature of immigration. One of these is through the initial conditions obtaining when a community begins its development. Slight differences in the original state can strongly affect the outcome of community establishment and therefore its final biodiversity. This idea of minor differences in initial conditions being highly influential in the ultimate outcome is known as **chaos theory**. It can be illustrated by considering a pencil balanced on the tip of its sharpened point. Any minute imperfection in the point, or in the hand of the operator, will determine the direction in which the pencil will eventually fall when it is released. The same is true of ecosystems; very small chance events, such as the occurrence of a certain set of weather conditions, wind direction or the time of year, can have a great influence on the ultimate outcome of a succession. The types and abundance of seeds already present in soils following a catastrophe can profoundly affect the development of the subsequent succession.

The fact that species continue to arrive and then vanish even within apparently stable ecosystems can be observed by monitoring this process of change. The tropical rainforests are generally regarded as both biodiverse and stable in their species content, but observations on the forests of Central America, northern South America and the Amazon Basin have all demonstrated that continual changes are proceeding within the composition of these ecosystems. In particular, the woody climbing plants (lianas) are generally becoming more abundant in all of these regions [86]. Neither the causes nor the consequences of these changes are yet clear. It may be a response to global atmospheric or climatic changes, or it may be an example of cyclic or random variation in species composition. The growth of lianas often impairs the performance of some tree species, so the outcome of this widespread vegetation change may prove considerable from the point of view of future biodiversity of these regions.

The entire question of why some parts of the world are richer in species than others, and what factors influence the assembling of communities, thus remains one of active debate and research. It is one of the most profound and important questions that is asked of biogeography, for the answer will be extremely valuable in predicting future changes and managing the threatened biodiversity of the planet. Explaining the global patterns of biodiversity is a process that requires a consideration of many factors. Some of these have been discussed here, but further understanding of the subject demands a knowledge of many other aspects of biogeography. How is it that many species manage to occupy the same habitat? What factors limit the geographical range of individual species? How have such ranges changed during the course of the Earth's history? How do new species evolve, and why do they evolve in particular ways? These are some of the questions that must be faced if the complex issue of biodiversity is to be further understood.

Summary

1 Biodiversity means the full range of life on Earth, including all the different species found, together with the genetic variation between populations and individuals, and the variety of ecosystems, communities and habitats present on our planet.

2 We are losing species at an unknown but undoubtedly accelerating rate. We need to know more about the variety of life on Earth before we can even appreciate how fast we are losing it.

3 Only about 1.8 million of the species that live on Earth have so far been described. This is a very small percentage, perhaps less than 5 per cent, of the likely total of species on the planet.

4 The tropics are generally richer in species than the high latitudes, possibly as a result of high productivity and food availability, high biomass and hence complex structure, past patterns of evolution, survival of fragments of habitats through the cold episodes of the last 2 million years, and also the degree of small-scale disturbance resulting in a mosaic of successional processes.

5 The term 'diversity' involves both species number (richness) and the pattern of allocation of numbers or biomass between the different species (evenness). It generally increases during the course of succession.

6 The species composition (and therefore the biodiversity) of any community is constantly changing. Chance is likely to play a part in the assemblage of communities and in subsequent changes, but there are also stabilizing forces at work.

7 There is a tendency for human populations to be dense in biodiversity hotspots. It is possible that people enhance biodiversity by diversifying habitats, but the destructive tendency of high population densities is a matter for conservation concern.

Further Reading

Boenigk J, Wodniok S, Glücksman E. *Biodiversity and Earth History*. Heidelberg: Springer, 2015.

Gaston KJ, Spicer JI. *Biodiversity: An Introduction*. 2nd ed. Oxford: Blackwell, 2004.

Groombridge B, Jenkins MD. *World Atlas of Biodiversity: Earth's Living Resources in the 21st Century*. Berkeley: University of California Press, 2002.

Hails RS, Beringer JE, Godfray HCJ. *Genes in the Environment*. Oxford: Blackwell, 2003.

Hubbell SP. *The Unified Neutral Theory of Biodiversity and Biogeography*. Princeton: Princeton University Press, 2001.

Lomolino MV, Riddle BR, Brown JH. *Biogeography*. 3rd ed. Sunderland, MA: Sinauer, 2006.

Smith FA, Gittleman JL, Brown JH. *Foundations of Macroecology*. Chicago: University of Chicago Press, 2014.

References

1. Taylor B, van Perlo B. *Rails: A Guide to the Rails, Crakes, Gallinules and Coots of the World*. Robertsbridge: Pica Press, 1998.

2. Sibley DA. *The Sibley Guide to Birds*. 2nd ed. New York: Alfred E. Knopf, 2014.

3. Garcia-Ramirez JC, Trewick SA. Dispersal and speciation in purple swamphens (Rallidae: *Porphyrio*). *The Auk* 2015; 132: 140–155.

4. May RM, Nee S. The species alias problem. *Nature* 1995; 378: 447–448.

5. Wilson EO. *Biodiversity*. New York: National Academic Press, 1988.

6. Erwin TL. Beetles and other insects of tropical forest canopies at Manaus, Brazil, sampled by insecticidal fogging. In: Sutton SL, Whitmore TC, Chadwick AC (eds.), *Tropical Rain Forest: Ecology and Management*. Oxford: Blackwell Scientific Publications, 1983: 59–75.

7. Stork N, Gaston K. Counting species one by one. *New Scientist* 1990; 127: 43–47.

8. Thomas CD. Fewer species. *Nature* 1990; 347: 237.

9. Pace NR. A molecular view of microbial diversity and the biosphere. *Science* 1997; 276: 734–740.

10. Holms B. Life unlimited. *New Scientist* 1996; 148: 26–29.

11. Fyfe WS. The biosphere is going deep. *Science* 1996; 273: 448.

12. O'Donnell AG, Goodfellow M, Hawksworth DL. Theoretical and practical aspects of the quantification of biodiversity among microorganisms. In: Hawksworth DL (ed.), *Biodiversity: Measurement and Estimation*. London: Chapman & Hall, 1995: 65–73.

13. Zettler LAA, Gomez F, Zettler E, Keenan BG, Amils R, Sogin ML. Eukaryotic diversity in Spain's River of Fire. *Nature* 2002; 417: 137.

14. May RM. A fondness for fungi. *Nature* 1991; 352: 475–476.

15. Groombridge B (ed.). *Global Biodiversity: Status of the Earth's Living Resources*. London: Chapman & Hall, 1992.

16. Morell V. New mammals discovered by biology's new explorers. *Science* 1996; 273: 1491.

17. Smith FA, Gittleman JL, Brown JH (eds.). *Foundations of Macroecology*. Chicago: University of Chicago Press, 2014.

18. Wright MG, Samways MJ. Gall-insect species richness in African fynbos and karoo vegetation: the importance of plant species richness. *Biodiversity Letters* 1996; 3: 151–155.

19. Lawton JH, Bignell DE, Bolton B, *et al.* Biodiversity inventories, indicator taxa and effects of habitat modification in tropical forest. *Nature* 1998; 391: 72–76.

20. Vellend M, Lilley PL, Starzomski BM. Using subsets of species in biodiversity surveys. *Journal of Applied Ecology* 2008; 45: 161–169.

21. Siemann E, Tilman D, Haarstad J. Insect species diversity, abundance and body size relationships. *Nature* 1996; 380: 704–706.

22. Aarssen LW, Schamp BS, Pither J. Why are there so many small plants? Implications for species coexistence. *Journal of Ecology* 2006; 94: 569–580.

23. Jackson JA. *In Search of the Ivory-Billed Woodpecker*. New York: HarperCollins, 2006.

24. da Silva W. On the trail of the lonesome pine. *New Scientist* 1997; 155: 36–39.

25. Chapin FS, Zavaleta ES, Eviner VT, *et al.* Consequences of changing biodiversity. *Nature* 2000; 405: 234–242.

26. Gaston KJ. Global patterns in biodiversity. *Nature* 2000; 405: 220–227.

27. Duellman WE. Patterns of species diversity in anuran amphibians in the American tropics. *Annals of the Missouri Botanical Garden* 1988; 75: 70–104.

28. Collins NM, Morris MG. *Threatened Swallowtail Butterflies of the World. IUCN Red Data Book*. Cambridge: IUCN, 1985.

29. Currie DJ, Paquin V. Large-scale biogeographical patterns of species richness of trees. *Nature* 1987; 329: 326–327.

30. Davies JT, Barraclough TG, Savolainen V, Chase MW. Environmental causes for plant biodiversity gradients.

Philosophical Transactions of the Royal Society of London B 2004; 359: 1645–1656.

31. Brown JH, Gillooly JF, Allen AP, Savage VM, West GB. Toward a metabolic theory of ecology. *Ecology* 2004; 85: 1771–1789.

32. Algar AC, Kerr JT, Currie DJ. A test of metabolic theory as the mechanism underlying broad-scale species-richness gradients. *Global Ecology and Biogeography* 2007; 16: 170–178.

33. Whittaker RJ, Nogues-Bravo D, Araujo MB. Geographical gradients of species richness: a test of the water-energy conjecture of Hawkins *et al.* (2003) using European data for five taxa. *Global Ecology and Biogeography* 2007; 16: 76–89.

34. Terborgh J, Petren K. Development of habitat structure through succession in an Amazonian floodplain forest. In: Bell SS, McCoy ED, Mushinsky HR (eds.), *Habitat Structure: The Physical Arrangement of Objects in Space.* London: Chapman & Hall, 1991: 28–46.

35. Blondell J. *Biogeographie et Ecologie.* Paris: Masson, 1979.

36. Dixon AFG. *Aphid Ecology: An Optimization Approach.* 2nd ed. London: Chapman & Hall, 1998.

37. Blackburn TM, Gaston KJ. A sideways look at patterns in species richness, or why there are so few species outside the tropics. *Biodiversity Letters* 1996; 3: 44–53.

38. Rosenzweig ML. *Species Diversity in Space and Time.* Cambridge: Cambridge University Press, 1995.

39. Rohde K. The larger area of the tropics does not explain latitudinal gradients in species diversity. *Oikos* 1997; 79: 169–172.

40. Rohde K. Latitudinal gradients in species-diversity – the search for the primary cause. *Oikos* 1992; 65: 514–527.

41. Crane PR, Lidgard S. Angiosperm diversification and paleolatitudinal gradients in Cretaceous floristic diversity. *Science* 1989; 246: 675–678.

42. Emerson BC, Kolm N. Species diversity can drive speciation. *Nature* 2005; 434: 1015–1017.

43. Colinvaux PA, De Oliveira PE, Moreno JE, Miller MC, Bush MB. A long pollen record from lowland Amazonia: forest and cooling in glacial times. *Science* 1996; 274: 85–88.

44. Colinvaux PA. *Amazon Expeditions: My Quest for the Ice-Age Equator.* New Haven: Yale University Press, 2007.

45. Klicka J, Zink RM. The importance of recent ice ages in speciation: a failed paradigm. *Science* 1997; 277: 1666–1669.

46. Stevens GC. The latitudinal gradient in geographical range: how so many species coexist in the tropics. *American Naturalist* 1989; 133: 240–256.

47. Rhode K. Rapoport's rule is a local phenomenon and cannot explain latitudinal gradients in species diversity. *Biodiversity Letters* 1996; 3: 10–13.

48. Ribas CR, Schoereder JH. Is the Rapoport effect widespread? Null models revisited. *Global Ecology and Biogeography* 2006; 15: 614–624.

49. Smith KF, Brown JH. Patterns of diversity, depth range and body size among pelagic fishes along a gradient of depth. *Global Ecology and Biogeography* 2002; 11: 313–322.

50. Gaston KJ, Blackburn TM, Spicer JI. Rapoport's rule: time for an epitaph? *Trends in Ecology and Evolution* 1998; 13: 70–74.

51. Blackburn TM, Gaston KJ. There's more to macroecology than meets the eye. *Global Ecology and Biogeography* 2006; 15: 537–540.

52. Brown JH. *Macroecology.* Chicago: University of Chicago Press, 1995.

53. Smith FA, Gittleman JL, Brown JH. Introduction: the macro of macroecology. In Smith FA, Gittleman JL, BrownJH (eds.), *Foundations of Macroecology.* Chicago: University of Chicago Press, 2014: 1–4.

54. Grytnes JA, Beaman JH. Elevational species richness patterns for vascular plants on Mount Kinabalu, Borneo. *Journal of Biogeography* 2006; 33: 1838–1849.

55. Vetaas OR, Grytnes JA. Distribution of vascular plant species richness and endemic richness along the Himalayan elevation gradient in Nepal. *Global Ecology and Biogeography* 2002; 11: 291–301.

56. Schworer C, Colombaroli D, Kaltenrieder P, Rey F, Tinner W. Early human impact (5000–3000 BC) affects mountain forest dynamics in the Alps. *Journal of Ecology* 2015; 103: 281–295.

57. Bhattarai KR, Vetaas OR, Grytnes JA. Fern species richness along a central Himalayan elevational gradient, Nepal. *Journal of Biogeography* 2004; 31: 389–400.

58. Kroemer T, Kessler M, Gradstein SR, Acebey A. Diversity patterns of vascular epiphytes along an elevational gradient in the Andes. *Journal of Biogeography* 2005; 32: 1799–1809.

59. Kueper W, Kreft H, Nieder J, Koester N, Barthlott W. Large-scale diversity patterns of vascular epiphytes in Neotropical montane rain forests. *Journal of Biogeography* 2004; 31: 1477–1487.

60. Ding T-S, Yuan H-W, Geng S, Lin Y-S, Lee P-F. Energy flux, body size and density in relation to bird species richness along an elevational gradient in Taiwan. *Global Ecology and Biogeography* 2005; 14: 299–306.

61. Kattan G, Franco P. Bird diversity along elevational gradients in the Andes of Colombia: area and mass effects. *Global Ecology and Biogeography* 2004; 13: 451–458.

62. Thompson BA, Kalas JA, Byrkjedal I. Arctic-alpine mountain birds in northern Europe: contrasts between specialists and generalists. In Fuller RJ (ed.), *Birds and Habitat: Relationships in Changing Landscapes.* Cambridge: Cambridge University Press, 2012: 237–252.

63. McCain CM. Could temperature and water availability drive elevational species richness patterns? A global case study for bats. *Global Ecology and Biogeography* 2007; 16: 1–13.

64. Nogues-Bravo D, Araujo MB, Romdal T, Rahbek C. Scale effects and human impact on the elevational species richness gradients. *Nature* 2008; 453: 216–219.

65. Myers N. The biodiversity challenge: expanded hotspots analysis. *The Environment* 1990; 10: 243–256.

66. Myers N, Mittermeier RA, Mittermeier CG, da Fonseca GAB, Kent J. Biodiversity hotspots for conservation priorities. *Nature* 2000; 403: 853–858.

67. Pimm SL, Raven P. Extinction by numbers. *Nature* 2000; 403: 843–845.

68. Cincotta RP, Wisnewski J, Engelman R. Human population in the biodiversity hotspots. *Nature* 2000: 404: 990–992.

69. Rackham O, Moody J. *The Making of the Cretan Landscape*. Manchester: Manchester University Press, 1996.

70. Cowling RM, Rundel PW, Lamont BB, Arroyo MK, Arlanoutsou M. Plant diversity in Mediterranean-climate regions. *Trends in Ecology and Evolution* 1996; 11: 362–366.

71. Araujo MB. The coincidence of people and biodiversity in Europe. *Global Ecology and Biogeography* 2003; 12: 5–12.

72. Orme CDL, Davies RG, Burgess M, *et al*. Global hotspots of species richness are not congruent with endemism or threat. *Nature* 2005; 436: 1016–1019.

73. Lamoreux JF, Morrison JC, Ricketts TH, Olson DM, Dinerstein E, McKnight MW, Shugart HH. Global tests of biodiversity concordance and the importance of endemism. *Nature* 2006; 440: 212–214.

74. Mora C, Chittaro PM, Sale PF, Kritzer JP, Ludsin SA. Patterns and processes in reef fish diversity. *Nature* 2003; 421: 933–936.

75. Pickett STA, Cadenasso ML, Meiners SJ. Vegetation dynamics. In: van der Maarel E, Franklin J (eds.), *Vegetation Ecology*. Chichester: Wiley-Blackwell, 2013: 107–140.

76. Bormann BT, Sidle RC. Changes in productivity and distribution of nutrients in a chronosequence at Glacier Bay National Park. *Alaska. Journal of Ecology* 1990; 78: 561–578.

77. Forcier LK. Reproductive strategies in the co-occurrence of climax tree species. *Science* 1975; 189: 808–810.

78. Plasse C, Payette S. Frost hollows of the boreal forest: a spatiotemporal perspective. *Journal of Ecology* 2015; 103: 669–678.

79. Kelly CK, Bowler MG, Fox GA. *Temporal Dynamics and Ecological Processes*. Cambridge: Cambridge University Press, 2013.

80. Perovic D, Gamez-Virues S, Borschig C, *et al*. Configuational landscape heterogeneity shapes functional community composition of grassland butterflies. *Journal of Applied Ecology* 2015; 52: 505–513.

81. Spellerberg IF, Fedor PJ. A tribute to Claude Shannon (1916–2001) and a plea for more rigorous use of species richness, species diversity and the 'Shannon-Wiener' Index. *Global Ecology and Biogeography* 2003; 12: 177–179.

82. Hubbell SP. *The Unified Neutral Theory of Biodiversity and Biogeography*. Princeton: Princeton University Press, 2001.

83. Bullock JM, Kenward RE, Hails RS (eds.). *Dispersal Ecology*. Oxford: Blackwell, 2002.

84. Ostling A. Neutral theory tested by birds. *Nature* 2005; 436: 635–636.

85. Clark JS, McLachlan JS. Stability of forest biodiversity. *Nature* 2003; 423: 635–638.

86. Phillips OL, Vásquez Martínez R, Arroyo L, *et al*. Increasing dominance of large lianas in Amazonian forests. *Nature* 2002; 418: 770–774.

(a)

(b)

Plate 1 (a) Desert scrub vegetation in eastern Iran. The most conspicuous feature is the deciduous shrub, *Zygophyllum eurypterum*, together with smaller plants of *Artemisia herba-alba*. This vegetation is typical of the drier regions of alluvial flats. (b) Desert scrub vegetation in eastern Iran in which the main scrub is *Haloxylon persicum*. This vegetation occupies the moister and more saline regions of the alluvial flats.

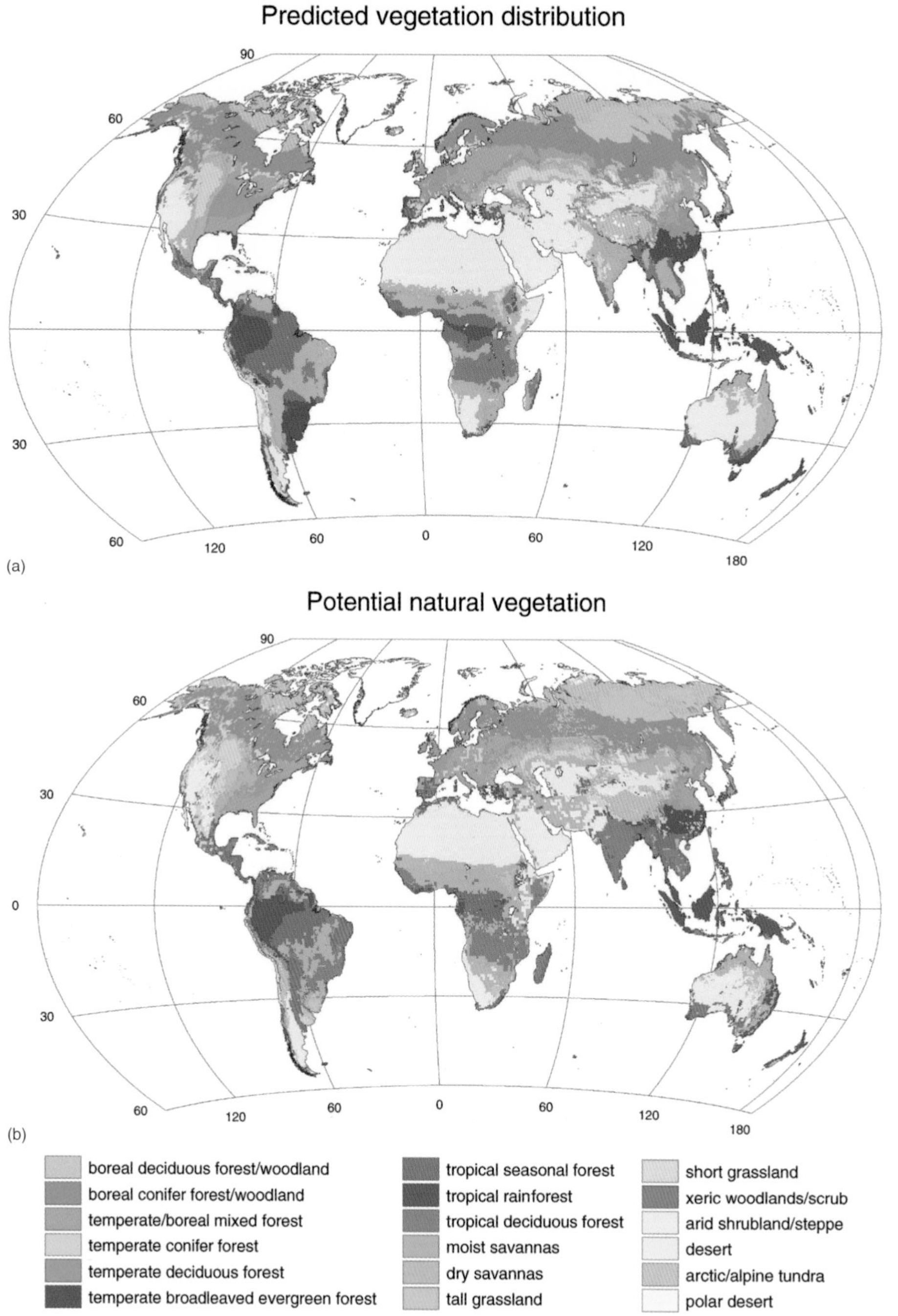
Predicted vegetation distribution
90
60
30
30
60
120
60
0
60
120
180
(a)
Potential natural vegetation
90
60
30
0
30
60
120
60
0
60
120
180
(b)
boreal deciduous forest/woodland
boreal conifer forest/woodland
temperate/boreal mixed forest
temperate conifer forest
temperate deciduous forest
temperate broadleaved evergreen forest
tropical seasonal forest
tropical rainforest
tropical deciduous forest
moist savannas
dry savannas
tall grassland
short grassland
xeric woodlands/scrub
arid shrubland/steppe
desert
arctic/alpine tundra
polar desert

Plate 2 (a) Global map of the present distribution of vegetation on Earth. This is somewhat idealized since it ignores the impact of human beings. It can be regarded as the potential vegetation if climate alone were the operative determinant. (b) Global map of vegetation derived from the predictions of a computer model, BIOME 3. As in the case of (a), this is a potential vegetation map that is based solely on climatic conditions. It can be seen that the two maps are very similar, which means that this model is a reliable predictor of vegetation under given climatic conditions. This close correspondence suggests that the model could be used to predict reliably the consequences of future climatic change for vegetation distribution patterns. From Haxeltine A, Prentice IC. BIOME 3: an equilibrium terrestrial biosphere model based on ecophysiological constraints, resource availability, and competition among plant functional types. *Global Biogeochemical Cycles* 1996; 10: 693–709.

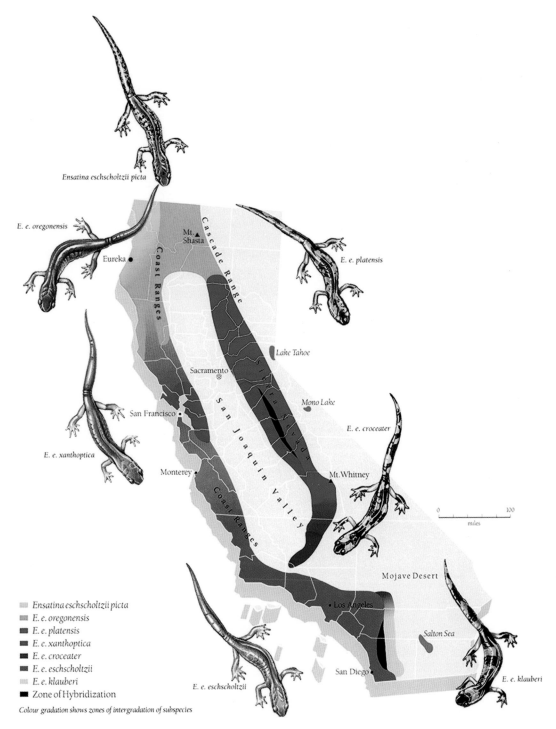

Plate 3 Ring species of the salamander *Ensatina* in the western United States. One species (*Ensatina oregonenis*) is found in Oregon, Washington and British Columbia. It then divides in northern California and forms a more or less continuous ring around the San Joaquin Valley. The salamanders vary in form from place to place, and consequently they have been given a number of taxonomic names. Where the coastal and island sides of the ring meet in southern California, the salamanders behave as good species at some sites (black zones on the map). From Thelander CG. Life on the Edge: A Guide to California's Endangered Natural Resources. Berkeley, CA: Ten Speed Press).

Legend:

- Ensatina eschscholtzii picta
- E. e. oregonensis
- E. e. platensis
- E. e. xanthoptica
- E. e. croceater
- E. e. eschscholtzii
- E. e. klauberi
- Zone of Hybridization

Colour gradation shows zones of intergradation of subspecies

Map labels: Ensatina eschscholtzii picta, E. e. oregonensis, Eureka, Mt. Shasta, Cascade Range, Coast Ranges, E. e. platensis, Lake Tahoe, Sacramento, Sierra Nevada, Mono Lake, San Francisco, E. e. croceater, San Joaquin Valley, E. e. xanthoptica, Monterey, Mt. Whitney, Coast Ranges, Mojave Desert, Los Angeles, Salton Sea, E. e. eschscholtzii, San Diego, E. e. klauberi

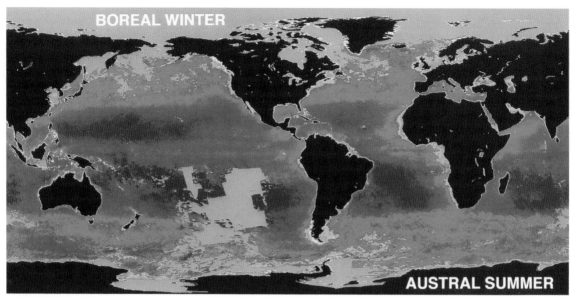

(a)

(b)

Plate 4 Density of plankton, as measured by chlorophyll concentrations from the sea surface to a depth of 25 m, averaged over 1978–1986. (a) Northern Hemisphere winter, December–February; (b) Northern Hemisphere spring, March–May; (c) Northern Hemisphere summer, June–August; (d) Northern Hemisphere autumn, September–November. Colours indicate concentrations on a log scale: extremes are purple (<0.06 mg/m^3), via dark blue, light blue and green to orange-red (1–10 mg/m^3). From Longhurst A. *Ecological Geography of the Sea*. 2nd ed. London: Academic Press, 2006. (Reproduced with permission of NASA/Goddard Space Flight Center.)

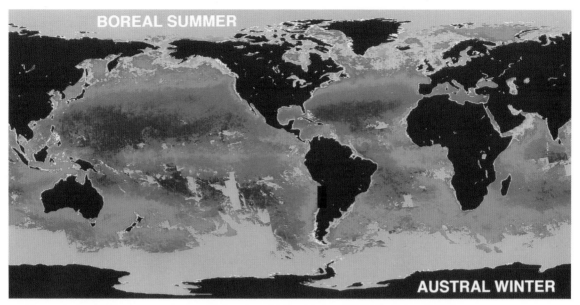

(c)

(d)

Plate 4 (*continued*)

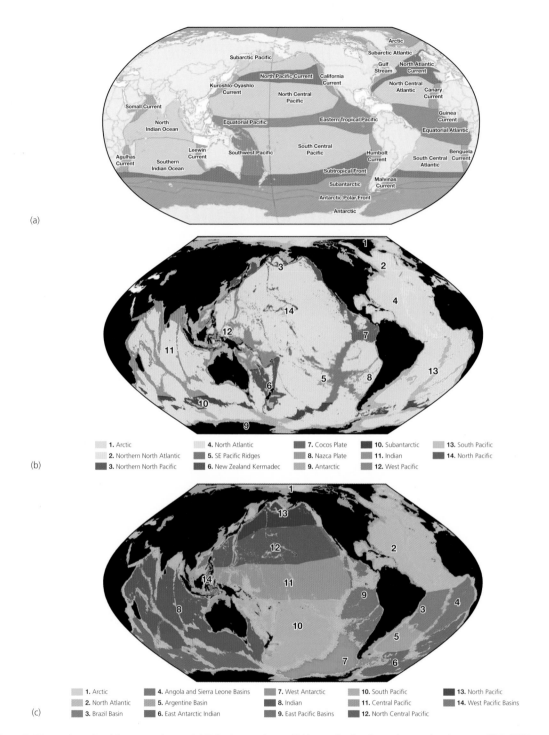

Plate 5 Maps of marine biogeoprovinces. (a) Pelagic provinces; (b) lower bathyal provinces, depth range 800–3000 m; (c) abyssal provinces, depth range 3500–6500 m. From Vierros M, Cresswell I, Escobar Briones E, Rice J, Ardron J (eds.). *Global Open Oceans and Deep Seabed (GOODS) Biogeographic Classification.* Intergovernmental Oceanographic Commission (IOC) Technical Series 84. Paris: UNESCO, 2009.

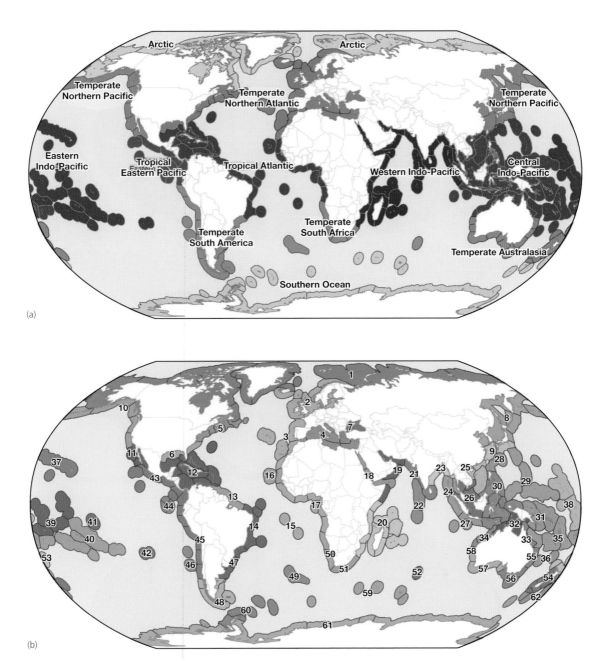

Plate 6 Marine coastal and shelf biogeographical realms (a) and provinces (b). From Spalding MD, Fox HE, Allen GR, *et al.* Marine ecoregions of the world: a bioregionalization of coastal and shelf areas. *Bioscience* 2007; 57: 573–583. (Reproduced with permission of Oxford University Press.)

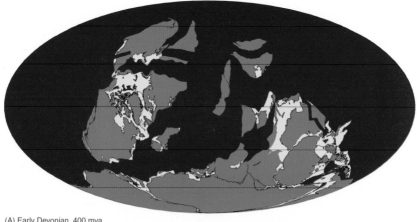

(A) Early Devonian, 400 mya

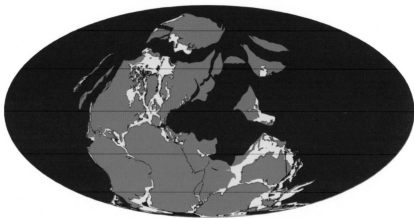

(b) Late Carboniferous, 300 mya

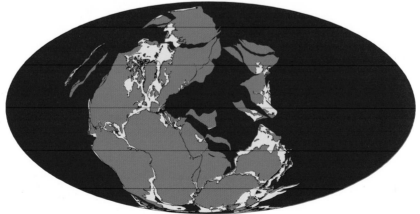

(c) Late Permian, 260 mya

Plate 7 Maps to show the arrangements of the land masses at different times in the past. Deep oceans are shown in dark blue. In all of the maps, the distribution of land (in green) and shallow epicontinental seas (in light blue) are shown as they are today, so that today's continents can be recognized in their former positions as parts of ancient continents. However, the extent of the shallow seas will have varied with time, as shown in Figures 10.1 and 10.3, which also identify some of the smaller continental fragments. Courtesy of Cambridge Paleomap Services, as amended by R. Hall. (Reproduced with permission.)

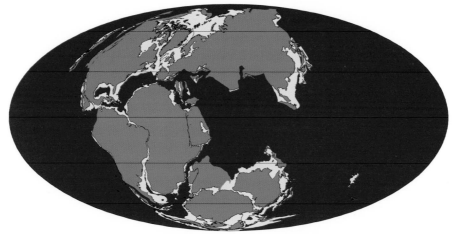

(d) Early Cretaceous, 140 mya

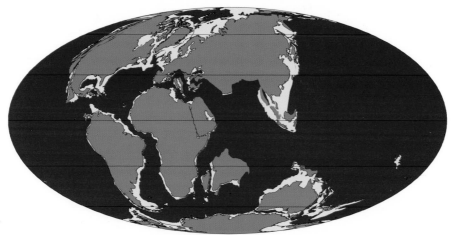

(e) Late Cretaceous, 90 mya

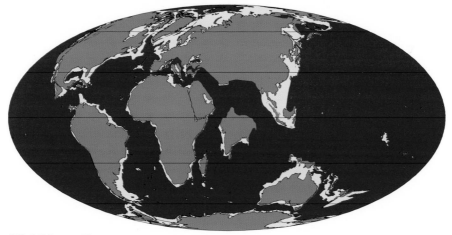

(f) Early Paleocene, 60 mya

Plate 7 (*continued*)

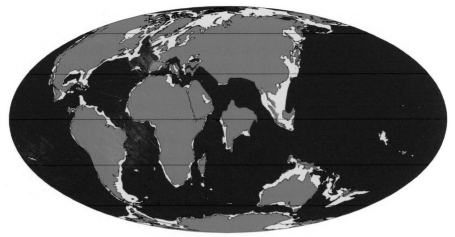

(g) Early Eocene, 55 mya

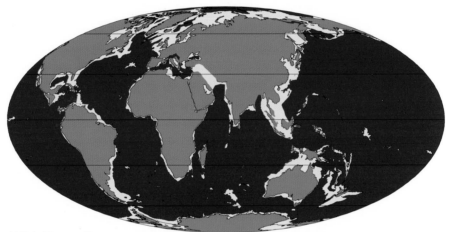

(h) Early Oligocene, 30 mya

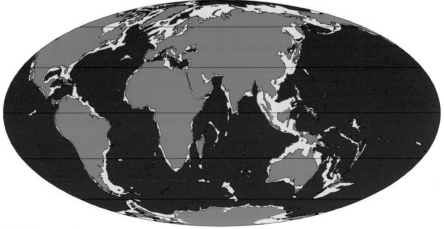

(i) Early Miocene, 20 mya

Plate 7 (*continued*)

The Engines of
the Planet

Section

II

Plate Tectonics

This chapter first explains the evidence for plate tectonics. It then describes how that process affects the patterns of life on the continents, in two ways. Firstly, it changes them directly by changing the patterns of interconnection of the continents. Secondly, it causes changes in the patterns of the continents, oceans, shallow seas, mountains and ocean currents, which have indirect effects on biogeography by altering climate patterns. It also produces different types of island, which may show different biotic histories. The engine of plate tectonics thus continually provides new challenges and opportunities for living organisms, to which they respond via the engine of evolutionary change, as explained in Chapter 6.

The Evidence for Plate Tectonics

As explained in Chapter 1, the idea that continents could fragment and move across the face of the planet was first suggested by the German meteorologist Alfred Wegener in 1912, but was rejected by scientists because he could not suggest any mechanism for such a phenomenon. It was only in the 1960s that new discoveries vindicated Wegener and revealed the motive force. The first breakthrough came with the invention of techniques that used the phenomenon of **palaeomagnetism**. This uses the presence of magnetized particles in many rocks to trace the movements of the rocks, and therefore also of the landmasses in which they lie. Obviously, if the continents had never moved, these 'fossil compasses' should all point to the present magnetic poles – but they

do not [1]. Instead, if a series of rocks of different ages from one continent are studied, and the positions of one of the magnetic poles at these different times are plotted on a map, it looks as though the pole has gradually moved across the Earth's surface (Figure 5.1). Of course, it is instead the continents that have moved across the poles. Furthermore, if similar paths of 'polar wandering' are constructed for each of today's continents, they also show that these have moved relative to one another. Finally, if we plot these paths on a globe and move the continents back along the paths that they have followed through time, we find that they gradually come together in a pattern very similar to that which Wegener first proposed. As he noted, other evidence for the positions of the continents can be gained from the types of rocks that were laid down within them (e.g. desert sandstones or glacial deposits).

The pattern of continental movements suggested by this palaeomagnetic research was supported and confirmed by study of the floor of the oceans. This study revealed a system of great submarine chains of volcanic mountains that rise up to 3000 m, and another system of deep troughs or **trenches** around the edges of the Pacific Ocean. In 1962, the American geophysicist Harry Hess [2] suggested that the volcanic chains were **spreading ridges**, where new seafloor is being formed as the regions on either side move apart, and that the **trenches**, in contrast, are where old ocean floor is consumed, disappearing downward into the Earth. He theorized that all this activity is the result of great convection currents that bring heated material to the surface from the hot interior of the Earth. The spreading ridges,

Biogeography: An Ecological and Evolutionary Approach, Ninth Edition. Edited by C. Barry Cox, Peter D. Moore, Richard J. Ladle.
© 2016 John Wiley & Sons, Ltd. Published 2016 by John Wiley & Sons, Ltd.

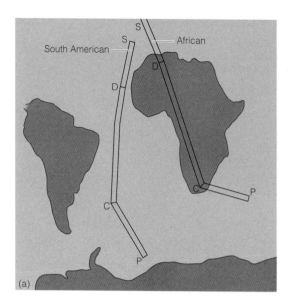

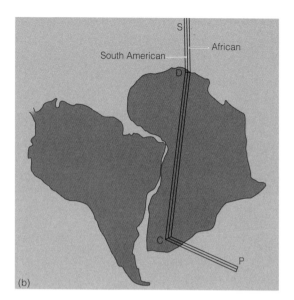

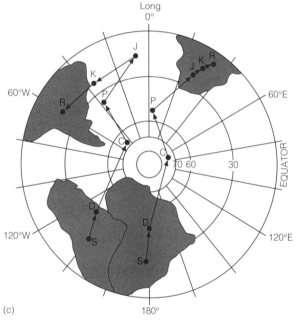

Figure 5.1 A simplified version of the polar wandering patterns of South America and Africa, to illustrate the concept. (a) The position of the South Magnetic Pole relative to each continent in its present position is shown for the Silurian (S), Devonian (D), Carboniferous (C) and Permian (P). (b) The continents are moved together until their polar wandering patterns overlap, proving that they moved as a single landmass during this period of time. (c) The palaeomagnetic data suggest that the continents moved across the South Magnetic Pole as shown here; their paths only diverge from one another during the Jurassic (J), to reach position K in the Cretaceous and R today, as the widening South Atlantic Ocean separated them.

which extend for 72 000 km, mark the positions where these upward currents reach the surface, while the trenches indicate where the corresponding downward currents return cooler material to the depths of the Earth.

Where spreading ridges lie within the oceans, their activity will cause continents to move apart by what is known as **seafloor spreading**, and ultimately may lead them to collide with one another. Sometimes a ridge extends under a continent; its activity will then cause the gradual drifting apart of those regions of the continent that lie on either side of the ridge (Figure 5.2). As these move apart, they become separated by a new, widening ocean, the floor of which is similarly expanding to one side or another, away from the spreading ridge that runs along the centre of the new ocean. The rate at which this takes place is variable: the Atlantic is widening about 2.5 cm a year (about the same speed as your fingernails grow), while the ridges in the Pacific spread five times faster. As a result, the surface of the Earth, known as the **lithosphere**, is occupied by a number of areas known as **tectonic plates**, which may contain continents and parts of oceans, or may consist only of ocean floor (see Figure 5.3). Because the moving elements therefore include the ocean floors as well as the continents, the study of their movements is known as **plate tectonics** rather than continental drift. The

different plates may also move past one another at regions known as **transform faults**, which are regions of active earthquake activity. Perhaps the best known of these is the San Andreas fault in California, where the eastern edge of the Pacific plate is rotating northward past the western edge of the North American plate.

When it first appears from the depths of the Earth, the new ocean crust is still hot and rich in iron minerals, which are sensitive to the prevailing direction of the Earth's magnetic field. Due to changes in the flows of material within the mantle of the Earth, this magnetic field reverses direction every 10^4 to 10^6 years. In 1963 the American geologists Fred Vine and Drummond Matthews [3] found that, as a result, there are strips of varying width, magnetized in opposite directions, in the new ocean crust on either side of the spreading ridge. The symmetry of these bands on either side of the ridge provides striking evidence of the reality of seafloor spreading (Figure 5.4). Because the ages of these stripes can easily be determined, these also provide direct evidence of the past positions of the continents. By removing from the map any ocean floor younger than, for example, 65 million years, one can return the continents to their positions at that time. However, all ocean floor older than 180 million years has disappeared by being returned to the Earth's interior at the great submarine trenches, and the

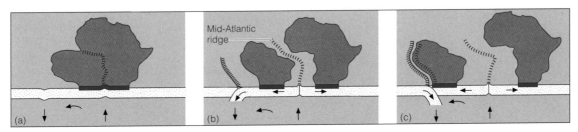

Figure 5.2 How and why South America and Africa drifted apart. (a) The two continents were originally part of a single continent, Gondwana (the rest of which is not shown). An upward convection current from the deeper layers of the Earth appeared under them, with the corresponding downward current further to the west, in the Pacific Ocean. (b) The two continents moved apart, separated by the new South Atlantic Ocean. Down the centre of this runs the mid-Atlantic spreading ridge, on either side of which new ocean crust is continually created. This extension movement is balanced by the appearance, in the Pacific, of an ocean trench where old ocean crust disappears into the Earth. (c) South America has moved westwards until it is adjacent to the ocean trench. To continue to balance the still-widening South Atlantic Ocean, the ocean crust that disappears into the trench is now derived from the west. This old crust now disappears below western South America, causing earthquakes and the rise of the volcanic Andes Mountains.

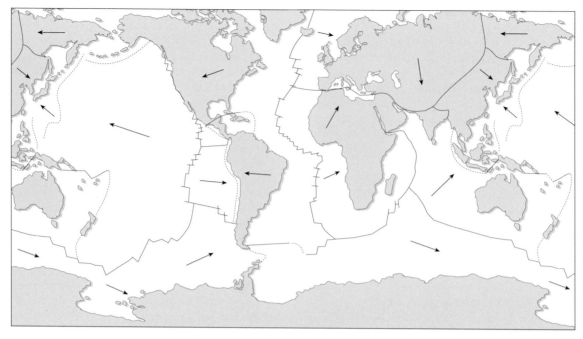

Figure 5.3 The major tectonic plates. Lines within the oceans show the positions of spreading ridges: dotted lines indicate the positions of trenches. Lines within the continents show the divisions between the different plates. Arrows indicate the directions and proportionate speeds of movement of the plates. The Antarctic plate is rotating clockwise.

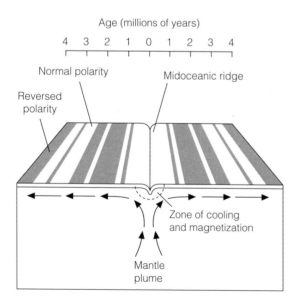

Figure 5.4 Diagram of a portion of seafloor across a mid-oceanic ridge, showing the symmetrical pattern of bands of varying width but of alternating polarity. Adapted from Stanley [13].

positions of the continents before that time therefore have to be deduced from palaeomagnetism. The former patterns of the union of continents can also be deduced from matching the sequences of rock types or rock ages and from dating the times of the rise of mountain chains that mark their collision. Of course, all of these geological changes took place over vast periods of time: the geological timescale is shown in Figure 5.5.

All this evidence for the theory of plate tectonics was so overwhelmingly convincing that it quickly gained general acceptance by biogeographers during the 1960s. Palaeobiogeography was particularly important in the formulation of Wegener's original theory of continental drift. Nevertheless, the degree of detail provided by the geophysical data is now on the whole more precise than that of the palaeobiogeographical evidence, so this biological evidence often has only a confirmatory role. The geophysical data, by establishing the times at which landmasses split or united, have also identified the units of time and geography within which it is

Era	Period	Epoch	Approximate duration in millions of years	Approximate date of commencement in millions of years BP	
Cenozoic	Quaternary	Pleistocene	2.6	2.6	Millions of years ago
	Tertiary	Pliocene	2.4	5	
		Miocene	18	23	
		Oligocene	11	34	— 50
		Eocene	22	56	
		Paleocene	10	66	
Mesozoic	Cretaceous		79		— 100
				145	— 150
	Jurassic		56		— 200
				201	
	Triassic		51		— 250
				252	
Palaeozoic	Permian		47		— 300
				299	
	Carboniferous		60		— 350
				359	
	Devonian		60		— 400
				419	— 450
	Silurian		24	443	
	Ordovician		42		— 500
				485	
	Cambrian		55		— 550
				540	
Proterozoic			c. 4000		— 4600
Formation of the Earth's crust about 4600 million years ago					

Figure 5.5 The geological timescale.

appropriate to make palaeobiogeographical analyses [4]. Until then, such analyses often made little sense, for they frequently combined units of time within which major changes of geography had taken place, or combined geographical areas in inappropriate patterns. The names that have been given to the different units of geological time, their times of beginning and ending and their durations are shown in Figure 5.5.

Today, we can see that the distribution patterns of plants 350–250 million years ago (mya) (Figure 1.3) and of vertebrates 300–270 mya (Figure 1.7) reflect the geography that appears when the continents are placed in the locations suggested by the geophysical data and the outlines of shallow seas are added. Fossil marine faunas provide evidence on the development of faunal provinces on either side of widening oceans (see Chapter 9). The shapes of the fossil leaves of flowering plants indicate the climatic regime of the areas they inhabited, as do the types of plant themselves, and their pollen also provides information on climatic change during the Ice Ages (see Chapter 12).

Although palaeobiogeography therefore often plays a subordinate role, in some situations its data provide more direct, detailed evidence than the geophysical record. For example, both the fossil marine faunas on either side of the Panama Isthmus (see Chapter 9) and the fossil mammals of North and South America (see Chapter 11) provide clearer and more reliable evidence of the date of the final linking of those two continents than do any geophysical data from the area.

For reasons explained in this chapter, sea levels have varied over time, and they have therefore covered varying amounts of the lower-lying margins of the continents; this submerged area is known as the **continental shelf**. In identifying the outlines of landmasses through time, it is important to remember that it is the edge of the continental shelf, and not the coastline, that marks the true edge of the continent. Between the continental shelves, the deep oceans separate the continental plates. Sometimes the whole of these plates has been above sea level. At other times, comparatively shallow 'epicontinental seas' have covered the edges of the continents (e.g. the North Sea today) or formed seas within the continents (e.g. Hudson Bay today). Because seafloor spreading does not provide data on the presence or spread of these seas, palaeobiogeography also provides crucial evidence on the times during which these subdivided areas of land.

Changing Patterns of Continents

The series of maps in Plate 7, and also Figures 10.1 and 10.3, show how the arrangement of the continents has changed through time. In the middle of the Palaeozoic, 400 mya, a large landmass that we call **Euramerica**, made up of today's North America plus Europe, lay across the equator. To its south lay a great continent known as **Gondwana**, a huge area of land that included five of today's landmasses – Antarctica (which was itself made up of two originally separate landmasses), South America, Africa, Australia and India. These two landmasses joined together about 340 mya and then, about 295 mya, the resulting supercontinent was joined by two other Northern Hemisphere continents, Siberia and Kazakhstan, the collision causing the rise of the Ural Mountains. Finally, about 260 mya, this was joined by a number of smaller fragments (including north China, south China, Tibet, Indochina and South-East Asia) that had split off from the northern edge of Gondwana and moved northward. The result was a single world continent we call **Pangaea** (Plate 7c). However, it was not long before Pangaea started to become divided. About 160 mya, Gondwana separated from the northern landmass, now made up of North America and Eurasia, that we call **Laurasia**. The new ocean between the two landmasses is known as the **Tethys Ocean** (Plate 7d).

From then on, the histories of what we see today as the Northern and the Southern Hemisphere continents were entirely different. That of the southern continents was one of continual fragmentation. From about 135 mya, Gondwana became progressively broken up into a number of separate tectonic plates, each bearing a separate continent. Plate 7 shows how Africa, India/Madagascar, Australia and South America in turn became separate from Antarctica as a seaway gradually extended clockwise around Africa, and Figure 11.7 shows this sequence in diagrammatic form. As explained in Chapters 10 and 11, some of these continents moved considerable distances northward, bearing their faunas and floras across zones of latitude with differing climates, to which they had to adapt. The final collision of India with Asia added a new element to the floras and faunas of Asia, and the approach of Australia to South-East Asia allowed a complex interchange between the biotas of those two areas. Even the comparatively small northern movement of South America led to its connection with North America about 3 mya, and an even more complex interchange between their faunas known as the Great American Interchange (see Chapter 11).

In contrast to this complex geographical history of the Southern Hemisphere, that of the Northern Hemisphere has been comparatively uniform. Its two continents, North America and Eurasia, have never been far apart, so that the dispersal of organisms between them has usually been fairly easy. However, three factors have, over the past 180 million years, subdivided or connected them in several different patterns (Figure 11.19): continental movements, the expansion or contraction of shallow seas and the rise or erosion of mountain chains. The resulting changes in the relationships between their floras and faunas are explained in Chapter 11.

How Plate Tectonics affects the Living World, Part I: Events on Land

Plate tectonics has overwhelmingly been the most important factor in causing major, long-term changes in the patterns of distribution of organisms. The movements of the plates are usually quite slow (only about 5–10 cm/year (2–4 inches), about the same rate as the growth of our fingernails), so that any resulting changes must have been extremely gradual. The most obvious effect has been the direct one, the splitting and collision of landmasses altering the patterns of land within which new types of living organism could evolve and spread.

But the changing positions of the continents indirectly affected life in several other ways by changing the patterns of climate. The most obvious of these changes was the result of their movement across the latitudinal bands of climate, thereby causing different areas of land to lie in cold polar regions, in cool, damp temperate regions, in dry subtropical regions or in the hot, wet equatorial regions. The distribution of land relative to the poles is also an important factor; it is noteworthy that during the two periods of time when there were great ice caps at the poles (first about 350–250 mya, as well as the more recent period that caused the Ice Ages), the poles were surrounded by land, not water. Because of this, any ice and snow that fell on the land formed a white surface that reflected back the light and heat of the sun, so that the land became progressively cooler. If, instead, the pole is surrounded by sea-water, the snow often melts, leaving the sea's dark surface free to absorb the Sun's rays.

Another indirect effect of plate tectonics on living things arises from the fact that new mountains, oceans or land barriers deflect the atmospheric and oceanic circulations, thus changing the climatic patterns on the landmasses. There are several ways in which this takes place, as we shall see.

Between the continents lie the deep oceans, with their spreading ridges (Figure 5.3). The amount of activity and the length of these ridges have varied over time. The ridges form chains of huge, undersea volcanic mountains. The more active and extensive these are, the more volume they occupy within the oceans. During periods of increased tectonic activity, sea levels therefore rise, and comparatively shallow **epicontinental seas** cover the lower-lying regions of the continents. These areas, such as the North Sea or Hudson Bay today, are known as continental shelves. Even though they are much shallower than the oceans, these seas form just as effective a barrier to the spread of terrestrial organisms; they were particularly extensive in the Mesozoic, when such seas covered much of North America and Eurasia. But these shallow seas not only affect life by forming barriers to distribution; they also affect the climate of the surrounding areas of land. The climate of any area largely depends on its distance from the sea, which is the ultimate source of rainfall. The central part of great supercontinental landmasses, such as Eurasia today or Gondwana in the past, is therefore inevitably dry, and the climate of such areas experiences great daily and seasonal changes of temperature. The breakup of a supercontinent, or the spread of shallow epicontinental seas into the interior of continents, would have brought moister, less extreme climates to these regions.

Mountain chains are another major influence on climate, and their appearance, location and orientation are all the result of plate tectonics. Mountain chains appear when two continents collide, as when the Ural Mountains resulted from the collision of Siberia and Euramerica in the Permo-Carboniferous. Another example is the collision between India and Asia in the Eocene, 55–45 mya, leading to the uplift of the huge, high Tibetan Plateau which covers over a million square kilometres and also, later, causing the rise of the Himalayan Mountains. As a result, air that had cooled during the Central Asian winter could no longer escape southward, and Central Asia also became isolated from any seas that could have been the source of rain-bearing winds, so that its climate became both cooler and drier. Later, towards the end of the Miocene, a large sea that had covered much of western Central Asia gradually shrank. This caused a further increase in summer temperatures there and increased seasonality in the Indian rainfall, so that savanna replaced the old tropical vegetation in parts of India.

Mountain chains also appear if a continent comes to lie next to an ocean trench, where the descent of lighter ocean-crust material below the edge of the continent causes the rise of volcanic mountains along its margin; this is the cause of the appearance of the Andes (Figure 5.2). Finally, if a continent comes to lie across the position of a spreading ridge, the presence of this heated area below the continent causes the appearance of a mountain range. This is the reason for the appearance of the Rocky Mountains in North America, which started to rise in the latest Cretaceous to the Middle Eocene; by 10–15 mya, they had reached half their present height. The uplift of the Sierra

Nevada Range began in the latest Oligocene, but most of it took place over the last 10 million years, whereas the Cascade and Coast Ranges only rose over the last 6 million years. All these mountains will have increased the seasonality of the climate of North America. It has also been suggested that, by diverting the westerly winds into a more northerly track, they may have played a part in initiating the climatic cooling that eventually led to the Ice Ages. The effects of the rise of the Andes in South America during the Late Cenozoic were less severe because the continent as a whole is narrower and is mainly at lower latitude.

New mountain ranges inevitably affect climate patterns, especially if they arise across the paths of the prevailing moisture-bearing winds, since areas in the lee of the mountains then become desert. Such deserts can be seen today in the Andes, to the east of the mountain chain in southern Argentina, and to the west along the coast from northern Argentina to Peru – the prevailing winds in these two regions blow in opposite directions.

How Plate Tectonics affects the Living World, Part II: Events in the Oceans

The global circulation pattern of winds also affects climate via its effects on the system of currents in the oceans. At present, the presence of

ice at the poles means that there is a very strong temperature difference between the poles and the equator. Because of this, there is a correspondingly strong system of winds transporting heat towards the poles by the system of cells that circulate air in the vertical dimension (see Figure 3.13), made up of a pair of cells either side of the equator, another pair of cells in the polar regions and an intermediate pair of cells in the northern and southern temperate regions. This system of atmospheric and oceanic heat transport has probably been stable since the Oligocene, 30 mya. But, as the American scientist Bill Hay [5] has pointed out, the whole picture was very different in the Cretaceous (Figure 5.6). At that time, there was no polar ice and therefore no permanent high-pressure systems at the poles. Because of the resulting lower temperature differential between the poles and the equator, the systems for transporting heat around the planet would have been less powerful and less stable, except for the easterly equatorial winds. The result, according to Hay, would have been wide arid zones in both the Northern and the Southern Hemispheres, and an oceanic circulation dominated by medium-sized eddies, moving generally from west to east.

Hay's findings suggest that we should be wary of attempting detailed reconstructions of the world's patterns of ocean currents too far back in time. But, if we concentrate on the stable, warm, westwardly directed Equatorial Current, it may be possible to suggest the outlines of the history of the major

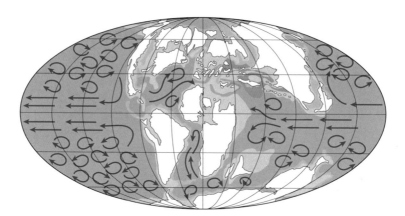

Figure 5.6 Hypothetical ocean surface currents on a Cretaceous ice-free Earth. From Hay [5]. (Reproduced with permission of the Geological Society of London.)

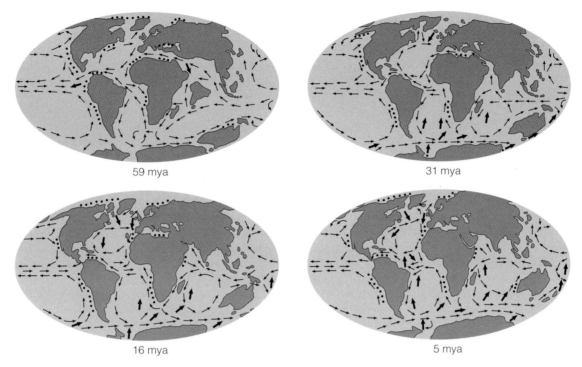

59 mya   **31 mya**

16 mya   **5 mya**

Figure 5.7 Reconstruction of the distribution of the continental landmasses and the inferred pattern of circulation of the ocean currents over the last 59 million years. The black dots indicate regions of upwelling, and the large arrows indicate possible regions of bottom-water formation. (In many cases, the formation of bottom water is by the sinking of warm saline water.) Adapted from Angel [14]; after Haq [15].

gyres, as the patterns of horizontal circulation of water in the ocean basins are known (Figure 5.7).

After Africa and India met Eurasia, closing the old Tethys Ocean to the south of that continent, the waters of the Equatorial Current would have had to find a new path. Some of them probably formed an anticlockwise gyre in the new Indian Ocean, while the rest may have found a path south of Africa to start a similar gyre in the widening South Atlantic (Figure 5.7). These would then also, via the Atlantic section of the Equatorial Current, have continued northward into the North Atlantic. Once in the North Atlantic, some of the waters would have continued westward between the two Americas. The remainder continued in a clockwise gyre, part of which forms the Gulf Stream that brings warm waters from the Caribbean region to western Europe.

The separation of Antarctica, first from Australia (by a shallow sea c. 50 mya and by deep ocean 35 mya) and then from South America 30–28 mya, was to have far-reaching effects. First, the northward movement of Australia progressively reduced the westward flow of warm Pacific water into the Indian Ocean and instead directed it anticlockwise along the north coast of Australia. This put an end to any movement of tropical marine faunas between the Atlantic and the Pacific, and added to the power of the South Atlantic gyre system. But the more important result of the separation of Australia and South America from Antarctica was the establishment of the cold, eastwardly directed Antarctic Circumpolar Current, which runs around the whole periphery of Antarctica. This, together with associated strong westerly winds, separated the weather system of Antarctica from that of its warmer, more northern neighbour and led to fundamental changes in the weather of the Southern Hemisphere. The major result was the progressive cooling of Antarctica: glaciers started to form there about 45 mya, had grown to at least 40% of their present size by 35 mya and covered nearly all of the continent by 29 mya [6]. There was some warming in the Late Oligocene 25 mya, so that less than 50% of Antarctica was ice-covered from 26 to 14 mya. Later, the collision of northward-moving Australia/New Guinea with South-East Asia 20 mya formed a new blockage in the path of the westward Pacific current, forcing some of its waters to flow southward along the eastern margin of Australia and then eastward along Antarctica. This reinforced the Antarctic Circumpolar Current and led to a Mid-Miocene expansion of the eastern Antarctic ice sheet.

The separation of Antarctica from Australia would also have led to a reduction in the amount of evaporation from the now cooler seas around Australia, reducing the rainfall on that continent and causing the arid climate that we see there today. The new continental pattern also added to the power of the gyres in the South Atlantic, including the cold Benguela Current up the western side of southern Africa and the warm clockwise Gulf Stream of the North Atlantic.

Through this period of time, the northward movement of Australia toward South-East Asia would also have gradually weakened the power of the Pacific Equatorial Current as it was forced to pass through the narrowing pattern of inter-island straits in the East Indies, which today is the only low-latitude connection between the world's oceans. In this area, known as the Indonesian Gateway or Indonesian Throughflow [7], the waters at more superficial levels are different from those that lie deeper. The surface water comes from the North Pacific, whose waters are cooler and less saline than those of the South Pacific. They provide most of the surface water just north of Australia, thereby reducing the amount of evaporation of water in the region and therefore also the rainfall in northern Australia. The deeper water in the Indonesian Gateway, on the other hand, is warmer and carries westward a substantial fraction of the heat absorbed by the equatorial Pacific. It is less well understood than the surface flow, but computer models suggest that variations in its power have major effects on the climates of both the Indian and the Pacific oceans. Such variations may well have been caused by the sea-level changes of the later Cenozoic, which united or subdivided the islands on the Sunda shelf (cf. Figure 11.9).

The meeting between Africa and Eurasia during the Oligocene turned the eastern part of the old Tethys Ocean into the ancestor of the Mediterranean Sea. Within this landlocked sea, the rate of water loss by evaporation is greater than its replacement by freshwater from the rivers, which led to an increase in its salinity. When its connection to the Atlantic was briefly lost at the end of the Miocene, this culminated in the drying up of part of the Mediterranean Sea (see Chapter 11). The closure of the Tethys Seaway also reduced the strength of the Equatorial Current through the Panama gap between the Americas, allowing an increase in the eastward flow of water from the Pacific into the northern Caribbean. This led to a cooling of the water there and may have been the reason for the decline in its coral reefs in the Miocene.

The final major change in the pattern of intercontinental connections was the closure of the Panama Seaway between North and South America by the Panama Isthmus, which finally separated the Pacific Ocean from any communication with the Atlantic Ocean. This took place gradually from 3.5 to 3.1 mya as both a narrowing and a shallowing of the seaway between the two continents, although there may have been a breakdown in the Isthmus 2.4–2.0 mya. It is thought that the formation of the Isthmus led to the appearance of the larger ice sheets in West Antarctica and the Northern Hemisphere. Cold, deep Arctic water has been able to enter the North Atlantic since the opening of the Norwegian Sea between Greenland and Scandinavia in the Late Eocene. However, the lands surrounding the North Atlantic have been affected by the warm surface waters of the Gulf Stream, channelled northward by the east coast of North America. The Kuroshio Current in the western North Pacific produces a similar effect, warming Japan.

The extent to which water is removed from the oceans to form continental ice caps inevitably affects sea levels, altering the extent to which they cover the lower, peripheral parts of the continents to form epicontinental seas. On the whole, sea levels have been much higher than they are today, but they have become lower since the Late Miocene 10 mya. But land absorbs less heat than does water and releases it more rapidly. So the draining of seawater from the continents increases the intensity of their seasonal cycles, and their climates

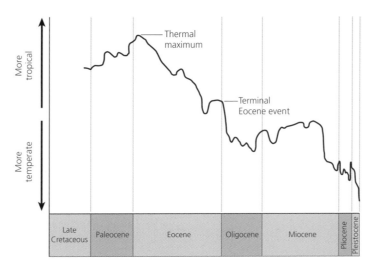

Figure 5.8 Changes in temperature during the Late Cretaceous and Cenozoic, as deduced from oxygen isotope analysis of benthic foraminifera from Atlantic deep-sea drilling sites. Adapted from Miller *et al.* [16].

become less equable, with hotter, drier summers and colder, wetter winters. But this can also lead to further climatic deterioration because, once the land has become cold enough for ice to form, this reflects the sun's rays back into space accelerating the cooling. The appearance of extensive ice caps in the Eocene is the probable cause of the Terminal

Eocene cooling of the planet (Figure 5.8). Similarly, the Pleistocene saw rapid and repeated changes between fully glacial and interglacial conditions, and the resulting declines and rises in sea levels (Figure 5.9).

The movements of the tectonic plates thus produced a variety of changes in the world's environ-

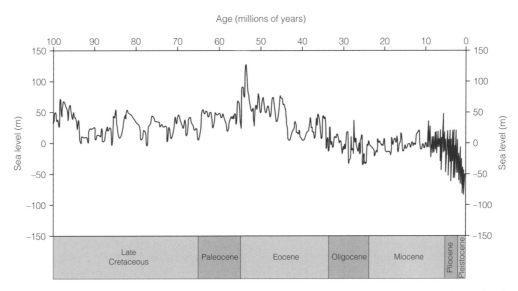

Figure 5.9 Changes in sea level over the last 100 million years; the 100 m level represents the sea level today. Modified from Miller *et al.* [17].

mental patterns, in a variety of interacting ways. We can now recognize these factors and can produce complex computer-generated models of the effects of the changes. However, it is still not possible to construct models that accurately match the high temperatures of the Cretaceous and Early Cenozoic, especially in the continental interiors. This at present can only be explained by making additional *ad hoc* assumptions, for example, of higher solar output of radiation or higher levels of atmospheric carbon dioxide in these earlier periods (some estimates place this at 10 times above current levels!), although little evidence of this has yet been found.

As explained in Chapters 10 and 11, all the processes that have been noted here can have extensive effects on biogeography. As a result of the environmental changes that they caused, organisms had to change their area of distribution or evolve new adaptations. New ecological barriers delimited new areas in which new endemic organisms might appear, or linked them so that the organisms could spread into areas that were previously not available. It should also not be forgotten that the appearance of a new link between two land areas (such as the Panama Isthmus between the Americas) at the same time produces a new barrier between the marine faunas on either side.

The complex interplay between plate tectonics and biogeography in the area between South-East Asia and Australia, known as **Wallacea**, is described in Box 5.1.

Plate tectonics and Wallacea

Professor Robert Hall, Earth Sciences Department, Royal Holloway College, University of London

Guest Author

Box 5.1

In the non-geological literature, one still occasionally reads interpretations of Earth history that dismiss plate tectonics, relying on supposed land bridges or an expanding planet to account for biogeographical patterns. These suggestions give the impression that plate tectonics is little more than a hypothesis that fits a few facts but overlooks others. Nothing could be further from the truth. Today, we have many more techniques of investigation into the structure of the Earth than were available to those who first suggested the theory of plate tectonics. For example, the present spreading centres, where new crust is forming, can now be mapped in great detail throughout the world's oceans, using satellite gravity measurements, as can the transform faults that segment the ridge crests. This helps us to trace the relative movements of plates and so provide increasingly complete models of past plate motions [11]. Above subducting plates, there is an almost continuous chain of volcanoes resulting from melting as fluids rise in the mantle. The movements of plates relative to one another also produce earthquakes, and the resulting **seismic** or energy waves can be used to map the sinking plates. Small-velocity variations in the mantle detected with these waves, using methods known as *seismic tomography*, identify slabs carried deep into the lower mantle, which provide evidence of former subduction episodes. There remain disagreements, but these are not about plate tectonics in general, but merely about details of plate movements, the nature of plate boundaries and similar limited issues.

Different plate tectonic models reflect problems such as difficulties in accurately dating rocks, the incomplete preservation of the rock record due to erosion, deformation and subduction, and the uneven knowledge of global geology. For example, North America and Europe have been studied for longer and by more people than South-East Asia, Australia and Antarctica. The oldest ocean crust on the Earth is about 160 million years old. The ocean-floor record of magnetic stripes and transform faults which provides the basis for reconstructing plate movements is therefore missing for the major part of the Earth's history. Models of the arrangements and movements of the plates in the early Earth therefore rely more on interpretations of an incomplete rock record and are consequently less detailed. Plate movements have caused the assembly and breakup of continents several times in the last 4000 million years. Wegener's interpretation of Pangaea, Laurasia and Gondwana has stood the test of time, although reconstructions of earlier supercontinents are more controversial.

Early plate tectonic thinking emphasized the rigidity of plates and the narrow zones of deformation between them. This concept works well for the oceans, but the continents are generally weaker and have a more complex structure than oceans. We now know that continents can deform over large areas, that they have grown by addition of material added during subduction (see the 'Terranes' section) and that continental crust may be subducted deep into the mantle – although it is buoyant and tends to return to the surface. Some rocks in mountain belts, containing diamonds and other minerals that form only under the high pressures of the Earth's interior, record deep subduction followed by subsequent reappearance at the surface during collision between tectonic plates. Furthermore, movements of subduction zones may lead to extension, and even creation of new oceans, within convergent settings. Subduction may not cease when continents collide, as suggested in early plate tectonic models, and mountain belts are much more complex than originally thought. As in other sciences, a simple paradigm has become more complex with time as our knowledge has increased. Nonetheless, the plate tectonic framework is at the heart of geological thinking, and some of these ideas are well illustrated by the history of East and South-East Asia [12].

One of the first people to recognize the fundamental importance of geology in influencing biogeographical patterns was Alfred Russel Wallace, who interpreted the great differences in the distribution of animals across the Malay Archipelago as reflecting differences in geological history. According to Wallace, the islands of Sumatra, Java and Borneo in the west once formed part of a single continent, which has since been divided into separate islands by changes in sea level, while to the east there was a wide ocean that included remnants of a former Australian and Pacific continent. Wallace, like most other scientists of the 19th and early 20th centuries, thought in terms of fixed continents, but now all Earth scientists recognize that the surface geography of the planet has changed with time, due to plate movements. No area better illustrates the complex way in which geology has influenced biogeography, and the importance of plate tectonics, than Wallace's Malay Archipelago, which broadly corresponds to today's Indonesia.

Asia has grown by closure of oceans between the continents of Gondwana and the Asian part of Laurasia (Plate 8). Continental blocks were rifted from the northern Gondwana margins, so that oceans south of the rifted blocks widened at spreading centres. To compensate for this, other oceans further north narrowed by subduction beneath Asia and finally closed, so that the continental blocks they carried collided with the southern margin of Asia. Asia is therefore a mosaic of continental fragments, separated by sutures containing the remnants of the oceans and the active volcanic margins above the subducting

plates. This general process was repeated several times during the last 400 million years and will no doubt continue, resulting in the separation of further fragments from Gondwana and their incorporation into a growing Asian continent, which may be the core of a supercontinent of the future. The most recent stages in this process have been the collision of India with Asia and the collision of Australia with South-East Asia. Both of these major events include numerous smaller events such as the formation of volcanic arcs, accretion of buoyant features on oceanic plates, island arc–continent collisions and fragmentation of larger blocks by faulting at active margins.

In Asia, from the Alpine-Himalayan mountain chain to South-East Asia, the mantle structure recorded by seismic tomography reveals this history of subduction and provides a test of tectonic reconstructions. Linear high-velocity anomalies in the lower mantle beneath India and north of India are the result of the closure of different oceans between India and Asia during the Mesozoic and Cenozoic. A broad high-velocity anomaly beneath Indonesia is evidence of a different subduction history north of Australia. Here, subduction ceased in the mid-Cretaceous after the collision of continental fragments in Sumatra and Java, and it resumed about 45 mya from the north, beneath Borneo, and from the south, beneath Sumatra, Java and the Sunda Arc.

As Wallace suggested, geology provides the basis for understanding the distributions of faunas and floras in South-East Asia, but only via a complex interplay of plate tectonic movements, palaeogeography, ocean circulation and climate. Plate movements and collisions were intimately linked to changing topography, ocean depths and land–sea distributions, which in turn influenced oceanic circulation and climate. The convergence of Australia and South-East Asia has almost closed a former deep, wide ocean. The remaining Indonesian Gateway between the Pacific and Indian oceans is the only low-latitude oceanic passage on Earth, has an important influence on local and probably global climate and is likely to have been just as significant in the past. As one ocean closed, smaller deep marine basins opened, volcanic arcs formed and mountains rose. The distribution of land and sea was altered, and changes in sea level further contributed to a complex palaeogeography. Understanding first the geology and then the palaeogeography, and their oceanic and climatic consequences, are vital steps on the way to interpreting the present distributions of plants and animals. But the plate tectonic engine is the driving force for this whole sequence of changes.

Islands and Plate Tectonics

The nature of islands and the biotas that they contain are also affected by the way in which the direct or indirect effects of plate tectonics created them. These islands may be of three different types (Figure 5.10). As explained in Chapter 7, different types of island may, as a result of their different histories, contain biotas with different characteristics.

The first type of island was originally a part of a nearby continent, but it became separated from it by rising sea levels (e.g. Britain, Newfoundland, Sri Lanka, Sumatra, Java, Borneo, New Guinea and Tasmania) or by tectonic processes that split them away from an adjacent continent (e.g. the larger islands of the Mediterranean, Madagascar, New Zealand and New Caledonia).

A second type of island is part of a volcanic **island arc**. The most obvious of these are the Kurile and Aleutian island arcs that lie along the edge of the Pacific. Here, old ocean crust is being forced into the depths of the crust, and the resulting stresses cause the appearance of volcanic islands. The Lesser Sunda Islands of the East Indies have similarly formed where the northward-moving Australian plate is undercutting South-East Asia.

Other islands form as a result of the activity of what geologists call **hotspots** (Figure 5.11). These are scattered, but fixed, locations over 700 km deep within the Earth, from which plumes of hot material rise to form volcanoes at the Earth's surface. Where this volcano lies within an ocean rather than within a continent, it may either remain submerged as a **seamount** or *guyot*, or grow to rise above the surface as a volcanic island. However, because the seafloor is part of a moving tectonic plate, the island is gradually carried away from the plume of hot material. Volcanic activity in the island then ceases, and the surrounding seafloor cools and contracts, while the island itself is subject to erosion. As a result, over thousands of years the island

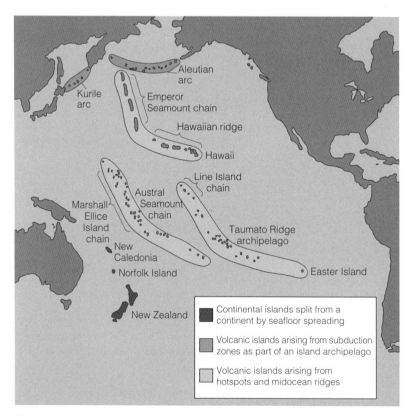

Figure 5.10 Map of the Pacific Ocean, showing examples of the three types of island. Adapted from Mielke [18].

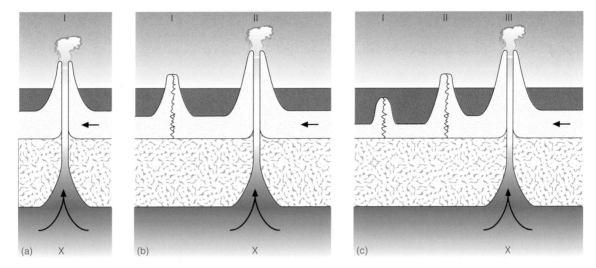

Figure 5.11 The formation of a chain of islands as the result of the activity of a hotspot. (a) A volcanic island, I, has formed above the position of a hotspot, X, deep within the Earth's crust. The more superficial layer of the crust is being carried westward, due to plate tectonics. (b) This movement carries the first island away from the hotspot, so that it is starting to erode away, and the seafloor around it is cooling and shrinking. A new volcanic island, II, now forms above the hotspot. (c) The continuing movement of the superficial layer of the crust has now carried the second island away from the hotspot, and it is therefore starting to erode. The seafloor around islands I and II is cooling and shrinking, so that the further eroded island I is now merely a seamount that does not appear above the surface. Yet another island, III, now forms above the hotspot.

gradually disappears below the surface of the ocean. Meanwhile, a new volcano is developing in that part of the seafloor that now lies above the hotspot. Over millions of years, repetition of this process causes the appearance of a chain of volcanoes (extinct, except for the youngest), the orientation of which clearly shows the direction of movement of the underlying seafloor. There are several such chains of islands in the Pacific Ocean, as shown in Figure 5.10 – the bend partway along each of these chains was caused by a change in the direction of motion of the Pacific plate about 43 mya.

The longest chain is that caused by the activity of the hotspot that today lies under the Hawaiian Islands (see Figure 5.10). The whole chain comprises 129 volcanoes, most of which eroded away and can only be detected as submerged seamounts; those visible today range in age from Hawaii in the east, the most recent, to the oldest, Kure in the west. The rocks that form the Hawaiian Islands can be accurately dated, and the modern technique of calculating the rates of molecular change in the lineages of animals and plants provides a similar system of dating the times of divergence of different genera and species. As the American workers Hampton Carson and David Clague [8] have shown, the ages of the species of animal and plant closely match the ages of the islands on which they are found. This matching integrates the gears of those two great engines of our planet: the plate tectonic engine that here leads to the appearance of new islands, and the evolutionary engine that responds to these opportunities by producing new forms of life. As plate tectonics is the paradigm of the earth sciences, while evolution is the paradigm of the biological sciences, this matching immensely strengthens our whole confidence in the validity of our theories to explain the phenomena of nature.

A hotspot may also cause the appearance of a larger area or **plateau** that, if exposed and colonized by animals and plants, could play a role in their dispersal between continents. An example of this is the Kerguelen plateau, which lay between Antarctica and India and could have allowed dinosaurs and other animals and plants to disperse between these two continents until the beginning of the Late Cretaceous.

Terranes

Where volcanic islands or their eroded, submerged remains reach the edge of a trench within the ocean, they are merely recycled back into the interior of the Earth – the oldest member of the Hawaiian seamount chain is about to disappear into the great trench that lies just east of the Kamchatka Peninsula of Asia. But where the trench lies adjacent to a continent, although the ocean crust will simply be subducted, any superficial islands, seamounts, reefs or other masses of volcanic material are scraped off against the edge of the continent. There, they form individual patches known as **terranes** – regions within which the rocks are quite different from those of the surrounding area. These are best known from the eastern margin of the Pacific, where a complex of 42 terranes makes up a strip 80–450 km wide along the western margin of the United States and Canada; another 48 terranes make up much of Alaska (Figure 5.12). Others lie along the western edge of South America – especially at the northern end, where their arrival from the Early Cretaceous onward caused the rise of the eastern, central and western ranges of the northern Andes.

Some biogeographers have suggested that such terranes might have carried living organisms that became incorporated into the biotas of the continents in which they are now found, and might therefore explain some examples of organisms that are found in both the western Pacific and in the Americas. However, nearly all of these terranes are of oceanic origin rather than continental

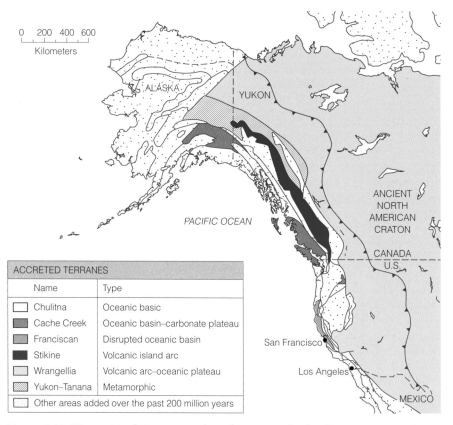

Figure 5.12 Western North America, to show the terranes that lie along its western edge. Over the past 200 million years, islands, seamounts and other superficial features on the seafloor of the Pacific Ocean have been scraped off and added to the western edge of the North American plate as the seafloor itself slipped beneath it. Adapted from Jones *et al.* [19].

fragments, originating from scattered locations in the Pacific rather than from some single notional mid-Pacific continent, and they became emplaced over 100 million years ago [9]. It is therefore very unlikely that they could have carried representatives of any modern groups, such as mammals or flowering plants, that are involved in these trans-Pacific patterns of distribution. The biota of the Pacific island of New Caledonia provides a good example of the incompatibility of the predictions of the two approaches, and it is discussed in Chapter 11.

As noted, the Pacific Ocean also contains a number of volcanic island arcs, and some of these appear to have become incorporated into the northern part of New Guinea and its neighbouring islands as the Australian plate on which they lie moved northward. Some panbiogeographers [10] believe that the distribution of some animals reflects this geological event and that these had already become distributed along the island arc before the collision. Their current distribution, according to this theory, is therefore the result of such a 'Noah's Ark' event, in which living organisms were carried as passengers on moving islands (Figure 5.13). However, these volcanic islands were mostly small and short-lived, and it seems very unlikely that such animals as cicadas and birds of

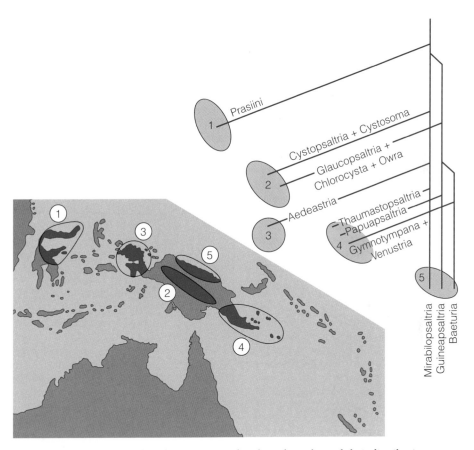

Figure 5.13 The relationships between some families of cicadas and their distribution on Pacific islands, as suggested by de Boer and Duffels [20], who believed that these families had earlier occupied areas 1–5 in an island arc (grey-shaded numbered areas, above right), which then collided with parts of New Guinea and neighbouring islands to form areas 1–5 (white circles) where these families are found today. From de Boer and Duffels [20]. (Reproduced with permission of Elsevier.)

paradise, which are very poor at dispersing, would have been able to disperse along such an oceanic chain. A more straightforward explanation is that their distribution reflects their ecological preferences, because the geology, soils and environment in the areas of New Guinea that were contributed by these former islands differed from those of the rest of New Guinea and therefore provided a more suitable environment for these animals.

All of the phenomena described in this chapter result from the processes of plate tectonics, and the evidence for them is drawn from many quite independent aspects of geology and biology. As a result, the 'theory' of plate tectonics is immensely strong. This is shown when, from time to time, an alternative explanation is put forward for some aspect of the geological phenomena involved – such as the idea that the opening of the oceans was due to expansion of the Earth as a whole. The problem that

arises for such theories is that they do not provide satisfactory solutions for other phenomena that are easily explained by the theory of plate tectonics. In this case, it is difficult to identify the origin of the vast quantity of water needed to fill these enlarging oceans, and the suggested changing diameter of the Earth is incompatible with the coherent pattern of palaeomagnetic results [9]. Such ideas may also raise problems with other aspects of our planet's history – for example, an expanding Earth would have caused a rapid reduction in its rate of rotation.

The theory of plate tectonics is based on a great variety of independent lines of evidence. Such a theory is known as a paradigm, and the theory of plate tectonics is the central paradigm of the earth sciences. The theory of evolution by natural selection, which is considered in Chapter 6, is similarly based on a great variety of lines of evidence and is the central paradigm of the biological sciences.

Summary

1 The fact that continents had moved across the face of the Earth was first proved in the 1960s by data from the magnetized particles preserved in their rocks. This was soon followed by data from the ocean beds, which showed that the Earth's surface is covered by a pattern of moving 'plates', bounded by spreading ridges where new material appears from the Earth's interior and by deep trenches where old material returns to its depths.

2 As a result of this process of plate tectonics, the pattern of the continents has changed greatly over the last 350 million years. At first there were three northern continents, plus a huge southern continent that has been called Gondwana. Later all these landmasses joined to form a single supercontinent, Pangaea, which later broke up into the continents that we see today. Palaeontologists have been able to show how these geographical changes affected the evolving faunas and floras that inhabited them.

3 The most direct effect of the movements of the tectonic plates on the climates of the different continents was caused by their movements across the latitudinal zones of climate. But the movements also affected the climate in other ways – for example, by changing sea levels, by the appearance of mountain chains and by changing the patterns of ocean currents – and also resulted in the Ice Ages. All of these phenomena affect the biota of the continents and islands.

4 Plate tectonics is also the cause of the appearance of volcanic islands, either as chains above geological hotspots deep in the Earth or as island arcs where ocean crust is disappearing. These islands provide interesting biogeographical studies and are home to many endemic species.

5 Small volcanic islands, reefs and other volcanic masses may also become attached to the edges of continental masses, where they are known as terranes, but there is no evidence that they contributed any living organisms to the continents into which they eventually became absorbed.

Further Reading

Edwards J. *Plate Tectonics and Continental Drift*. North Mankato, MN: Smart Apple Media, 2005.

References

1. Runcorn SK. Paleomagnetic comparisons between Europe and North America. *Proceedings of the Geological Association of Canada* 1956; 8: 77–85.

2. Hess HH. History of ocean basins. In: EngelAE, James HL, Leonard BF (eds.), *Petrologic Studies: A Volume in Honour of A.F. Buddington*. Denver, CO: Geological Society of America, 1962: 599–620.

3. Vine FD, Matthews DH. Magnetic anomalies over a young ocean ridge. *Nature* 1963; 199: 947–949.

4. Cox CB. Vertebrate palaeodistributional patterns and continental drift. *Journal of Biogeography* 1974; 1: 75–94.

5. Hay WW. Cretaceous oceans and ocean-modelling. In: Cretaceous oceanic red beds. *Society for Sedimentary Geology Special Publication* 2009; 9: 243–271.

6. Crowley TJ, North GR. Palaeoclimatology. *Oxford Monographs on Geology and Geophysics* 1991; 8: 1–330.

7. Schneider N. The Indonesian Throughflow and the global climate system. *Journal of Climate* 1998; 11: 676–689.

8. Carson HL, Clague DA. Geology and biogeography of the Hawaiian Islands. In: Wagner WL, Funk VA (eds.), *Hawaiian Biogeography: Evolution on a Hot-Spot Archipelago*. Washington, DC: Smithsonian Institution Press, 1995.

9. Cox CB. New geological theories and old biogeographical problems. *Journal of Biogeography* 1990; 17: 117–130.

10. Heads M. Regional patterns of diversity in New Guinea animals. *Journal of Biogeography* 2002; 29: 285–294.

11. Hall R. Cenozoic geological and plate tectonic evolution of SE Asia and the SW Pacific: computer-based reconstructions and animations. *Journal of Asian Earth Sciences*. 2002; 20: 353–434.

12. Hall R. SE Asia's changing palaeogeography. *Blumea* 2009; 54: 148–161.

13. Stanley S. *Earth and Life through Time*. New York: W.H. Freeman, 1986.

14. Angel MV. Spatial distribution of marine organisms: patterns and processes. In: Edwards PJR, May NR, Webb NR (eds.), *Large Scale Ecology and Conservation Biology*. British Ecological Society Symposium no. 35. Oxford: Blackwell Science, 1994: 59–109.

15. Haq BU. Paleoceanography: a synoptic overview of 200 million years of ocean history. In: Haq BU, Milliman HD (eds.), *Marine Geology and Oceanography of Arabian Sea and Coastal Pakistan*. New York: Van Nostrand Reinhold, 1984: 201–231.

16. Miller KG, Fairbanks RG, Mountain GS. Tertiary oxygen isotope synthesis, sea level history, and continental margin erosion. *Paleoceanography* 1987; 2: 1–19.

17. Miller KG, Kominz MA, Browning JV, *et al.* The Phanerozoic record of global sea-level change. *Science* 2005; 310: 1293–1295.

18. Mielke HW. *Patterns of Life: Biogeography in a Changing World*. Boston: Unwin Hyman, 1989.

19. Jones DL, Cox A, Coney P, Beck M. The growth of western North America. *Scientific American* 1982; 247 (5): 70–84.

20. de Boer AJ, Duffels JP. Historical biogeography of the cicadas of Wallacea, New Guinea and the West Pacific. *Palaeogeography, Palaeoclimatology, Palaeoecology* 1996; 124: 153–177.

Evolution, the Source of Novelty

Chapter 6

This chapter explains how evolution by natural selection works and presents the evidence for it. The genetic mechanism that leads to natural variations in the characteristics of organisms is explained. Also discussed are the definition of the species and the way in which new species appear, as is the role of isolation in enabling this process of speciation. Studies of 'Darwin's finches' in the Galápagos Islands have shown the effectiveness of natural selection. The technique known as cladistics provides a reliable method of discovering the patterns of evolution and relationship between different species.

The background to Charles Darwin's discovery of the mechanism of evolution is explained in Chapter 1. Published in 1858 in his great book *On the Origin of Species*, his explanation is now an almost universally accepted part of the basic philosophy of biological science. Darwin realized that any pair of animals or plants produces far more offspring than would be needed simply to replace that pair. For example, many fish produce millions of eggs each year, and many plants produce millions of seeds. Nevertheless, the number of individuals normally remains almost unchanged from generation to generation. It follows that they must be competing with one another in order to survive. They must survive disease, parasites, predators, shortages of food and nutriments or hostile weather conditions, simply in order to reach the age of reproduction. Then they must succeed in the competition to attract a mate, or the complete uncertainty as to which pollen grain will pollinate and fertilize the plant ovum, so that they may reproduce in their turn. To do so, they must employ the multitude of characteristics that they possess and have inherited from their parents.

But the offspring of any single pair of parents are not identical to one another. Instead, they vary slightly in their inherited characteristics. Inevitably, some of these variations will prove to be better suited than others to the hazards of life that the individual experiences. The offspring that have these favourable characteristics will then have a natural advantage in the competition of life, and will tend to survive at the expense of their less fortunate relatives. By their survival and eventual mating, this will lead to the persistence of the favourable characteristics into the next generation, while the less advantageous characteristics will gradually disappear. This process of differential survival is known as **natural selection**.

Scientists refer to Darwin's explanation of evolution by natural selection as a 'theory'. But this does not imply that it is merely a working hypothesis that could be overturned at any time. Scientists commonly use the word *theory* because, technically, no hypothesis can be proved. However many times an experiment is repeated and leads to the same result, there is always a possibility, however remote, that the next try will lead to a different result. So we still refer to the *theory of evolution* and the *theory of gravity*. To a scientist, a 'theory' is an explanation that is compatible with *all* the known evidence. But the important aspect of any theory is not that it cannot be proved, but that it is always possible to disprove it, to subject it to tests that may show it to be untrue, faulty or inadequate. The great strength of Darwin's theory is the huge variety of evidence for it (Box 6.1), so that new evidence from any of these fields of enquiry could disprove his theory. So far, all the new research into aspects of Darwin's theory has only served to confirm it in ever-increasing detail.

Biogeography: An Ecological and Evolutionary Approach, Ninth Edition. Edited by C. Barry Cox, Peter D. Moore, Richard J. Ladle.
© 2016 John Wiley & Sons, Ltd. Published 2016 by John Wiley & Sons, Ltd.

Each chromosome is composed of a pair of strands that spiral around one another (Figure 6.1a) to form a double helix. Each strand is made up of a string of molecules called **nucleotides**,

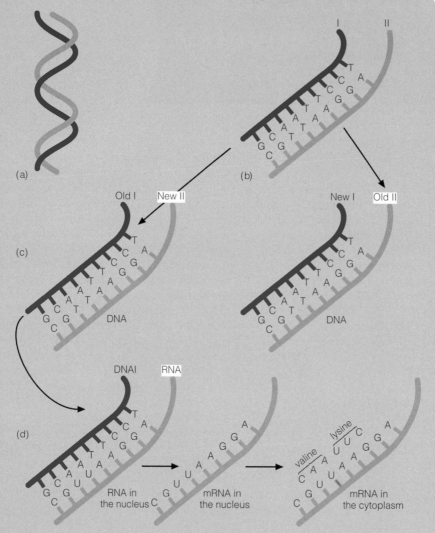

Figure 6.1 How RNA and proteins are formed. (a) The DNA thread is made up of two strands that spiral around one another in a double helix. (b) The two strands are connected to one another via their nucleotides, which bind together in pairs: A (adenosine) and T (thymine), or C (cytosine) and G (guanine). (c) At cell division, the two strands separate, and each goes to one of the new cells, where it builds a new partner identical to the old partner. (d) During the manufacture of proteins, a thread of DNA first builds a thread of RNA, in which U (uracil) has replaced the T (thymine). This thread of RNA detaches itself to form messenger RNA (mRNA), which travels from the nucleus to the ribosome in the cytoplasm. There, the nucleotides in the mRNA act as triplets, each of which forms a template for the formation of a particular amino acid – here, valine and lysine.

of which there are just four types present in the string: adenine (A), cytosine (C), guanine (G) and thymine (T). The nucleotides of one strand link with those of its partner in the double helix, but only in pairs: A with T, and C with G (Figure 6.1b). When a cell divides to produce two new body cells, each thread of DNA separates from its partner, and each of the daughter cells receives one of these two threads. Within each new cell, the thread of DNA then builds a new partner, but because of the unique pairing system of A and T versus C and G, each new partner is an exact replica of the original partner before cell division took place (Figure 6.1c). This precision of replication is essential if the new body cells are to retain the nature of the parent cell. The chromosomes themselves are also paired, one chromosome in each pair having originated from the mother and one from the father; this is known as the **diploid** condition, and the total set of genes within it is known as the **genotype**. In the process of sexual reproduction, each sperm or ovum contains only one of each pair of chromosomes; this is known as the **haploid** condition, and the total set of genes within this is known as the **haplotype**. During the process of production of these sperm and ova, there is some exchange of material between the maternal and the paternal chromosomes, a process known as **recombination**. This is an important part of the genetic system because it allows the next generation to have some characteristics of each of its parents, rather than being a simple replica of one of them.

As already mentioned, DNA is also responsible for all the characteristics of the individual. But to do this, it has to pass through several intermediate stages. First, it is modified within the nucleus into a slightly different molecule called ribonucleic acid, or RNA (Figure 6.1d) This, in a form known as messenger RNA, then travels out of the nucleus into the **cytoplasm** that surrounds the nucleus, where it manufactures proteins (Figure 6.1d). These are large molecules made up of units called **amino acids**, which are responsible for controlling all the metabolic processes within the cell. In manufacturing proteins, the nucleotides of the RNA act as triplets: each successive group of three nucleotides forms a template to which particular amino acids can attach, thus creating a chain of linked amino acids. In the example shown in Figure 6.1d, the amino acids valine and lysine are forming.

There are about 20 different common amino acids, and each type of protein is composed of a unique sequence of them, which may be thousands of amino acids long. The precise sequence of amino acids in each protein controls its properties. Some proteins are structural – for example, muscle tissue is made up of a protein called myosin. But the most important function of proteins lies in their activity as catalysts, or **enzymes**, in which they moderate and control all cell processes. The expression of the DNA code, therefore, is ultimately effected through the activities of these enzymes.

Evolution is therefore possible because of competition between individuals that differ slightly from one another. But why should these differences exist, and why should each species not be able to evolve a single, perfect answer to the demands that the environment makes on it? All the flowers of a particular species of plant would then, for example, be of exactly the same colour, and every sparrow would have a beak of precisely the same size and shape. Such a simple solution is not possible because the demands of the environment are neither stable nor uniform. Conditions vary from place to place, from day to day, from season to season. No single type can be the best possible adaptation to all these varying conditions. Instead, one particular size of beak might be the best for the winter diet of a sparrow, while another,

slightly different size might be better adapted to its summer food. Since, during the lifetimes of two sparrows differing in this way, each type of beak is slightly better adapted at one time and slightly worse adapted at another, natural selection will not favour one at the expense of the other. Both types will therefore continue to exist in the population as a whole.

Because we do not normally examine sparrows very closely, we are not aware of the many ways in which the individual birds may differ from one another. In reality, of course, they vary in as many ways as do different individual human beings. In our own species, we are accustomed to the multitude of trivial variations that make each individual recognizably unique: the precise shape and size of the nose, ears, eyes, chin, mouth and teeth, the colour of the eyes and hair, the type of complexion, the texture and waviness of the hair, the height and build, and the pitch of voice. We know of other, less obvious characteristics in which individuals also differ, such as their fingerprints, their degree of resistance to different diseases and their blood group. All of these variations are, then, the material on which natural selection can act. In each generation, those individuals with the greatest number of advantageous characteristics will be more likely to survive and breed, at the expense of those with fewer.

The Mechanism of Evolution: The Genetic System

The mechanism that controls the characteristics of each organism and their transmission to the next generation lies within the cell, in a rather opaque object called the nucleus. Inside this lie a number of thread-like bodies called **chromosomes**, made of a complex molecule called deoxyribonucleic acid, or **DNA** (see Box 6.1). Each characteristic of the organism is the result of the activity of a particular part of the DNA, which is referred to as a **gene**. It is the biochemical activity of these genes that is responsible for the characteristics of every cell of an individual, and thus for the characteristics of the organism as a whole. There might, then, be a particular gene that determines the colour of an individual's hair, while another might be responsible for the texture of the hair and another for its waviness. (There are just over 20 000 genes in human beings.) Each gene exists in a number of slightly different versions, or **alleles**. Taking the gene responsible for hair colour as an example, one allele might cause the hair to be brown while another might cause it to be red. Many different alleles of each gene may exist, and this is the main reason for much of the variation in structure that Darwin noted. The total of all the genes, which makes up the total genetic inheritance of an organism, is known as its **genotype**. The activity of the genotype produces the characteristics of the individual (its morphology, physiology, behaviour etc.); this is known as the **phenotype**. But in some cases, the environment can modify the phenotypic expression of the genes. For example, identical twins develop from the splitting of a single developing egg and therefore have exactly the same genotype. But they nevertheless may come to differ from each other if they are brought up experiencing differing environmental conditions such as different amounts of sunlight, etc. This slight plasticity of the genotype is valuable from an evolutionary point of view, for it makes it possible for a single genotype to survive in slightly different habitats.

An individual, of course, inherits characteristics from both its parents. This is because each cell carries not one set of these gene-bearing chromosomes, but two: one set derived from the individual's mother, and the other derived from its father. Both parents may possess exactly the same allele of a particular gene. For example, both may have the allele for brown hair, in which case their offspring would also have brown hair. But very often, they may hand down different alleles to their offspring; for example, one might provide a brown-hair allele, while the other provides a red-hair allele. In such a case, the result is not a mixing or blurring of the action of the two alleles to produce an intermediate such as reddish-brown hair. Instead, only one of the two alleles goes into action, and the other appears to remain inert. The active allele is known as the **dominant** allele, and the inert one as the **recessive** allele. Which allele is dominant and which is recessive normally is firmly fixed and unvarying; in the above example, the brown-hair allele is usually dominant, and the red-hair allele is usually recessive.

This, then, is the genetic system that provides two vital properties of the organism. First, it provides the stability that ensures that its complex systems will function and be adapted to the demands of the environment. Second, it provides the plasticity that allows it to respond to minor changes in that environment. But how do modifications of its characteristics take place?

The genes themselves are highly complex in their biochemical structure. Although normally each is precisely and accurately duplicated each time a cell divides, it is not surprising that from time to time – due to the incredible complexity of the molecules involved – there is a slight error in this process. This may happen in the cell divisions that lead to the production of the sexual gametes (the male sperm or pollen, and the female ovum or egg). If so, the individual resulting from that sexual union may show a completely new character, unlike that of either parent. In the example just given, such an individual might have completely colourless hair. Such sudden alterations in the biochemical structure of the genes are known as **mutations**.

This genetic system can lead to changes in the characteristics of an isolated population, in two ways. First, new mutations may appear and, if they are advantageous, spread through the population. Second, since each individual carries several thousand genes, and each may be present in any one of its several different alleles, no two individuals (unless they are identical twins) carry exactly the same genetic constitution. Even if no mutations have taken place, so that they carry the same sets of characteristics, these may be present in different combinations. Inevitably, therefore, each isolated population will come to differ from the others in its genetic content, some alleles being rarer or, perhaps, being absent altogether. As mating goes on in different populations, new combinations of alleles will appear haphazardly in each, and this will lead to further differences between them.

Whether they are new mutations or merely new recombinations of existing alleles, new characteristics will therefore appear within an isolated population. Any of these that confer an advantage on the organism are likely to spread gradually through the population and so change its genetic constitution. However, it is important to realize that chance, as well as its genetic constitution, plays a role in determining whether a particular individual survives and breeds. Even if a new, favourable genetic change appears in a particular individual, it may by chance die before it reproduces, or all of its offspring may similarly die, so that the new mutation or recombination disappears. Nevertheless, however rare each genetic change may be, each is likely to reappear in a certain percentage of the population as a whole. In a larger population, each mutation or recombination will therefore reappear sufficiently often that the effects of random chance are reduced. The underlying advantages or disadvantages that they confer will then eventually show themselves as increased or decreased reproductive success. For this reason, it is the population, and not the individual, that is the real unit of evolutionary change. In smaller populations, however, chance will play a greater role in controlling whether a particular allele becomes common or

rare or disappears; this effect is known as **genetic drift** because it is not controlled by selective pressures. Smaller populations therefore contain less genetic variability and are less closely adapted to their environment; they are therefore more likely to become extinct than larger populations. (This can be a particular problem in island populations; see Chapter 7.)

All that is now required for the appearance of a new species is that some of the new characteristics of this isolated population fit it for a way of life that is in some fashion different from that of the ancestral population from which it became isolated.

From Populations to Species

However great may be the area of land (or water) within which a particular species is found, it is not present everywhere there. Any area is a patchwork quilt of differing environments, of meadow, pond and woodland, of dense forest or of regrown forest – and even the meadow or woodland is itself made up of a myriad of differing habitats, as we saw in Chapter 4. As a result, the species is broken up into many individual populations that are separate from one another. Furthermore, no two patches of woodland, no two freshwater ponds, will be absolutely identical, even if they lie in the same area of country. They may differ in the precise nature of their soil or water, in their range of temperature or their average temperature, or in the particular species of animal or plant that may become unusually rare or unusually common in that locality. Each population independently responds to the particular environmental changes that take place in its own location. The response of each population is also dependent on the particular pattern of new mutations and of new genetic combinations that have taken place within it. Each population will therefore gradually come to differ from the others in its genetic adaptations.

Provided that the barriers between the two populations are great enough to prohibit genetic exchange between them, the foundations for the appearance of a new species have now been laid. If two such divergent populations should meet again when the process of divergent adaptive change has not gone very far, they may simply interbreed and merge with one another. If they have become significantly different in their adaptations, they may still be able to mate and have fertile offspring, known as **hybrids**; these are likely to have a mixture of the characters of their two parents. Because each of the parents had already become adapted to its individual environment, these hybrid offspring, not particularly adapted to either of the environments, will not be favoured by natural selection. From the point of view of each of the well-adapted parent populations, this hybridization is disadvantageous because it merely leads to the production of poorly adapted individuals that will not survive. So, evolution will then favour the appearance of any characteristics that reduce the likelihood of hybridization. These are known as **isolating mechanisms**, and they can take two different forms – systems that prevent mating between related species from taking place and systems that lead to reduced fertility if such mating *does* take place.

Pre-mating isolating mechanisms are common in animals such as birds and insects that have a complicated courtship and mating behaviour. This is because small differences in these rituals may in themselves effectively prevent interbreeding. In the case of Darwin's finches, related species recognize one another because they have different songs, and they do not mate with an individual that sings the 'wrong' song [1]. Sometimes the preference for the mating site may differ slightly. For example, the North American toads *Bufo fowleri*

and *Bufo americanus* live in the same areas but breed in different places [2]. *B. fowleri* breeds in large, still bodies of water such as ponds, large rainpools and quiet streams, whereas *B. americanus* prefers shallow puddles or brook pools. Interbreeding between species is also hindered by the fact that *B. americanus* breeds in early spring and *B. fowleri* in the late spring, although there is some midspring overlap. However, where the group concerned is rapidly evolving, such pre-mating barriers may not have had sufficient time to become effective. For example, in ducks, where colour pattern and behaviour are the pre-mating barriers, 75% of the British species are known to hybridize.

Many flowering plants are pollinated by animals, which are attracted to the flowers by their nectar or pollen. Hybridization may then be prevented by the adaptation of the flowers to different pollinators. For example, differences in the size, shape and colour of the flowers of related species of the North American beard-tongue (*Penstemon*) adapt them to pollination by different insects or, in one species, by a hummingbird (Figure 6.2). In other plants, related species have come to differ in the time at which they shed their pollen, thus making hybridization impossible. Even if pollen of another species does reach the stigma of a flower, in many cases it is unable even to form a pollen tube because the biochemical environment in which it finds itself is too alien. It cannot therefore grow down to fertilize the ovum. Similarly, in many animals the spermatozoa

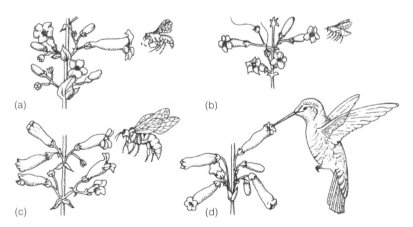

(a)

(b)

(c)

(d)

Figure 6.2 Four species of the beard-tongue (*Penstemon*) found in California, together with their pollinators. Species (a) and (b) are pollinated by solitary wasps, species (c) by carpenter bees, and species (d) by hummingbirds. Adapted from Stebbins [21].

of a different species cause an allergic reaction in the walls of the female genital passage, and the spermatozoa die before fertilization.

Post-mating isolating mechanisms do not prevent mating and fertilization, but instead ensure that the union is sterile. In some cases, the two species may mate together, but they will have no offspring – the mating is sterile. In other cases, the two species may mate and have offspring, but these descendants are themselves sterile. An example is the case of the horse and the ass. Although these are separate species, they do sometimes breed together, but the resulting mule (male) or hinny (female) is sterile. A third category of cases comprises those in which the two species may breed together and have offspring, but the fertility of the offspring is reduced. The hybrids therefore soon become extinct in competition with the more fertile results of mating within each species.

The basic cause of all these incompatibilities is usually to be found in the genetic system. Sometimes the structure and arrangement of the genes on the chromosomes are so different that the normal processes of chromosome splitting and pairing that accompany cell division are disrupted. Other genetic differences may disrupt the normal processes of embryological development, or the growth and maturation of the hybrid individual. Whatever it may be, the final result is the same – the hybrid mating is sterile or, if offspring are produced, they are themselves sterile or have reduced fertility.

The final biogeographical step towards the appearance of a new species occurs when hybrids between the two independent populations are found only along a narrow zone where the two populations meet. Such a situation suggests that, although continued interbreeding within this zone can produce a population of hybrids, these hybrids cannot compete elsewhere with either of the pure parent populations. But, of course, if two closely related populations have diverged in their adaptations but are not in contact with one another, it is quite possible that isolating mechanisms will not have appeared. The two groups may then be able to mate together, although it is likely that their offspring will be less well adapted to the environment than either of their parents. Nevertheless, 25% of plant species and 10% of animal species are known to hybridize, and the number of cases known in animals is growing rapidly [3]. It now seems likely that this

process may be quite important in the evolution and maintenance of biodiversity.

The boundaries between different biogeographical provinces, within each of which many species-pairs may have diverged from one another, are the areas most likely to show hybridization, and recent research [4] provides a good example of this in the marine environment. Reef fishes, because their systems of species-recognition often depend on easily visible patterns of colouration, provide an easily detectable test of the existence of hybrids. The area near the southern end of the Red Sea is at the meeting point of three different marine provinces. In that area, merely six days of collecting identified seven different fishes, belonging to four different families of reef fish, whose colour patterns suggested that they might be intermediate between other individuals of related species. Genetic testing of their DNA showed that four of these were indeed hybrids. It is probably significant that one or both of the parental species of these hybrids were at the periphery of their geographical range, so that the fishes may have found the process of potential mate identification and selection unusually difficult.

Sometimes, where a species has been extending its range and, in the process, has had to adapt to new environments, we can see the resulting pattern of evolutionary change laid out on the landscape as a pattern of biogeography. In a few cases, the end-products of that process have come into contact and demonstrated the extent of the genetic change by refusing to mate with one another: they have evolved into separate species, known as **ring species**.

One of the best examples of a ring species is the pattern of distribution of the salamander *Ensatina* in the western United States [5]. The story appears to have begun with the species *Ensatina oregonensis* living in Washington State and Oregon, which spread into northern California, where it formed the new species *Ensatina eschscholtzii*. As this species continued to spread southward, encircling the hot lowlands of the San Joaquin Valley, it developed populations with different genetic constitutions. One result of this was that the populations had different colour patterns (Plate 3): the populations on the western side of the valley became lightly pigmented, while those on the eastern side developed a blotchier pattern. Biologists therefore gave them different names but, because these populations were able to interbreed with one another, so that hybrids were found, they were recognized as merely separate subspecies of the single species *E. eschscholtzii* (*E. e. picta*, *E. e. platensis* etc.).

In two areas, these two sets of subspecies came into contact. At some time in the past, the subspecies that lives in the San Francisco area, *E. e. xanthoptica*, colonized the eastern side of the valley, where it met one of the 'blotched' subspecies, *E. e. platensis*. Here, the amount of genetic divergence between these two subspecies is not enough to prevent them from interbreeding (Plate 3). The other meeting point is below the southern end of the San Joaquin Valley, in southern California, where *E. e. eschscholtzii* met *E. e. croceater* and *E. e. klauberi*. (There is nowadays a gap in the ring of subspecies in this area, perhaps because of climatic change, but populations of *E. e. croceater* are found both northwest and east of Los Angeles, showing that the chain was once complete [6].) Over much of this area of overlap, the two subspecies hybridize to a limited extent; approximately 8% of the salamanders of the area are hybrids. However, at the extreme southerly end of the chain, the length of time since the two types of salamander diverged from their common northern ancestors is at its maximum. As a result, the genetic differences between them are so great that they behave as completely separate species, *E. eschscholtzii* and *E. klauberi*.

There is no general rule as to the length of time that it will take for the descendants of one original species to diverge so far from one another in their genetic constitution that they have become separate species. The most important factor in determining the rate of genetic change is the speed at which the environment changes. If it changes rapidly, the organism must also change rapidly or else become liable to extinction. But the rate at which an organism can respond is also dependent on population size. In a small population, the random effect of genetic drift may by chance produce a new mixture of genetic characteristics that match the new requirements of the environment. This is less likely to happen in a larger population, where the sheer size of the gene pool makes rapid evolutionary change of this kind less likely. One of the best examples of the age of species comes from the volcanic Hawaiian Islands, on which ero-

sion has carved narrow valleys in the lava flows. These valleys have been colonized by the fruit fly *Drosophila*, and many of the small populations of the fly belong to separate species. These can only have started to diverge from one another after the erosion of the lava flows, which can be dated to only a few thousand years ago. But even that comparatively short period of time is longer than biologists have studied any species, so we have never been able to watch while a new species emerges (although see later in this chapter, 'A Case Study: Darwin's Finches').

Sympatry versus Allopatry

In all the examples considered so far, it has been assumed that the divergence between the populations as they evolved into separate species took place in isolation from one another, a situation known as **allopatric speciation**. Until comparatively recently, it was generally accepted that, with the exception of polyploidy (Box 6.2), this was the almost unvarying way in which speciation takes place. But there is now a lively discussion as to whether new species can also arise by **sympatric speciation**, *within* the area of distribution of the ancestral species. There is even evidence for this in Darwin's finches (discussed in this chapter), where two types of *Geospiza fortis*, with different beak sizes, live alongside one another. They prefer to mate with individuals with a similarly sized beak, and genetic analyses show that there is reduced genetic flow between the populations of the two types, which also have different songs. This could well be an example of sympatric speciation.

The Great Lakes of East Africa show dramatic examples of what appears to be sympatric speciation, for they are large enough to provide a great diversity of environments and old enough for these ecological opportunities to have become realized through evolutionary change. The cichlid fishes, in particular, have been able to take advantage of this (Figure 6.3) and are the most rapidly speciating organisms known. These fish are extremely good at changing their diet, and the different jaw shapes and diets shown in Figure 6.3 have evolved repeatedly and independently in several of these lakes. The result is the presence of hundreds of different types of cichlid in each of the Great Lakes: Tanganyika, Victoria and Malawi [7].

A quite different method by which new species can appear is by **polyploidy** – the doubling of the whole set of chromosomes in the nucleus of a fertilized egg cell. The result is that each chromosome automatically has an identical partner, with which the normal processes of pairing and cell division can take place. This may occur in the development of a hybrid individual (in which case it can overcome any genetic isolating mechanisms), or in the development of an otherwise normal offspring of parents from a single species. In either case, the new polyploid individual will be unlikely to find another similar individual with which to mate, and the origin of new species by polyploidy has therefore been important only in groups in which self-fertilization is common. Only a few animal groups fall into this category (e.g. flatworms, lumbricid earthworms and weevils), but in these groups an appreciable proportion of the species probably arose in this way.

In plants, however, in which self-fertilization is common, polyploidy is an important mechanism of speciation. Perhaps three-quarters of all plant species have probably arisen in this way, including many valuable crop plants such as wheat (see Chapter 13, 'The Human Intrusion'), oats, cotton, potatoes, bananas, coffee and sugarcane. Polyploid species are often larger than the original parent type, and also hardier and more vigorous; many weeds are polyploids. But there is another, even more important aspect of polyploidy: the additional set of chromosomes also provides an additional set of genes that are available for adaptation to new needs. They may also contain normal genes that may mask the effects of harmful recessive alleles that may have appeared as a result of mutation. These additional sources of flexibility may be the reason why, as we shall see in Chapter 10, flowering plants were far less affected by the great pattern of extinctions in the life of the planet after the great meteor strike at the end of the Cretaceous.

An example of a pest species resulting from such polyploidy is the cord-grass, a robust plant of coastal mudflats around the world. There are several species of this plant, but none of them was a serious pest until two species met in the waters around the port of Southampton, United Kingdom, in the latter half of the 19th century. An American species of cord-grass, *Spartina alterniflora*, was brought into the area, probably carried in mud on a boat, and was able to hybridize with the native English species, *Spartina maritima*. The hybrid was first found in the area in 1870 and was named *Spartina x townsendii*. It contained 62 chromosomes in its nucleus but, because these chromosomes were derived from two different parent species, they were unable to join together in compatible pairs before gamete formation; the hybrid therefore did not produce fertile pollen grains or egg cells. It was nevertheless able to reproduce vegetatively, and it is still found along the coasts of western Europe. But in 1892 a new fertile cord-grass appeared near Southampton and was named *Spartina anglica.* This has 124 chromosomes. As a result of this doubling of the number found in the sterile hybrid, the chromosomes could once again form compatible pairs and fertile gametes could be produced. This new species, formed by natural polyploidy, has been extremely successful and has spread around the world, often creating problems for shipping by forming mats of vegetation within which sediments are deposited, and so contributing to the silting-up of estuaries. Polyploidy can also be artificially produced, for example by the use of colchicum, an extract of the meadow saffron plant, *Colchicum autumnale*. Techniques of this kind have been used to provide new strains of commercially valuable plants, such as cereals, sugar beet, tomatoes and roses.

Recent research on the cichlids of Lake Victoria [8] has thrown considerable light on the mechanism of this rapid speciation. Genetic mechanisms for post-mating isolation are frequently weak or absent in these fishes. However, they are mainly brightly coloured, show obvious differences in characteristics that allow the two sexes to recognize one another and depend on vision for their selection of mates. The waters of the lake contain suspended particles that absorb particular wavelengths of light. As a result, blue colours that are clearly visible in the surface waters become increasingly unclear at greater depths, where red colours instead are more visible. The female cichlids have mating preferences for conspicuously coloured males, and their colour preference is correlated with a genetic difference, for they have evolved different alleles of the visual pigments in

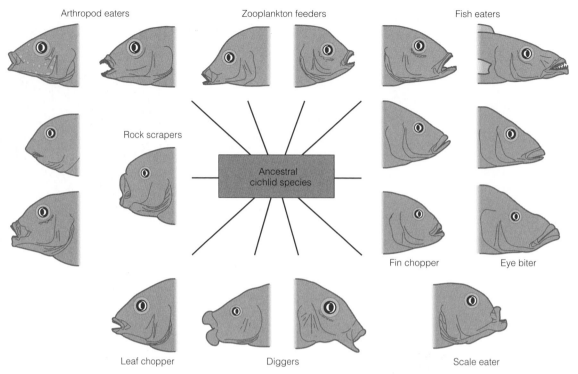

Arthropod eaters

Zooplankton feeders

Fish eaters

Rock scrapers

Ancestral
cichlid species

Fin chopper

Eye biter

Leaf chopper

Diggers

Scale eater

Figure 6.3 A selection of the different shapes of head and mouth that have evolved for different feeding habits in the cichlid fishes in Lake Malawi. Adapted from Fryer and Iles [7].

their eyes. So, in fish that live close to the surface, where blue is more visible, the fish themselves are blue and their visual pigment is more sensitive to that colour. In those that live at greater depths, where red is more visible, the fish are red and their visual pigment is more sensitive to that colour. As long as the rate of change in the relative visibility of the colours changes gradually with depth, these differences in the environment and in the genetics and mating preferences of the fish are sufficient to keep the different populations genetically separate, so that they are potential new species. However, where there is a more rapid change in relative visibility, so that the two populations encounter each other more frequently, they interbreed with one another, forming a single population within which both types of visual pigment are to be found. This type of evolutionary change, associated with differences in the sensory systems and behaviour, has been called **sensory drive** [9].

However, none of this could lead to speciation unless the different populations involved lived in

different environments, each of which provided opportunities for adaptive evolution. In Lake Victoria lie a number of small islands, some of which rise steeply from the bed of the lake, while others rise more gradually. These slopes also vary in the size of the boulders that litter them, and also in the clarity of the water. All this provides gradients in the physical environment, including that of the spawning sites, temperature, oxygen concentration, the intensity and composition of illumination and the amount of growth of the algae on which some of the fish feed, providing microenvironments to each of which the fish can become adapted. This can also only take place if the fish show shoaling behaviour, which would be necessary to restrict gene flow between the different populations as they adapt to the different depth-related microenvironments. So, although the evolution of the different species may seem sympatric to us, at the level of the perceptions of the fish themselves this evolution is **parapatric** (i.e. the distributions of the populations are adjacent to one another but only overlap very narrowly); it is not allopatric.

There is an unhappy ending to this tale. The fish need good visibility in order for the females to be able to see the colour of male that they prefer, but recent changes in their environments are making this more and more difficult. Because of increasing human activity around Lake Victoria, nutrients and silt are draining into the lake from surrounding farmland, together with sewage from urban settlements, causing turbidity and poor light penetration in the waters. Human activity could thus bring the diversification of fish in this lake to a halt [10].

Now that we have established how new species arise, we can turn to the methods we use to define just what a species is, how different species are related to one another and where the species 'lives'.

Defining the Species

There is a great variety of definitions of the **species** which focus on a variety of different aspects of the organisms, so let's first of all consider what type of data are *normally* the concern of the biogeographer.

Practical constraints usually lead the scientist, in analysing the distribution of species, to rely on those that are easy to observe and collect, and on morphological characteristics that can be measured. This in turn normally results in the use of data from such groups as the vertebrates, the larger invertebrates and macroscopic plants. (Studies on other groups, such as mosses, lichens or worms, show that they often exhibit similar biogeographical patterns.)

As we have seen, the most important factor in preserving the identity and nature of a species is isolation, and this aspect is emphasized in the **biological species concept**, which focuses on its reproductive isolation: 'the species is a group of natural populations whose members can all breed together to produce offspring that are fully fertile, but that in the wild do not do so with other such groups' (so the horse and the ass belong to separate species because, although they can breed together to produce a mule, the mule is always sterile).

In deciding whether a particular group of populations should be recognized as a separate species, biologists in practice usually start by finding that they can recognize the group by particular aspects of their appearance, that is, by their physical characteristics. In defining the species, the biologist will then try to use characteristics that appear to be important in its way of life, perhaps because they are obviously used by the members of the species when they mate – as is the case in the cichlid fishes described in this chapter. This may be fairly easy to detect if the habits of the species are not too different from ours (i.e. it is a terrestrial vertebrate and uses its eyes in social interaction), but it becomes progressively more difficult as one examines groups that are more and more different from us in structure and environment. Species recognition must be quite different in, for example, a blind burrowing worm, and in all groups that merely release their eggs and sperm into the water, or that release their pollen into the air. Here, one can only try to identify the characteristics of the species that are connected to its ecology.

This takes us to the second type of definition of the species, the **ecological species concept**, which arises from the organism's need to find its own place in the natural world. Because it must do so in the face of competition from other organisms, it must develop a set of characteristics (morphological, behavioural, physiological etc.) that give it an advantage over those competitors. (These features are, of course, genetically based, so we are also describing the species as a distinct genotype.) These specializations define its ecological niche, which may be quite narrow if competition is intense, as in the case of the tanagers and monkeys described in Chapter 2. If there is less competition, or if conditions are so changeable that the species must remain more variable in order to be able to survive during these changing environmental demands, then the species may show more variation in its own characteristics. (A good example is found in Darwin's finches, discussed in this chapter.) Here, as always, the organism has adapted to the demands of the environment.

It should never be forgotten, however, that evolution is a continuing, dynamic process. So, from time to time, detailed research will reveal related populations that are in the process of becoming separate species. They have therefore not yet reached the point at which their differing adaptations, and the necessary isolating mechanisms, have become completely established. This evidence of evolution in progress provides support for our concept of speciation by natural selection, rather than posing a problem for it.

These definitions obviously cannot be used in the case of fossil species, where we have no knowledge of their abilities to interbreed with one another. Here, morphological data are the only type of evidence available. However, again, it is best to try to find those features that are connected to the way of life or reproduction of the species, and to try to find an assemblage of characteristics that appear to be reliably associated with one another. Where we have an extremely good fossil record, as in the case of some invertebrates and our own species (see Chapter 13), we may find a series of linked forms, evolving and changing over a long period of time. We can then only subdivide them into a sequence of intergrading 'species'.

Any scientific definition reflects the nature and detail of the information available at the time. That is why definitions of the species have changed over time, as we know more and more about the genetic and ecological nature, as well as the population structure, of the species we observe.

A Case Study: Darwin's Finches

Until recently, it was thought that evolution took place too slowly for it to be detectable over the timescale of scientific studies of living organisms, so that it could only be detected in the fossil record. However, it is now clear that this perception was wrong. One of the most detailed and fruitful studies of evolution in action has been carried out precisely where Darwin first noted evidence of it – in the Galápagos Islands in the eastern Pacific Ocean, 960 km from the nearest land (Figure 6.4), and on the birds now known as **Darwin's finches**, which are found only in those islands. Modern evidence suggests that these finches are descended from a flock of South American finches that entered the Islands 2–3 million years ago and evolved into 15 different species belonging to five different genera. A number of them feed on buds or insects, cactus seeds, flowers and pollen, while six species, belonging to the genus *Geospiza*, feed on the seeds of some two dozen species of plant, which they find on the ground.

For many years, the 'Finch Unit' led by the British workers Peter and Rosemary Grant at Princeton University have studied the geospizid finches on the tiny Galápagos island called Daphne Major,

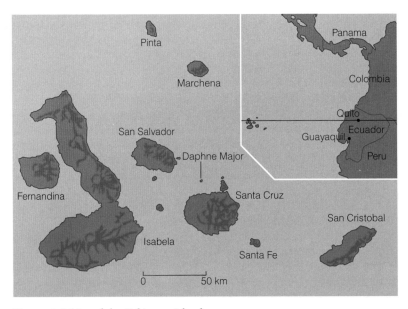

Figure 6.4 Map of the Galápagos Islands.

Figure 6.5 Galápagos ground finches with beaks of different size: from left to right, *Geospiza magnirostris*, *Geospiza fortis*, *Geospiza fuliginosa* and *Geospiza difficilis*.

only 34 ha in area and 8 km from the nearest large island. The largest of them is *Geospiza magnirostris* and the smallest is *Geospiza fuliginosa*, while *Geospiza fortis* is of intermediate size (Figure 6.5). The size of the beak is the only difference in the appearance of these three species, and it is a vital aspect of their lives, for it determines the size of the seed that the bird is best suited to eat. Both the size of the finch's body and that of its beak are strongly dependent on those of its parents; that is, there is a strong inherited factor in their size.

The work of the Finch Unit has been vividly described in Jonathan Weiner's excellent book, *The Beak of the Finch* [11]. Since 1973, they have measured and weighed some 20 000 of these birds, belonging to 24 generations. Their populations have ranged from fewer than 300 birds in the hardest, driest year, to approximately 1000 in the best year. Since 1977, the Finch Unit has recognized, measured and ringed every single bird, and recorded with which others it has mated, how many offspring they had, how many survived, how many in turn mated and so on. They have recorded the abundance of each type of food plant and its seeds, the hardness of the seeds and the pattern of temperature and rainfall. Analysis of their results showed with startling clarity the extent of the differences in the environment from year to year, and the immediacy of the impact of these changes on the populations of the finches.

When food is plentiful, all three species prefer to eat the softest types of seed. In drier seasons, there are fewer of these favourite seeds because they are produced by smaller plants, which tend to wither and die during a drought. As a result, each species has then to become more specialized, spending more of its time feeding on those seeds to which its beak size is best adapted. For example, *G. magnirostris* has the biggest and strongest beak,

and is the best at cracking the hardest seeds, which belong to a little creeping plant called *Tribulus cistoides*. The fruit of this plant (Figure 6.6) breaks up into hard, spiny 'mericarps', each of which is about 7 mm long and contains up to six seeds. *G. magnirostris* can crack two mericarps in less than a minute and extract at least four seeds, while *G. fortis* obtains only three seeds in over 1.5 min. Therefore, in this particular, tiny example of competition, *G. magnirostris* is gathering food at 2.5 times the rate of *G. fortis*. Furthermore, not all the *G. fortis* birds can even try to compete for this food, for only those with beaks at least 11 mm long can crack a mericarp; those whose beaks are only 10.5 mm long cannot do so. Meanwhile, little *G. fuliginosa* has to feed on smaller, softer seeds. This provides a very precise example of the nature and results of the link between morphology (beak size) and ecology (availability of different types of food).

But evolution is not merely about relative ease of existence; natural selection is about life and death. Lying in the Pacific Ocean, the climate of the Galápagos Islands is greatly affected by the cyclical changes in ocean temperature known as El Niño (during which there is heavy rainfall) and La Niña (during which the rainfall is much reduced). The Finch Unit was fortunate that their studies covered a period when the adaptations of the finches were tested to their fullest, for it included two of the most extreme years in the century – both the driest year, 1977, and the wettest year, 1983.

The drought of 1977 (Figure 6.7a) first affected the amount of food available for all the ground finches. During the wet season of 1976, there had been more than 10 g of seeds per square metre of ground. Through 1977, as the drought struck harder and harder, the plants failed to flower and set the new season's seeds. The finches therefore had to continue to eat the seeds produced in 1976:

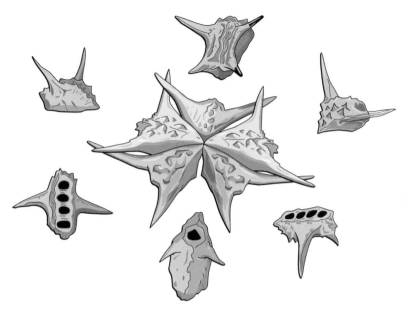

Figure 6.6 The fruit of *Tribulus*, to show how it breaks up into mericarps. Some of these mericarps have been opened by finches, which have removed the seeds, leaving holes.

by June, there were only 6 g/m^2 of seeds, and by December only 3 g/m^2 (Figure 6.7c). The food shortage hit first the new generation of finches because of the shortage of the small seeds on which all the youngest, smallest birds feed. As a result, all the fledglings of 1977 died before they were 3 months old. But the drought also hit the adults. In June 1977 there had been 1300 ground finches on Daphne Major; in December, there were fewer than 300 – only about a quarter of the population had survived. But death hit harder among the finches that could only eat softer seeds. The numbers of *G. fortis* dropped by 85%, from 1200 birds

down to 180 (Figure 6.7b), but little *G. fuliginosa* suffered most, for its population dropped from a dozen to only a single bird [12], so that the population could only recover by immigration from the neighbouring larger island of Santa Cruz. Finally, the drought had been kindest to birds with big beaks, who could feed on the larger, harder seeds. The average size of the *G. fortis* birds that survived was 5.6% greater than the average size of the 1976 population, and their beaks were correspondingly longer (and stronger) – 11.07 mm long and 9.96 mm deep, compared with 1976 averages of 10.68 and 9.42 mm.

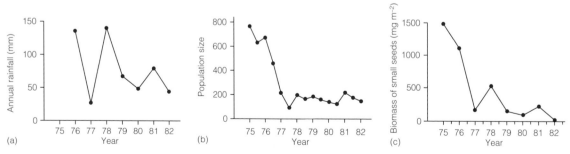

Figure 6.7 Changes on the island of Daphne Major, 1975–1982. (a) Annual rainfall (that of 1975 is unknown). (b) Population size of *Geospiza fortis*. (c) Biomass of small seeds. Adapted from Grant [22].

As if that opportunity to watch and document the harsh workings of natural selection were not enough, 1983 allowed the Finch Unit to see a total reversal in the demands of the environment of Daphne Major. That was the year of the strongest 20th-century El Niño event; the rainfall was 10 times the previously known maximum. The island was drenched, and its plants grew rampantly – by June, the total mass of seeds was almost 12 times greater than in the previous year. This time, small soft seeds predominated – they formed up to 80% of the total mass of seeds, up to 10 times more than the previous maximum. That was partly because the smaller plants grew luxuriously and produced many more seeds, and partly because the growth of *Tribulus* had been hampered by smothering grasses.

As a result of these lush conditions, the finch population spiralled upward. By June, there were more than 2000 ground finches on the island, the numbers of *G. fortis* having increased by more than four times. But in 1984 there was only 53 mm of rain, and in 1985 only 4 mm. Now there was a new episode of drastic selection, but in the opposite direction from that which had followed the drought of 1973. Now it was selecting smaller birds, with smaller beaks, more suited to eating the plentiful smaller seeds [13].

The moral of this second part of the finch story is that conditions, and selection, can oscillate violently. The Finch Unit had witnessed and documented a total reversal of the selection pressures within 6 years – a timespan that would have been totally invisible in the fossil record, had there been one.

But there was still another aspect of the action of evolution on Daphne Major to be found in the Finch Unit's records. Normally, any species has its own niche, and hybridization between species is therefore disadvantageous, for the hybrid young are less well adapted than either of their parents and therefore cannot compete successfully with either of them. That had indeed been true on Daphne Major before the flood year. Interspecies mating was rare, and the occasional hybrid was unsuccessful and did not find a mate. But when there is so much food available that competition has for the moment disappeared, hybridization is no longer necessarily a disadvantage – and that was the situation on Daphne Major in the Year of Abundance,

1983. After that, hybrids between *G. fortis* and *D. fuliginosa* in fact did better than pure-bred members of either species and, by 1993, about 10% of the finches of the island were hybrids [14]. It would now seem that the three supposed species are in reality merely members of one species that shows an unusual degree of variation. But such a flexibility is advantageous in an environment that shows the immense oscillations in climate and environment that the El Niño system brings to the Galápagos Islands. So, the process and results of evolution are closely tailored to the situation in which they take place.

The Finch Unit has therefore seen evolution in action. Their records show precisely what Darwin's explanation had predicted and what biologists have long accepted. Different species show adaptations (here, of beak size and power) that fit them to live on different resources in the environment, thereby reducing competition with other species living alongside them. But that adaptation is not immutable: it cannot afford to be, for the environment itself is fluid and changeable. Thus, as from year to year the environment makes new demands and provides new opportunities, so the adaptations of the species will change in harmony with those demands and opportunities. It is changes of this kind in the characteristics of a species that have led to the evolution of strains of bacteria that are resistant to commonly used antibiotics, to strains of insects that are resistant to DDT and other chemicals designed to control pests, and to strains of malarial parasite that are resistant to modern drugs. Adaptation by evolution is never far behind our efforts to control the biological world.

Controversies and Evolution

There is now a vast amount of evidence for Darwin's explanation of evolution by natural selection. Nevertheless, controversy still exists about some details of the circumstances in which new species evolve or the rate at which this happens. For example, some biologists believe that evolutionary change normally takes place at a steady, gradual rate – a concept known as **gradualistic evolution**. Others instead believe that, even if genetic alterations gradually accumulate within a population, this may not be reflected in detectable morphological or physiological changes until they are so numerous as to shift the balance of the whole genotype. At this point, a comparatively large number of changes are seen to take place at the same time; this is known as the **punctuated equilibrium** model of evolutionary change. Each group of theorists provides examples that may support their view – and, at times, each interprets the same example as supporting their own view!

It is also difficult to isolate such underlying patterns from the more direct effects of the environment. For example, a study of the fossil shells of gastropod (spiral-shelled) molluscs that lived in northern Kenya over the last few million years shows long periods during which their structure and size remained unchanged, interrupted by shorter periods (5000–50 000 years) during which they changed rapidly (Figure 6.8). This was interpreted as an example of punctuated equilibrium [15]. However, the fact that the periods of change took place in several lineages at about the same time suggests that they were the result of external events that affected all of them, rather than resulting from some inherent evolutionary mechanism.

The main difficulty with such studies is that, in general, the fossil record is not sufficiently detailed for us to be able to be certain whether gradualistic evolution or punctuated evolution was involved. In

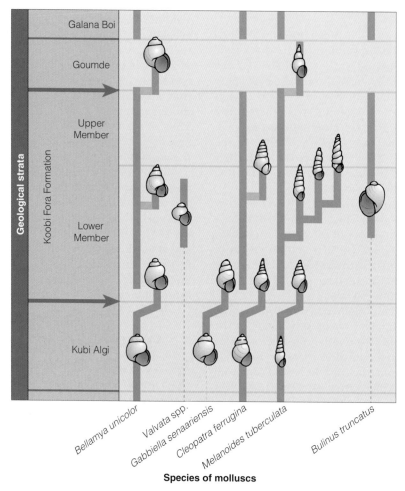

Figure 6.8 Evolutionary changes in fossil gastropod molluscs in northern Kenya. The arrows indicate the levels at which sudden evolutionary changes took place simultaneously in several different species. From Dowdeswell [23].

any case, we have no reason to believe that either style of evolution systematically prevails over the other. Therefore, the real point of interest should instead be to identify the circumstances under which one or another type of evolution would be more likely to take place.

These controversies, and others like them, are common in any area of science, as new observations provoke new theories or suggest modifications of existing theories. But the protagonists in these disputes are only arguing about details of evolution by natural selection: all accept that the explanation itself is correct and is indeed the only one that makes sense of the phenomena of the living world [16]. It is particularly important to realize this, as some groups within society are basically opposed to the idea of evolution, which they view as conflicting with their belief that the human race was created by divine action. One such group is the **Creation Science** organization, who believe an interpretation of the Bible that holds that everything on the planet was created in a short burst of divine activity a few thousand years ago, and insist that all the 'supposed evidence for evolution' is spurious [17]. Another group, who believe in **Intelligent Design**, consider that the process of evolution by natural selection is incapable of producing the degree of adaptation that they see in the organic world, and therefore that it must have appeared by divine action. But these aspects of the biological world are the result of millions of years of continual improvement, as each organism competes with its neighbours. In any case, to the observant scientist, many organisms are less than perfect: even our own species would be better off without wisdom teeth or an appendix and, like all mammals, would find life easier if our system of getting rid of nitrogenous waste was like that of birds and reptiles, which demands much less water than our own great volumes of urine! Such groups try to present these academic controversies as symptoms of widespread and fundamental scepticism of the validity of Darwin's explanation. They are, of course, nothing of the kind. (Darwin's explanation is in any case not incompatible with religion. It is perfectly possible that the biological phenomena that scientists have discovered are merely documenting the gradual way in which the world and all its fauna and flora were created – a view that is accepted by many biologists.)

These groups also sometimes suggest that Darwin's explanation is founded on a circular argument, as follows: 'Darwin suggests that it is the fittest that survive. But how do we know which is the fittest? Why, because those are the ones that survive, of course!' But the catch-phrase 'survival of the fittest' was not Darwin's, and in any case, as noted earlier, the central points of his reasoning are straightforward, as follows:

1. All organisms normally produce many more offspring than are required to replace them in the population, but the population remains approximately constant in its numbers – so only a small proportion of the offspring survive.

2. The individuals are not identical in their characteristics; many of these differences are inherited from one generation to the next, and some of these differences will affect how well the individual is adapted to the environment.

3. It is therefore likely that the offspring that survive will be those whose characteristics make them better adapted.

4. It is thus more likely that these individuals will reproduce and that their characteristics will be more frequent than the others in the next generation. This relative increase in numbers lies at the heart of the concept of natural selection, and it does not involve any circular argument.

Another misunderstanding concerns the appearance of 'new' characters. Some critics object that natural selection can only affect existing characters and does not explain the appearance of new characters, such as an eye or a limb. However, of course these did not begin like the complex structures we see today, which are the result of many millions of years of gradual evolutionary change. In its essentials, the eye is merely a light-sensitive organ, and its basic requirement is the appearance, by chance mutation, of a molecule that is changed when exposed to a ray of light. Mutations can then gradually add a whole repertoire of simple additions to this, such as using energy to return the molecule to its original state, and so making it available for re-use. One can see in the animal kingdom a complete spectrum of different kinds of eye, from those that merely provide an indication of the direction from which light is coming, to the complicated vertebrate eye that provides an accurate, colour image of the world. All that natural selection requires in order to improve any system is that it is slightly better than the other systems around it.

Similarly, all that is required for a whole variety of structures in the vertebrates is the appearance of bone, which may have started as a way of depositing an unwanted excretory product in the body. From this, gradual evolution could lead to the appearance of the bony scales of fishes, which in turn led to the bony shell of tortoises, the scales of reptiles, the feathers of birds and the hair of mammals.

The five-fingered limbs of land vertebrates are merely an elaboration of the bony skeleton of the fin of early bony fishes. This can be demonstrated by the fossil record, which documents the history of such changes through geological time, and by comparative embryology, which shows how these evolutionary changes were achieved by the process of development. So, for example, the appearance of such a 'new' type of mammal as the horse is shown in the fossil record to be merely gradual changes in the limbs, skull and teeth in a lineage of vertebrates that began with a little dog-like creature that lived over 50 million years ago. The embryology of the horse similarly shows how the skeleton of its limbs gradually changed in proportions and structure to produce the elongate, simplified limbs that we see today (see Chapter 11).

The idea that natural selection of genetic characteristics is the mechanism of evolution is, therefore, deeply rooted in biological science today and, like the idea of evolution itself, is supported by a great deal of evidence (Box 6.3). This support from a wide variety of independent areas of scientific inquiry makes natural selection an immensely strong explanation of the method by which evolution takes place. It is therefore the central **paradigm** of the biological sciences, just as the theory of plate tectonics is the central paradigm of the earth sciences. The two systems are based on quite independent and different sets of evidence – biological evidence in the case of evolution and physical evidence in the case of plate tectonics. The fact that the two systems provide similar dates for origins of the individual islands in groups such as the Hawaiian Islands (see the 'Evolutionary Radiations within the Hawaiian Islands' section of Chapter 7) and for the organisms that are found on these islands therefore gives the two paradigms even greater strength as explanations of the causes of the phenomena of the world.

The evidence for natural selection

Concept Box 6.3

The sequence in which different organisms appear in the geological record is strong evidence for evolution. If each new type evolved from other, pre-existing organisms, we should see them appearing in turn, the simpler before the more advanced. The palaeontological record provides varied and detailed evidence of this, for fish first appeared over 450 million years ago (mya), followed by amphibians over 360 mya, the first reptiles about 320 mya and the first mammals about 200 mya. The great radiation of the mammals did not start until 65 mya, and the first primitive types of human being only appeared about 6 mya.

If this sequence in time is because each new lineage has evolved from another, we would expect that the gradual diversification of successful new lineages should be reflected in a hierarchy of characteristics. This in turn should make it possible to construct a hierarchy of groupings into which the organisms can be placed. We see this in our classification of animals and plants. For example, all mammals have hair and produce milk to feed their young. There are

many types of mammal but one group, known as the perissodactyls, has limbs in which the main axis runs through the third digit; they include the tapirs, with their short, fleshy trunk, the horned rhinoceros, and the equids (horses and zebras). So the equids have all the preceding characters, but all also have their own specializations, including high-crowned teeth and limbs ending in a single hoofed digit. They have an excellent fossil record (see Chapter 11), in which we can trace the gradual appearance of these characteristics, which evolved in response to the spread of grasslands, which permitted fast running but required teeth that could withstand the heavy wear caused by the silica in grass.

Similarly, if these similarities in adult structure are the result of evolutionary relationships, we would expect to be able to find similar relationships in the embryology of the groups. Again, that is precisely what we find. For example, in the early stages of their development, mammals still show traces of the gill clefts of fish. They also show relics of the system of membranes that reptiles employ, while they are in the egg, in order to use their yolky food supply and to obtain oxygen. Similarly, the early larval stages of many marine organisms betray evidence of their relationships. These are sometimes surprising – for example, the larvae of barnacles show clearly that they are crustaceans, related to crabs, rather than being limpet-like molluscs that their adults closely resemble.

The patterns of biogeography are strong evidence for evolution, for they show that each type of organism that appears is restricted to the part of the world in which it evolved. For example, the great variety of Australian marsupials is restricted to that continent, in which they radiated, and the unique and strange mammals that evolved in South America remained confined to that continent until the Panama land bridge formed (see Chapter 11). This is also true of the relationship between the patterns of distribution of extinct animals and the geography of the times in which they lived, such as the way in which the distribution of dinosaurs conforms to the patterns of landmasses during the Cretaceous Period.

The techniques of modern genetics, which can analyse the structure of their molecules such as DNA (see Figure 6.1 and Box 6.2), allow us to document the details of the genetic make-up of any organism, to compare it with that of others and to deduce the pattern of relationship between them and the times of their divergence from one another. It also allows us to identify the precise genes that are responsible for each characteristic and process in the organism. As a result, we can now see the genetic basis for the varied lines of evidence for evolution. So, to take the evidence from geographical distribution, we can identify individual genes that link together the apparently very different mammalian lineages that evolved from a common ancestor in Africa, known as the Afrotheria. (This was a complete surprise, for there had not previously been any strong indication of some of these relationships.) Similarly, a particular stretch of DNA is found only in the group of mammalian lineages found only in North America and Eurasia (see Figure 10.6). We also find that the sequence of dates of divergence of related groups suggested by the fossil record is reflected in the degree of difference of their DNA. The evidence from human genetics shows that the first divergence of our lineage took place when our ancestors migrated from Africa into Eurasia, and also documents our gradual expansion across Asia, into North America and down into South America, and from Asia across the island groups of the

Pacific Ocean. Genetics also supports the classification of organisms that has been built up by taxonomists. For example, it confirms that the different types of anteater found in South America, Africa and Australia are, despite many similarities in their structure, related to other lineages of mammal in those continents rather than to one another.

The genes are also intimately involved in the processes of embryology, in which a single-celled fertilized egg is transformed into a complex organism composed of many types of tissue and a variety of organs of different function. Some studies of the involvement of genes in this have provided a very great surprise. It turns out that the same linked groups of genes, known as *homeoboxes*, are responsible for the segmentation of the body in such diverse and ancient groups as the annelid worms, crustaceans, insects and vertebrates. So these groups of genes have been in existence over all the hundreds of millions of years since these groups diverged from one another. At a very different level, genetics also shows us the DNA basis for the different embryological processes that led to the quite different structural features found in the various breeds of dog or pigeon, whose domestication by human selection Darwin had cited as one of his proofs of similar selection in nature.

In addition to the evidence from genetics, modern biological science has provided convincing examples of natural selection in action. We now know many examples in which different populations of a single species live under slightly different circumstances and show particular adaptations to these circumstances. If natural selection works, we would expect that, if we exchange populations of these species, each of them would gradually evolve the adaptations needed in their new environment. For example, some populations of the fish known as the guppy (*Poecilia reticulata*) live in localities in which predation pressure is intense, while others live where that pressure is lower. Those in the high-predation localities mature and breed at a smaller size and earlier than those in low-predation localities, so as to minimize the chances of being eaten before they have reproduced. In one experiment, populations from high-predation localities were introduced to low-predation sites – and, within a few years, each had evolved the new strategy appropriate to its new environment [18].

Every one of the above areas of biological knowledge is evidence for the action of evolution. It follows that any alternative that purported to be an alternative explanation for any one of these would also have to provide a new explanation for all the others.

Charting the Course of Evolution

When we observe a number of related species or genera, it at first seems a hopeless task to try to establish precisely how the different taxa are related to one another. (Whether the group in question is a species, a genus, or some larger unit, it is known as a **taxon**, (plural **taxa**), whose interrelationships are studied in **taxonomy**.) How can we start to put some order into the puzzle and simplify the task of understanding the course that their evolution had taken? The solution to all this was worked out in 1950 by the German taxonomist Willi Hennig [19], and it is called the cladistic method (Box 6.4).

Hennig's method seems clear and simple, but the problem arises when deciding which characters to use. The most readily available are aspects of the

---

Cladistics and parsimony

Concept Box 6.4

First of all, Hennig tried to identify a group of taxa that were all related to one another, sharing a common ancestor and including all its descendants. Such a lineage is known as a **clade**. He then treated the process of evolutionary change in this clade as a series of branching events, or 'dichotomies', at each of which a single group divides into two daughter groups. At each dichotomy, known as a *node*, one or more of the characteristics of the group changes from the original ancestral or **plesiomorphic** state into a derived or **apomorphic** state. The plesiomorphic characters are recognized by comparison with an *outgroup*, which is closely related to the lineage being studied but not a part of it. The evolutionary history of the group can then be portrayed as a branching **cladogram**. Thus, in Figure 6.9, characters a–g evolved after the divergence between group 1, which is the outgroup, and groups 2–5. They are therefore derived, or apomorphic, relative to the characters of group 1 (in which these characters have remained primitive, or plesiomorphic), but plesiomorphic for groups 2–5. Other new, apomorphic characters then evolved at different points within the evolutionary history of groups 2–6 and can therefore be used to analyse their patterns of relationship.

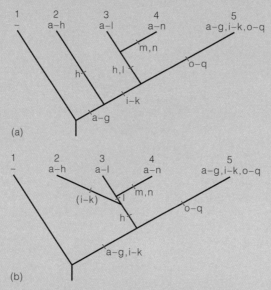

Figure 6.9 Cladogram of the relationships between five groups, using characteristics a to q. The positions at which characters were lost are shown in brackets.

In constructing a cladogram, the characters shown by the different taxa are listed, and the taxa are then arranged so that those that show a similar set of characters are placed in adjacent positions on the branching 'tree'. As far as possible, it is assumed that each apomorphic evolutionary event only occurred once in the history of each group of related taxa (a concept known as economy of hypothesis, or **parsimony**), and the taxa are arranged on the cladogram in such a way as to minimize the number of parallelisms. For example, in Figure 6.9 it is most parsimonious to believe that character h, which is absent in group 5, has evolved twice, because that involves the assumption of only that single additional evolutionary event (Figure 6.9a). The alternative is to transfer the origin of group 2 to near the base of groups 3–4, with the consequent need to assume that characters i–k had been lost in the evolution of group 2 (shown in brackets) – an assumption of three additional evolutionary events, instead of only one (Figure 6.9b).

Of course, species show an enormous variety of characteristics, some of which are linked to one another by such aspects as function, development or behaviour. In constructing the cladogram, the taxonomist therefore has to be careful to try to avoid selecting more than one of each set of linked characters. Even then, it is still possible to select different arrays of characteristics that may give rise to different cladograms. Although the use of a greater number of characteristics may minimize the importance of these difficulties, it becomes progressively more difficult to arrange the taxa and analyse the results. This has led to the introduction of computer programmes such as PAUP (Phylogenetic Analysis Using Parsimony) that calculate the cladogram showing the most parsimonious arrangement of the taxa. Even then, it is not rare for more than one cladogram to show equally parsimonious, and therefore equally likely, possible patterns of evolutionary relationship. The relationship between a particular set of taxa may also be so uncertain that they have to be shown as converging to a single point; this is known as a *polychotomy*, and the cladogram is said to be not fully 'resolved'. An example of these problems comes from the study of the Hawaiian plants known as silverswords (Figure 6.10).

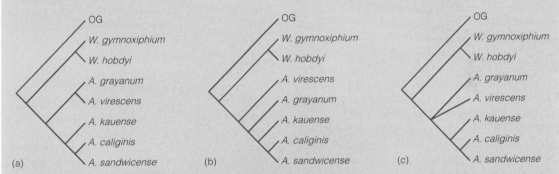

Figure 6.10 Three equally parsimonious cladograms of the interrelationships of species of the Hawaiian silversword genera *Argyroxiphium* and *Wilkesia*, showing different interpretations of the relationships between *Argyroxiphium virescens* and *Argyroxiphium grayanum*. OG is the outgroup used in the analysis. From Funk and Wagner [24].

morphology of the organism, but the danger is that similar ways of life can lead different groups that are not closely related to show similar characteristics. There are many instances of this in the fossil record, but usually a careful study of the complete organism will reveal other features showing that the similarity is the result of *convergent evolution*. A good example is found in the extinct litoptern mammal herbivores of South America, whose limb adaptations to rapid movement on the open, grassy pampas show a remarkable similarity to the limbs of horse ancestors – but many features of their skulls show that the litopterns are really part of a great radiation of South American herbivores, and are not at all closely related to horses. But where the taxa are more closely related, as when one is investigating the interrelationships of species or genera, and the characteristics being used are much finer points of detail, the results of convergent evolution may be far more difficult to detect. In any case, the problem is not confined to the fossil record. Until the advent of molecular methods, some 'groups' of living organisms that biologists had assembled on the basis of what seemed to be quite reasonable features were in fact not closely related at all. For example, a number of lineages of songbird, such as wrens, warblers, nuthatches, treecreepers, robins, shrikes, crows, magpies and jays, seemed to have representatives in Australia, and it was thought that they had originated in the Old World and spread from South-East Asia through the East Indies to Australia. But molecular studies [20] showed that in fact the songbirds originated in Australia. Only after their dispersal out of Australia in the Eocene did they produce the huge worldwide radiation that today includes nearly half of all living species of bird.

But the new revelations of molecular methods can also work the other way around by showing that taxa that had previously been thought to be quite unrelated are in fact part of a single radiation. The most spectacular example is the demonstration that the huge elephants, the tiny elephant-shrews, the insectivorous tenrecs and long-snouted aardvarks, the rabbit-like conies, the Cape golden mole, and the aquatic sea-cows and manatees are all the descendants of a single early radiation of mammals that took place in Africa some 55 million years ago. As a result, they are now placed in a single group known as the Afrotheria (see 'The Biogeography of the Earliest Mammals' in Chapter 10) and provide a completely new example of endemic biogeography, instead of being a rather puzzling array of diverse mammals whose history and biogeography had all to be analysed separately.

These new insights into relationships depend on analysis of the DNA of the organisms (see Box 6.1). The constitution of the DNA of any individual is virtually unique, depending on the combination of molecular structures received from its parents, together with any changes that have taken place in the process. The more closely related two individuals are, however, the more similar their DNA profiles are likely to be.

DNA is also found in other parts of the cell as well as the nucleus, such as in the **mitochondria** (which are responsible for the control of respiration in cells) and the **chloroplasts** (the green structures in plant cells where the conversion of energy from sunlight takes place); these types of DNA are not involved in cell reproduction. They are therefore not subjected to the recombination process (during which some genetic material is exchanged between the maternal and paternal chromosomes), seen at the commencement of each new generation, and they are consequently more stable. They are also haploid and therefore contain a smaller variety of genetic information than the chromosomes in a diploid cell. A further great advantage of the use of extranuclear DNA in such research is that the rate of change of its molecule is far greater than that of nuclear DNA, so that differences in this show up in a much smaller number of generations. It also survives better than nuclear DNA in dead tissues, so it has proved more useful in analysing preserved and dried specimens from museum collections. Studies of extranuclear DNA found in the mitochondria, known as mtDNA, have therefore become increasingly important in phylogenetic research, and that found in the chloroplasts of plants is also proving useful in the study of their interrelationships.

When taxonomists study the evolutionary relationships between organisms, either within a species or when comparing species, the DNA profile is therefore an ideal source of information, since it is not confused by superficial resemblances of structure caused by such phenomena as convergent evolution. Techniques are now available for the direct study of DNA sequences of living organisms, so relationships can be worked out using such molecular methods. The application of these methods to fossil material is limited by the gradual breakdown and eventual loss of DNA in the process of fossilization.

Starting with an explanation of the discovery and mechanism of evolution, we have now followed the process of speciation, explained the role of isolation, defined the nature of the species and explained how its course can be simply and unequivocally portrayed. In the next chapter, we show how advances in understanding the relationships of species to one another in space and time have at long last provided us with a reliable method of revealing the history of lineages, biota, biomes and areas.

1 The organisms that make up the living world are separated into distinct species, each of which has a particular assemblage of adaptations – characteristics that enable it to survive in a particular environment, utilizing a particular source of nourishment.

2 Evolution, the gradual modification of one type of organism or structure into another, is the fundamental phenomenon that underlies most of the processes of the living world. It was Darwin who first put forward convincing evidence for its occurrence, and suggested that natural selection is the mechanism of evolution.

3 Natural selection arises from the fact that each pair of individuals produces far more descendants than are needed to replace them in the population, so that there is competition for survival between these descendants. This leads to the natural selection of those individuals that have the greatest number of favourable characters. In the short term, this ensures that the species remains adapted to its existing environment. In the longer term, if the environment changes, natural selection then acts on the population to change its characteristics, so that it remains adapted.

4 These characteristics are controlled by a genetic mechanism, so they are inherited from generation to generation. This mechanism involves the continual production of slight variations in the characteristics of the species, due to the recombination of existing characteristics, and also the appearance of new features by mutation.

5 Natural selection takes place independently in each population of a species, adapting it to local conditions. Continued isolation between these populations can lead them to gradually become so different from one another that they have become new species. However, under some circumstances, a new species can arise in the same area as its parent species.

6 Cladistics, a method that presents the pattern of relationship between species as a series of bifurcations, provides a clear way of presenting those relationships. Although it can be misleading if based only on morphological characteristics, in which convergent evolution due to similar ways of life may take place, this danger can now be avoided by the use of molecular methods.

Further Reading

Coyne JA, Orr HA. *Speciation*. Sunderland, MA: Sinauer, 2004.

Desmond A, Moore J. *Darwin (A Biography of Charles Darwin)*. London: Michael Joseph, 1991.

Futuyama DJ. *Evolution*. 3rd ed. Sunderland, MA: Sinauer, 2013.

Grant PR, Grant BR (eds.). *In Search of the Causes of Evolution*. Princeton: Princeton University Press, 2010.

Schluter D. *The Ecology of Adaptive Radiation*. Oxford: Oxford University Press, 2000.

Weiner J. *The Beak of the Finch*. London: Vintage Books, 1994.

References

1. Grant PR, Grant BR. Species before speciation is complete. *Annals of the Missouri Botanical Garden* 2006; 93: 94–102.

2. Blair AP. Isolating mechanisms in a complex of four species of toads. *Biology Symposium* 1942; 6: 235–249.

3. Mallet J. Hybrid speciation. *Science* 2007; 446: 279–283.

4. Di Battista JD, Rocha L, Hobbs JA, *et al.* When biogeographical provinces collide: hybridization of reef fishes at the crossroads of marine biogeographical provinces in the Arabian Sea. *Journal of Biogeography* 2015; 42: 1601–1614.

5. Kuchta SR, Parks DS, Mueller RL, Wake DB. Closing the ring: historical biogeography of the salamander ring species *Ensatina eschscholtzii*. *Journal of Biogeography* 2009; 36: 982–995.

6. Jackman TR, Wake DB. Evolutionary and historical analysis of protein variation in the blotched forms of salamanders of the *Ensatina* complex (Amphibia: Plethodontidae). *Evolution* 1994; 48: 876–897.

7. Fryer G, Iles TD. *The Cichlid Fishes of the Great Lakes of Africa: Their Biology and Evolution*. Edinburgh: Oliver & Boyd, 1972.

8. Seehausen O, Magalhaes IO. Geographical mode and evolutionary mechanism of ecological speciation in cichlid fish. In: PR Grant, Grant BR (eds.), *In Search of the Causes of Evolution*. Princeton: Princeton University Press, 2010: 282–308.

9. Seehausen O, Terai Y, Magalhaes IS, *et al.* Speciation through sensory drive in cichlid fish. *Nature* 2008; 4552: 620–626.

10. Seehausen O, van Alphen JJM, Witte F. Cichlid fish diversity threatened by eutrophication that curbs sexual selection. *Science* 1997; 277: 1808–1811.

11. Weiner J. *The Beak of the Finch*. London: Vintage Books, 1994.

12. Grant PR, Boag PT. Rainfall on the Galápagos and the demography of Darwin's finches. *The Auk* 1980; 97: 227–244.

13. Gibbs HL, Grant PR. Ecological consequences of an exceptionally strong El Niño event on Darwin's finches. *Ecology* 1987; 68: 1735–1746.

14. Grant PR. Hybridization of Darwin's finches on Isla Daphne, Galápagos. *Philosophical Transactions of the Royal Society of London B* 1993; 340: 127–139.

15. Williamson PG. Palaeontological documentation of speciation in Cenozoic molluscs from Turkana Basin. *Nature* 1981; 293: 437–443.

16. Coyne JA. *Why Evolution Is True*. New York: Viking Press, 2010.

17. Scott EC. *Evolution vs. Creationism*. Berkeley: University of California Press, 2004.

18. Reznick DN, Shaw FH, Rodd FH, Shaw RG. Evaluation of the rate of evolution in natural populations of guppies (*Poecilia reticulata*). *Science* 1997; 275: 1934–1937.

19. Hennig W. *Grunzüge einer Theorie der phylogenetischen Systematik*. Berlin: Deutscher Zentralverlag. 1950. (English translation of *Phylogenetic Systematics*, trans. DD Davis, R Zanderl. 3rd ed. Urbana: University of Illinois Press, 1966.)

20. Barker KF, Cibois A, Schikler P, Feinstein J, Cracraft J. Phylogeny and diversification of the largest avian radiation. *Proceedings of the National Academy of Science USA* 2004; 101: 11040–11045.

21. Stebbins GL. *Variation and Evolution in Plants*. New York: Columbia University Press, 1950.

22. Grant PR. *Ecology and Evolution of Darwin's Finches*. Princeton: Princeton University Press, 1986.

23. Dowdeswell WH. *Evolution: A Modern Synthesis*. London: Heinemann, 1984.

24. Funk VA, Wagner WI. Biogeographic patterns in the Hawaiian Islands. In: WI Wagner, VA Funk (eds.), *Hawaiian Biogeography: Evolution on a Hotspot Archipelago*, Washington, DC: Smithsonian Institution Press, 1995.

Island Biogeography

Section

III

Life, Death and Evolution on Islands

The limited area and biota of islands provide three unique fields of biogeographical research. The first focuses how isolation and the unusual, unbalanced biota provoke changes in mainland colonists. The second uses the unique characteristics of islands to make statistical analyses of the processes of colonization, extinction, isolation and island area. The third field is the study of colonization of islands that have been devastated of life, generating invaluable insights into the ways in which ecosystems develop and change.

Continental biogeography is quite different from that of islands. Those great areas of land have, over long periods of time, changed in their positions and interconnections, allowing new types of organism to evolve, compete with one another and change their patterns of distribution. Their complex co-evolved ecosystems, with a great variety of interacting species, make their study and interpretation challenging. Much biogeographical research on continental biota has therefore focused on these long-term patterns and the factors that underlie them.

Oceanic islands, in contrast, have a limited biota and comparatively simple ecosystems. Such islands provide unique insights into three quite different areas of research. The first concerns the ways in which the environment affects and controls the evolutionary process. Islands provide one of the essential necessities of evolution, and especially speciation – isolation. Another important point is that new colonists from the mainland are confronted by a very different environment. While it may be released from the pressures exerted by former competitors, predators and parasites, the smaller area and different patterns of climate of the island environment provide new limitations. This aspect of island biogeography is dealt with in the first part of this chapter, culminating in an examination of how these factors have operated in forming the biota of the Hawaiian Islands.

The second area of research is the relevance of island biogeography to evolution and speciation. For example, what controls the number of species on an island? The huge diversity and number of islands (there are over 20 000 in the Pacific alone) provide the equivalent of an enormous, ongoing natural experiment from which we can make comparisons between the biotas of islands of different age, history, climate, size or topography, or of islands that lie at different latitudes or at different distances from their source of colonists. In particular, this research attempts to identify and quantify the factors that control three phenomena: the rate at which new species reach an island, the rate at which species become extinct on an island and the number of species that an island can support (known as its **carrying capacity**).

The third area of study is the role of biogeography in the creation of new island ecosystems. The great volcanic explosion of Krakatau in 1883 created new islands such as Rakata that were initially completely denuded of life. Research on the colonization of Rakata and similar islands provides unique opportunities to study the processes of ecological change and assemblage formation, and it has afforded biogeographers unique opportunities to document how biological complexity incrementally increases within a precisely delimited arena.

Biogeography: An Ecological and Evolutionary Approach, Ninth Edition. Edited by C. Barry Cox, Peter D. Moore, Richard J. Ladle.
© 2016 John Wiley & Sons, Ltd. Published 2016 by John Wiley & Sons, Ltd.

Types of Island

As we have seen, our planet contains three different types of island, and their differing origins lead to their having biotas with distinct characterristics (see Chapter 5). Islands formed from fragments of continent have biotas that were originally made up of a limited subset of species from the continent itself, but that have changed because of independent evolution and extinction within the new island. Some of that evolutionary change will have been a response to the different conditions of life on an island compared with those on the mainland (as will be discussed in this chapter). In addition, if the island gradually moves farther from the parent mainland, its biota will start to be influenced by transoceanic dispersal.

In contrast to islands that were formed from fragments of a continent, all of the biotas of island arcs and chains caused by geological hotspots arrived by transoceanic dispersal. These biotas will subsequently have become changed by evolution, but they may also show a progressive ecological change as the island ecosystem changes and offers new opportunities.

The islands of an arc all appear more or less simultaneously, while those of a hotspot island chain appear (and disappear) in turn, but in both cases a number of islands exist at the same time (note that *hotspot* is used here in a geophysical sense rather than a biological one). This provides the potential for inter-island dispersal and for a more complex pattern of cladogenesis – a pattern that is sometimes called **archipelago speciation**. This is an important driver of species diversity on islands which, due to their relatively small area, have very little potential for intra-island speciation. Even very large islands with complex topographies appear to be poor areas for speciation, as was recently shown by American ornithologist Nicholas Sly and his colleagues [1]. They used molecular techniques to reconstruct the history of speciation of 11 endemic bird species on Hispaniola, a large Caribbean island noted for its high numbers of endemic birds and fantastic mountain ranges. The results clearly showed that genetic divergence (speciation) of closely related species was always associated with the presence of ancient sea barriers that once divided Hispaniola into several smaller paleo-islands.

Getting There: The Challenges of Arriving

Oceans are the most effective barrier to the distribution of land animals. Because few terrestrial or freshwater organisms can survive for any length of time in seawater, organisms can normally only reach an island if they possess special adaptations for transport by air or water. Dispersal to islands is therefore by a sweepstakes route, the successful organisms sharing adaptations for crossing the intervening region rather than for living within it. This greatly restricts the diversity of life that is capable of dispersing to an island.

Some flying animals, such as birds and bats, may be capable of reaching even the most distant islands unaided, using their own powers of flight, especially if, like water birds, they are able to alight on the surface of the water to rest without becoming waterlogged. Smaller birds and bats and, especially, flying insects may reach islands by being carried passively on high winds. These animals may, in their turn, carry the eggs and resting stages of other animals, as well as the fruits, seeds and spores of plants.

Most land animals cannot survive in seawater for long enough to cross oceans and reach a distant island, but it seems possible that some may occasionally make the journey on masses of drifting debris. Natural rafts of this kind are washed down the rivers in tropical regions after heavy storms, and entire trees may also float for considerable distances. Such a floating island could carry small animals such as frogs, lizards or rats, the resistant eggs of other animals, and specimens of plants not adapted to oceanic dispersal. It is rare for such raft-aided dispersal to be seen and documented, but there was a good example of this in the West Indies in September 1995 [2]. Soon after two hurricanes had passed through the Caribbean, a mass of logs and uprooted trees, some over 10 m long, was found on the beach of the island of Anguilla, and at least 15 individuals of the green iguana lizard, *Iguana iguana*, were seen on logs offshore and on the beach. They included both males and females in reproductive condition, and specimens were still surviving on the island (where the species was previously unknown) over 2 years later. Judging by the track of the hurricanes, the lizards probably originated on the island of Guadeloupe, some

250 km away, and the journey probably lasted for about a month.

A few plants have developed fruits and seeds that can be carried unharmed in the sea. For example, the coconut fruit can survive prolonged immersion, and the coconut palm (*Cocos nucifera*) is widespread on the edges of tropical beaches. However, since the beach is as far as most seaborne fruits or seeds are likely to get, only species that can live on the beach are able to colonize distant islands in this way. The fruits or seeds of plants that live inland would be less likely to reach the sea and, even if they were able to survive prolonged immersion and were later cast up on a beach and germinated, they would be unable to live in a beach environment. They are therefore only able to reach island habitats by evolving different methods of dispersal, by wind or animals.

It is potentially far easier for a plant to adapt to long-distance dispersal. Many plants show adaptations to ensure that the next generation is carried away from the immediate vicinity of the parent [3]. It requires little elaboration of some of these dispersal devices to make it possible for them to traverse even wide stretches of ocean. In addition to this, successful colonization requires only a few fertile spores or seeds, whereas in most animals it requires the dispersal of either a pregnant female or a breeding pair. The spores of most ferns and lower plants are so small (0.01–0.1 mm) that they are readily carried considerable distances by winds; they therefore tend to have very wide patterns of dispersal. Some plants have seeds that are specially adapted to being carried by the wind. Orchid seeds, for example, are surrounded by light, empty cells, and some have been known to travel over 200 km. Liriodendron and maple seeds have wings, and the seeds of many members of the Asteraceae (daisies and their relatives) have tufts of fluffy hairs; those of thistles have been carried by the wind for 145 km.

Many fruits and seeds have special sticky secretions or hooks to make them adhere to the bodies of animals. Examples are the spiny fruits of burdocks and beggarticks, and the berries of mistletoe, which are filled with a sticky juice so that the seeds that they contain stick to birds' beaks. The seeds of many plants can germinate after passing through a bird's stomach, and those of some (e.g. *Convolvulus*, *Malva* and *Rhus*) can germinate after

up to 2 weeks there. In the Canary Islands off the western coast of Africa grows the plant *Rubia*, whose seeds are carried in fleshy fruits that are eaten by gulls. One study [4] showed that the seeds remained in the bird's digestive system for 9–17 hours, during which time the gulls could have flown for 300–677 km. Furthermore, over 80% of the seeds germinated after passing through the bird. Another interesting aspect of this example is that gulls are omnivorous rather than specialist seed eaters, which tend to be smaller with less ability to fly between islands. Likewise, predatory birds may also contribute to seed dispersal by consuming frugivorous lizards (common on islands) and seed-eating birds that, in turn, carry viable seeds in their stomachs [5].

Dying There: Problems of Survival

Like any other population, an island population of a species must be able to survive periodic variations in its environment. But island life is more hazardous than that on the mainland, for several reasons. Critically, the small, isolated nature of many oceanic islands means that resident species do not have the option to move elsewhere when conditions deteriorate. Consequently, island populations are particularly sensitive to natural catastrophes, such as volcanic eruptions. Such events have longer lasting effects in an island situation because if a species does go extinct, reinvasion may be difficult. On the mainland, by contrast, chance extirpation of a species in a particular area can soon be made good by immigration from elsewhere. An island will therefore contain a smaller number of species than an equivalent mainland area of similar ecology. For example, study of a 2 ha plot of moist forest on the mainland of Panama showed that it contained 56 species of bird, while a similar plot of shrubland contained 58 species. In contrast, the offshore Puercos Island, 70 ha in area and ecologically intermediate between the two mainland plots, contained only 20 of these species [6].

Since breeding success and survival are the only measures of an organism's degree of adaptation to its environment, the fact that a species has become extinct also demonstrates that it was not able to adapt to the biotic or climatic stresses to which it was exposed. The island environment is inevitably different from that of the mainland, which was the source of the colonists, and adaptation to it is not easy. First, if the colonists are few in number, they can include only a very small part of the genetic variation that provided the mainland population with the flexibility to cope with environmental change; this is sometimes known as the **founder principle**.

Founding populations are necessarily small, and they may remain that way if the island itself is small. This causes an additional suite of problems because when a population becomes very small, it is thrown to the mercy of chance. Random factors such as fluctuations in sex ratios or age distribution, disease outbreaks or unusual weather events may wipe out the entire population. If a population stays small, sooner or later some essentially random event will wipe them out [7]. If that is not bad enough, individuals in small populations inevitably end up breeding with close relatives – if they can even find a mate. This causes the phenomenon of *inbreeding depression*, resulting in lower fertility and less viable offspring. Even if a group avoids inbreeding depression, the effects of genetic drift (the process by which certain genes may disappear from a population, thereby reducing genetic diversity) are greater in small populations [8]. Like founder effects, genetic drift affects the ability of a population to evolutionarily adapt to new challenges such as outbreaks of disease or a changing climate.

The extent of this added risk to survival in islands is shown by the fact that, although only 20% of the world's species and subspecies of bird are found only on islands, they have contributed 155 (90%) of the 171 taxa that are known to have become extinct since 1600 – an extinction rate about 50 times as great as on the continents. The influence of area on extinction rate is underlined by the fact that 75% of these extinctions took place on small islands [9].

Because of these obvious risks in island life, it has been assumed that there was a one-way traffic of colonists from the continents outward, where they island-hopped across the ocean. However, here, as in so many fields of biogeography, molecular studies have shown us otherwise. A recent study [10] of the molecular genetics of the Monarch flycatcher birds, which are found in nearly every Pacific archipelago including the Hawaiian Islands, has shown

that, after an initial single colonization from the mainland, there was a single radiation into six genera and 21 species wholly within the islands, followed by a reverse colonization of Australia within the last 2 million years.

A species which can make use of a wide variety of food is therefore at an advantage on an island, for its maximum possible population size will be greater than that of a species with more restricted food preferences. The advantage of this will be especially great in a small island, in which the possible population sizes are in any case smaller. This is probably the reason why, for example, although on the larger islands of the Galápagos group in the eastern Pacific both medium-sized and small-sized *Geospiza* finches can coexist, on some of the smaller islands of the group there is only a single type, of intermediate size [11].

Chance extinction is a particular danger for predators, since their numbers must always be far lower than those of their prey. As a result, island faunas tend to be unbalanced in their composition, containing fewer predators and fewer varieties of predator than a similar mainland area. This in turn reinforces the fundamental lack of variety of the animal and plant life of an island that is due to the hazards involved in entry and colonization. The complex interactions of continental communities containing a rich and varied fauna and flora act as a buffer that can cope with occasional fluctuations in the density of different species, and even with temporary local extirpation of a species. This resilience is typically lacking in the simple island community. As a result, the chance extinction of one species may have serious knock-on effects on other species that ecologically interact with it, sometimes leading to domino extinction effect. All these factors increase the rate at which island species may become extinct.

There are clearly several possible reasons why a particular organism may be absent from a particular island. It may be unable to reach it; it may reach it, but be unable to colonize it; it may have colonized it but later have become extinct; or it may simply as yet, by chance, not have reached the island [12]. It is often very difficult to decide which of these possible reasons is causing an absence in any particular case.

On the other hand, there is another side to all the above problems. Once a species has arrived in an island and found a niche in the island biota, it is to some extent protected from many of the problems that may plague its continental relatives. The surrounding ocean waters often protect islands from the more extreme episodes of climatic change experienced by the continents, while the comparatively small number of species on an island reduces competition, and the frequent absence of the parasites and predators also makes life easier for the successful immigrant. Finally, the island species is also protected from the appearance of new, more competent species that have evolved on the continents. All of this explains the fact that species may survive on islands, as what Cronk has called 'relict endemics', long after their continental relatives have disappeared [13]. For example, the plant *Dicksonia* is found both as living plants and as 9 million-year-old fossils on the little (122 km^2) island of St Helena, isolated in the South Atlantic, although none of its relatives survive in Africa, and its closest known living relatives are found in New Zealand.

Adapting and Evolving

Colonists may encounter many difficulties when they first enter an island, but there are rich opportunities for those species that can survive long enough for natural selection to adapt them to the new environment. Opportunity for alterations in behavioural habits, diet and way of life provides in turn opportunity for the organism to become permanently adapted, through evolutionary change, to a new way of life. This process requires a longer period of time and is therefore unlikely to take place except on islands that are large and stable enough to ensure that the evolving species does not become extinct. But if an island does provide these conditions, then remarkable evolutionary changes may occur as colonizing species become modified to fill vacant niches. This can be seen at a whole range of different levels, from the trivial to the comprehensive.

A good example of this can be found in the Dry Tortugas, the islands off the extreme end of the Florida Keys, which only a few species of ant have successfully colonized [14]. One species, *Paratrechina longicornis*, on the mainland normally nests only in open environments under, or in the

shelter of, large objects; but on the Dry Tortugas it also nests in environments such as tree trunks and open soil, which on the mainland are occupied by other species. Not every species, however, is capable of taking advantage of such opportunities in this way. On the Florida mainland, the ant *Pseudomyrmex elongatus* is confined to nesting in red mangrove trees, occupying thin, hollow twigs near the tree top. Although it has managed to colonize the Dry Tortugas, it is still confined there to this very limited nesting habitat.

At a higher level of opportunity, a whole category of niches may be unoccupied. For example, many islands are treeless because the seeds of trees are usually much larger and heavier than those of other plants, and therefore do not often disperse over long distances. The valuable adaptive property of the tree niche is that its height allows it to shade out its competitors and also to live longer [15]. The modifications needed to produce a tree from a shrub that already possesses strong, woody stems are comparatively slight – merely a change from the many-stemmed, branching habit to concentration on a single, taller trunk. For example, although most members of the Rubiaceae are shrubs, this family has produced on Samoa the tree *Sarcopygme*, which is 8 m tall and with a terminal palm-like crown of large leaves. Although more comprehensive changes are needed to produce a tree from an herb, many islands show examples of this. In many cases, the plants involved are members of the Asteraceae, perhaps because they have unusually great powers of seed dispersal, are hardy and often already have partly woody stems. (It has been suggested that the arborescent habit may be primitive in the Asteraceae.) To this family belong both the lettuces, which have evolved into shrubs on many islands, and the sunflowers. Five different types of tree, 4–6 m high, found on the isolated island of St Helena in the South Atlantic have evolved there from four different types of immigrant sunflower (Figure 7.1). Two of these (*Psiadia* and *Senecio*) are endemic species of more widely distributed genera, while the other three (*Commidendron*, *Melanodendron* and *Petrobium*) are recognized as completely new genera.

At an even higher level of opportunity, even birds may find it difficult to colonize oceanic islands. As a result, any successful bird colonist, originally adapted to a particular, limited diet and way of

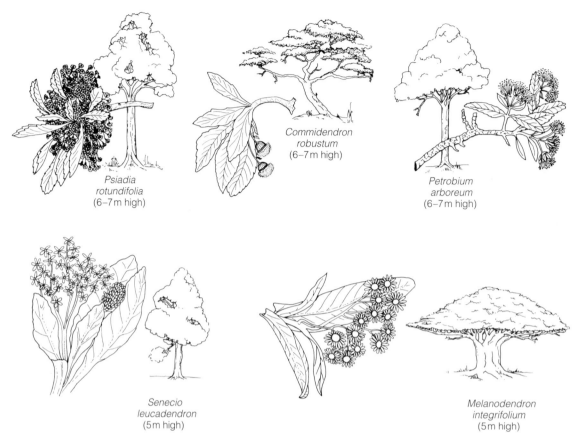

Psiadia
rotundifolia
(6–7m high)

Commidendron
robustum
(6–7m high)

Petrobium
arboreum
(6–7m high)

Senecio
leucadendron
(5m high)

Melanodendron
integrifolium
(5m high)

Figure 7.1 The varied trees that have evolved from immigrant sunflowers on St Helena Island. From Carlquist [15].

life, may find a whole range of other avian niches empty and available to it. The most spectacular example of this is the radiation of the cardueline finches on the Hawaiian Islands, from an original colonist that probably had the standard finch adaptation for cracking seeds using a stout beak to birds with an array of beak shapes and feeding methods that parallel many of those of the whole range of songbirds (discussed further in this chapter).

Examples of such evolution in isolation are also provided by the Great Lakes of East Africa ('islands' of water in an expanse of land), large enough to provide a great diversity of environments and old enough for these ecological opportunities to have become realized by evolutionary change.

Another tendency is for island species (especially those on small islands) to lose the dispersal mechanisms that originally allowed them to reach their new home. Once on the restricted area of the island, the ability for long-distance dispersal is no longer of value to the species: in fact it is a disadvantage, for the organism or its seeds are now more likely to be blown out to sea. Seeds tend to lose their 'wings' or feathery tufts, and many island insects are wingless – 18 out of the 20 endemic species of beetle on the island of Tristan de Cunha have reduced wings. Flightlessness and fear of flying over water, seen in some island birds, may also be partly for this reason and partly because there are often no predators from which they need to escape. A few of many examples are the kiwi and moa of New Zealand, the elephant birds of Madagascar and the dodo of Mauritius (the last three are extinct, but only because they were killed by humans).

Due to the strong selection pressures, especially on small islands, dispersal traits can disappear remarkably quickly. This is illustrated by a classic study [16], performed by Martin Cody and Jacob Overton, who sampled 240 small islands off the Pacific coast of Canada ranging in size from a few square metres to 1 km². They studied members of the Daisy family (Asteraceae), all of which possess a characteristic two-part dispersal structure consisting of an achene (a tiny seed within a covering) and a pappus (a large ball of fluff connected to the achene). The bigger the pappus and the smaller the achene, the longer the diaspore stays aloft, and the farther it is likely to be carried. Cody and Overton performed an 8-year census that illustrated many of the principle characteristics of island life. First, population turnover was very high due to small population size. Second, the youngest populations (e.g. founders) had the smallest achenes (15% smaller than on the mainland), illustrating that only the 'best' dispersers (those with the lightest seeds) were able to reach these islands. Second, achene size increased with population age, reaching mainland proportions after 8 years due to selection against high dispersers. Likewise, pappus volume decreased with population age, becoming smaller than mainland population after only 6 years. What makes these results so remarkable is that the plants are biennial, meaning that evolutionary change took place in only 1–5 generations.

It is not unusual to find that island species are different in size from their mainland relatives, and this phenomenon has been discussed by the American ecologist Ted Case [17]. Sometimes an island lacks a particular type of predator, because the size of the population of its prey is not large enough to provide a reliable source of food. This may decrease the death rate of the prey species and allow it to grow more rapidly, for it can feed at times and places that were previously dangerous. The predator might also have preferred to catch larger specimens of the prey species. For all these reasons, the average size of island herbivores, such as some rodents, iguanid lizards and tortoises, is often greater than that of their mainland relatives.

Another reason for a change in size in island species is the absence of competitors. For example, if a smaller competitor is absent, then an island species may become smaller to colonize that vacant niche. The converse may be true if its competitor was larger, as in the case of the Komodo dragon (*Varanus komodoensis*), a giant lizard which lives on Komodo Island and nearby Flores Island in the East Indies. These animals have dramatically increased in size to occupy niches which on the mainland are filled by much larger animals.

All these organisms have evolved in islands to fill habitats normally closed to them. But other evolutionary changes frequently found on islands are the direct result of the island environment itself, not of the restricted fauna and flora. We have seen how serious the effect of a small population may be. But the same island will be able to support a larger population of the same animal if the size of each individual is reduced. This evolutionary tendency on islands is shown by the find of fossil pygmy elephants that once lived on islands in both the Mediterranean and the East Indies, and by the discovery of what seems to be a dwarf species of human being, *Homo floresiensis*, only 1 m high and weighing only 25 kg, on Flores Island [18].

The tendency for small island species to become larger and for large island species to become smaller is sometimes called the **island rule**, and relevant data have recently been reviewed and analysed by the New York biogeographer Mark Lomolino [19]. He found that the rule was true for mammals in general, as well as for examples in some birds, snakes and turtles, although another study [20] suggested that it did not apply to the Carnivora. It has also been suggested that the island rule also applies to deep-sea species, perhaps because, like islands, this environment suffers from decreased total food availability [21].

The Hawaiian Islands

As has been seen, there are many aspects of island life that are unique, and many others that differ only in degree from life on the continental landmasses. The result of the action of all these different factors can be seen by examining the flora and fauna of one particular group of islands. The Hawaiian Islands provide an excellent example, for they form an isolated chain, 2650 km long, lying in the middle of the North Pacific, just inside the Tropics (Figure 7.2). Sherwin Carlquist has provided an interesting account of the islands and of their fauna and flora, pointing out the significance of many of the adaptations found there [22]. A collection of papers on Hawaiian biogeography [23] contains many fascinating contributions, and

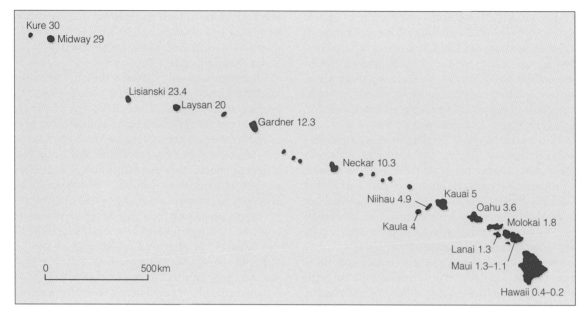

Figure 7.2 The Hawaiian island chain. The figures indicate the age of each island, in millions of years.

others are to be found in a collection of papers on Pacific island biotas in general [24].

The Hawaiian Islands are a part of the most outstanding example of a volcanic island chain generated by a geological hotspot (see Chapter 5). They rise from a sea floor that is 5500 m deep, and that is moving north-westward at 8–9 cm per year. The whole chain comprises 129 volcanoes; 104 of these were originally high enough to reach the surface and form islands (or add to an existing island). The youngest and biggest island, Hawaii, is only 700 000 years old and is made up of six volcanoes, two of which are still active. The most westerly island still visible above the sea, Kure, is nearly 30 million years old, but the Emperor Seamount chain, consisting of other submerged islands and volcanoes, extends farther westward and then northward to near the Kamchatka Peninsula of Siberia. There, the Meiji Seamount is perched on the edge of the Kurile Trench and will eventually disappear back into the depths of the earth.

Meiji was formed by the hotspot 85 million years ago, at the present-day location of Hawaii, so islands and submerged volcanoes have been forming in the central North Pacific Ocean for at least that length of time. However, there have been periods in the past when there were no visible islands, or only low islands that were soon submerged. The most recent of these periods lasted for 18 million years, and it ended with the appearance of Kure Island about 30 million years ago. Kure is therefore the oldest island that was a potential home for colonists that might be ancestral to the biota of the Hawaiian Islands of today. Although most later islands (until the last 5 million years) were small, there continued to be the occasional larger, higher island that might have provided a home for the descendants of the biota of Kure.

The rocks that form the Hawaiian Islands can be accurately dated, and the modern technique of calculating the rates of molecular change in the lineages of animals and plants provides a similar system of dating the times of divergence of different genera and species. As the American workers Hampton Carson and David Clague [25] have shown, it is therefore now possible, in the case of the Hawaiian Islands, to produce a fascinating integration between the gears of those two great engines of our planet – the plate-tectonic engine that here leads to the appearance of new islands, and the evolutionary engine that responds to these opportunities by producing new forms of life. As will be seen in this chapter, this integration strongly suggests that the ancient islands, which appeared between

30 million and 10 million years ago, were in fact the sites of evolution of some of the ancestors of today's Hawaiian animals and plants. Once they had arrived on any particular island, these animals and plants were able to diversify and, often, to spread to the newly formed additions to the chain as the older islands steadily eroded away and disappeared beneath the sea. The high volcanic peaks seem to be islands within islands, for it is difficult for their alpine plants to disperse from one island to another. Instead, they have evolved from the adjacent lower-lying flora independently within each island: 91% of the alpine plants of Hawaii are endemic, a far higher proportion than that of the island flora as a whole (16.5%) or even of the angiosperms alone (20%) [26].

The closest relatives of many of the Hawaiian animals and plants live in the Indo-Malaysian region. For example, of the 1729 species and varieties of Hawaiian seed plants, 40% are of Indo-Malaysian origin but only 18% are of American origin; also, nearly half of the 168 species of Hawaiian ferns have Indo-Malaysian relatives, but only 12% have American affinities. This is not surprising, for the area to the east is almost completely empty, while that to the south and west of the Hawaiian chain contains many islands, which can act as intermediary homes for migrants. Organisms adapted to life in these islands would also be better adapted to life in the Hawaiian Islands than would those from the American mainland. However, about half of the 21 lineages of birds found on the Hawaiian Islands appear to have originated in North America.

Mechanisms of Arrival

The way in which the Hawaiian birds reached the islands is obvious enough. One of the plants which probably came with them is *Bidens*, a member of the Compositae, whose seeds are barbed and readily attach themselves to feathers; about 7% of the Hawaiian non-endemic seed plants probably arrived in this way [26]. The Hawaiian insects, too, arrived by air. Entomologists have used aeroplanes and ships to trail fine nets over the Pacific at different heights and have trapped a variety of insects, most of which, as would be expected, were species with light bodies. Both birds and insects have tended to colonize from the east, the typical direction of arrival of the strongest storms. Although

small, light insects predominate in the Hawaiian Islands (an indication of their airborne arrival), heavier dragonflies, sphinx moths and butterflies are also found there.

Similarly, because ferns have spores that are much smaller and lighter than the seeds of angiosperms, they provide a greater proportion of the flora in the Hawaiian Islands than in the rest of the world. Of the non-endemic seed plants of the Hawaiian Islands, about 7.5% almost certainly arrived carried by the wind, while another 30.5% have small seeds (up to 3 mm in diameter) and may also have arrived in this way. For example, the unusually tiny seeds of the tree *Metrosideros* have allowed it to become widely dispersed through the Pacific Islands. It is a pioneering tree, able to form forests on lowland lava rubble with virtually no soil – a great advantage on a volcanic island. *Metrosideros* shows great variability in different environments, from a large tree in the wet rainforest, a shrub on wind-swept ridges, to as little as 15 cm high in peatlands; it is therefore the dominant tree of the Hawaiian forest. Although these differences are probably at least partially genetically based, the different forms are not distinct species, and intermediates are found where two different types (and habitats) are adjacent to one another.

The carriage of seeds within the digestive system of birds is an important mechanism for the arrival of plants; about 37% of the non-endemic seed plants of the islands (e.g. blueberry and sandalwood) probably arrived in this way. Significantly, many plants that succeeded in reaching the islands are those that, unlike the rest of their families, bear fleshy fruits instead of dry seeds (e.g. the species of mint, lily and nightshade found in Hawaii). An exception to this appears to be *Viola*, but it is significant that the Hawaiian members of this genus are most closely related to those of the Bering region of the northern Pacific, from which some 50 species of bird regularly overwinter in the Islands.

Dispersal by sea accounts for only about 5% of the non-endemic Hawaiian seed plants. As well as the ubiquitous coconut, the islands also contain *Scaevola toccata*; this shrub has white, buoyant fruits and forms dense hedges along the edge of the beach on Kauai Island. Another seaborne migrant is *Erythrina*; most species of this plant genus have buoyant, bean-like seeds. On Hawaii, after its arrival on the beach, *Erythrina* was unusual in adapting to an island environment, and a new endemic species, the coral tree *Erythrina sandwichensis*, has evolved on the island. Unlike those of its ancestors, the seeds of the coral tree do not float – an example of the loss of its dispersal mechanism often characteristic of an island species.

The successful colonists of the Hawaiian Islands are the exceptions; many groups have failed to reach them. There are no truly freshwater fish and no native amphibians, reptiles or mammals (except for one species of bat), while 21 orders of insect are completely absent. As might be expected, most of these are types that seem in general to have very limited powers of dispersal. For example, the Formicidae (ants), which are an important part of the insect fauna in other tropical parts of the world, were originally absent. They have, however, since been introduced by humans, and 57 different species from 24 genera have now established themselves and filled their usual dominant role in the insect fauna. This proves that the obstacle was reaching the islands, not the nature of the Hawaiian environment.

Evolutionary Radiations within the Hawaiian Islands

As ever, the absence of some groups has provided greater opportunities for the successful colonists. Several insect families, such as the crickets, fruitflies and carabid beetles, are represented by an extremely diverse adaptive radiation of species, each radiation derived from only a few original immigrant stocks. For example, the fruitflies, belonging to the closely related genera *Drosophila* and *Scaptomyza*, have undergone an immense radiation in the Hawaiian Islands; of the over 1300 species known worldwide, over 500 have already been described from the Hawaiian Islands, where there are probably another 250–300 species awaiting description. The abundance of species of fruitflies in the islands is probably due partly to the great variations in climate and vegetation to be found there, and also to the periodic isolation of small islands of vegetation by lava flows, each island providing an opportunity for independent evolution of new species. But another major factor has been that the Hawaiian fruitflies, in the absence of the normal inhabitants of the niche,

have been able to use the decaying parts of native plants as a site in which their larvae feed and grow. This change is probably also due to the fact that their normal food of yeast-rich fermenting materials is rare in the Hawaiian Islands.

Molecular studies indicate that the common ancestor of all the drosophilids of the Hawaiian Islands diverged from the Asian mainland type about 30 million years ago; comparison with the ages of the different islands (see Figure 7.2) suggests that this must have happened on Kure Island when that island was young and nearly 900 m high. Similarly, these studies indicate that *Scaptomyza* diverged from *Drosophila* 24 million years ago, which suggests that this event took place on Lisianski, which was at one time an island about 1220 m high. Detailed studies of the chromosome structure of the Hawaiian 'picture-winged' fruitflies have made it possible to reconstruct the sequence of colonizations that must have taken place [27]. As might have been expected, the older islands to the west in general contain species ancestral to those in the younger islands in the east (Figure 7.3). The youngest, Hawaii itself, has 19 species descended from species in older, more westerly islands, but none of its species appears to be ancestral to those in the older islands. (The present-day islands of Maui, Molokai and Lanai together formed a single island until the postglacial rise in sea levels; their *Drosophila* species are therefore treated together as a single fauna.) Within Hawaii itself, molecular studies show that the species on the more southern part of the island, which is up to 200 000 years old, evolved from those on the more northern part, which is 400 000–600 000 years old.

The same phenomenon of a great adaptive radiation has taken place in other groups of animals and plants. In general, therefore, although the islands contain comparatively few different families, each contains an unusual variety of species, nearly all of which are unique to the islands. In fact, out of the whole of the Hawaiian flora, over 90% of the species are endemic to the islands.

There are many other examples of Hawaiian adaptive radiations, but three are of particular interest: the silverswords and lobeliads among the plants [28,29], and the honey-creepers among the birds [30]. The silverswords are descended from the tarweeds of southwest North America (members

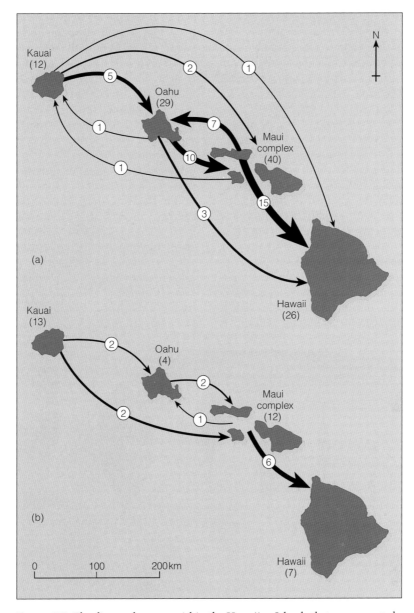

Figure 7.3 The dispersal events within the Hawaiian Islands that are suggested by the interrelationships of the species of 'picture-winged' *Drosophila* flies (above) and of silversword plants (below). The width of the arrows is proportional to the number of dispersal events implied, and the number of species in each island is shown in parentheses. From Carr *et al.* [27].

of the family Compositae, which includes sunflowers and daisies), and probably arrived on the islands as sticky seeds attached to the feathers of birds. They have produced only three genera in the islands (*Dubautia*, *Argyroxiphium* and *Wilkesia*),

but these have colonized a variety of habitats. For example, on the bare cinders and lava of the 3050 m high peak of Mt Haleakala on the island of Maui, two of the few plant species that can survive are silverswords. *Dubautia menziesii* is adapted to this

arid environment by its tall stem and stubby suc-
culent leaves, and *Argyroxiphium sandwichense* is
covered by silvery hairs that reflect the light and
heat. A few hundred metres below the bare volcanic peaks, conditions are at the other extreme,
because most of the rain falls at heights of from
900 to 1800 m; these regions receive 250–750 cm
of rain per year. The upper regions of 1770 m high
Mt Puu Kukiu on Maui are covered by mire in
which thrives another silversword, *Argyroxiphium
caliginii*. On the island of Kauai, the heavy rainfall
has led to the development of dense rainforest, in
which *Dubautia* has evolved a tree-like species,
Dubautia knudsenii, with a trunk 0.3 m thick, and
large leaves to gather the maximum of sunlight in
the dim forest. Kauai bears another silversword
which shows the tendency for island plants to
become trees. In the drier parts of this island grows
Wilkesia gymnoxiphium, with a long stem which
carries it above the shrubs that compete with it
for light and living space. This species also shows

another example of the loss of the dispersal mechanism that first brought the ancestral stock to the
island: the seeds of *Wilkesia* are heavy and lack the
fluffy parachutes usually found among the Compositae. The pattern of dispersal of the 28 species
in the three silversword genera within the Hawaiian Islands is very similar to that of the drosophilid
flies (Figure 7.3) [27]. Molecular evidence suggests
that these three genera had a common ancestor
that arrived on Kauai.

Lobeliads are found in all parts of the world, but
they have undergone an unusual adaptive radiation in the Hawaiian Islands, because their normal
competitors, the orchids, are rare. The Hawaiian
lobeliads include 150 endemic species and varieties, making up six or seven endemic genera.
Over 60 species of the endemic genus *Cyanea* are
known, showing an incredible diversity of leaf
form (Figure 7.4) [22]. The plants range from the
tree *Cyanea leptostegia*, which is 9 m tall (similar in appearance to the tarweed *Wilkesia*), to

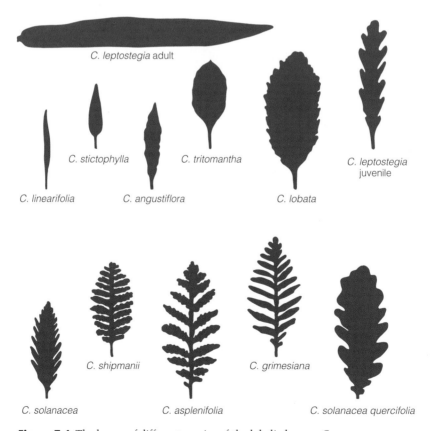

Figure 7.4 The leaves of different species of the lobeliad genus *Cyanea*.

soft-stemmed *Cyanea atra* at only 0.9 m tall. The species of another genus, *Clermontia*, are less varied in overall size but are very varied in the size, shape and colour of their flowers. These are mainly tubular and brightly coloured, a type of flower that is often associated with pollination by birds. On isolated islands such as Hawaii, the adaptation of larger flowers to bird pollination may be because the large insects that would normally pollinate such flowers on the mainland are absent. It is no coincidence that the adaptive radiation of the Hawaiian lobeliads has been accompanied by the adaptive radiation of a nectar-eating type of bird, the honey-creepers [30].

The ancestors of these birds were probably finch-like immigrants from Asia [31] which fed on insects and nectar. From the original immigrants, adaptive radiation has produced 11 endemic genera comprising the endemic family Drepanididae (Figure 7.5), but another circa 30 extinct species are known. Many of the genera, such as *Himatione, Vestiaria, Palmeria, Drepanidis*, many species of *Loxops* and one species of *Hemignathus* (*Hemignathus obscurus*), are still nectar eaters, feeding from the flowers of the tree *Metrosideros* and the lobeliad *Clermontia*. Since insects, too, are attracted to the nectar, it is not surprising to find that many nectar-eating birds are also insect-eaters, and it is a short step from this to a diet of insects alone. *Hemignathus wilsoni* uses its mandible, which is slightly shorter than the upper half of its bill, to probe into crevices

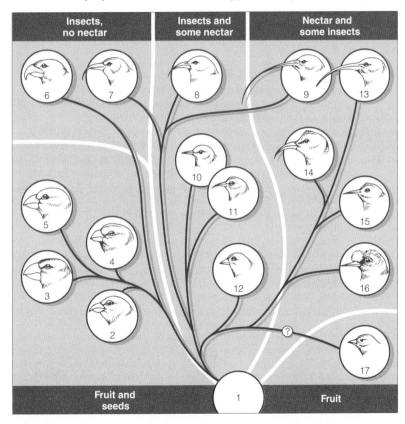

Figure 7.5 The evolution of dietary adaptations in the beaks of Hawaiian honey-creepers. (1) Unknown finch-like colonist from Asia; (2) *Psittirostra psittacea*; (3) *Chloridops kona*; (4) *Loxioides bailleui*; (5) *Telespyza cantans*; (6) *Pseudonestor xanthophrys*; (7–11) *Hemignathus munroi, H. lucidus, H. obscurus, H. parvus,* and *H. virens*; (12) *Loxops coccineus*; (13) *Drepanis pacifica*; (14) *Vestiaria coccinea*; (15) *Himatione sanguinea*; (16) *Palmeria dolei*; (17) *Ciridops anna*. Taxonomy adapted from Pratt *et al.* [73].

in bark for insects, and *Pseudonestor xantho-phrys* uses its heavier bill to rip open twigs and branches in search of insects. Other types have heavy, powerful beaks, which they use for cracking open seeds, nuts or beans. The light bill of the recently extinct *Ciridops* was used for eating the soft flesh of the fruits of the Hawaiian palm *Pritchardia*.

It is interesting to speculate why these Hawaiian birds should have radiated into so many more species than did Darwin's finches on the Galápagos Islands (cf. Chapter 6). Possible reasons include the facts that the Hawaiian Islands are older, contain a far greater variety of environments and lie farther apart than the islands in the Galápagos archipelago.

Studies of the Hawaiian avifauna also show the unreliability of the modern biota as a basis for estimates of rates of biotic change or of the relationship between island area and the number of species. It has long been known that about a dozen Hawaiian bird species became extinct after the arrival of Europeans and the animals that they introduced. But studies by the American biologists Storrs Olson and Helen James [32,33] revealed at least 50 now-extinct species of Hawaiian bird – more than the entire avifauna of the islands today. These included flightless types of ibis, rail and goose-like ducks, six hawks and goshawk-like owls, and nearly two dozen species of drepanidid finches, mostly of the insectivorous type. Most of these species were still alive when the Polynesians arrived in the Islands in about AD 1500, but had become extinct before Europeans arrived 300 years later. Similar evidence for bird extinctions has been reported from many other Pacific islands, and it is clear that the patterns of distribution, endemism and numbers of species on individual islands that are seen today are wholly unreliable as a basis for generalizations about bird populations on islands.

Living as they do on fruits, seeds, nectar and insects, it is not surprising that none of the drepanidids shows the island fauna characteristic of loss of flight. However, both on Hawaii and on Laysan to the west, some genera of the Rallidae (rails) have become flightless. (This is particularly common in this particular family of birds; palaeontological work has shown that nearly every Pacific island once had at least one species of flightless rail; and, before extinction due to human activity, it is pos-sible that this was true of more than 800 of the islands, compared with the 27 surviving flightless species!) The phenomenon of flightlessness is also common in Hawaiian insects: of the endemic species of carabid beetle, 184 are flightless and only 20 are fully winged. The Neuroptera or lacewings are another example – their wings, usually large and translucent, are reduced in size in some species, while in other species they have become thickened and spiny.

Now that the island scene in general has been set in the above sections, it is time to turn to the challenging task of trying to find general rules that might underlie the immense diversity of islands and their biota. This is comprehensively reviewed by Whittaker and Fernández-Palacios (see Further Reading). The realization that dispersal, rather than vicariance, is the normal source for the biota of oceanic islands (see Chapter 7), together with the demonstration by molecular studies of the extent of allopatric speciation within them, have given a renewed impetus to the analytical study of island biotas [34,35].

Integrating the Data: The Theory of Island Biogeography

One of the most obvious characteristics of the biota of islands is that it is strongly affected by the degree of isolation of the island. However diverse the habitats that it offers, the variety of the island life depends, in the short term, very much upon the rate at which colonizing animals and plants arrive. This, in turn, depends largely upon how far the island is from the source of its colonizers, and upon the richness of that source. If the source is close and if its biota is rich, then the island in its turn will have a richer biota than another, similar island which is more isolated or which depends upon a source with a more restricted variety of animals and plants. Each sea barrier further reduces the biota of the next island, which in turn becomes a poorer source for the next. For example, the data provided by Van Bal-gooy [36] make it possible to map the diversity of conifer and flowering plant genera in the Pacific island groups (Figure 7.6). This clearly shows that diversity is much lower in the more isolated island groups of the central and eastern Pacific.

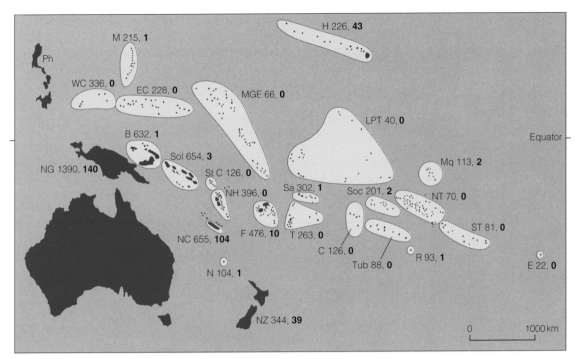

Figure 7.6 The distribution of conifers and flowering plants in the Pacific Islands. The first number beside each island group is the total number of genera found there; the second is the number of endemic genera found there. B, Bismarck Archipelago; C, Cook Islands; E, Easter Island; EC, East Carolines; F, Fiji Islands; H, Hawaiian Islands; LPT, Line, Phoenix and Tokelau Island groups; M, Marianas; MGE, Marshall, Gilbert and Ellis Islands; Mq, Marquesas; N, Norfolk Island; NC, New Caledonia; NG, New Guinea; NH, New Hebrides; NT, Northern Tuamotu Islands; NZ, New Zealand; Ph, Philippines; R, Rapa Island; Sa, Samoa group; Soc, Society Islands; Sol, Solomon Islands; ST, Southern Tuamotu Islands; StC, Santa Cruz Islands; T, Tonga group; Tub, Tubai group; WC, West Carolines. Data from Van Balgooy [36].

However, in several of the more westerly island groups, the diversity is much higher than their geographical position alone would lead one to predict. A logarithmic graph of the relationship between the number of genera and the area of the islands (Figure 7.7) clearly shows that, in most cases, the generic diversity is simply dependent on island area. (The fact that some islands therefore have more genera than do the islands from which most of their flora is derived suggests that other genera were once also present in these latter islands, but have since become extinct there.) Nearly all of the more isolated island groups (shown as triangles in Figure 7.7) have, as would be expected, a much lower diversity than would be predicted from their areas alone. The number of land and freshwater bird species in each island shows a similar relationship to island area (Table 7.1), but this is probably due, not to island area directly, but instead to the resulting higher floral diversity.

As we have seen above, the number of species found on an island depends on a number of factors: not only on its area and topography, its diversity of habitats, its accessibility from the source of its colonists, and the richness of that source, but also on the equilibrium between the rate of colonization by new species and the rate of extinction of existing species. Many individual observations and analyses of such phenomena have been made over the past 160 years. As explained in Chapter 1, a quantitative theory to explain these relationships was put forward in 1967 by the American ecologists Robert MacArthur and Edward Wilson in their book, *The Theory of Island Biogeography* [14]. This

Table 7.1 The relationships between island area and the diversity of bird genera and non-endemic flowering plant genera in some Pacific islands. Data from Van Balgooy [36], Mayr [71] and MacArthur and Wilson [72].

	Area (km^2)	Angiosperm genera	Bird genera
Solomon Islands	40 000	654	126
New Caledonia	22 000	655	64
Fiji Islands	18 500	476	54
New Hebrides	15 000	396	59
Samoa group	3100	302	33
Society Islands	1700	201	17
Tonga group	1000	263	18
Cook Islands	250	126	10

made two main suggestions: that the changing and interrelated rates of colonization and immigration would eventually lead to an equilibrium between these two processes, and that there is a strong correlation between the area of an island and the number of species it contains.

To explain the relationship between rates of colonization and of immigration, MacArthur and Wilson took the case of an island that is newly available for colonization. They pointed out that the rate of colonization will at first be high, because the island will be reached quickly by those species that

are adept at dispersal, and because these will all be new to the island. As time passes, immigrants will increasingly belong to species that have already colonized the island, so that the rate of appearance of new species will drop (Figure 7.8). The rate of immigration will also be affected by the position of the island, for it will be higher for islands that are

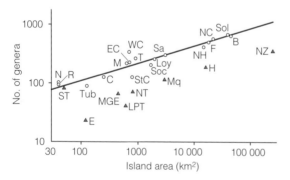

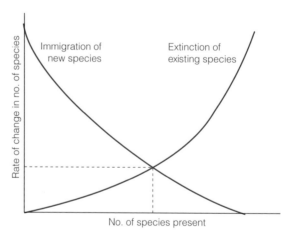

Figure 7.7 The relationship between island area and the diversity of conifer and flowering plant genera in the Pacific Islands. The more isolated islands are indicated by triangles. The data from the other islands lie very close to a straight line (the regression coefficient), suggesting that generic diversity in these islands is almost wholly controlled by island area – the correlation coefficient is 0.94, indicating a very high degree of correlation. For abbreviations, see the legend of Figure 7.6, plus: Loy, Loyalty Islands. Data from Van Balgooy [36].

Figure 7.8 Equilibrium model of the biota of an island. The curve of the rate of immigration of new species and the curve of the rate of extinction of species already on the islands intersect at an equilibrium point. The interrupted line drawn vertically from this point indicates the number of species that will then be present on the island, while that drawn horizontally indicates the rate of change (or turnover rate) of species in the biota when it is at equilibrium. Adapted from MacArthur and Wilson [72].

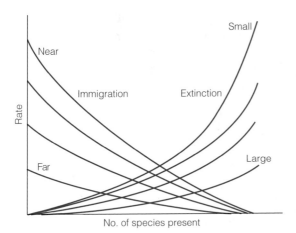

Figure 7.9 The interrelationship between isolation and area in determining the equilibrium point of biotic diversity. Increasing distance of the island from its source of colonists lowers the rate of immigration (left). Increasing area lowers the rate of extinction (right). Adapted from MacArthur and Wilson [72].

close to the source of their colonists, and lower for those that lie farther away (Figure 7.9).

The rate of extinction, in contrast, will start at a low level but gradually rise. This is partly because, since every species runs the risk of extinction, the more that have arrived, the more species there are at risk. In addition, as more species arrive, the average population size of each will diminish as competition increases – and a smaller population is at greater risk of extinction than a larger population.

The **Theory of Island Biogeography** (**TIB**) suggests that the two curves that represent these two conflicting processes (immigration and extinction) will intersect at a point where the rates of immigration and extinction are equal, known as the **turnover rate**, so that the number of species is constant at this equilibrium number.

At first, the few species present can occupy a greater variety of ecological niches than would be possible on the mainland, where they are competing with many other species. For example, in the comparison mentioned (in the 'Dying There: Problems of Survival' section) between the Panama mainland and Puercos Island, the smaller number of bird species in the island were able, because of reduced competition, to be far more abundant: there were 1.35 pairs per species per hectare in Puercos Island, compared with only 0.33 and 0.28,

respectively, for the two mainland areas [9]. This effect of release from competition was especially noticeable in the antshrike (*Thamnophilus doliatus*). On the mainland, where it competed with over 20 other species of ant-eating bird, there were only eight pairs of antshrike per 40 ha; on Puercos Island, where there was only one such competitor, there were 112 pairs per 40 ha.

This effect of release from competition will be reversed if the island is later colonized by a new species whose diet overlaps with that of one of the earlier immigrants. This may result in the extinction of one of them. This may be because they compete too closely with one another, so that they cannot coexist. Alternatively, it may be because the competition between them leads to a reduction in the population size of each – because each species has to become more specialized in its ecological requirements. This in turn renders the species more vulnerable to extinction. In either case, the rate of extinction upon the island will have increased.

In all these theoretical cases, the number of species present in the biota will obviously be the result of the balance between the rate of immigration and the rate of extinction. MacArthur and Wilson suggested that the biota will eventually reach an equilibrium, at which the rates of immigration and of extinction are approximately equal, and that this equilibrium level is comparatively stable.

Many immigrant island species, now with less competition than they had on the mainland, will be able to expand into new habitats. But if other competing new colonists now arrive, they may later find their distribution and evolutionary expansion reduced. This concept of alternating expansion and construction gave rise to the theory of the taxon cycle (Box 7.1).

The TIB was widely welcomed, for it gave biogeographers a theoretical background with which to compare their own individual results, and therefore encouraged a more structured and less ad hoc approach to biogeographical studies. Its methodology was also extended into types of isolation other than that of dry land surrounded by water, such as mountain peaks (Figure 7.10), caves and individual plants, and the theory was even extended to evolutionary time, with individual host-plant species being regarded as islands as far as their 'immigrant' insect fauna was concerned.

In addition to his attempt to establish general principles in the Theory of Island Biogeography, Edward Wilson had earlier [67] suggested that the distribution and ranges of individual species in island communities went through stages of expansion and contraction, which he named the **taxon cycle**. The methodology of this theory is to establish categories for island species that have different ecological and distributional characteristics, and then infer that these differences are the result of their having colonized the island at different times and interacted with one another.

The concept is best understood by visualizing a time sequence, imagining the history of a species from its first dispersal from the mainland and arrival in an island. The species is likely to arrive in one of the ecologically marginal habitats, such as shore communities, grassland or lowland forest. Here, they were at first generalists, with a wide and continuous range of distribution. However, taking advantage of the lack of its normal competitors, predators or parasites, the species may extend its ecological range into other environments, such as inland forests or montane rainforest. The species is in the expansion phase of the taxon cycle.

The next event is the arrival of other species, whose competition expels the original species from its original marginal habitat, so that its distribution is restricted to the more specialized, inland habitats. It is now in a contraction phase of the taxon cycle, with a patchier, less continuous pattern of distribution. The original species will have become genetically adapted to life in these habitats, and so may have become recognizable as a new, endemic species. This process may take place independently on more than one island but, if so, each of these new species will be most closely related to the original colonist from the mainland, rather than to each other.

The taxon cycle theory has attracted a great deal of criticism. At the most fundamental, theoretical level, the problem is that the diversity of patterns of distribution to be found within and between island faunas is so varied that it is easy to find examples that will fit within almost any set of categorizations. Furthermore, it is impossible to prove that the suggested linkages between distribution, adaptation and relationship are cause and effect, or that they form a sequence in time. However, the American biogeographers Robert Ricklefs and Eldredge Bermingham, in a review of the taxon cycle concept [68], have shown that molecular phylogenetic analyses of the times of divergence of 20 lineages of West Indian birds conform to the assumptions of the hypothesis, and phylogenetic analyses of lineages of West Indian anolid lizards [69] have similarly supported it. More recently, Danish biologist Knud Jønsson and his colleagues used molecular methods to analyse the evolution of the avian genus *Pachycephala* on Indo-Pacific islands [70]. They demonstrate that relict species persist on the largest and highest islands, whereas recent archipelago expansions resulted in colonization of all islands in a region. Moreover, earlier colonists tended to be found in the interior and highest parts of an island and rarely mixed with later colonists. These studies strongly suggest that, despite much variation, many island taxa continuously pass through phases of expansions and contractions.

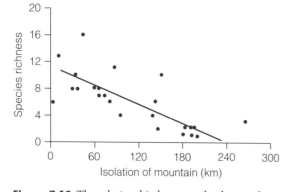

Figure 7.10 The relationship between the degree of isolation of mountain peaks in the southwestern United States and the number of species of mammal found in each. From Lomolino *et al.* [74].

Modifying the Theory

The story of how the TIB came to dominate this whole area of research has been outlined in Chapter 1 and will not be repeated here. Even though this delayed its critical evaluation, its shortcomings were eventually realized. Many studies that had been extensively quoted as supporting the TIB were really far too imprecise, as pointed out by the American ecologist Dan Simberloff [37] and in a review by the British biologist Francis Gilbert [38], while the statistical procedures of many earlier studies were criticized by the American ecologists Edward Connor and Earl McCoy [39]. It was also pointed out that the TIB treats species as simple numerical units of equal value to one another, so that their possible biological inter-actions, such as competitive or co-evolutionary effects, are therefore ignored. However, this was an essential part of the methodology of the TIB, for it was expressly designed to try to rise above the inevitable complexity that results from the anal-ysis of individual species in order to see whether this might reveal general rules against which indi-vidual species or circumstances might be judged. The general point that arises here is the difficulty of transforming the essentially gradual nature of many biological phenomena into the individual points of defined, quantified data that mathemati-cal treatments require.

There has been considerable criticism of work that had seemed to support the TIB's prediction that the number of species will come to, and remain at, an equilibrium, as long as the envi-ronment remains constant. Jared Diamond [40] had compared the numbers of bird species found in the Californian Channel Islands in a 1968 sur-vey with those recorded in a 1917 review, and he concluded that there had been an equilibrium. However, Lynch and Johnson pointed out [41] that fundamental changes in the environment of the Channel Islands had taken place between 1917 and 1968, rendering the example worthless. Lynch and Johnson found similar flaws in other studies of bird faunas. Similarly, Simberloff [42] analysed the records of the bird faunas of two islands and three inland areas over 26–33 years, and found that none of them showed any evidence of regulation towards an equilibrium.

Apart from these specific studies, can we in any case be sure that any biota that we see today is in a state of stable equilibrium? The level of that equi-librium will alter if the environment changes – and the environment has changed a very great deal over both the comparatively distant past and the recent past. These changes include alterations in climate and sea level that may result in changes in island area or in the union or subdivision of islands. We ourselves have caused the extinction of endemic island species and introduced new species to them. It has been estimated that human activities over the last 30 000 years have led to the extinction of thousands of populations of birds and as many as 2000 bird species [43]. Under these circumstances, it is difficult to be confident that any island biotas are at stable equilibrium. If so, they cannot provide

a database for estimates of the actual numerical equilibrium level in any existing situation, nor for any prediction of where such an equilibrium might emerge in the future. In any case, the lag-time of many shot, the biological characteristics still being in the process of adjusting to an earlier event of climatic or sea-level change. As Shafer [44] has commented, no aspect of the concept of an equilib-rium level in species numbers can now be consid-ered as established, and this would in any case be of limited value in an environment that is subject to both cyclical and occasional change.

When considering the rate of turnover of species, the longevity of the dominant species is also impor-tant. Case and Cody [45] have pointed out that the life spans of forest trees are so great that turnover is inevitably very slow. This is shown by the fact that the species structure of the forest at Angkor in Cambodia, which started to grow when the ancient Khmer capital was abandoned 560 years ago, has still not become identical to that of the surround-ing older forest.

Other research has been directed at trying to mod-ify and improve the TIB. MacArthur and Wilson's first point, that larger islands contain more species, has met general acceptance. But there has been considerable research to try to identify the factors that produce this effect, and their relative impor-tance, and to examine the impact of other factors that the TIB ignores, such as climate and evolu-tionary history. For example, the American bioge-ographer David Wright suggested that the funda-mental aspect of enlarging the island area is that it increases the amount of energy that falls upon the island – a concept known as **species-energy theory** [46]. Wright proposed that islands are essentially energy collectors, but that the amount of energy that an island can collect will also vary according to its climate: those with a warm, wet climate are more productive and therefore have more species than those that are cool and dry. Wright showed that the species richness of flowering plants on 24 islands world-wide, and of land birds on 28 islands of the East Indies, were better explained by his the-ory than by the TIB. More recently, the Canadian biogeographers Attila Kalmar and David Currie have extended Wright's work to find out whether the degree of isolation of the islands affected their species richness [47]. Using data on non-marine birds from 346 islands (Figure 7.11), they found

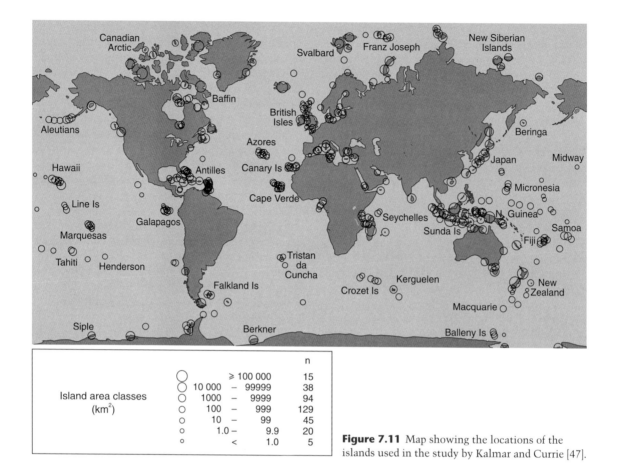

Island area classes (km^2)			n
		≥ 100 000	15
	10 000	– 99999	38
	1000	– 9999	94
	100	– 999	129
	10	– 99	45
	1.0	– 9.9	20
		< 1.0	5

Figure 7.11 Map showing the locations of the islands used in the study by Kalmar and Currie [47].

that species richness of an island is correlated to its distance from the nearest island, but far more strongly correlated to its distance from the nearest mainland. Impressively, they showed that a combination of average annual temperature, total annual rainfall and distance from the nearest continent together explained 87.5% of the bird species richness of these islands. However, in contrast to the TIB, their analysis showed that the slope of the species-richness versus island-area curve depends upon climate, not upon isolation.

The General Dynamic Model for Oceanic Island Biogeography

Islands such as the Hawaiian chain, which result from the activities of an oceanic hotspot, have their own cycle of birth, growth, subsidence and disappearance. Professor of Biogeography at Oxford University Robert Whittaker and his colleagues have recently combined the patterns of this geological life history with some aspects of the TIB to produce a new **general dynamic model** (GDM) that predicts the patterns of biodiversity, endemism and diversification through the lifetime of such an island [48,49].

Whittaker and colleagues provide a graph that illustrates the general pattern of the changes in the area, altitude and topographical complexity of a hotspot island. Its area and altitude at first steadily increase, but then these begin to be reduced by erosion, which, together with subsidence of the cooling ocean floor as the island moves away from the centre of the hotspot, and major landslips resulting from the erosion of the steep volcanic slopes, will also lead to the eventual disappearance of the island below the sea (Figure 7.12a). Erosion also produces a more complex topography – although this complexity reaches a maximum a little after the time of maximum elevation. This in turn increases the habitat diversity and therefore the number of vacant ecological niches on the island (Figure 7.12b), and so also increases the potential carrying capacity of the island (K in Figure 7.12c). These vacant niches may be filled either by new colonists or by the radiation of existing colonists, which together increase the extent to which the potential species' carrying capacity is realized (R in Figure 7.12c). While the gap between these two figures remains high, the ecological isolation between existing species will be high, and opportunities for the appearance of such new species will be correspondingly high (Figure 7.12b). As these niches are filled (R gradually coming closer to K)

and competition between the species increases, the rate of appearance of new species will decrease (S in Figure 7.12c).

To begin with, nearly all the vacant niches will be filled by immigrants from neighbouring islands. But as the habitat diversity and vacant niche space increase, more and more of these niches are likely to be filled by the evolution of new species from existing immigrants. Over time, an increasing proportion of the biota will be single-island endemics (SIEs) resulting from such within-island evolution and the fact that these clades will themselves become increasingly species-rich. Although later the number of SIEs is likely to be reduced by, for example, their dispersal to other islands or their extinction as competition increases, Whittaker and his co-workers [48] point out that the number

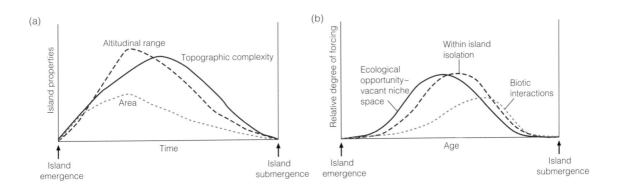

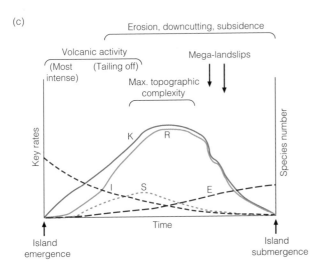

Figure 7.12 (a) The geology of the GDM: how the area, height and topographical complexity of an island may change over its lifetime.
(b) The biology of the GDM: how the number of vacant ecological niches, within-island speciation and competition may change over its lifetime.
(c) Integrating the geology and the biology of the island's lifetime. The rise in the carrying capacity of the island (K) is followed by an increase in the speciation rate (S) and, after a delay, by a consequent rise in its realized species richness, R. Also shown are the rates of immigration (I) and extinction (E). The kinks in the K and R curves represent mega-landslips, which would lead to an increase in the rate of extinction. Adapted from Whittaker et al. [48].

of these SIEs should vary in a predictable manner through the lifetime of the island. The chain of islands resulting from a hotspot, similar in their location and the nature of the animals and plants available, should provide a particularly suitable test of this. Accordingly, the authors use 10 sets of data for organisms from the Hawaiian, Canary, Galápagos, Azores and Marquesas chains to test their model, and find that they do support it. They comment that their model should in principle apply to other island archipelagos, such as island arcs and those that were affected by Pleistocene changes in climate and sea level.

The GDM model has now been tested in various archipelagos with various groups of organisms [e.g. 50], but its importance is probably less to do with precise quantitative predictions and more to do with an inclusive conceptual framework for understanding the evolutionary biogeography of oceanic islands.

Nestedness

Within-archipelago species often show distinct patterns of arrangement, with smaller islands containing a subset of the species present on the largest island. This pattern, known as *nestedness*, is driven by the interaction between the physical differences between different sized islands and the familiar processes of immigration, extinction and speciation. It is often the case that smaller islands only contain some of the habitats available on the largest island and therefore cannot support many of the species present on the larger islands, even if they are able to reach it. Moreover, extinction is higher on smaller islands, and they therefore lose some of the species that are retained on the largest island. Finally, the greater number of habitats (niches) available on the largest island means that there are more opportunities for speciation, creating endemic species that are not found on the smaller islands. Of course, speciation can also occur on smaller islands, reducing nestedness.

Another factor that can reduce nestedness is when a small island contains a unique habitat. In this case, species that arrive there may evolve into new (endemic) species that are no longer able to survive on other islands in the archipelago. Moreover, isolation can reduce nestedness by

making speciation more likely (by reducing gene flow between islands) and by curtailing the rate of exchange of species between islands. It is also worth noting that very small islands typically show an extreme version of nestedness because their area can only sustain viable populations of a tiny number of very small species (e.g. some insects). Indeed, very small islands don't even conform to the species–area relationship – there appears to be a size threshold after which the diversity of species drops dramatically, known as the **small island effect** [51].

As with the TIB, one of the most important applications of nestedness research has been to the conservation of habitat fragments (e.g. forest fragments in a 'sea' of agricultural land). In this case, nestedness is mainly caused by the rapid extirpation of species from smaller fragments. Clearly, the optimal group of fragments to conserve biodiversity in a highly nested system will be different from an equivalent system with a low degree of nestedness. From this perspective, the development of metrics to precisely capture the degree of nestedness is essential. Various methods have been proposed [52], the commonest of which is *nestedness temperature* [53], a measure of disorder or unexpectedness in the species presence–absence matrix. A *nestedness temperature calculator* has been developed by Wirt Atmar and Bruce Patterson [54] which is able to calculate how much individual islands (or fragments) deviate from the overall nestedness of the entire system, allowing scientists to easily identify outlying islands that may be of special interest to conservation.

Living Together: Incidence and Assembly Rules

Understanding how biological communities form is one of the main objectives in ecology. However, it is difficult to study for at least two reasons. First, assembly takes place over long time periods, making direct study almost impossible. Second, community assembly is a complex process that involves interactions at different scales of organization, space and time. Once again, islands have proven to be invaluable to overcome these limitations due to their discrete and replicated nature and because new islands are essentially 'clean sheets' upon which the processes of immigration, establishment, adaptation, speciation and local extinction form and re-form communities [55].

Building on the work of MacArthur and Wilson, Jared Diamond took the idea of islands as 'clean sheets' one stage further, elaborating a set of assembly 'rules' that determine the composition of communities on oceanic islands. Diamond derived his rules from his work on Pacific island birds. Specifically, he documented the occurrence, or **incidence**, of 513 species on thousands of islands near New Guinea (Figure 7.13) [56]. He found that, on the one hand, some species (which he called **sedentary species**) are present only on the largest, most species-rich islands. On the other hand, a few others (which he called **supertramps**) are absent from these and are usually found in the smaller or most remote and species-poor islands. In between these two extremes, Diamond arbitrarily defined four other categories of lesser tramps. The data suggested that the combinations of bird species to be found on the different islands were not random but depended on many factors, including the size of the island, number of species, availability of suitable habitats and extent of the required habitat in area or duration in time.

Analysing the different assemblages of species found in the islands, Diamond drew up a number of generalizations (**assembly rules**). He found that some pairs of species were never found together, apparently because they competed directly with one another, so that the first to arrive was able to exclude the other. This appears to be the case in the colonization of central Pacific islands by wind-dispersed spiders [57]. Some combinations therefore appear to be 'permissible', while others are 'forbidden'. Other such pairs can coexist, but only if they are part of a larger assembly. But precisely what combination of species is permissible varied according to the size and species richness of the island. The data also suggested that potential invaders were unable to colonize islands if their presence there would produce a 'forbidden' combination (presumably because the species that were already there did not leave a suitable niche vacant for the invader). However, the invader might be able to enter such a combination on a larger or more species-rich island.

Over the last 30 years, Diamond's suggestions have stimulated a wide range of criticism and support. Perhaps the greatest problem was the sheer diversity of factors that Diamond invokes in his incidence functions and assembly rules (distribution versus island area, species richness, pattern of habitats, history of changes in climate, area, connection with the mainland, human intervention, and interaction between the species). It could therefore fairly be claimed that some explanation could

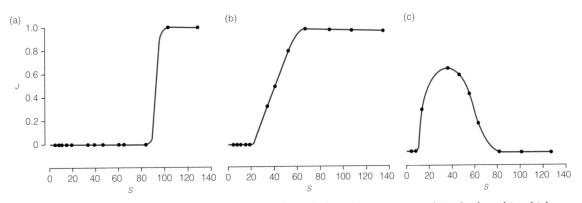

Figure 7.13 Examples of Diamond's incidence rules. Each graph shows the percentage of 50 islands within which each species is found (*J*), plotted against the number of bird species found on each island (*S*). Each point represents grouped data for between 3 and 13 islands, except that the two largest values each represent only one island. (a) The distribution of a sedentary species, the cuckoo *Centropus violaceus*. (b) The distribution of an intermediate tramp, the pigeon *Ptilinopus superbus*. (c) The distribution of a supertramp, the pigeon *Macropygia mckinlayi*. Adapted from Diamond [56].

be found for any island biota, but that one could not rule out the possibility that it was merely the result of chance (the **null hypothesis**), an approach supported by a study of woody flowering plants in 27 islands west of Vancouver Island, Canada [58]. However, as Robert Colwell and David Winkler have shown [59], it is in fact extremely difficult to design a null model that does not itself suffer from serious biases. It is equally difficult to use a real-world mainland biota as a comparison, because its composition has been conditioned by competition within a much more varied biota. Given the variety and complexity of island biotas, areas and history, it is unlikely that any single theory will explain them all, and it is pragmatic to retain a variety of approaches and to see which is the most helpful in each case.

Another line of research has been to try to identify what characteristics are found in earlier or later colonists. The earliest are found to be **r-selected**: highly mobile, capable of rapid growth to early maturity and therefore with a high potential rate of population increase. They contrast with the later arrivals, known as **K-selected**: slower at dispersing and reproducing, but with a greater ability to sustain their population when this is getting close to the carrying capacity of the island.

Although in the very early stages of development, a new approach to biogeography has the potential to provide more quantitative answers to questions of coexistence, competition and 'permissible' levels of morphological similarity. The nascent sub-discipline of *functional biogeography* seeks to understand the geographical distribution of organismal trait diversity across organizational levels [60]. Robert Whittaker and his colleagues [61] recently applied such an approach to spiders and beetles of the Azores archipelago in an attempt to understand how the suites of traits possessed by native and exotic species may affect the ability of the latter to invade an island. They found that functional diversity of both taxa increases with species richness, which in turn scales with island area – strongly suggesting that the island is not 'saturated' for either of these groups. Interestingly, exotic spiders added novel 'trait space' to a greater degree than exotic beetles, probably due to a greater historical loss of spider competitors leaving more empty niches that the colonizing species could fill.

Building an Ecosystem: The History of Rakata

In most cases, we have little knowledge of the history that lies behind the complex assemblage of animals and plants that inhabit a particular island. We may attempt to compare different islands, and to place their biota in a series, or several series, that might represent an historical process, but such an enterprise is fraught with the difficulties of subjective interpretation. We are on surer ground only where the history of the biota of a single island has been documented over an extended period of time – and we are now fortunate in having such a documentation for one island, Rakata, whose biota is in the process of being reassembled after total destruction. It has been studied since 1979 by Robert Whittaker and his associates and since 1983 by the Australian zoologist Ian Thornton and his co-workers. Much of the information in this section is taken from Thornton's enjoyable and stimulating book *Krakatau* [62], plus other data from Whittaker *et al.* [63]. The data accumulated (and still being accumulated) by these research programmes allow us to analyse the sequence of colonization of the island, and how different methods of colonization contribute to sequence. Another valuable aspect of these studies, as we shall see, is that it has been possible to extend them to other neighbouring new islands, giving us a rare opportunity to make comparative studies.

Rakata lies in the East Indies, between the major islands of Java (40 km away) and Sumatra (35 km away), which act as the main sources of its colonists (Figure 7.14). With an area of 17 km^2 and up to approximately 735 m high, Rakata is the largest remaining fragment of the island of Krakatau, which was destroyed by an enormous volcanic explosion in 1883. Two other islands, Sertung (13 km^2, 182 m) and Panjang (3 km^2, 147 m), are fragments of an older, larger version of Krakatau, and a new island, Anak Krakatau, appeared in 1930. All life on the islands was extinguished by the eruption, which covered them with a layer of hot ash 60–80 m deep on average and up to 150 m deep in places. Surveys of the biota of Rakata were made intermittently from 1886, with a gap between 1934 and 1978 (apart from a little work in 1951), and intensive work has been carried out since the centenary of the eruption, in 1983. (In the following

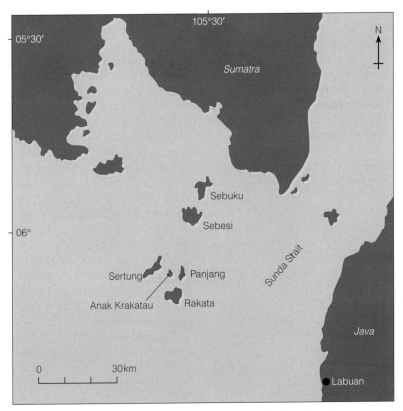

Figure 7.14 Location of Rakata and neighbouring islands. Adapted from Whittaker *et al.* [75].

account, the dates of these surveys are indicated as the length of time that had elapsed since the 1883 eruption; thus 1908 = E + 25. See also the scale at the base of Figure 7.15.) These surveys show that the patterns of colonization and extinction are not smooth, but are heavily influenced by the times of emergence of new ecosystems, and by the linkage between plants and animals due to food requirements or mechanisms of dispersal.

The Coastal Environment

The development of the biota of Rakata is best understood by analysing the beach and near-beach environment separately from the inland environment. There are two reasons for this. First, the beach environment is largely unchanged by the establishment of living organisms there, and is also itself ceaselessly being destroyed and re-created as ocean currents and storms erode one part of the

beach and redeposit its materials elsewhere. This causes frequent local extinction of the biota and simultaneous recolonization elsewhere. Second, these environments are primarily colonized by plants that have evolved methods of dispersal by sea, and for whom the ceaseless tides provide daily opportunities to colonize the beach.

There were already nine species of flowering plant on the beaches of Rakata by E + 3 (including two species of shrub and four species of tree). Eleven years later (E + 14), this had risen to 23 species of flowering plant, including three species of shrub and 10 species of tree. The flora already included three distinct communities. Along the beach itself lay the strand-line creepers such as *Ipomoea pes-caprae*. The trees and shrubs grew a short distance inland, and were made up of stands of the Indian almond tree, *Terminalia catappa*, or of the tree *Casuarina equisetifolia*. All of the species in these woodlands are widely distributed on the beaches of South-East Asia and the western Pacific, showing that they are good at dispersal by sea. The number of trees in these early beach-arrival figures may at first seem surprising, but the larger size of the fruits or nuts of trees makes it easier for them to have flotation devices and not be overwhelmed by the waves. (Not all of these species necessarily arrived as solitary individuals; in 1986, the beach on the neighbouring island of Anak Krakatau bore a mass of vegetation 20 m², including complete palm trees 3–4 m tall.) Once they had grown, the early trees and shrubs also provided food in the form of fruit, as well as perching places, for birds and bats whose droppings were the probable source of other trees, such as two species of fig. Therefore, early colonists themselves provide a beach-head for other arrivals. The seeds and fruits of trees (such as *Terminalia*) that had arrived on the beach may also have been taken farther inland by both fruit bats and crabs; the lack of mammals other than bats and rats on Rakata may explain why this inland spread seems to have been comparatively slow on the island.

By E + 25, 46 species of plant had arrived in Rakata by sea (Figure 7.15c), but thereafter the seaborne component of its flora started to level off. Potential colonists are continually arriving by sea – a 2-month-long survey of the beach of Anak Krakatau on 2 successive years found the fruits, seeds or seedlings of 66 species of plant. Because it is so easy for these species to colonize the beach, most of them soon do so, and thereafter there will be few arrivals of new seaborne species. So, for example, between E + 41 and E + 106, the diversity of the beach flora increased by only six new species (from 53 to 59 species), and the beach community is now relatively stable, with few new gains or losses.

Life Inland

The history of the colonization of the inland areas of Rakata is more complex, because the environment did not remain constant. Instead, the presence and activities of each wave of colonists not only changed the environment, but also produced a greater variety of habitats, some of which were suited to a new selection of colonists. Therefore, both the complexity and the variety of the inland ecosystems of Rakata steadily increased. Figure 7.15b is an attempt to give an impression of the

timing and rate of these ecological changes, using the descriptions given in the early surveys, so that they can be compared with the changes in some of the animals and plants that were colonizing the island through this period of time.

At first, by E + 3, the devastated, ash-covered inland areas bore a gelatinous film of *blue-green algae*, or cyanobacteria. This had provided a moist environment in which the spores of 10 species of fern and two species of moss had already germinated and grown, as well as four species of herb. (This early preponderance of ferns may be because their spores are lighter than the seeds of flowering plants.) By E + 14, many new species of flowering plant had arrived, so that the ferns were less dominant, although they still covered much of the upland areas. By that time, definite plant associations could be distinguished at different levels in the island. The interior hills and valleys were covered by a 'grass steppe', up to 3 m high, consisting of the wild sugar cane *Saccharum spontaneum*, together with other grasses and scattered trees. Interestingly, some of the creeping plants, and a shrub, that are normally confined to the beach had been able to extend inland in this still-impoverished Rakata flora.

Grassland also covered the higher slopes of Rakata, but here it was dominated by *Imperata cylindrica*, a grass that is usually the first to colonize fire-cleared areas in the East Indies, together with the bamboo grass *Poganotherum*. Although ferns were still present in these grassland floras, they contributed only 14 species, compared with 42 species of flowering plants, and dominated only the higher regions.

By E + 25, new species of animal-dispersed tree such as figs and *Macaranga* were arriving in the lowlands. As a result, the interior grasslands were being replaced by mixed woodland and forest, and they had almost completely disappeared by E + 45. The forest grew denser, and its canopy gradually closed between E + 36 and E + 51. This caused a progressive change in the physical habitat and microclimate of the forest floor: the wind velocity, light intensity and temperature all decreased, while the humidity increased. The graphs of the immigration rates of ferns, flowering plants, butterflies and birds (Figure 7.15a) show interesting changes that appear to be results of these ecological changes. (The immigration rate is the rate of addition of new

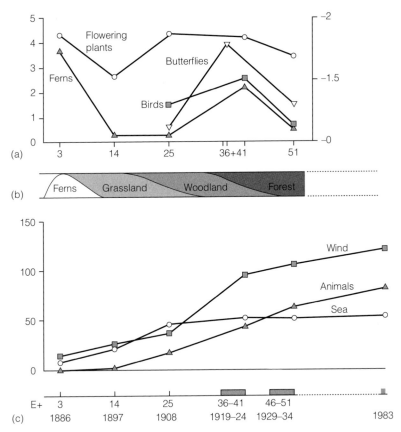

Figure 7.15 The changing biota of Rakata, 1883–1934. The lower axis shows the dates, and the periods elapsed, since the eruption (E+x); the blocks indicate the periods at which collections were made. (a) Changes in the immigration rates of ferns and flowering plants (left-side scale) and of butterflies and resident land birds (right-side scale; these were not censused before 1908). Adapted from Thornton [62]. (b) A subjective interpretation of the rate and nature of the environmental changes, deduced from the descriptions given by scientific investigators. (c) Methods of initial colonization by ferns and flowering plants. Beyond the figures for 1934, the level has been extrapolated towards the figures from the 1983 census. Adapted from Bush and Whittaker [76].

species to the biota, per year. The addition of 10 new species over the space of 5 years between one survey and the next would therefore be an immigration rate of two species per year.)

For both ferns and flowering plants, the grasslands that replaced the early fern phase do not seem to have provided an environment that encouraged a diversity of new colonists, and the immigration rate of both groups fell. The fall was greater for the ferns, because their immigration rate had earlier been particularly high during the formation of the fern phase. The immigration rate of the flowering plants had improved when they were next sampled, at E + 25, but had fallen by E + 51; this may be because the increasing woodland at first provided a greater variety of habitats for them, but the later closure of the forest canopy restricted the light. The immigration rate of the ferns, in contrast, continued to increase even during the early stages of canopy formation, probably because the moist forest provided an ideal environment for a second group of ferns, mainly shade-demanding species, many of which were epiphytes, living on the trunk and branches of the forest trees. In both butterflies and resident land birds, the immigration rate rose as the forest started to form, but had fallen by the time the canopy had closed. Thus, closure of the canopy took place at the same time as a reduction in the immigration rate of all these groups. Nevertheless, because extinction rates still remained comparatively low, the total number of species increased slightly in the case of the flowering plants, and remained approximately constant in the other groups. (The fact that extinction rates remained low during the period of closure of the canopy suggests that patches of open ground or woodland must have remained, either around the edges of the forests or within them, perhaps where trees had fallen and provided a continuing opportunity for the survival of species that preferred an open habitat.) Analysis of the species that were lost suggests that some of them had never become properly established, that some had lived in habitats that had disappeared or had been transformed and that others had had a very restricted distribution. The higher plant flora of the interior is still gaining new species, so that the forest succession is still continuing and the balance of species in the canopy is still changing.

Analyses of the data on the floras of Rakata and its neighbouring islands provide interesting and important results [63]. Although earlier analyses suggested that many extinctions had been due to the chance interactions of a complex of variables, this was the result of sampling errors. Instead, a relatively high proportion of the extinctions were the inevitable results of the loss or transformation of habitats as part of processes of change that affected whole islands or communities.

As noted here, the biota of the interior of Rakata also differs from that of the coast in that nearly all of it arrived by air, not by sea. That was not a very difficult journey, for winds from Java and Sumatra have an average speed of 20–22 km/h (12–14 mph), so that wind-blown seeds could arrive in Rakata in about 2 hours. However, although some of the airborne arrivals came on the wind, others came in or upon the bodies of animals. To begin with, the ash-covered interior of Rakata was totally uninviting to animals. Thus, as can be seen from Figure 7.15c, animal-aided dispersal only became a significant contributor to the biota from the time of the E + 25 survey. From then on, the graphs of the arrival of wind-dispersed species and of animal-dispersed species move in parallel. But they also

rise more steeply, because of a positive-feedback effect between the plants and the animals. The woodlands that had developed by E + 25 provided an environment that other species of flowering plant could colonize (Figure 7.15b). The increasing diversity of these plants in turn provided food for an increasing diversity of animals, as can be seen from the fact that the immigration rates of both butterflies and birds increased at that time. This effect became especially evident as the forest canopy formed. But the increasing numbers and diversity of animals arriving in the growing forests also brought the seeds of other new species of plant, either within their alimentary canal or adhering to their bodies. The resulting increase in plant diversity in turn encouraged more animal diversity, and so on.

The animal-dispersed component has been the most important one ecologically, because the seeds of nearly all the tree species of the inland forests arrived in this way. Wind dispersal, on the other hand, was particularly important in the addition of other plant species, providing all of the forest ferns and many of the herbs and shrubs. Of these, 17% belong to the Compositae and 13% to the Asclepiadaceae (milkweeds, whose seeds bear silky hairs rather like those of the Compositae and are easily carried in the wind), while over 50% of these species belong to the Orchidaceae, some of which are dispersed by wind and others by animals. (Orchid seeds have no food reserve and need root fungi in order to germinate and grow, which suggests that the original colonists, at least, may have arrived on the legs of birds, in mud that also contained the fungus.) For the flora as a whole, species with small, wind-dispersed seeds form a much higher proportion of the flora of Rakata than they do on neighbouring Java.

Rakata is the largest and highest of the three islands that were the surviving, initially lifeless fragments of Krakatau. It might have been expected that the forests that eventually appeared on these three islands would be similar to one another and to those on neighbouring Java and Sumatra – but they are not. The lowland forests of Rakata are unique, because they are dominated by the wind-dispersed tree *Neonauclea* (Rubiaceae), which grows up to 30 m high. The forests of Panjang and Sertung, in contrast, are instead dominated by the animal-dispersed trees *Dysoxylum* (Meliaceae) and

Timonius (Rubiaceae). Why should the forests of these islands be unique and also different from one another? Various theories have been put forward.

One of the differences between the trees is that *Neonauclea* is less tolerant of shade than are the other two species. It may therefore be 'shaded out' if all three species arrive on an island at about the same time. However, *Neonauclea* arrived on Rakata in 1905, nearly 25 years before the others, and this may have given it a head start and allowed it to become dominant on that island. It also seems that it requires overturned soil or fresh ash in order to establish itself. Few scientific visits were made to the other two islands, so the history of their forests is not well known – only that all three species were present by 1929. The Japanese ecologist Hideo Tagawa and his colleagues [64]

have suggested that, if they all arrived at about the same time, the shade produced by the growing *Dysoxylum* and *Timonius* trees might have made it more difficult for *Neonauclea* to flourish. Rob Whittaker and his colleagues initially pointed out [63] that, unlike Rakata, the other islands had been partially covered by up to 1 m of ash during the eruption of Anak Krakatau in 1930, and that this might well have affected the floras of the islands. They later [65] suggested that there is a very strong chance element determining which species happens to form the dominant element, its success depending on such factors as the time of year, the prevailing climate, what elements of the previous vegetation had survived and which of them were fruiting, which dispersal agents were available and so on (Figure 7.16).

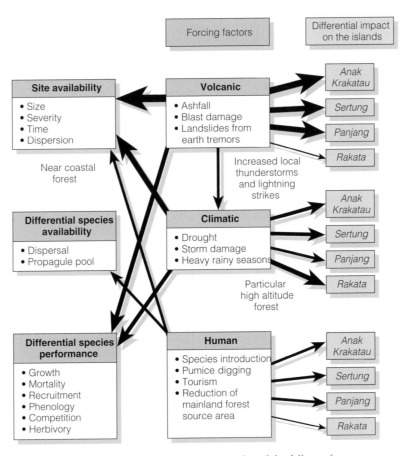

Figure 7.16 The relative importance, and hierarchy, of the different factors affecting plant succession in the Krakatau Islands. From Schmitt and Whittaker [77].

It is interesting to note that the recolonization of Anak Krakatau following occasional eruptions from 1932 to 1973 showed a similar pattern to that following the main Krakatau eruption of 1883. In particular, the coastal and pioneer inland colonists were very similar, and it will be very interesting to see what light the interior forest colonists, only now starting to appear, will throw on the variations shown by the forests of Rakata, Panjang and Sertung.

All of this is a good illustration of how difficult it is to interpret ecological biogeography, even in the apparently simple situation of the colonization of a lifeless island environment.

Fig trees, which are a particularly important part of the flora of tropical forests, provide another interesting problem in the colonization of Rakata. They are a major component of the forests; by E + 40, the 17 fig species found on Rakata, Panjang and Sertung made up nearly two-thirds of the total number of tree species there. Figs are also important because they are used as food by many animals. (In Malaya, one individual fig tree was visited by 32 species of vertebrate, and 29 species of fig tree were used by 60 species of bird and 17 species of mammal.) Just as important, however, is the fact that few of these animals eat only figs, and therefore those fig-eating animals (especially bats) arriving in Rakata were quite likely to bring, within their digestive system, the seeds of other trees. (Bats can retain viable seeds in their gut for over 12 hours, which gives them plenty of time to feed on the mainland and then fly to Rakata.) But figs also provide a problem in colonization, for each species of fig requires the services of its own species of pollinating wasp in order to produce fertile seeds – and the fig is similarly necessary for the life cycle of the wasp. Therefore, for this symbiotic system to become effective and established, there has to be a sufficiently large population of both the fig tree and the wasp, each component having arrived in Rakata independently, the figs having been borne by animals and the wasps having arrived by air. (However, some fig species have solved this problem by hybridizing with one another, thus making up for the absence of one of the required species of wasp [66].)

Finally, one can learn something about the processes and difficulties of island colonization by noting what types of animal or plant have so far not been able to colonize Rakata. Small, non-flying mammals such as cats, monkeys and most rodents are absent; they are incapable of crossing sea barriers of more than 15 km. The exception is the country rat *Rattus tiomanicus*, which has been known to swim 35 km and which had colonized Rakata by E + 45. Because there is no supply of running or standing fresh water on the island, there are no mangrove trees, no freshwater birds, no insects that have aquatic larvae and no freshwater molluscs. There is as yet no mature forest on Rakata, and therefore none of the birds that require that environment, such as trogons, parrots, nuthatches, hornbills, pittas and leafbirds. Finally, although some trees (e.g. dipterocarps) and bushes have winged seeds, these do not normally travel in the wind for more than a mile (1.6 km), and these species are absent from Rakata.

There are also some interesting interactions at a more detailed level. For example, the bird fauna of Rakata includes the flowerpecker *Dicaeum*, which distributes the seeds of the plant family Loranthaceae, which are epiphytic plant parasites of the trees of the forest canopy. However, the Rakata forest is not yet old enough to contain the mature and dying trees that the parasite can attack. As a result the Loranthaceae are absent, together with the butterfly *Delias* that feeds upon these plants, even though the butterfly itself is highly migratory and a competent potential colonist.

The complexity of all these ecological and successional changes shows that the colonization history of an island will not follow the simple path predicted by the TIB. MacArthur and Wilson themselves pointed this out in their book, in which they gave an example of a single-peaked colonization curve and suggested that the sequence of invasion might affect the nature of colonization and the equilibrium number [14]. The replacement of one plant community by another causes pronounced irregularities in the graphs of immigration and extinction, not only of the plants themselves but also of the associated fauna. The integrated nature of the successional ecosystems therefore makes it likely that the colonization history of an island like Rakata will show pronounced waves of change. The simple monotonic curves predicted by the TIB therefore apply only to the situation after the pattern of communities in an island has already become firmly established. There can be no doubt that the detailed studies of Whittaker and his colleagues will continue to provide fundamental data and insights into the varied processes of the establishment of island communities, and of ecological interactions between the different species of the island biota.

We now turn from the islands to study the oceans that surround them, whose biogeography involves a new dimension, depth, but whose organisms provide fundamental problems in the identification of species and their areas of endemicity.

Summary

1 The islands' impoverished faunas and floras are the ideal situation for rapid evolutionary modification and adaptive radiation of colonists. The events of colonization and subsequent adaptation to the new environment, sometimes taking advantage of major new ecological opportunities, provide many fascinating examples of evolution in action. These processes are illustrated with a detailed study of the biogeography of the Hawaiian Islands.

2 Island life is unusually hazardous, so that there is a complex interaction between the processes of immigration, colonization and extinction. However, attempts to construct a predictive theory of the numbers of species that would be found on islands of different sizes and locations have proved to be unreliable, and these theories also provide only limited help in the design of nature reserves (see Chapter 14).

3 Research on the recolonization of the once-lifeless island of Rakata in the East Indies is providing unique insights into the interaction between environmental factors and the arrival of new animals and plants in a developing ecosystem. Comparison between the results of these processes in Rakata and two other neighbouring islands poses interesting questions on the role that chance plays in these ecological developments.

Further Reading

Grant P (ed.). Evolution on Islands. Oxford: Oxford University Press. 1988.

Quammen D. The Song of the Dodo – Island Biogeography in an Age of Extinction. London: Pimlico/Random House, 1996.

Whittaker RJ, Fernández-Palacios JM. Island Biogeography: Ecology, Evolution and Conservation. 2nd ed. Oxford: Oxford University Press, 2007.

References

1. Sly ND, Townsend AK, Rimmer CC, Townsend JM, Latta SC. Ancient islands and modern invasions: disparate phylogeographic histories among Hispaniola's endemic birds. *Molecular Ecology* 2011; 20 (23): 5012–5024.

2. Censky EJ, Hodge K, Dudley J. Over-water dispersal of lizards due to hurricanes. *Nature* 1998; 395 (6702): 556–556.

3. Van der Pijl L. *Principles of Dispersal.* Berlin: Springer, 1982.

4. Nogales M, Medina FM, Quills V, González-Rodríguez M. Ecological and biogeographical implications of yellow-legged gulls (*Larus cachinnans* Pallas) as seed dispersers of *Rubia fruticosa* Ait. (Rubiaceae) in the Canary Islands. *Journal of Biogeography* 2001; 28 (9): 1137–1145.

5. Nogales M, Heleno R, Traveset A, Vargas P. Evidence for overlooked mechanisms of long-distance seed dispersal to and between oceanic islands. *New Phytologist* 2012; 194 (2): 313–317.

6. MacArthur RH, Diamond JM, Karr JR. Density compensation in island faunas. *Ecology* 1972: 330–342.

7. Boyce MS. Population viability analysis. *Annual Review of Ecology and Systematics* 1992; 23: 481–506.

8. Lande R. Genetics and demography in biological conservation. *Science* 1988; 241 (4872): 1455–1460.

9. Diamond JM. Historic extinctions: a Rosetta Stone for understanding prehistoric extinctions. In: Martin PS, Klein RG (eds.), *Historic Extinctions: A Rosetta Stone for Understanding Prehistoric Extinctions.* Tuscon: University of Arizona Press, 1984.

10. Filardi CE, Moyle RG. Single origin of a pan-Pacific bird group and upstream colonization of Australasia. *Nature* 2005; 438 (7065): 216–219.

11. Lack D. Subspecies and sympatry in Darwin's finches. *Evolution* 1969: 252–263.

12. Simberloff D. Using island biogeographic distributions to determine if colonization is stochastic. *American Naturalist* 1978: 713–726.

13. Cronk Q. Islands: stability, diversity, conservation. *Biodiversity and Conservation* 1997; 6 (3): 477–493.

14. MacArthur RH. *The Theory of Island Biogeography*. Princeton: Princeton University Press, 1967.

15. Carlquist SJ. *Island Life: A Natural History of the Islands of the World*. New York: Natural History Museum Press, 1965.

16. Cody ML, Overton J. Short-term evolution of reduced dispersal in island plant populations. *Journal of Ecology* 1996: 53–61.

17. Case TJ. A general explanation for insular body size trends in terrestrial vertebrates. *Ecology* 1978: 1–18.

18. Brown P, Sutikna T, Morwood MJ, *et al.* A new small-bodied hominin from the Late Pleistocene of Flores, Indonesia. *Nature* 2004; 431 (7012): 1055–1061.

19. Lomolino MV. Body size evolution in insular vertebrates: generality of the island rule. *Journal of Biogeography* 2005; 32 (10): 1683–1699.

20. Meiri S, Dayan T, Simberloff D. The generality of the island rule reexamined. *Journal of Biogeography* 2006; 33 (9): 1571–1577.

21. McClain CR, Boyer AG, Rosenberg G. The island rule and the evolution of body size in the deep sea. *Journal of Biogeography* 2006; 33 (9): 1578–1584.

22. Carlquist SJ. *Hawaii: A Natural History*. New York: Natural History Press, 1970.

23. Wagner WL, Funk VA. *Hawaiian Biogeography*. Washington, DC: Smithsonian Institute Press, 1995.

24. Keast A, Miller SE. *The Origin and Evolution of Pacific Island Biotas, New Guinea to Eastern Polynesia: Patterns and Processes*. Amsterdam: SPB Academic Publishing, 1996.

25. Carson H, Clague D. Geology and biogeography of the Hawaiian Islands. In: WagnerWL, FunkVA (eds.), *Hawaiian Biogeography: Evolution on a Hot Spot Archipelago*. Washington, DC: Smithsonian Institution Press, 1995: 14–29.

26. Stone BC. A review of the endemic genera of Hawaiian plants. *The Botanical Review* 1967; 33 (3): 216–259.

27. Carr GD *et al.* Adaptive radiation of the Hawaiian silversword alliance (Compositae-Madiinae): a comparison with Hawaiian picture-winged Drosophila. In: GiddingsLY, KaneshiroKY, AndersonWW (eds.), *Genetics, Speciation and the Founder Principle*. New York: Oxford University Press, 1989: 79–97.

28. Givnish TJ. Adaptive radiation, dispersal, and diversification of the Hawaiian lobeliads. In: KatoM (ed.), *The Biology of Biodiversity*. Berlin: Springer, 2000: 67–90.

29. Baldwin BG, Sanderson MJ. Age and rate of diversification of the Hawaiian silversword alliance (Compositae). *Proceedings of the National Academy of Sciences* 1998; 95 (16): 9402–9406.

30. Raikow RJ. The origin and evolution of the Hawaiian honeycreepers (Drepanididae). *Living Bird* 1977; 15: 95–117.

31. Sibley CG, Ahlquist JE. The relationships of the Hawaiian honeycreepers (Drepaninini) as indicated by DNA–DNA hybridization. *The Auk* 1982; 99: 130–140.

32. Olson SL, James HF. Descriptions of thirty-two new species of birds from the Hawaiian Islands: Part I. Non-passeriformes. *Ornithological Monographs* 1991; 46: 1–88.

33. James HF, Olson SL. Descriptions of thirty-two new species of birds from the Hawaiian Islands: Part II. *Passeriformes. Ornithological Monographs* 1991; 46: 1–88.

34. Cowie RH, Holland BS. Dispersal is fundamental to biogeography and the evolution of biodiversity on oceanic islands. *Journal of Biogeography* 2006; 33 (2): 193–198.

35. Heaney LR. Is a new paradigm emerging for oceanic island biogeography? *Journal of Biogeography* 2007; 34 (5): 753–757.

36. Van Balgooy MMJ. Plant-geography of the Pacific as based on a census of phanerogam genera. *Blumea Supplement* 1971; 6: 1–122.

37. Simberloff D. Species turnover and equilibrium island biogeography. *Science* 1976; 194 (4265): 572–578.

38. Gilbert F. The equilibrium theory of island biogeography: fact or fiction? *Journal of Biogeography* 1980; 7: 209–235.

39. Connor EF, McCoy ED. The statistics and biology of the species-area relationship. *American Naturalist* 1979: 791–833.

40. Diamond JM. Avifaunal equilibria and species turnover rates on the Channel Islands of California. *Proceedings of the National Academy of Sciences* 1969; 64 (1): 57–63.

41. Lynch JF, Johnson NK. Turnover and equilibria in insular avifaunas, with special reference to the California Channel Islands. *Condor* 1974; 76: 370–384.

42. Simberloff D. When is an island community in equilibrium? *Science* 1983; 220 (4603): 1275–1277.

43. Steadman DW. *Extinction and Biogeography of Tropical Pacific Birds*. Chicago: University of Chicago Press, 2006.

44. Shafer CL. *Nature Reserves: Island Theory and Conservation Practice*. Washington, DC: Smithsonian Institution Press, 1990.

45. Case TJ, Cody ML. Testing theories of island biogeography. *American Scientist* 1987; 75: 402–411.

46. Wright DH. Species-energy theory: an extension of species-area theory. *Oikos* 1983; 41: 496–506.

47. Kalmar A, Currie DJ. A global model of island biogeography. *Global Ecology and Biogeography* 2006; 15 (1): 72–81.

48. Whittaker RJ, Triantis KA, Ladle RJ. A general dynamic theory of oceanic island biogeography. *Journal of Biogeography* 2008; 35 (6): 977–994.

49. Whittaker RJ, Triantis KA, Ladle RJ. A general dynamic theory of oceanic island biogeography: extending the MacArthur–Wilson theory to accommodate the rise and fall of volcanic islands. In: Losos JB, Ricklefs RE (eds.), *The Theory of Island Biogeography Revisited*. Princeton: Princeton University Press, 2010: 88–115.

50. Borges PA, Hortal J. Time, area and isolation: factors driving the diversification of Azorean arthropods. *Journal of Biogeography* 2009; 36 (1): 178–191.

51. Triantis K, Vardinoyannis K, Tsolaki EP, et al. Re-approaching the small island effect. *Journal of Biogeography* 2006; 33 (5): 914–923.

52. Ulrich W, Almeida Neto M, Gotelli NJ. A consumer's guide to nestedness analysis. *Oikos* 2009; 118 (1): 3–17.

53. Atmar W, Patterson BD. The measure of order and disorder in the distribution of species in fragmented habitat. *Oecologia* 1993; 96 (3): 373–382.

54. Atmar W, Patterson BD. The nestedness temperature calculator, a visual BASIC program, including 294 presence/absence matrices. *Chicago: AICS Research*, 1995.

55. Warren BH, Simberlof D, Ricklefs RE, et al. Islands as model systems in ecology and evolution: prospects fifty years after MacArthur-Wilson. *Ecology Letters* 2015; 18 (2): 200–217.

56. Diamond JM. Assembly of species communities. In: Cody ML, Diamond JM (eds.), *Ecology and Evolution of Communities*. Cambridge, MA: Harvard University Press, 1975: 342–444.

57. Garb JE, Gillespie RG. Island hopping across the central Pacific: mitochondrial DNA detects sequential colonization of the Austral Islands by crab spiders (Araneae: Thomisidae). *Journal of Biogeography* 2006; 33 (2): 201–220.

58. Burns K. Patterns in the assembly of an island plant community. *Journal of Biogeography* 2007; 34 (5): 760–768.

59. Colwell R, Winkler D. A null model for null models in biogeography. In: StrongDRJr (ed.), *Ecological Communities: Conceptual Issues and the Evidence*. Princeton: Princeton University Press, 1984: 344–359.

60. Violle C, Reich PB, Pacala SW, Enquist BJ, Kattge J. The emergence and promise of functional biogeography. *Proceedings of the National Academy of Sciences* 2014; 111 (38): 13690–13696.

61. Whittaker RJ, Rigal F, Borges PAV, et al. Functional biogeography of oceanic islands and the scaling of functional diversity in the Azores. *Proceedings of the National Academy of Sciences* 2014; 111: 13709–13714.

62. Thornton I. *Krakatau: The Destruction and Reassembly of an Island Ecosystem*. Cambridge, MA: Harvard University Press, 1997.

63. Whittaker RJ, Bush MB, Partomihardjo T, Asquith NM, Richards K. Ecological aspects of plant colonisation of the Krakatau Islands. *GeoJournal* 1992; 28 (2): 201–211.

64. Tagawa H, Suzuki E, Partomihardjo R, et al. Vegetation and succession on the Krakatau Islands, Indonesia. *Vegetatio* 1985; 60 (3): 131–145.

65. Whittaker RJ, Field R, Partomihardjo T. How to go extinct: lessons from the lost plants of Krakatau. *Journal of Biogeography* 2000; 27 (5): 1049–1064.

66. Parrish TL, Koelewijn HP, van Dijk PJ. Genetic evidence for natural hybridization between species of dioecious ficus on island populations. *Biotropica* 2003; 35 (3): 333–343.

67. Wilson EO. Adaptive shift and dispersal in a tropical ant fauna. *Evolution* 1959; 13: 122–144.

68. Ricklefs RE, Bermingham E. The concept of the taxon cycle in biogeography. *Global Ecology and Biogeography* 2002; 11 (5): 353–361.

69. Losos JB. Phylogenetic perspectives on community ecology. *Ecology* 1996: 1344–1354.

70. Jønsson KA, Irestedt M, Christidis L, Clegg SM, Holt BG, Fjeldså J. Evidence of taxon cycles in an Indo-Pacific passerine bird radiation (Aves: Pachycephala). *Proceedings of the Royal Society B: Biological Sciences* 2014; 281 (1777): 20131727.

71. Mayr E. Die Vogelwelt Polynesiens. *Mitteilungen aus dem Zoologischen Museum in Berlin* 1933; 19: 306–323.

72. MacArthur RH, Wilson EO. An equilibrium theory of insular zoogeography. *Evolution* 1963; 17: 373–387.

73. Pratt HD, Bruner PL, Berrett DG. *A Field Guide to the Birds of Hawaii and the Tropical Pacific*. Princeton: Princeton University Press, 1987.

74. Lomolino MV, Brown JH, Davis R. Island biogeograrahy of montane forest mammals in the American Southwest. *Ecology* 1989; 70: 180–194.

75. Whittaker RJ, Jones SH, Partomihardjo T. The rebuilding of an isolated rain forest assemblage: how disharmonic is the flora of Krakatau? *Biodiversity and Conservation* 1997; 6 (12): 1671–1696.

76. Bush MB, Whittaker RJ. Krakatau: colonization patterns and hierarchies. *Journal of Biogeography* 1991; 18: 341–356.

77. Schmitt S, Whittaker RJ. Disturbance and succession on the Krakatau Islands, Indonesia. In: Newbery DM, Prins HH, Brown ND (eds.), *Dynamics of Tropical Communities: The 37th Symposium of the British Ecological Society, Cambridge University*, 1996. Oxford: Blackwell Science, 1998: 515–548.

Patterns of Life

From Evolution to Patterns of Life

The previous chapter explained how the process of evolution leads to the appearance of new species. As time goes by, that may lead to the appearance of its radiation into several other new species. Modern methods allow us to establish precisely how the different species are related to one another, when each of them diverged from its relatives and how they came to spread from their places of origin to other locations. This in turn enables us to trace the biogeographical history of lineages of organisms, of biotas and of biomes, and of the areas in which they are found.

If the evolutionary process continues after a new species has evolved, the many descendants of the original species may eventually spread over large areas of the planet. Sometimes the areas in which related species are found today may be widely separated from one another, a situation known as a *disjunct distribution*. Trying to understand how and why these patterns appeared and changed makes us confront a series of questions – and the more of these questions we can answer, the more confidence we can have in the correctness of our analysis of the reasons for these patterns of life.

1. On the geographical/geological side, when did two areas in which we see the organisms living today become separate from one another?

2. Was this the result of geological processes such as plate tectonics, which might have caused the areas to split apart or to become separated by mountain building, or was it the result of environmental events such as changes in sea level or climate?

3. On the biological side, how are the taxa related to one another?

4. Did their appearance in separate areas take place before or after the appearance of a barrier between them?

5. When did they start to become separate due to evolutionary change?

Until comparatively recent years, we were unable to answer many of these questions. As explained in Chapter 1, it was only in the 1960s that understanding of the processes of plate tectonics enabled us to answer questions 1 and 2. Only a few years after that, in the late 1960s, the advent of molecular analysis of the proteins of living organisms, together with Hennig's cladistic method, allowed us to obtain reliable answers to question 3. Most methods of analysis in historical biogeography therefore start with data that provide answers to these three questions, so we can now turn to the problems posed by questions 4 and 5.

Dispersal, Vicariance and Endemism

Let us imagine a species that has recently evolved. It is likely, to begin with, to extend its area of distribution or **range** until it meets barriers of one kind or another to its further spread; this is known as **range extension** (Figure 8.1a). Sometimes this barrier later disappears, so that the species is able to extend its range into the area previously not available; this is known as **dispersion**, but it has also been called **geodispersal** because it is the result of a geological rather than a biological event (Figure 8.1b). Sometimes the species is eventually able to disperse across the barrier, which now provides the necessary isolation for the population on

Biogeography: An Ecological and Evolutionary Approach, Ninth Edition. Edited by C. Barry Cox, Peter D. Moore, Richard J. Ladle.
© 2016 John Wiley & Sons, Ltd. Published 2016 by John Wiley & Sons, Ltd.

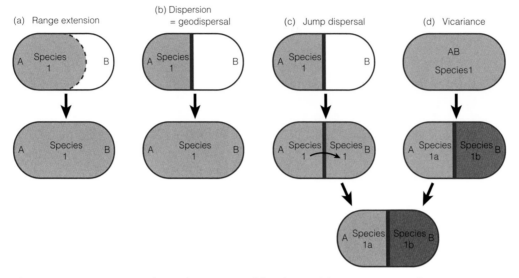

Figure 8.1 How species can change their patterns of distribution. (a) Range extension: the species is at first found only in area A. Later, it gradually extends its range of distribution into the neighbouring area B. (b) Dispersion or geodispersal: the barrier between the two areas disappears, so that the species can extend its range into area B. (In these two cases, because there is no barrier between the two areas, it cannot differentiate into a new separate species.) (c) Dispersal or jump dispersal: the species is at first restricted to area A by a barrier that separates it from area B, but later disperses across the barrier. (d) Vicariance: the species originally occupies the whole of area A and B, but these two areas become separated from one another by a barrier. The original species then differentiates into two separate species separated by the barrier. (In both c and d, because the barrier remains, the two populations of the species can now differentiate into two separate species.) Note: The results of dispersal and of vicariance are identical.

either side to differentiate into separate species; this is known as simple dispersal or **jump dispersal** (Figure 8.1c). Sometimes the barrier may appear within the area of a distribution of an existing species, subdividing it into separate populations, which could then diverge from one another into separate species; this process is known as **vicariance** (Figure 8.1d),

After it has arisen, the new species will gradually extend its area of distribution until it reaches barriers of one kind or another (physical, ecological etc.) beyond which it cannot readily spread. It is said to be **endemic** to that particular area, being found there and nowhere else. Various definitions of **areas of endemicity** have been proposed, the criteria required ranging from the biological (relatively extensive sympatry of the taxa involved) to the physical (areas delimited by barriers). No two organisms live in *exactly* the same area; even those that live in the same lake or island will have at least slightly different ecological preferences and

will therefore not live in precisely the same set of locations. This is a particular problem at smaller scales of study, where local ecology is important. At the other end of the scale, where major continental or subcontinental areas are the subjects of study, even though the areas are easy to define, problems of tectonic subdivision or fusion are more likely to arise. This is then more likely to lead to successive patterns of subdivision and subsequent reunion into different patterns – a phenomenon known as a reticulate pattern (see the 'Reticulate Patterns' section).

Methods of Analysis

Phylogenetic Biogeography

The analysis of the significance of the endemism patterns found in related species lies at the heart of much of modern biogeography, but there are two different approaches to it. The first starts with the selection of a group of biological taxa that are assumed to have diverged from a common ancestor, and then uses their pattern of distribution to study the implications for the history of the areas to which they are endemic. The earliest and clearest example of this method is the 1966 work of the Swedish entomologist Lars Brundin, who was the first to realize the potential of cladistics (see the 'Cladistic, or "Pattern-Based" Biogeography' subsection) as a tool for analysing the distribution patterns of areas of endemism [1]. He referred to his method as **phylogenetic biogeography**. Studying the distribution of three subfamilies of chironomid midges in the Southern Hemisphere, he first produced a cladogram of the evolutionary relationships of all the species. In place of the name of each species in the cladogram, he then instead inserted the name of the continent on which it is found, and where it is therefore endemic, transforming the phyletic cladogram into a **taxon–area cladogram** (Figure 8.2). The result was a consistent pattern in which the African species appeared to have diverged first, followed in turn by those of New Zealand, South America and Australia. So, cladistics plus information on endemicity produce

a biological taxon–area cladogram in which information on a number of lineages is now also providing information on their sequence of association into the biotas of the different continents.

This sequence, based on the evolutionary relationships of the midges, was independently supported by geophysical data on the sequence of breakup of the Gondwana supercontinent. Again, Africa was the first to break away, followed in turn by New Zealand, South America and Australia. (India and Antarctica do not appear in this analysis because they do not contain these midges.) So, the biological taxon–area cladogram was paralleled by a **geological area cladogram**. This in turn had useful implications as to the apparent geological ages of the different groups of midges, because the dates of separation of the continents were known from the geophysical data.

Cladistic or 'Pattern-Based' Biogeography

The second type of approach to the relationship between patterns of endemism and the history of the areas concerned starts with the identification of areas that are assumed to have had a simple

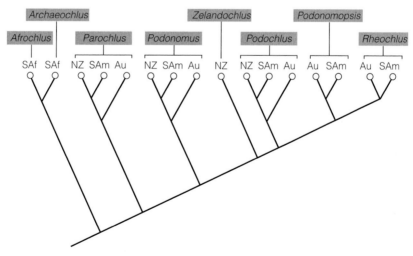

Figure 8.2 Simplified taxon–area cladogram of some of the Gondwana genera of podonominine chironomid midges studied by Brundin. The names in italics are those of the genera involved, while the circles represent individual species. The initials indicate the continent in which each species is found: Au, Australia; NZ, New Zealand; SAf, South Africa; SAm, South America. The African genera appear to have diverged first. In each of the other genera, the divergence of the New Zealand species preceded the divergence between the South American and the Australian species.

history of successive subdivision from an original single unit. So these did not subsequently fuse together, nor were they colonized more than once by any individual taxon, whereas any geological subdivision was always accompanied by speciation. The presumed history of the subdivision is shown as a geological area cladogram, and its accuracy is then tested by seeing the extent to which the biological cladograms of the organisms that are found in these areas conform with the geological area cladogram. Because it starts with the areas, this type of approach is known as **area biogeography**, but most of the examples of its use are usu-

ally known as **cladistic biogeography**. Because they rely on the identification of patterns of relationship among areas of endemism, such methods are also termed **pattern-based**.

It is useful to have two sets of data, one on geological history and the other on biological history, for it enables us to compare the patterns of each, which may show interesting parallels or differences. When the underlying assumptions of the cladistic biogeography method are fulfilled, there is a neat concordance between the geological and the biological cladograms (Figure 8.3a). However, there are many practical difficulties. For example,

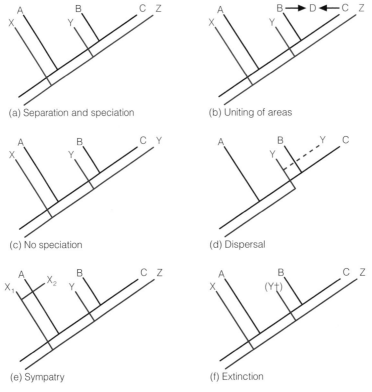

Figure 8.3 The possible relationships between a cladogram of the history of the separation or fusion of areas (in black) and the corresponding cladogram of the evolution of the taxa (in blue). (a) Areas A, B and C have separated from one another, and this has been accompanied by the evolution of separate taxa (X, Y and Z). (b) Areas B and C have become united, so that the resulting area D contains taxa Y and Z. (c) Areas B and C have separated, but taxon Y has not separated into two species, and is therefore present in both areas. (d) Taxon Y has dispersed from area B to area C, and is therefore present in both areas. (e) The results of sympatric evolution of two taxa, X_1 and X_2, in area A. (f) The results of the extinction of taxon Y in area B.

the results of range extension and geodispersal are identical, and it may be impossible to decide which was involved unless there is clear evidence of a geological event. The results of plate tectonic changes may gradually change what was originally a simple range extension across continuous land to becoming progressively more difficult because a gap dividing the area into two becomes steadily wider until dispersal across it becomes impossible. Another difficulty is when the splitting of a single area into two parts is not accompanied by biological speciation, so that the same taxon is present in more than one area (Figure 8.3c). The same result is found if one of the areas resulting from the geographical split is colonized by recent dispersal (Figure 8.3d). In contrast, sympatric speciation may take place within a single area, which therefore contains more than one related taxon (Figure 8.3e), which will also alter the apparent taxon–area cladogram (Figure 8.3f). Finally, the same geological event may have quite different implications for the distributions of biological groups of different ecologies.

To begin with, attempts to explain the more ancient patterns therefore tended to accept and use this increasingly detailed geological information, then try to fit the biological information to it in order to identify the sequence of patterns of distribution that are most likely to have led to that which we see today. That pattern may have arisen from the occurrence of four different types of event: vicariance, **duplication**, dispersal or extinction. (Where we are dealing with large areas such as continents in the distant past, it is impossible to estimate whether new species that appear within them arose by dispersal or vicariance at the local level. Their appearance can therefore only be ascribed to a process of 'duplication' that makes no assumptions as to which of these processes was involved. This is the result of allopatric speciation in response to the appearance of a transient geographical barrier; unlike vicariance, duplication affects only a single lineage.) (Range extension cannot be included as a separate type of event, because its result, the fact that the same species is now found in an additional area B (cf. Figure 8.1) is the same as the result of geodispersal or jump dispersal.)

In the case of vicariance and duplication, the new species is still found in the area in which it originated. Its geographical relationships to related species, especially to the ancestral species, have therefore not changed, and consequently these provide firm data on biogeographical relationships. By contrast, where a new species arises by dispersal, it is now found in an area where its ancestor did not occur. Similarly, in the case of extinction, part of the ancestral area of endemism has been lost. In both of these cases, the pattern of ancestor–descendant geographical relationships has been broken. These phenomena therefore do not provide information on the biogeographical history of the groups involved and do not help us in trying to choose between alternative hypotheses. To try to avoid this uncertainty, modern techniques therefore try to discover systems that minimize the extent to which we have to invoke dispersal or extinction in order to produce the final biogeographical patterns [2]. It is only if the resulting proposed pattern of sequential biogeographical changes still does not make sense that we have to find out whether the occurrence of dispersal or extinction may have been the cause of this problem.

Cladistic biogeography therefore relied on finding patterns of relationship between different areas that were shared by a number of lineages. A fundamental difficulty, however, was that this would only be obvious if the patterns of distribution were the result of range extension, geodispersal or vicariance (cf. Figure 8.1). That is because, in each of these cases, all the lineages in the area have been affected by the same event and are likely to provide congruent patterns. In contrast to this, jump dispersal is confined to a single lineage, whose pattern of relationship would therefore be different from that of others, providing 'noise' in the system. Cladistic biogeographers were therefore driven to assume that jump dispersal was a rarity and could be ignored. These methods also ignore any information that may be available from other sources, such as knowledge of the sequence of plate tectonic events that led to changes in the physical relationships between the areas of endemism, even though these are clearly relevant to the problem.

As a result of all these problems, cladistic biogeographers had to develop increasingly complicated statistical methods for analysing the history of biotas, which have been well reviewed and explained by the Mexican biogeographer Juan Morrone [3]. Methods such as **Brooks parsimony analysis (BPA)** [4] and **parsimony analysis for comparing trees (PACT)** [5] first try to find a common pattern of relationships, or **general area cladogram (GAC)** that shows the biotic history of the areas of endemism involved, which presumably reflects the history of biotic connections between them. Any data that do not fit this GAC are assumed to have been the result of jump dispersal, extinction or speciation, and the computer programme then attempts to find the explanation that involves the smallest number of such events. However, no attempt is made to identify which of these processes is involved in each event, so this is left to the investigator to evaluate, nor is any estimation made of the times at which these events may have taken place. This vagueness makes it very difficult to evaluate and compare the results of these techniques [6].

Event-Based Biogeography

In response to all these problems, biogeographers developed new methods that are not based on the identification of patterns but instead focus on the events that led to these patterns. They are therefore known as **event-based** methods. These help us to identify the most accurate GAC by assigning a cost to each type of event. As already noted, only two types of event (vicariance and duplication) help us to identify the patterns of endemism of the ancestral species and of their descendant species, because they leave these unchanged. In contrast, extinction and jump dispersal lead to a break between these patterns. So, because anything that damages our ability to identify the most accurate GAC must be penalized, these are given higher costs than vicariance and duplication. Randomization trials have shown that a cost difference such as 2.0 for jump dispersal, 1.0 for extinction and 0.01 for vicariance and duplication provide a greatest likelihood for success in this. Other figures can be used as long as the general magnitude of the penalties is unchanged.

A good example of such a method is **parsimony-based tree fitting**, which starts by constructing a GAC, which is our hypothesis on the relationships among the areas analysed. Next, the phylogeny and species distributions of the group in question are fitted to the GAC to find the biogeographical reconstruction that has the minimum cost in terms of the biogeographical events needed to explain the observed distribution pattern. The GAC can also use geological evidence, for example to investigate to what extent geological vicariance could better explain the observed distributions, instead of other events such as dispersal and extinction. It can be also estimated directly from the phylogeny, using standard phylogenetic methods to search for the cladogram that minimizes the cost of the events, for example by using the computer programme TreeFitter. The cost of the fit of phylogeny and GAC can be compared with the cost that is expected by chance under the null hypothesis that no relationship exists between the taxon cladogram and the area cladogram. (A **null hypothesis** is one that assumes that the statistical relationship between two phenomena is due to chance alone.) In this case, the null hypothesis can be produced by randomizing the distributions of the taxa in the phylogeny, so that they cannot provide any information on evolutionary relationships. As long as the costs of the suggested GAC are less than those of the null hypothesis, it is reasonable to continue to use this as a representation of what actually occurred.

The Spanish biogeographer Isabel Sanmartín has given a good example of the use of this method to explain the distribution patterns of the species of the southern beech tree *Nothofagus* in the Southern Hemisphere [6]. Over the past 30 years, attempts to explain this have perhaps provided a greater number of research papers in historical biogeography than for any other problem. Analysis of the distribution of 23 of its 35 species allows us to compare the advantages and disadvantages of parsimony-based tree fitting and BPA.

The phylogeny and distribution of the species are shown in Figure 8.4d. Figure 8.4a shows the general area cladogram that results from a BPA analysis of the distribution of the characters. The numbers shown on the tree represent ancestral nodes in the phylogeny: some ancestors such as the most recent common ancestor of *Nothofagus* (no. 45) or the ancestor of subgenera *Fucospora*, *Nothofagus* and *Brassospora* (no. 44), were present in the common area from which all the later areas were derived by subdivision; other more derived ancestors (e.g. nos 33 and 41) evolved in more restricted areas that had appeared later in this process of subdivision. The analysis suggests that the Southern South America

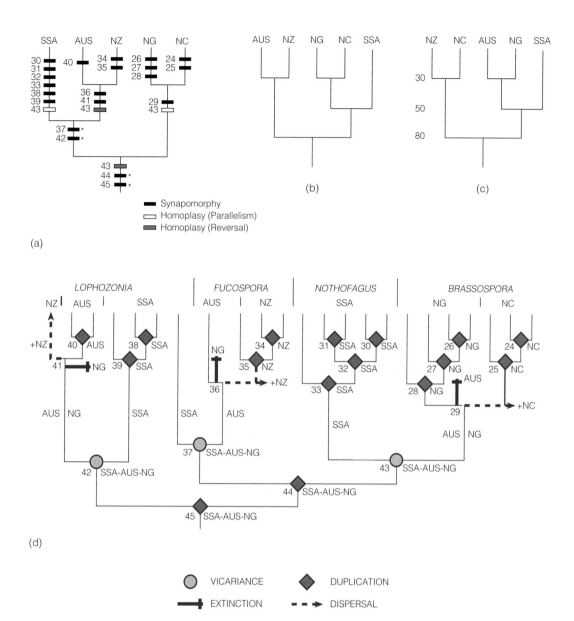

Figure 8.4 Analyses of the historical biogeography of *Nothofagus*. (a) General area cladogram suggested by the pattern-based Brooks parsimony analysis. (b) Alternative general area cladogram suggested by TreeFitter. (c) Geological area cladogram (the figures indicate the age of the events in millions of years). (d) Detailed TreeFitter solution using an event-based analysis, mapped onto the geological area cladogram. The numbers on (a) and (d) represent the ancestral nodes of the different taxa and subgenera. AUS, Australia; NC, New Caledonia; NG, New Guinea; NZ, New Zealand; SSA, Southern South America. Adapted from Sanmartín [6].

(SSA) clade is closest to an Australia (AUS)–New Zealand (NZ) clade, the New Guinea (NG)–New Caledonia clade being separate from these. The TreeFitter programme finds that the total 'cost' of this area cladogram, which involves one extinction and one dispersal, is 3.21. However, the TreeFitter programme also finds an alternative area cladogram (Figure 8.4b) with the same cost, but in which NG–New Caledonia is shown as closest to SSA, whereas AUS–NZ is now the more isolated clade.

This shows the extent to which these methods can provide significantly different results from the same data. However, in the case of problems involving the landmasses of the Southern Hemisphere, geological data can provide crucial diagnostic help. This is because the sequence of breakup of the various areas and the dating of these events are known with considerable confidence (Figure 8.4c). If this geological area cladogram is fitted to the phylogeny of *Nothofagus* using TreeFitter, the result is the event-based reconstruction of the historical biogeography of the genus shown in Figure 8.4d. This suggests that the ancestor of *Nothofagus* evolved in Gondwana before that supercontinent split up, and also diversified into the subgenera *Lophozonia* and *Fucospora* and the ancestor of the subgenera *Brassospora* and *Nothofagus*. *Lophozonia* and *Fucospora* later became extinct in New Guinea (cost, 2.00) but dispersed from Australia to New Zealand (cost, 4.00), while *Brassospora* became extinct in Australia (cost, 1.00) but dispersed from New Guinea to New Caledonia (cost, 2.00). There were also three 'cheap' vicariance events (cost, 0.03) and 16 duplication events (cost, 0.16). The higher total cost (9.19) shows the extent to which this reconstruction of the biogeographical history of *Nothofagus* differs from the geological history of the breakup of Gondwana – which is not a surprising conclusion. (In fact, palaeontological evidence supports some aspects of this reconstruction because it suggests that the southern beech tree became extinct in New Zealand – probably because of the Oligocene marine flooding of the island. If so, the presence of *Lophozonia* and *Fucospora* there must be the result of dispersal, as the reconstruction suggests. Similarly, fossil evidence shows that *Brassospora* was originally present in Australia, but became extinct there – just as the reconstruction suggests.)

Several observations can be made. First, TreeFitter provides a very clear portrayal of its solutions because these directly show the location and relative timing of the events, as well as the direction of the implied dispersals, unlike the BPA model. Second, use of the reliable geological cladogram provides an input from the real world, for it introduces data that arise from the quite independent phenomena of plate tectonics, defining not only the set of events that led to this pattern of biogeographical distributions, but also their relative timing. Finally, this geology-aided TreeFitter solution also suggests that dispersals and extinctions are considerably more common than is likely to appear in unaided TreeFitter, with its relative heavy cost penalties for these events. Nevertheless, as will be seen (see Chapter 10), the frequency with which molecular data indicate that dispersal, rather than vicariance, is the cause of many patterns in historical biogeography makes one suspect that this is the correct interpretation. The fact that TreeFitter would be unlikely to suggest this solution if the geological evidence were not available might lead one to question whether these heavy penalties are realistic.

Isabel Sanmartín and the Swedish biogeographer Fredrick Ronquist [7] have shown that this technique can also be used to infer a general biogeographical pattern for a set of different groups of Southern Hemisphere organisms. They used the phylogenies of 54 animal groups and 19 plant groups, and found interesting differences between the two. The patterns of relationship between the animal groups (Figure 8.5a) showed a close relationship between Australia and South America. This is presumably a result of the long, ancient link between the two areas before their separation – both tectonically, by plate tectonic rifting, and biologically, by the glaciation of Antarctica and consequent extinction of most of its biota. The animal data show a weak relationship between Australia and New Zealand – so low that it is not statistically significant. In contrast, the plant data (Figure 8.5b) show a stronger link between Australia and New Zealand. Apparently, the 2000 km distance between these two areas is not sufficiently great to exclude long-distance dispersal of the plants, aided by the strong winds of the West Wind Drift. In a recent paper [8], Muellner *et al.* comment that the resistant properties of seeds make it much easier for plants to get from place to place, and therefore plants are less affected by the details of physical geography than are animals.

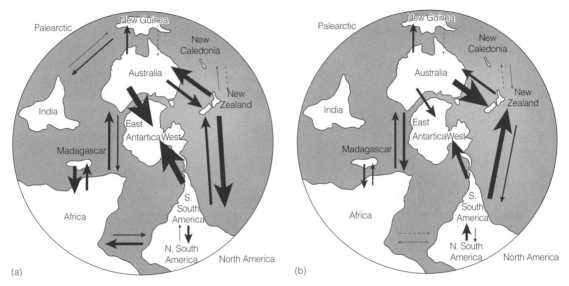

Figure 8.5 Polar view of the Southern Hemisphere, showing the frequency of dispersal (number of events inferred by Treefitter) between the landmasses in (a) the animal data set and (b) the plant data set during the late Cretaceous–Cenozoic period. The thickness of the arrows is proportional to the frequency of the dispersal events. From Sanmartín and Ronquist [7].

Sanmartín and Ronquist point out that such data contradict the strong contention of the vicariance biogeographers, who held that dispersal was always a rare, random event, which could never lead to coherent, systematic patterns of distribution, and should therefore be ignored as mere 'noise' in the biogeographical record. The Australian biogeographers Lyn Cook and Michael Crisp have suggested methods that can take account of such non-random, asymmetric factors in evaluating species trees [9]). Recent phylogeographical work also shows that the distributions of some southern temperate marine macroalgae, starfish and gastropods are the result of dispersal by the ocean currents driven by the West Wind Drift, rather than being the relics of ancient vicariance [10].

It must not be forgotten, however, that the same area cladogram may have arisen more than once but at different times, as a result of a repetition of the same geographical change – for example, the separation between western North America and eastern North America during periods B, C and E in Figure 11.19, a phenomenon known as **pseudo-congruence**. Event-based methods, like cladistic

approaches, do not integrate dating information into the biogeographical analysis and are therefore sensitive to this source of error, because they infer a common GAC from the phylogenies of several groups that were affected by the same barrier, even though they took place at different times (unless geological information is considered).

Reticulate Patterns

All of the preceding methods involve the assumption that there is a single area cladogram that describes the relationships between the areas involved. However, in many cases, especially where a long period of time is involved, some of the areas may have had more than one type of relationship with one another, leading to a network-like or **reticulate pattern** of relationships. This is prevalent in the historical biogeography of the Northern Hemisphere where, unlike the simple, tectonically driven, progressive subdivision and scattering of land areas in the Southern Hemisphere, the areas of land came into several different patterns of relationship, mainly due to changes in sea level (see Figure 11.19). Both BPA and TreeFitter assume that the biogeographical history of the organisms being analysed was the result of a series of splitting events that can be portrayed as a single branching tree, the general area cladogram. They would therefore not be able to analyse a reticulate pattern, in which a group might appear in an area that had previously separated from the area in which the organism is found. To cope with such a situation, Ronquist [11] has suggested a variant of parsimony-based tree fitting, known as **dispersal–vicariance analysis (DIVA)**, in which no general area cladogram is assumed. Instead, the area distributions are mapped directly onto the phylogeny.

DIVA can also be used [12] to find general patterns of distribution. Sanmartín, Enghoff and Ronquist [12] inferred frequencies of dispersal, duplication and vicariance events in 57 monophyletic animal groups in the Northern Hemisphere, using geological, fossil and molecular information to sort events into four periods of time. The DIVA analysis showed a satisfying degree of correspondence between the events in each period of time and the pattern of intercontinental relationship present at that time. Among other things, it pointed out the importance of the North Atlantic bridge for the historical migration of animals between Europe and North America in the Paleogene. The same type of study was repeated in plants by the American plant biogeographers Michael Donoghue and Stephen Smith, who used DIVA to analyse patterns of distribution in the temperate forests of the Northern Hemisphere [13]. Their results suggested that many of their plant groups originated and diversified in eastern Asia and later spread into North America across the Bering region (rather than via Europe). Furthermore, these dispersals took place much later, and perhaps more frequently, than the dispersals of animal taxa suggested by Sanmartín, Enghoff and Ronquist [12]. This implies that the insect fauna already resident in these forests must have had to adapt to the newly immigrant plants by switching hosts or foods. Such a degree of detail also enables investigation of the relationship between the times of arrival of new taxa, the climatic events that took place at that time and such aspects of the ecology of the migrants as methods of dispersal and whether they were deciduous or evergreen.

A further use of DIVA is to find where and when the elements of a particular biome appeared, and how they spread during the gradual evolution of the biome by its assembly from different elements. DIVA thus holds out the promise of rewarding inquiries into details of the evolution of the community of a biome and of coevolution within it – aspects of historical biogeography that had previously seemed inaccessible. These inquiries may also reveal how, when and why some of the unusual biomes that we see in the past were transformed into the more familiar ones that we see today.

However, as frequently happens, the now-widespread use of DIVA is revealing some aspects of its use that require caution or modification, such as those pointed out by Kodandaramaiah [14]. For example, it does not distinguish between range expansion and dispersal unless molecular-based estimates of the divergence times of the lineages are also taken into account, and dispersal is given a lower cost than vicariance. The technique's option to limit the number of areas allowed in the analysis can also lead to errors, so that it is preferable to run the analysis several times, using different levels of constraint. It does also fail to incorporate extinction in a realistic way; in fact, extinction events would not be inferred by DIVA unless cost constraints based on geological information are used in the analysis [2].

As we have seen in this chapter, there are a number of problems in the use of event-based methods. The cost of each type of event has to be fixed in advance, rather than arising from an analysis of the data itself. Most importantly, the need to find the most parsimonious solution leads to the penalization of dispersal and extinction, which will therefore tend to be underestimated, while vicariance will tend to be overestimated. It is also impossible to incorporate information from the dating of the divergence of lineages, the fossil record and palaeogeography directly into the calculation; they can only be considered later, in interpreting the results.

However, the whole problem has now been transformed by the development of a new class of methods without these limitations. These are known as **model-based** or **parametric** methods (Box 8.1). For example, it has now become possible to analyse information on the ranges of fossils in time and space, and to integrate the results directly into their inferred biogeographical history. Furthermore, data on their distribution and inferred ecological niches can be used to reconstruct the climatic preferences and potential patterns of distribution of lineages over time, revealing the role of climatic corridors (such as the Bering region) and climatic zones in limiting or permitting changes in their ranges. Such methods allow us to use every aspect of an organism's history and ecology in revealing its biogeographical history, enabling the final integration of ecological and historical biogeography.

Breaking the chains of parsimony: the development of parametric methods in historical biogeography

Guest Author

Box 8.1

Dr Isabel Sanmartín, Real Jardín Botánico–CSIC, Madrid, Spain

Both event-based and cladistic methods rely on the principle of parsimony as the criterion for estimating ancestral areas and the frequency of biogeographical processes. As seen in this chapter, this poses several problems in biogeographical inference; for example, it is not possible to estimate the rate of processes from the data, and that of dispersal and extinction tends to be underestimated. Parsimony also imposes other limitations: it cannot incorporate the uncertainty associated with the biogeographical reconstruction itself, since only scenarios (reconstructions) implying a minimum number of changes in the biogeographical range are evaluated, even though there could be alternative reconstructions that are almost as likely as and may be more compatible with other aspects of the problem [15]. The uncertainty in estimating phylogenetic relationships is also ignored: biogeographical history is often reconstructed on a single most parsimonious tree, which must be fully resolved (i.e. it should not contain polytomies). It could be the case, however, that some clades in the phylogeny are better supported than others by the data (i.e. they receive higher values of *clade support*), and this difference in the level of certainty or confidence in phylogenetic relationships should be accounted for in biogeographical inference. There have been some attempts to accommodate phylogenetic uncertainty in parsimony-based reconstructions. For example, one could use DIVA to analyse a sample of trees, in which each tree is weighted according to its sampling frequency in the phylogenetic analysis, and then average the frequency of ancestral ranges over all reconstructions [16].

Probably the most important limitation of parsimony-based methods is that they cannot integrate the temporal dimension into the biogeographical analysis. Methods such as TreeFitter can be used to map a geological area cladogram onto an organism phylogeny, allowing us to infer the relative timing of biogeographical events [7]. For example, in the event-based reconstruction of *Nothofagus* (Figure 8.4), dispersal to New Zealand takes place after the divergence of the South American and Australian clades. Similarly, if the original phylogeny is a time-calibrated tree, it is possible to use DIVA to classify the inferred biogeographical events into time classes, and thus estimate changes over time in the frequency of these events [12]. Nevertheless, these are all indirect ways of incorporating time into biogeographical reconstructions; time itself is not used as evidence in the analysis.

In recent years, a whole new class of biogeographical methods has been developed to overcome these perceived limitations of the parsimony-based approach [17,18]. They are termed *model-based* or *parametric* methods because they use probabilistic models of range evolution whose variables or parameters are quantifiable biogeographical processes such as dispersal, extinction or range expansion. These probabilistic models, termed *Markov chain processes*, are inspired by models used in phylogenetic studies to trace the evolution of a character in a phylogeny, but here the character states are the geographical ranges of the species (e.g. A, B and C in Figure 8.6a). The change or transition between character states (here, alterations in the geographical range) is governed by a probabilistic matrix (Q) that defines the rate of moving from one state to another. In the example shown in Figure 8.6a, for example, the rate of moving between A and B (p) is higher than between A and C (q). The likelihood of change in the geographical range of a taxon along the phylogeny (from ancestors to descendants) is determined by this probabilistic matrix of rates and is also a function of time. Integrating the Markov chain biogeographical model, a phylogenetic tree with branch lengths calibrated in units of time, and associated terminal distributions (Figure 8.6a, centre), allows us to make a statistical estimate of ancestral geographical ranges at each node in the phylogeny, as well as the rate of biogeographical processes that alter the geographical range of a taxon (*p, q* and *r*). Thus, unlike event-based methods, the probability of occurrence of dispersal or extinction is not fixed *a priori,* but is instead estimated from the data in parametric approaches [17,18].

Model-based methods offer several additional advantages [18]:

1 *Time.* Its ability to take the timing of events into account is undoubtedly the most important contribution of parametric methods to historical biogeography. Time is important as a measure of the probability of occurrence of a biogeographical event, because it affects the likelihood that a biogeographical event will take place. In Figure 8.6b, branch lengths represent estimates of the time since the evolutionary divergence between Species 1 and Species 2. Both species are currently present in area A. However, in the phylogeny on the left, very little time has elapsed since the divergence of the two species, so it is likely that the ancestor is still in area A (the 'full' grey pie chart). In the phylogeny on the right, the much longer periods of time involved make it more likely that changes in the geographical range (such as jump dispersal, range expansion or extinction) took place. The result is that the inference of the ancestral area for the two species in the right phylogeny is much more uncertain than in the left phylogeny: it was probably area A, but there is a degree of possibility that they started with B (the blue wedge in the pie chart in Figure 8.6b).

Thus, by integrating the length of phylogenetic branches (i.e. time) into the analysis, parametric methods are able to take account of the fact that the likelihood of biogeographical change is higher along long branches than along shorter ones: this also has implications for the degree of error in the reconstruction. Additionally, by integrating absolute divergence times, parametric approaches can be used to disentangle pseudocongruence from real shared biogeographical history. For example, the 'duplication' event that gave rise to the two species in the left phylogeny is not the same as the one

in the right phylogeny; they are not temporally congruent.

2 *Uncertainty.* Another advantage of parametric methods over parsimony-based approaches is that they allow us to estimate the relative probabilities of alternative ancestral areas at phylogenetic nodes. This is possible because they evaluate all possible biogeographical scenarios during the inference, rather than only the most parsimonious ones [18]. Also, phylogenetic uncertainty can be accounted for through the use of *Bayesian inference.* This is a method of statistical inference in which evidence from one source such as phylogenetic relationships (tree topology, branch lengths) can be combined with evidence from other sources, such as ancestral range reconstruction, in such a way as to improve the precision of any estimate based on any one of them [15,19].

3 *Model assessment.* Model-based approaches provide a more rigorous statistical framework for the testing of alternative biogeographical hypotheses than do event-based methods. For example, two different biogeographical scenarios can be formulated in terms of alternative models and compared on the basis of how well they fit the data. Since each biogeographical scenario is described in terms of processes such as dispersal, range expansion or extinction, we can identify the processes that best explain the data by identifying the best-fitting model.

4 *Additional evidence.* Parsimony-based approaches were limited in their inference of biogeographical scenarios to the use of the topology of the phylogeny and the distribution of the terminal species. Parametric methods, because they are defined in terms of probabilistic models with biogeographical parameters, can easily incorporate additional sources of evidence. This can be done either in the form of new parameters in the Q matrix, or by scaling the rate or likelihood of processes according to independent sources of evidence. For example, in the biogeographical model in Figure 8.6a, rates of dispersal could be made (inversely) dependent on the geographical distance between the areas. Similarly, in an island system one could scale the dispersal rate to incorporate the strength of wind currents facilitating dispersal [19]. Similarly, in a continental setting, land migration could be made reliant on the availability of a geological corridor ('land bridge') connecting two continents [20].

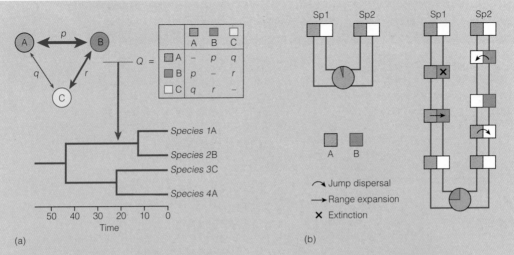

Figure 8.6 Examples of the use of parametric methods in biogeography. Biogeographical evolution is modelled as a probabilistic Markov chain process. See text for explanation.

The first parametric methods developed were very different (see [17,18] for a detailed explanation of them). The Dispersal–Extinction–Cladogenesis model (DEC) by Richard Ree and collaborators [21,22] implements a complex biogeographical model that allows geographical ranges comprising two or more areas (*widespread ranges*). It does this by modelling range evolution as being the result of two processes: range expansion from one area into another (e.g. from A to B in Figure 8.6b) and extinction within one area (X in area B; see Figure 8.6b) – direct dispersal between two areas (*jump dispersal* in Figure 8.6b) is not allowed. Instead, dispersal is modelled as range expansion – the ancestor moves into a new area but also retains its original distribution – followed by extinction in the original area ($D_{A\ to\ B} + E_A$). DEC uses maximum likelihood inference to estimate the relative probability of ancestral ranges at nodes, and the global rate of dispersal (range expansion) and extinction given a time-calibrated phylogeny and associated distributions. The Bayesian Island Biogeographical model (BIB) of Sanmartín, van der Mark and Ronquist [19] implements a simpler biogeographical parametric model that restricts geographical ranges to single areas, and models changes in geographical range only as jump dispersal. (*Jump dispersal* is when a lineage disperses from one area into another and is immediately followed by allopatric speciation into two disjunct lineages.)

Each model has its own advantages and weaknesses. Undoubtedly, the DEC model is more realistic in that it allows widespread ancestral distributions. However, this complexity comes with analytical limitations. For example, the size of the model increases exponentially with the number of areas – so, there would be three possible states for two areas (A, B and AB), seven possible states for three areas (A, B, C, AB, AC, BC and ABC), 15 for four areas and so on. The presence of ancestral widespread ranges implies the need to model biogeographical change at speciation nodes – in other words, the different ways (*speciation modes*) in which a widespread ancestral range could be divided into two descendant ranges. For example, the range AB may be divided by allopatry into non-overlapping subsets (A/B), by peripheral speciation in which one descendant inherits the entire range and the other only one area (AB/B). Alternatively, by sympatry, it could be the result of the two descendants inheriting the entire widespread range (AB/AB) – although this latter is not allowed in DEC [22]. The fact that dispersal is modelled exclusively as range expansion implies that DEC is more appropriate for the analysis of continental biogeographical scenarios. This is because, in such a scenario, areas are adjacent to one another, so that it is likely that lineages would expand their ranges and later speciate by vicariance or range division. In contrast, the simpler biogeographical model implemented in BIB does not allow widespread ancestral ranges and the Markov chain model has only one type of biogeographical process, *jump dispersal*. It is therefore more suitable for modelling scenarios in which areas are isolated by barriers that need to be crossed (e.g. oceanic islands) [19]. On the other hand, unlike DEC, where a phylogeny is generated first and is then used as input for the biogeographical inference, BIB uses Bayesian inference to simultaneously estimate phylogenetic relationships, ancestral geographical ranges and rates of biogeographical change from DNA sequence data and species geographical distributions. This implies that biogeographical parameters estimated in BIB are not dependent on a particular phylogeny as in DEC. This makes BIB more suitable for inferring general biogeographical patterns across a group of unrelated organisms that differ in their biological characteristics (dispersal ability, age of origin, rate of evolution etc.) but that inhabit the same set of areas – for example, an archipelago of oceanic islands [19].

There is one more difference between the BIB and DEC models [17]. The Markov chain stochastic models used in parametric biogeographical inference (and those used in modelling the process of nucleotide substitution in phylogenetics) are termed *time-homogeneous* or *stationary Markov models*. This is because the rates of change or transition between states (geographical ranges) are assumed to be constant over time and across lineages. They are also often assumed to be time-reversible, that is, the rate of change from state A to B is the same as that from B to A. In time-reversible, homogeneous Markov models, the rates of change between character states can be broken down into two parameters: the relative exchange rate parameters (r_{AB}), and the stationary frequencies of the character states (π_A). The latter correspond to the equilibrium frequencies of the process; over time, the state frequencies of a time-homogeneous Markov process converge to these values regardless of the starting point.

In the BIB model, the rates in the Q matrix can be also broken down into two parameters: the relative dispersal rate between two islands, and the **carrying capacity** of each island – the stationary frequencies or the number of lineages expected in each island at equilibrium conditions. This implies that a BIB model with three states such as the one in Figure 8.6a would include three carrying capacity parameters (π_A, π_B and π_C) and three (or six, if it is time-irreversible) relative dispersal rates (r_{AB}, r_{Ac} and r_{BC}). In contrast,

the standard DEC model assumes equal rates of range expansion (*D*) and extinction (*E*) among geographical ranges, and does not include a parameter for stationary frequencies. To obtain good estimates of model parameters in as complex a model as BIB, between-island dispersal rates and island carrying capacities are estimated from biogeographical data (geographical states) that are shared across multiple lineages. This increases the number of data points while allowing for lineage-specific differences in the rate of molecular evolution, age and ability to disperse [19]. The BIB model has so far been used to infer patterns of colonization in oceanic islands [19] and to study historical patterns of floristic exchange between isolated regions in the continental margins of Africa [23]. Later, it has been implemented in a phylogeographical context, for the inference of geographical evolution at the level of populations or individuals, for example to study patterns of viral spread [24].

Clearly, the biggest challenge of parametric methods is computational feasibility and learning how to balance this with increasing realism of the biogeographical scenarios [18]. To reduce the number of potential geographical ranges or states in the DEC model without decreasing the number of initial areas, one could use models in which movements between certain ancestral ranges are prohibited according to some biological, geographical or geological criteria. For example, in biogeographical continental scenarios, one could restrict widespread states (geographical ranges comprising two or more areas) to combinations only of areas that are geographically adjacent to one another. All other combinations would imply an extinction event in the intervening area or the crossing of a barrier between the two areas, so these widespread states are disallowed [18]. For island systems such as Hawaii, in which dispersal proceeds along the island chain, one can limit dispersal in BIB to follow a *stepping-stone model*, by making the rate of dispersal between nonadjacent islands in the chain equal to zero [19].

Another possible solution is to integrate additional sources of evidence. For example, fossil information can be used in DEC to limit the set of possible ancestral ranges of a phylogenetic clade to which the fossil is assigned in order to include the geographical distribution of the fossil [20].

Changing continental configurations (the collision and splitting of landmasses in reticulate scenarios) can also be modelled in DEC. This is done by dividing the phylogeny into time intervals, and assigning to each of them a different set of scaling values for the dispersal rate, depending on the availability of bridges facilitating land migration between areas in that particular time period [22]. Equally, dispersal rates could be made dependent on the existence of climatic dispersal corridors between two continents, based on the tolerances of the group analysed in the present and in the past (i.e. using fossil occurrences) [20]. This has opened an exciting area of research designed to integrate the ecological and evolutionary side of the discipline into a common research framework [2].

Recently, there have been methodological extensions of the DEC and BIB models. For example, the BioGeoBEARS model [25] extends DEC to introduce a third parameter, a rate of jump dispersal, and also allows for higher flexibility in modelling range inheritance scenarios at speciation nodes. The Bay-Area model [26] uses a Bayesian 'data augmentation approach' to increase the size of the parameter space, allowing for a larger number of geographical states. In the case of the BIB model, extensions have gone in the direction of relaxing the time homogeneity of the Markov process, by allowing the relative dispersal rates to vary over time or across time intervals in a stratified phylogeny [27]. The boundaries between the time intervals may be estimated from the molecular and phylogenetic data alongside the dispersal parameters. An exciting new approach is the modelling of non-equilibrium models, in which the island carrying capacities or equilibrium frequencies are allowed to change at a given point in time. This might be, for example, the result of an extinction event that partly wipes out the biota of an island or area, decreasing its standing diversity and changing the fundamental properties of the dispersal process. Bayesian inference could be used then to estimate the point in time when there is a change in equilibrium frequencies (carrying capacities) and also the intensity of the extinction event, which might have affected some areas more intensely than others.

Finally, there have been new developments towards the implementation of so-called *range-dependent diversification models*, in which there is a causal relationship between range evolution and lineage diversification. In all models described here, although the pattern of geography involved may change, this does not affect the rate or nature of the evolutionary processes that are simultaneously taking place in the phylogeny. However, this is unrealistic because, for example, we know that the dispersal of a lineage into a new area might increase its rate of speciation because it can now invade new niches and displace other species [28]. Similarly, widespread species that occupy more than one area are more likely to speciate by allopatry and less likely to become extinct [29]. Coupling rates of speciation and extinction to the process of range evolution allows us to address important questions such as whether migration into a new area increases diversification rate, and whether extinction is historically higher in a particular region rather than in another. Maximum likelihood and Bayesian Inference can be used to estimate the speciation and extinction rates associated with being or moving into a particular area. These models, however, are computationally demanding – they have more parameters than a standard parametric model [29] – and their inferential power remains to be tested with real complex datasets.

Despite these challenges, parametric models of biogeographical evolution represent an exciting new area of research in trying to reveal how evolutionary and ecological patterns and processes can be integrated in the reconstruction of the biogeographical history of lineages and biotas. We thus foresee an increase in the number and range of evolutionary questions that can be tackled with these approaches.

The Molecular Approach to Historical Biogeography

The investigation techniques described in the 'Reticulate Patterns' section have allowed great improvements in our understanding of biogeographical events of the distant past, because the use of geological evidence has enabled us to estimate when the taxa involved started to differentiate from one another. But if we consider the more recent past, our interest turns to a more detailed level of geographical and taxonomic change that such now-distant and large-scale processes as plate tectonic movements and changes in sea level are rarely of any help. Fortunately, advances in our ability to perceive rates of change in biological processes have now made it possible for us to interpret biogeographical changes that took place more recently.

As described in Chapter 6, each of the characteristics of an organism is controlled by genes, made up of highly complex molecules, which duplicate themselves at each cycle of cell division. Inevitably, from time to time there is a slight error in this process, and the resulting, slightly different gene is known as a *mutation*. Most of them either disappear from the genotype because they are harmful to the organism, or become accepted into the genotype by natural selection because they are advantageous to the organism. However, in the late 1960s the Japanese geneticist Motoo Kimura suggested that the vast majority of the mutations that become permanently established in the DNA molecule are neutral from the point of view of fitness – they neither harm nor help the organism concerned. They are therefore not affected by selective pressures, and so they steadily accumulate through time. As a result, the greater the number of molecular differences between two organisms, the more distant in time was their evolutionary separation from one another. We can thus place a timescale against the phylogenetic tree, so that its points of branching or 'nodes' are dated, transforming it into what is now called a *timetree* (see Chapter 10).

In the early days of testing Kimura's theory, relatively few proteins had been subjected to detailed structural analysis. But when these (e.g. haemoglobin or cytochrome c) were checked, they did indeed display a steady change in the course of time, which provided support for the neutral theory of molecular evolution. A variety of other biochemical molecules, such as the DNA, found in the mitochondria or chloroplasts, messenger RNA and ribosomal RNA (see Chapter 6), are now used in these studies. Although it soon became clear that the molecular clock's rate of ticking varies between evolutionary lineages, such variations can now be recognized and allowance can be made for them, for example by using Bayesian inference techniques that employ probability distributions to account for variation in the rate of molecular evolution, such as the *relaxed clock models* implemented in the software BEAST (Bayesian Evolutionary Analysis by Sampling Trees) [30]. Fur-

thermore, such variations are more important in studying smaller assemblages of DNA nucleotides; larger assemblages show more consistent rates of sequence evolution. Finally, the dating of nodes is usually based on at least one non-molecular datum, from palaeontological (fossil record) or geological (plate tectonic) data, or when this is missing, the rate of molecular evolution inferred from a higher level lineage including the study group can be used to calibrate the molecular clock (e.g. the family that encompasses the genus). Again, the use of Bayesian inference allows incorporating the uncertainty on these calibration points.

At the other extreme, these problems are minimized when we are studying the evolution of closely related groups, such as species or subspecies, and this has given rise to the new approach known as **phylogeography,** developed by the American biogeographer John Avise [31]. Its methodology is similar to that of phylogenetic biogeography because it, too, uses cladistics. However, its analyses are based on data from mitochondrial DNA (mtDNA) of animals, which evolves rapidly, is passed down only via maternal inheritance and does not undergo the complex genetic changes and exchanges of meiosis. As a result, it can reveal relationships of animals at the level of species or species complexes, and can therefore document distributional changes that have taken place in the more recent evolutionary past.

Study of the details of the precise sequence of genes within the mtDNA of different populations within the range of a given species has shown that these differ from one another, that each sequence is found in a restricted area and that closely similar sequences are normally found in localities that are close together. Furthermore, where the differences are more pronounced, suggesting that they had taken place further back in time (in order to allow for the accumulation of these differences), the corresponding geographical ranges were also much greater. By combining such results from clades from within different groups occupying the same areas, comparative phylogeography can also reveal previously unexpected geographical relationships.

For example, the American biogeographers Brian Arbogast and Jim Kenagy [32] have studied the patterns of distribution of four different mammals that live in the forests of North America (Figure 8.7). Two new insights resulted from this research.

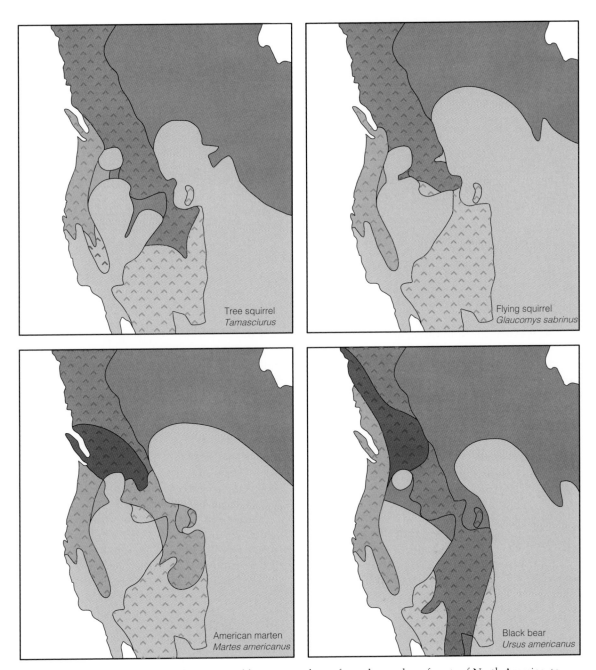

Figure 8.7 The phylogeographical structure of four mammal taxa from the northern forests of North America, to show the difference between the current taxonomy and the results of molecular phylogeography. The range of the mammals of Pacific Coast lineage is shown in light blue, the range of continental lineage in medium blue and the area of overlap of the two lineages in dark blue. Adapted from Arbogast and Kenagy [32]. (Reproduced with permission of John Wiley & Sons.)

First, they all show a genetic difference between a Pacific *coastal* lineage and an interior *continental* lineage. In three of the cases shown in Figure 8.7, this difference is present within the distribution of what had previously been considered a single species. Only in the case of the tree squirrel *Tamasciurus* had the difference been recognized as distinguishing two species, *Tamasciurus douglasi* and *Tamasciurus hudsonicus*. Second, the geographical position of the break between the two lineages is similar in all the groups. This suggests that they all responded similarly, by vicariant evolution, to an episode of fragmentation of their ranges. Both the fossil record and the genetic evidence suggest that this may have been connected with Quaternary episodes of cyclical fragmentation, contraction and expansion of the boreal forest. The later expansion allowed the coastal and continental lineages to come into renewed contact, which sometimes led to an overlap of their ranges. Neither of these insights was obvious from earlier analyses of these groups.

Inevitably, this method works better with groups whose dispersal ability is limited and that are therefore found within a smaller geographical area. However, the molecular genetics of more readily dispersed animals that are found on either side of an impenetrable barrier can also provide interesting and valuable phylogenetic information. For example, the populations of many marine animals became divided on either side of the Panama Isthmus when that formed. Since we know that this took place about 3 million years ago, this provides very useful data on the rate of change of their mtDNA.

Comparative phylogeography is now playing an increasingly important part in unravelling very detailed aspects of the patterns of biogeography that have resulted from events in the last few million years. In order to interpret the detailed information on the geographical distribution and genetic constitution of minor clades that comparative phylogeography can sometimes provide, it has now become necessary to develop complex computer methods such as **nested clade analysis (NCA)**.

The NCA method was developed by Alan Templeton of Washington University, and it illuminates the evolutionary history of a species over space and time by using information from its molecular structure. It first defines a series of haplotypes of a given molecule, first arranging these into groups that differ from one another by only a single mutational difference – the '1-step' clades. These groups are then treated as the units for an identical process that produces '2-step' clades, which similarly differ from one another by only a single mutation. This process is continued until there is only a single clade unit that includes all the haplotypes.

This technique can be illustrated by an investigation into the phylogeography of the Greek populations of the European chub *Leuciscus cephalus*, a fish that lives in rivers [33]. Figure 8.8 shows the area in which different populations of the fish were sampled, together with illustrations (Figure 8.8a and 8.8b) of two different theories as to how their distributions might have come about. Bianco's hypothesis (Figure 8.8a) suggests that the Danubian population spread to both eastern and central Greece via the Black Sea, while the western Greek populations were the result of a southward spread down the west coast. Economidis and Banarescu (Figure 8.8b) instead suggest that the spread of the Danubian chub via the Aegean only reached eastern Greece, whereas other Danubian chub spread southward by changes in the patterns of drainage of rivers (1–4) first to central Greece and then westward to the coastal regions.

The NCA of the chub populations is shown in Figure 8.8c and shows that they fall into three groups. The existence of a Danubian/Central Greek clade and of a separate eastern Greek clade supports those aspects of Economidis and Banarescu's theory. On the other hand, the presence of a quite separate western Greek clade supports Bianco's hypothesis that the chub of this region originated directly from the north and not from the Danubian populations. Other analyses of the data allow one to estimate the relative times, and therefore sequence, of these events and therefore their relationship to other geological or climatic phenomena.

Comparative phylogeography is now starting to throw light on an increasingly varied range of topics within the biogeography of the last few million years. These range from the expansion of a species out of the refugia to which it became restricted during the Ice Ages [34] and the direction of gene flow between geographically separate populations of sister species [35], to the relative contributions of historical and recent processes in causing the

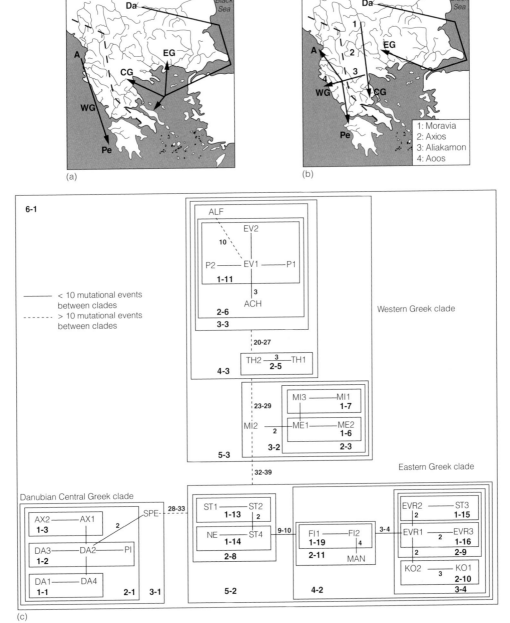

Figure 8.8 (a, b) Two hypotheses as to the biogeographical history of populations of the chub, *Leuciscus cephalus*, in Greece. From Durand *et al.* [33]. (c) Nested clade analysis of the interrelationships of these populations. Letters (e.g. ALF) represent abbreviations of the different locations in which they were found. Numbers indicate the hierarchy of clades (e.g. 3–3 indicates a three-step clade that includes 2–6, a two-step clade that in turn includes a one-step clade, 1–11). Each solid line without a number represents a single mutational change from one of the two genotypes to the other. Any numbers beside the lines indicate the number of mutations that took place between the two genotypes linked by the lines (e.g. there were between 32 and 39 mutational differences between the western Greek clade, 5–3, and the eastern Greek clade, 5–2). See text for further explanation.

distribution of genes in a population today [36]. Inevitably, it is starting to overlap with investigations into processes that took place earlier in the Tertiary [37]. A paper [38] by three biogeographers (Richards, Carstens and Knowles) working in the University of Michigan has suggested techniques that allow for the integration of genetic data and distributional data with palaeoclimatic modelling, and so testing alternative hypotheses about the timing and pattern of divergence of the populations.

Avise has outlined [31] the new perspectives that phylogeography has enabled; he also suggests some potentially fruitful directions that new research may take, such as comparative phylogeographical appraisals of multispecies regional biotas to identify shared patterns resulting from major historical events. It has proved to be a veritable revolution in population genetics.

Investigations of the genetics and distribution of populations today have to cope with a very large amount of data, and this is provoking the development of complex computer programmes. The variety, complexity and rate of appearance and modification of these are such that it is not appropriate to try to explain them in a book such as this, and students are instead advised to follow them up in the current relevant research literature.

Molecules and the More Distant Past

Although the difficulties in the use of molecular data, mentioned in this chapter, are minimized when studying more recent and more closely related groups, such data are still very valuable in studying events in the more distant past [39]. The molecular data can be used to set up a phyletic tree of relationship with the implied dates of divergence between the different lineages. A similar tree of relationship is set up using data from the fossil record, which also gives dates of divergence based on the ages of the fossils. This in its turn introduces the possibility of errors. The most fundamental and inevitable of these is that the age of any fossil, placed on a given lineage, is a minimum age, because other, undiscovered, older fossils may lie on this same lineage before its divergence from its closest relatives. Other errors may arise from

incorrect estimates of relationship, or if the age estimates of the fossil-bearing rocks are wrong. But recent techniques reduce the impact of any such individual error by calculating the effect that each datum has on the overall degree of difference between the molecular tree and the fossil-based tree [40]. Those that produce the greatest degree of inconsistency can then be removed from the data set and/or be reappraised. The results can then also be compared with different models of the rates of substitution in the molecular data tree in order to find a cross-correlated set that minimizes the differences between the two independent systems.

The extent to which plate tectonic, geological and timescale information has been integrated into modern methods of analysis of historical biogeography has created major problems for the proponents of pattern-based methods. The response of some of them has been to object to the use of any geological information, claiming that 'biogeography should not be subordinate to geology', and to point out that the geological information might be wrong and that there are other theories of the history of the continents, such as the expanding earth theory [41]. However, the value of using this information is not that it makes the biological data dependent on the geological data and theories, but that it provides an independent picture of events, with which the biological picture can be compared and evaluated. This is normal good practice in any part of science and, if an expanding-earth theory seemed likely to provide a better match for the biological scenario, biogeographers would be quite prepared to consider it. At present, that is not the case.

Comparisons with the ages of plate tectonic events are not, however, the only examples of the uses of molecular data in relating biological events to environmental events. For example, ancient patterns of river drainage have been revealed as causative agents in the biogeography of freshwater crabs in the Cape of South Africa [42] and of freshwater molluscs in western North America [43]. Perhaps the most complex set of factors is that being revealed in the history of the biota of the Amazon Basin (see Chapter 11). Here, molecular phylogeny has not only shown that the accepted taxonomy of such groups as parrots was incorrect, but has also revealed that the diversification patterns were much older than had been thought and had been influenced by the location and timing of episodes of mountain building. Other variations in the climate of the Basin seem to have been linked to changes in the location of the low-level jet stream, itself affected by the long-term precessional changes in the Earth's orbit [44]. Faced with ever more detailed biological data, biogeographers are now having to become aware of ideas and information in a greater variety of other branches of science than ever before.

In this chapter, we have shown how a species is defined, how it evolved and is related to other species, where it lives and how we can use our knowledge to interpret its history. That understanding will now help us to reconstruct how the patterns of life on the planet came to appear through time. This, the study of historical biogeography, makes up Chapters 10 and 11 of this book. We have already examined the patterns of life on islands in Chapter 7, which vividly illustrate evolution in action, and shall now turn to discuss the patterns of biogeography in the oceans that surround them.

Summary

1 If two closely related species are widely separated from one another, did the original species disperse across a pre-existing barrier, or did the barrier arise later, separating them by a process known as vicariance? The two possible answers to this question, as causes of patterns of disjunct patterns of distribution, have been supported by two opposing, disputing schools of biogeographers.

2 Cladistics, together with information on the areas of endemism of the taxa, can produce a taxon–area cladogram. Comparison of this with the plate tectonic history of the areas sometimes shows a relationship between the two processes. Further results may be obtained by comparing the taxon–area cladograms of different groups.

3 The many uncertainties involved in interpreting the results of these methods led to the development of complicated statistical analyses to try to resolve the problems. More recently, event-based methods have been developed. These specify whether dispersal, vicariance, extinction or duplication of taxa were involved at each point in the cladogram, and greatly improved the resolving power of the interpretation. Event-based methods can also analyse the biogeographical history of areas that have had more than one set of relationships with one another over time – known as a reticulate pattern.

4 In all these methods, the use of parsimony distorts the interpretation of the likelihood of the different biological processes and limits the variety of data that can be employed. Recently, new model-based or parametric methods have been developed that avoid these limitations.

5 All of these methods have become far more effective since the development of molecular methods of estimating the dates of divergence of taxa. These can then be compared with the dating of geological or climatic events, often unequivocally indicating whether the patterns were caused by dispersal or vicariance. In most cases, this has shown that dispersal, rather than vicariance, was the agent.

6 Using these methods, we can now analyse details of the evolution of the community of a biome, the coevolution within it and the processes of genetic change and dispersal of populations in response to environmental change. All of this is leading to an exciting integration of historical and ecological biogeography.

Further Reading

Lomolino MV, Heaney LR (eds.). *Frontiers of Biogeography*. Sunderland, MA: Sinauer Associates, 2004.

Pennington RT, Cronk QCB, Richardson JA. Plant phylogeny and the origin of major biomes. *Philosophical Transactions of the Royal Society of London B* 2004; 359: 1453–1656.

References

1. Brundin LZ. Phylogenetic biogeography. In: Myers AA, Giller PS (eds.), *Analytical Biogeography*. London: Chapman & Hall, 1988: 343–369.

2. Sanmartín I. Historical biogeography: evolution in time and space. *Evolution: Education and Outreach* 2012; 5: 555–568.

3. Morrone JJ. *Evolutionary Biogeography. An Integrative Approach with Case Studies.* New York: Columbia University Press, 2000.

4. Crisci JV, Katinas L, Posadas P. *Historical Biogeography: An Introduction.* Cambridge, MA: Harvard University Press, 2000.

5. Wojcicki M, Brooks DR. PACT: an efficient and powerful algorithm for generating area cladograms. *Journal of Biogeography* 2005; 32: 755–774.

6. Sanmartín I. Event-based biogeography: integrating patterns, processes and time. In: Ebach MC, Tangney RS (eds.), *Biogeography in a Changing World.* London: CRC Press, 2007: 135–159.

7. Sanmartín I, Ronquist F. Southern hemisphere biogeography inferred by event-based models: plants versus animal patterns. *Systematic Biology* 2004; 53: 216–243.

8. Muellner, AN, Pannell CM, Coleman A, Chase MW. The origin and evolution of Indomalesian, Australasian and Pacific island biotas: insights from Aglaieae (Meliaceae, Sapindales). *Journal of Biogeography* 2008; 35: 1768–1789.

9. Cook LG, Crisp M. Directional asymmetry of long-distance dispersal and colonization could mislead reconstructions of biogeography. *Journal of Biogeography* 2005; 32: 741–754.

10. Waters JM. Driven by the West Wind Drift? A synthesis of southern temperate marine biogeography, with new directions for dispersalism. *Journal of Biogeography* 2008; 35: 417–427.

11. Ronquist F. Dispersal-vicariance analysis: a new biogeographic approach to the quantification of historical biogeography. *Systematic Biology* 1997; 46: 195–203.

12. Sanmartín I, Enghoff H, Ronquist F. Patterns of animal dispersal, vicariance and diversification in the Holarctic. *Biological Journal of the Linnean Society of London* 2001; 73: 345–390.

13. Donoghue MJ, Smith S. Patterns in the assembly of temperate forests around the Northern Hemisphere. *Philosophical Transactions of the Royal Society of London B* 2004; 359: 1633–1644.

14. Kodandaramaiah U. Use of dispersal-vicariance analysis in biogeography – a critique. *Journal of Biogeography* 2010; 37: 3–11.

15. Ronquist F. Bayesian inference of character evolution. *Trends in Ecology and Evolution* 2004; 19: 475–471.

16. Nylander JAA, Olsson O, Alström P, Sanmartín I. Accounting for phylogenetic uncertainty in biogeography: a Bayesian approach to dispersal-vicariance analysis of the thrushes (Aves: *Turdus*). *Systematic Biology* 2008; 57: 257–268.

17. Ronquist F, Sanmartín I. Phylogenetic methods in biogeography. *Annual Review of Ecology, Evolution, and Systematics* 2011; 42: 441–464.

18. Ree RH, Sanmartín I. Prospects and challenges for parametric models in historical biogeographical inference. *Journal of Biogeography* 2009; 36: 1211–1220.

19. Sanmartín I, van der Mark P, Ronquist F. Inferring dispersal: a Bayesian approach to phylogeny-based island biogeography, with special reference to the Canary Islands. *Journal of Biogeography* 2008; 35: 428–449.

20. Meseguer AS, Lobo JM, Ree R, Beerling DJ, Sanmartín I. Integrating fossils, phylogenies, and niche models into biogeography to reveal ancient evolutionary history: the case of *Hypericum* (Hypericaceae). *Systematic Biology* 2015; 64 (2): 215–232.

21. Ree RH, Moore BR, Webb CO, Donoghue MJ. A likelihood framework for inferring the evolution of geographic range on phylogenetic trees. *Evolution* 2005; 59: 2299–2311.

22. Ree RH, Smith SA. Maximum likelihood inference of geographic range evolution by dispersal, local extinction, and cladogenesis. *Systematic Biology* 2008; 57: 4–14.

23. Sanmartín I, Anderson CL, Alarcon M, Ronquist F, Aldasoro JJ. Bayesian island biogeography in a continental setting: the Rand Flora case. *Biology Letters* 2010; 6: 703–707.

24. Lemey P, Rambaut A, Drummond AJ, Suchard MA. Bayesian phylogeography finds its roots. *PLoS Computational Biology* 2009; 5: e1000520.

25. Matzke NJ. Model selection in historical biogeography reveals that founder-event speciation is a crucial process in island clades. *Systematic Biology* 2014; 63: 951–970.

26. Landis MJ, Matzke NJ, Moore BM, Huelsenbeck JP. Bayesian analysis of biogeography when the number of areas is large. *Systematic Biology* 2013; 62: 789–804.

27. Bielejec F, Lemey P, Baele G, Rambaut A, Suchard MA. Inferring heterogeneous evolutionary processes through time: from sequence substitution to phylogeography. *Systematic Biology* 2014; 63: 493–504.

28. Moore BR, Donoghue MD. Correlates of diversification in the plant clade Dipsacales: geographic movement and evolutionary innovations. *American Naturalist* 2007; 170: S28–S55.

29. Goldberg EE, Lancaster LT, Ree RH. Phylogenetic inference of reciprocal effects between geographic range evolution and diversification. *Systematic Biology* 2011; 60: 451–465.

30. Drummond AJ, Rambaut A. BEAST: Bayesian evolutionary analysis by sampling trees. *Evolutionary Biology* 2007; 7: 214.

31. Avise JC. Phylogeography: retrospect and prospect. *Journal of Biogeography* 2009; 36: 3–15.

32. Arbogast B, Kenagy GJ. Comparative phylogeography as an integrative approach to historical biogeography. *Journal of Biogeography* 2001; 28: 819–825.

33. Durand JD, Templeton AR, Guinand B, Imsiridou A, Bouvet Y. Nested clade and phylogeographic analysis of the chub, *Leuciscus cephalus* (Teleostei, Cyprinidae) in Greece: implications for Balkan Peninsula biogeography. *Molecular Phylogenetics and Evolution* 1999; 13: 566–580.

34. Hewitt GM. Genetic consequences of climatic oscillations in the Quaternary. *Philosophical Transactions of the Royal Society of London B* 2004; 359: 183–195.

35. Zheng XJ, Arbogast BS, Kenagy GJ Historical demography and genetic structure of sister species: deermice (*Peromyscus*) in North American temperate rain forest. *Molecular Ecology* 2003; 12: 711–724.

36. Templeton AR, Routman E, Phillips CA. Separating population structure from population history: a cladistic analysis of the geographical distribution of mitochondrial DNA haplotypes in the tiger salamander, *Ambystoma tigrinum. Genetics* 1995; 140: 767–782.

37. Riddle BR, Hafner DJ. A step-wise approach to integrating phylogeographic and phylogenetic perspectives on the history of a core North American warm desert biota. *Journal of Arid Environments* 2006; 65: 435–461.

38. Richards CL, Carstens BC, Knowles LL. Distribution modelling and statistical phylogeography: an integrative framework for generating and testing alternative biogeographical hypotheses. *Journal of Biogeography* 2007; 34: 1833–1845.

39. Donoghue PCJ, Benton MJ. Rocks and clocks: calibrating the Tree of Life using fossils and molecules. *Trends in Ecology and Evolution* 2007; 22: 425–431.

40. Near TJ, Sanderson MJ Assessing the quality of molecular divergence time estimates by fossil calibrations and fossil-based model selection. *Philosophical Transactions of the Royal Society of London B* 2004; 359: 1477–1483.

41. McCarthy D. Are plate-tectonic explanations for trans-Pacific disjunctions plausible? Empirical tests of radical dispersalist theories. In: Ebach MC, Tangney RS (eds.), *Biogeography in a Changing World.* New York: CRC Press, 2007: 177–198.

42. Daniels SR, Gouws G, Crandall KA. Phylogeographic patterning in a freshwater crab species (Decapoda: Potamonautidae: Potamonautes) reveals the signature of historical climatic events. *Journal of Biogeography* 2006; 33: 1538–1549.

43. Liu H-P, Hershler R. A test of the vicariance hypothesis of western North American freshwater biogeography. *Journal of Biogeography* 2007; 34: 534–548.

44. Bush MB. Of orogeny, precipitation, precession and parrots. *Journal of Biogeography* 2005; 32: 1301–1302.

Patterns in the Oceans

Chapter 9

Life clothes the land; it merely stains the seas.

The oceans and seas of the world contain three very different types of habitat: the vast volumes of the open oceans, the floor of the deep ocean far from the light of the surface, and the far richer life of the shallow seas around the continents and oceans. Our understanding of the biogeography of these environments has been limited by two factors. First, it is far more difficult for us, as air-breathing creatures, to survey and sample them – particularly in those cases where they are far from land and at great depth. But it has also become clear that, unlike terrestrial species, marine species often cannot be distinguished by differences in morphology, so their taxonomy has to be based on more subtle genetic comparison. It is also more difficult to recognize the barriers that lie between the areas of distribution of marine species. As a result, our understanding of marine biogeography is still at an early stage of development.

The biogeography of the continental masses and that of the sea are similar in one way: both involve analysis of the biotas of vast areas of the surface of the globe. However, because their environments are very different, the marine biotas are much more difficult to study. As a result, we know far less about the composition and ecology of marine organisms than we do about those of the land. In many areas of marine research, we are therefore still at the stage of constructing and evaluating hypotheses at a comparatively basic level. This fact is of more than merely academic and technical importance, in view of our desire and need to conserve the world's present diversity of organisms.

To do this, we must first understand the fundamental patterns of distribution, both of the ecosystems and of the organisms they contain. Only then can we identify those that are threatened because of their rarity or because of their vulnerability to ecological change – whether natural or the result of human activities.

The terrestrial biogeographical regions are, effectively, the different continents. The interplay of the topography of the land and of the seasonal cycles in climate within each region also produces a considerable variety of physical environments. Those regions are usually separated from one another by barriers of ocean, mountain or desert that make it difficult for organisms to disperse from one to another. The geographical boundaries of the regions are therefore easy to define. Their inhabitants live in air, which has a very low density. As a result, it is impossible for terrestrial organisms to be permanently airborne, and it does not provide much help in their long-range dispersal. On the other hand, it allows plants to become structurally complex. In most parts of the continents (excluding the tundra, steppe and desert), the plants therefore dominate the environment. In addition to the variety of physical environments, the plants therefore add their own living architecture (grassland, woodland, forest, etc.) within which the animals exist. These habitats provide a framework for biogeographical analysis at a finer level of detail, and our investigations are helped by the fact that we, too, are terrestrial.

The liquid world of the oceans is quite different. The major oceans are all interconnected, so that their geographical boundaries are less clear than those of the continents. As a result, their biotas cannot show such clear differences as those on land. The oceans are also far larger than the continents, for they make up 71% of the surface of our planet and, because of their depth, they account for 97% of its habitable volume. The oceans themselves are continually moving because the water within each ocean basin slowly rotates. These moving waters carry marine organisms from place to place, and also help the dispersal of their young or larvae. Furthermore, the gradients between the environments (and therefore between the different faunas) of different areas of ocean water mass are very gradual and often extend over wide areas that are inhabited by a great variety of organisms of differing ecological tolerances. There are no firm boundaries within the open oceans.

Many terrestrial animals have evolved flight, but due to the low density of air it is impossible to remain aloft permanently. This constraint does not apply to ocean-dwelling organisms, for whom the high density of water enables a fully pelagic lifestyle. Over 90% of the marine environment's living space is pelagic (the three-dimensional layer from the water's surface through to the deepest trenches or bathypelagic zones). It is therefore not surprising that pelagic systems support the majority of marine biomass. Although pelagic areas may seem like a boundary-less continuum, water chemistry, salinity, depth/pressure, currents and variations in primary productivity create different regions within these systems with distinct biotas and geographies.

Because of the density and power of the waters, however, there are no large, complex plants to provide the equivalent of the terrestrial biomes. Photosynthesis in the sea is carried out mainly by tiny, single-celled organisms known as **phytoplankton**. The density of plant life in the ocean is therefore far less than that on land, and the primary productivity per unit area is only one-fifteenth of that in tropical rainforest, so that much less solar energy is being fixed in the system.

Nevertheless, there is one way in which the oceans are more complex than the land: they have an important extra dimension, that of depth. The physical conditions of light, temperature, density and pressure – and often also the concentrations of nutrients and oxygen – change much more rapidly with depth in the seas than they do with altitude on land. These lead to corresponding changes in the biotas. In fact, quite unlike the situation on land, the most important environmental gradients and discontinuities take place in the vertical dimension, not the horizontal, and are far more abrupt. These resulting vertical patterns of distribution also interact with the horizontal patterns.

Because we ourselves are terrestrial and air-breathing, it is difficult for us to study and census the life of the sea, even in the near-shore or surface-water regions, and even more so in its depths. Our knowledge of the fauna of the deep-ocean seabed, which covers an area of 270 million km^2 (10 sq. miles), is derived from cores totalling only about 500 m^2, together with the areas sampled from dragging a number of trawls and deep-sea sleds over the bottom! But our analysis of the life in the sea may well be blinkered by a far more fundamental error: the nature and recognition of species in the aquatic environment. One study comments on the extent to which many 'species' in fact contain more than one real species (and perhaps even a complex of several species); the reality only becomes apparent after phylogeographical analysis [1]. Inevitably, we base our identification of a species on those aspects of its appearance and behaviour that we can see and understand. But species recognition in the waters may well be based on biochemical signals that we would find very difficult to identify and monitor. This lack of comprehension may also underlie our problems with speciation in the waters of the African Great Lakes (see Chapter 6).

The history of the changing interconnections between the oceans in the past are shown in Plate 7 and described in Chapter 5. Today, all the oceans interconnect at high southern latitudes, and the Panama Isthmus was only completed in comparatively recent times, about 3 million years ago. Because of this lack of physical barriers in the oceans and their lower productivity, there may have been less opportunity for evolutionary diversification. As presently described, marine families contain fewer genera than terrestrial families, and these genera contain fewer species, so that many fewer species are known from the sea: only about 210 000 species of marine organisms have been described, compared with about 1.8 million from the land (but see the 'Faunas and Barriers in the

Oceans' section). Similarly, there are probably over 250 000 species of land plant, but only 3500–4500 species of phytoplankton. It has been estimated that we have recorded only about 10% of the biota of the oceans. However, even our very limited sampling of the seafloor shows a quite surprising variety of life, which has led the American oceanographers Frederick Grassle and Nancy Maciolek to estimate that the deep sea may contain between 1 million and 10 million species [2]. We have explored less than 5% of the deep sea, and we know less about it than about the dark side of the moon.

As we shall see in this chapter, although our understanding of the distribution of marine organisms is still limited, it would appear that, with the exception of those that live in the intertidal or shallow-water regions, they usually have a much wider distribution than those on land – at least at the level of the family or genus. Thus, while most families of mammal are found in a single zoogeographical region, most families of marine organisms are cosmopolitan or widespread throughout the world's oceans. As a result, marine faunas differ from one another in containing different genera or species rather than different families. These genera or species are not known by different English-language names, so we can only refer to them by their Latin names. Thus, although it is easy to explain that, for example, the South American zoogeographical region contains the endemic families of armadillo, anteater and sloth, one can only explain the differences between the biotas of marine regions by giving lists of Latin names.

The great ocean basins are equivalent to the continents as the major biogeographical units. Even though their patterns are more diffuse, they exhibit the same phenomena and raise the same problems of explanation by dispersal or vicariance, or of the extent to which differences are due to different histories of enlargement, fusion or subdivision, or to different evolutionary events. But since the marine faunas are still much less well known, we need to be much more cautious in coming to conclusions, or in assuming that particular deductions and generalizations associated with continental biogeography are necessarily valid for marine biogeography. Nevertheless, a study of the biogeography of the hake *Merluccius* by the South

African marine zoologists Stewart Grant and Rob Leslie [3] has shown some interesting parallels with the more familiar maps of dispersal on the landmasses (Figure 9.1). Fossils and the phylogeny suggested by molecular data make it likely that the genus originated in the Early Oligocene in the shallow epicontinental seas of the northeastern Atlantic–Arctic. From here, Old World and New World lineages diverged down the continental shelves of the eastern and western coasts of the Atlantics. The Old World lineage penetrated beyond the Cape of Good Hope into the Indian Ocean, with subsequent episodes of back-dispersal to and from West Africa. The New World lineage spread through the still-open gap between North and South America into the Pacific Ocean, within which it spread northward, southward and westward, with an episode of back-dispersal around Cape Horn into the South Atlantic.

Zones in the Ocean and on the Seafloor

We need to understand the biogeography of the oceans in order to use this information for more effective conservation and sustainable use of their biodiversity in areas outside those under national jurisdiction. (The latter areas, at depths of 300–800 m, are therefore excluded from the UNESCO report cited in [4].) To that end, an international group of experts has therefore published such an environmental classification of the open oceans and deep seabed [4], the maps, definitions and information of which are used here. Our knowledge of the biota of the provinces defined in that work is still very limited, and there is little doubt that many years of work, and of international collaboration over the results of the collection of specimens, will be needed before the biota can be properly characterized.

Around the deep ocean that makes up just over 50% of the marine environment are areas of shallow sea adjacent to the continents. The distinction between the two is a result of the basic structure of our planet. The continents rise above the level of the ocean floor because the rocks of which they are made are less dense and therefore lighter than those of the ocean floor. Seas cover the lower-lying parts of the continents, which are known as the

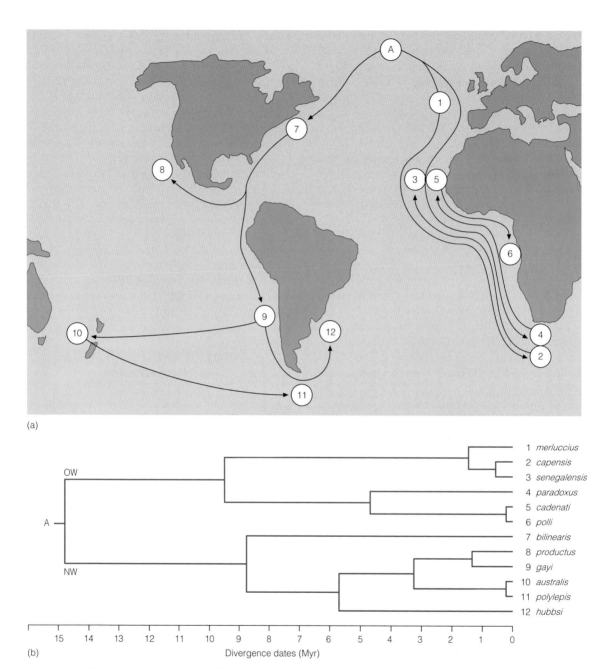

(a)

(b)

1	*merluccius*
2	*capensis*
3	*senegalensis*
4	*paradoxus*
5	*cadenati*
6	*polli*
7	*bilinearis*
8	*productus*
9	*gayi*
10	*australis*
11	*polylepis*
12	*hubbsi*

Divergence dates (Myr)

Figure 9.1 The dispersal of 12 species of hake, *Merluccius*, as suggested by fossil and molecular data. (a) Map showing distributions and dispersal. The numbers refer to the individual species shown in the cladogram of phylogenetic relationships (b), which also shows the dates of divergence between the different lineages, in millions of years. The numbers refer to the individual species as shown in the cladogram. A, ancestral *Merluccius*; NW, New World lineage; OW, Old World lineage. Data from Grant and Leslie [3].

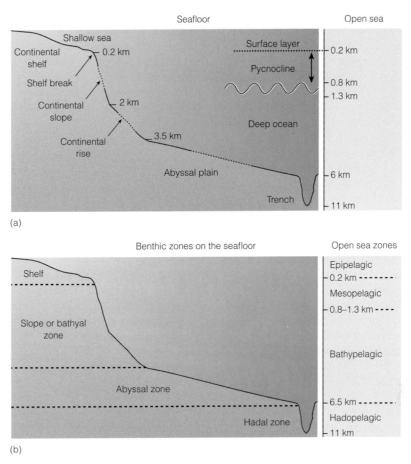

Figure 9.2 (a) Diagram of the vertical divisions of the seafloor and of the open sea. (b) Life zones on the seafloor and in the open sea.

continental shelves (Figure 9.2a). The depth of the water in these seas, which are known as *epicontinental seas* or shallow seas, varies according to how much of it has gone to form ice sheets and glaciers on the continents. At present, they have a maximum depth of 200 m (except in the Antarctic, where the great weight of ice has depressed the continent, so that the edge of the continental shelf lies at about 500 m). At this point, known as the **shelf break**, which is at an average depth of 135 m, the slight gradient of the seafloor increases sharply. From here onward, the continental slope descends with relative steepness until it reaches the deep ocean floor, which is known as the **abyssal plain**,

which lies at depths of 3500–6500 m. Sediments cover all of these ocean floors. Upon the continental shelf and continental slope are thick layers of sediment that are derived primarily from the erosion of the rocks of the continents. These sediments spill down over the edge of the continental slope and onto the adjacent margin of the abyssal plain to form a wedge known as the **continental rise**. The continental slope and continental rise together form the **bathyal** depth **zone** (Figure 9.2b). At an average depth of 4 km, the abyssal plain covers 94% of the area of the oceans and 64% of the surface of the world, and is therefore the most extensive of all environments. In the Indian and

Pacific oceans, the abyssal plain is also fringed by a system of trenches, reaching depths of up to 11 000 m, where old ocean floor disappears back into the depths of the earth (see Chapter 5).

Because the physical conditions change as one moves downward, away from the light and warmth of the surface, a number of different depth zones can be distinguished in the waters of the sea, while the shape of the seafloor similarly defines several different regions there (Figure 9.2).

Perhaps the most important physical characteristic of the sea is that conditions do not change uniformly from those at the surface, where the water is warmer and therefore less dense, to the colder denser conditions at greater depths. Instead, there is a comparatively narrow zone of rapid change in density, known as the **pycnocline**. (This is most often caused by a rapid change of temperature, in which case it is also known as a **thermocline**. But it can also be caused by a rapid change in salinity, in which case it is known as a **halocline**. Both can exist in the same area, as in some parts of the western tropical Pacific, where a near-surface halocline lies about 110 m above a thermocline.)

The surface layer of the sea, down to a depth of about 200 m, is the most liable to extremes of temperature. This can be as low as 21.9°C (the freezing point of seawater) below the ice cover found in high latitudes, or over 30°C in enclosed low-latitude waters such as the Red Sea. The rapid fall in temperature associated with the pycnocline commences at a depth of between 25 and 250 m and continues until, at a depth of 800–1300 m, it reaches about 4°C. At greater depths, the temperature decreases slowly until, at most latitudes, it reaches a minimum of 2–3°C at around 3000 m. In polar seas, however, it may be as low as 20.5°C in the Norwegian Sea, or 22.2°C in parts of the Southern Ocean. Below 3000 m, there is only a very slight temperature gradient.

As we shall see, the pycnocline is the most important ecological boundary in the ocean, and it is therefore useful to outline the major features that control its characteristics. The depth at which it begins, as well as its thickness, varies according to latitude, region and season. Closer to the equator, between about 20°N and 20°S, the pycnocline commences at a shallower depth, and is therefore within the lighted zone: it therefore shows a greater

rate of change with depth and is more resistant to mixing with the adjacent waters. There is also little change in temperature through the year in this near-equatorial band, so that the pycnocline is a permanent feature and lies at a constant depth. Closer to the poles, the depth at which the pycnocline begins lies at a deeper level but also changes according to the time of year, being deeper in winter than in summer. It is only rarely and temporarily in the lighted zone, shows a slower rate of change with depth and is less resistant to mixing. In near-polar regions, the pycnocline may disappear altogether in winter.

Basic Biogeography of the Seas

The shape of the ocean basins is the cause of the most basic divide within marine biogeography – between the shallow-sea (or **neritic**) environment and the open-sea environment. The major difference between these two environments is that of scale. The shallow seas occupy a smaller area than the oceans, and each of them is also far smaller than any ocean. But the individual shallow seas also differ more from one another than do the oceans, and each shallow sea contains within itself a greater variety of conditions than any equivalent area of ocean. As we shall see, the patterns of life in the oceanic environment are the results of external factors, such as the Earth's rotation and the heat and light of the sun. The resulting patterns can be seen at the largest scale, for example at the level of the North Pacific or the South Atlantic Ocean. Although these oceanic patterns also affect the shallow seas, they are less obvious there than local influences such as the nature of the seafloor, the contribution of sediments and freshwater by local rivers and streams, or the pattern of tides. In addition to these physical differences, the various shallow seas are also isolated from one another by wide stretches of ocean. As a result, their biotas are similarly isolated, giving the opportunity for independent evolution and endemicity.

A different divide within marine biogeography is between the patterns of distribution of **pelagic** organisms, which swim or float within the waters themselves, and those of **benthic** organisms, which live on or in the seafloor. (Pelagic organisms that swim are known as **nektonic**, while those that float are known as **planktonic**.) The pelagic world is dynamic, its environment in both the horizontal and vertical planes being dominated by the systems of oceanic circulation. The positions of different features of its regions are predictable only on scales of a few tens of kilometres and a few weeks' duration, and they are connected with one another on relatively short time scales compared to those of the life cycles and evolutionary changes of a region's biota. In contrast to this, the life of the benthic community is dominated by geomorphological features such as the topography and nature of the substrate, and the depth; these are stable for centuries, down to a scale of a metre or less. The resulting communities have higher levels of local endemism and are therefore more heterogeneous than those of the pelagic world, because they are less connected with one another and have slower rates of dispersal.

The Open-Sea Environment

Below the upper few tens of metres of water, which is called the **euphotic zone**, there is not enough sunlight to support photosynthesis. The heating effects of the sunlight, too, are mainly restricted to the upper, wind-mixed layer, termed the **epipelagic zone**, which varies in thickness from a few tens of metres to 200 m (Figure 9.2b). This warmer layer contains a high concentration of living organisms, whose activity reduces the nutrient content of the water. Below it lies the pycnocline, a zone in which the temperature drops rapidly and where the density of the water therefore also increases rapidly. As a result, the pycnocline is the most important boundary within the ocean waters. However, it is not merely a zone of rapid transition but also an environment in its own right, with its own characteristic biota that, in this poorly lit zone, is comparable to the shade flora of terrestrial ecology [5].

In the twilight zone below the pycnocline lies the **mesopelagic zone**, which extends to depths of approximately 1000 m. Many of the fish that live in this zone during the day move up towards the surface at night, to feed on the richer fauna

there. Below the mesopelagic zone lies the **bathypelagic zone**, which extends to a depth of 6500 m; it is a zone of total darkness and almost unchanging cool temperatures. The fish of this zone are too far from the surface to migrate there daily. Finally, the **hadopelagic** faunas of the deep trenches are thought to be mainly endemic because many of the trenches are isolated from one another.

The biogeography of the open-sea environment is best approached by first describing the circulation patterns within the oceans, for these have led to differences in their nutrient concentrations, which in turn have affected the patterns of distribution of life there. Much of this has been reviewed by the British oceanographer Martin Angel [5,6].

Dynamics of the Ocean Basins

The patterns of the waters' movement in the oceans are dominated by stable **gyres** – huge masses of water, filling a large proportion of an ocean basin, which rotate horizontally, with a periodicity that is usually 19–20 years. Their rotation is caused by wind patterns, which themselves result from the uneven distribution of solar energy on the surface of the Earth, as well as from the eastward rotation of the Earth due to the action of the Coriolis force (see Chapter 3). Heat from the equatorial regions is distributed towards the poles by patterns of wind movement that rotate clockwise in the northern mid-latitudes and anti-clockwise in the southern mid-latitudes, forming the subtropical high-pressure systems. As a result the winds, and therefore the warm waters, of the equatorial region move westward, and, as they reach the landmasses to their west, they are deflected towards the poles. The pattern of circulation within the gyre is completed by cold currents that flow away from the poles and towards the equator along the eastern edges of the oceans (see Figure 9.5). The resulting worldwide pattern of ocean currents is a major element in the transfer of heat around the world (discussed further in this chapter).

Major gyres lie either side of the equator in both the Atlantic and the Pacific oceans. In both these oceans there is also a minor gyre to the north, driven by the low Arctic low-pressure system.

This does not happen in the Southern Hemisphere because the southern ends of the continents lie far from the Antarctic continent, and therefore allow the development of a strong circum-Antarctic current.

The winds in the Indian Ocean show a seasonal reversal, the 'monsoon' (cf. Figure 3.14): the winds blow from the northeast from November to March or April, and from the southwest from May to September. As a consequence, the ocean current patterns in the Indian Ocean are similarly seasonal. The Arctic Ocean is almost totally enclosed by land, so that there is only a little southward dispersal of its waters, mainly to the east and west of Greenland.

In addition to these horizontal movements of the ocean waters, there is also a vertical circulation, driven by differences in water temperature and salinity. As seawater in the polar regions freezes into virtually salt-free ice, the excess salt enriches the water layer immediately below the ice. These waters are therefore unusually saline and dense, causing them to sink downward. In two regions (along the eastern side of Greenland and near the Antarctic Peninsula; see Figure 3.15), this water sinks down to the ocean floor as a coherent body of water known as **bottom water**. Because it is cold, this water is also rich in dissolved oxygen and carbon dioxide, so that it can be recognized by these characteristics as well as by its salinity. As it spreads through the oceans, it therefore brings oxygen to even the deepest waters, far from the oxygen of the surface. This in turn displaces water upward, producing what is known as a **thermohaline circulation**. The time it takes for the ocean waters to complete a cycle of vertical circulation is estimated at 275 years in the Atlantic, 250 years in the Indian Ocean and 510 years in the Pacific [7].

A similar phenomenon arises from the fact that, near the centres of the great ocean gyres in the North Pacific and North Atlantic, where the weather is normally clear and sunny, the surface loses more water by evaporation than it gains by rainfall. It, too, therefore becomes more saline and dense so that, where it meets neighbouring waters to the north and south, it sinks below their lighter, fresher waters along lines known as **convergences** (shown as dashed lines in Figure 9.5). Convergences are also found where two sets of

ocean currents converge, as in the North Pacific Polar Front Convergence and the Southern Subtropical Convergence (shown by a dark tint in Figure 9.5).

Another cause of vertical water movement is the winds that blow offshore along the western coasts of parts of the Americas, Africa and Australia. These winds blow the warm surface waters away from the coast, and they are replaced by an upwelling of deeper water. A similar upward movement of water is found at regions known as **divergences** in the Atlantic and Pacific, where there is a shear between currents going in different directions. The Pacific Equatorial Divergence (PEqD) lies to the south of the zone of meeting between the westward equatorial current and the eastward North Pacific Countercurrent (NPEqC). The Antarctic Divergence (the line of small circles in Figure 9.5) lies between the eastward Antarctic current and the narrow ribbon of the westwardly directed Austral Polar Current that lies adjacent to the coastline of Antarctica.

The circulation patterns of the ocean waters, both horizontally and vertically, also create patterns in the concentration of such major nutrients as nitrate, phosphate and silicate. The patterns of availability of these nutrients in space and time have profound effects on the marine organisms. For example, silicate is vital for the production of the skeleton of the phytoplankton known as diatoms, which are the main source of the rain of organic material that descends into the deeper layers of the ocean. Therefore, a close correlation exists between the pattern of regions where waters rise to the surface at upwellings and divergences (see Figure 9.5), the availability of nutrients in the surface waters (Figure 9.3) and the patterns of productivity in the oceans (Figure 9.4). Thus, although upwelling areas account for only 0.1% of the ocean surface, these areas provide 50% of the world's fisheries catch. (It must be emphasized that *all* the primary productivity in the oceans is confined to the uppermost few tens of metres as, below this level, the sunlight is absorbed and scattered by the water.) There has long been a debate as to whether the levels of phosphate, or those of nitrate, are the limiting factor in determining productivity. One model [8] suggests that each of these has a role, with nitrate being more important in the short

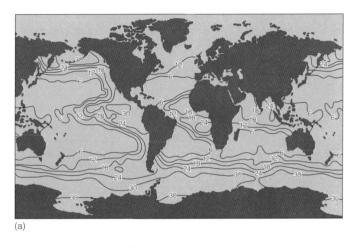

(a)

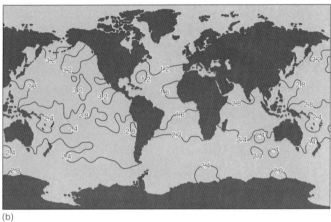

(b)

Figure 9.3 The distribution of nutrients in the oceans. (a) Mean dissolved nitrate at a depth of 150 m; (b) mean dissolved phosphate at a depth of 1500 m. From Levitus [61]. (Reproduced with permission of Elsevier.)

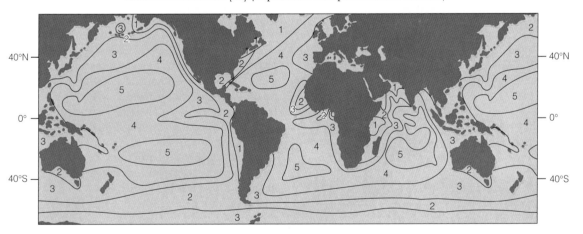

Figure 9.4 Patterns of annual primary production in the oceans. Numbered areas refer to annual productivity of carbon in g/m²/year, as follows: (1) 500–200 g; (2) 200–100 g; (3) 100–60 g; (4) 60–35 g; (5) 35–15 g. Adapted from Berger [62].

term of hours or days, while phosphate is more important over the longer term in regulating total ocean productivity.

Another study shows the importance of sea-surface temperature [9]. Satellite measurements of this temperature show that it accounts for nearly 90% of all the geographical variation in the density of the type of plankton known as foraminifera in the Atlantic. Also, because it is closely correlated with temperatures at 50, 100 and 150 m, this is almost as adequate an explanation of the diversity patterns at these depths. The results of this study also show that the diversity of planktonic foraminifera does not simply diminish from the equator to the poles, but instead peaks at middle latitudes in all oceans. This appears to be controlled primarily by the thermal structure of the near-surface levels of the ocean and may be determined by the thickness and depth of the pycnocline. A pycnocline with a gradual temperature change and a deep base may provide more niches per unit of surface area than one with a shallow base and more rapid temperature change. So, in high latitudes, where the pycnocline is nearly absent, there can be little partitioning of niches, and a consequent low diversity of plankton. In intermediate latitudes, where a higher sea-surface temperature and a thick permanent pycnocline prevail, this leads to a maximum diversity. In the tropics, however, although the surface temperature is higher, the pycnocline has a shallow base and a rapid rate of temperature change, so that it provides fewer niches and diversity is lessened.

Patterns of Life in the Surface Waters

We saw earlier in this chapter how the patterns of the continents and climate made it possible to recognize biogeographical provinces in the surface waters. Although the patterns of movement in these waters have long been known, it is only since satellite observations became available that it has been possible to monitor the annual changes in the patterns of the life that lies within them. Importantly, the chlorophyll content of the water can now be measured in this way (Plate 4). This allows us to deduce the density of the phytoplankton, the depth of the euphotic zone

and the seasonal cycles in the balance between phytoplankton productivity and loss, which may or may not lead to a seasonal increase in phytoplankton biomass known as a **bloom**. The British oceanographer Alan Longhurst has integrated this biological data with data on the movements of the ocean waters to identify and define what he calls *marine biomes and provinces*. He recognizes three biogeographical biomes in the oceans (the **polar**, **westerly winds** and **trade winds biomes**), plus a **coastal biome** that comprises the shallow seas; these biomes are divided into ecological provinces [10]. Figure 9.5 is a simplified, redrawn version of Longhurst's map, showing most of his 33 provinces of the oceanic biomes. (The 'trade' winds were so named, not because they were more used by traders than other winds, but from an archaic use of the word to mean *steadily and regularly*.)

Longhurst's biomes are not really comparable to those on land, however. This is because a terrestrial biome is a *biological* entity, recognized by its plant life, which creates a characteristic environment to which its animal component has become adapted (see Chapter 3). Except in the deserts and tundra, the physical environment has been profoundly modified by the biological world. In contrast, in the open oceans the *physical* environment everywhere dominates the life that it contains. Longhurst's biomes differ from one another only in the patterns of their winds and currents and in the ways these vary throughout the year. Nevertheless, these offer different annual patterns of opportunity for growth to the plants and animals that are carried in their waters. Another difference between the two systems is that, unlike those on land, the boundaries between Longhurst's marine provinces vary from year to year and from season to season – although the underlying pattern remains stable. So, although Longhurst's usage and system are given here, they should not be taken as suggesting any close similarity to the terrestrial systems.

Both the North Atlantic and the North Pacific are enclosed oceans with major gyres. The more polar half of each has westerly winds and ocean currents and belongs to Longhurst's westerlies biome, while the more equatorial half has easterly winds and currents and belongs to his trades

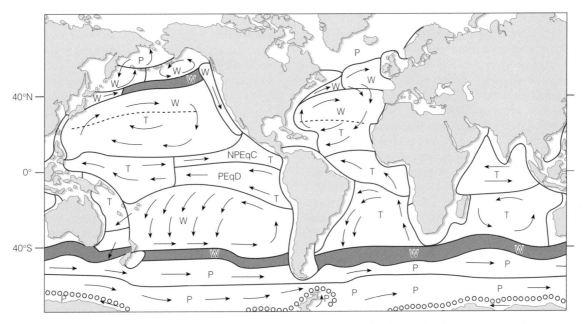

Figure 9.5 The oceanic biogeographical biomes and provinces. Letters in blue (P, T, W) indicate provinces that are placed in the Polar, Trade Winds or Westerly Winds Biomes, respectively. NPEqC, North Pacific Equatorial Counter-Current Province; PEqD, Pacific Equatorial Divergence. The North Pacific Polar Front Convergence and Southern Subtropical Convergence are shown as a dark tint. The Antarctic Divergence is shown as a line of small circles. From Longhurst [10]. (Reproduced with permission of Elsevier.)

biome (Figure 9.5). Elsewhere, the pattern is more complicated. In the South Atlantic, the whole oceanic gyre is placed in the trades biome because it is all under the influence of southeasterly winds; full westerlies only develop south of the latitude of Cape Horn. Both this and the South Pacific gyre are still very poorly known, and Longhurst therefore does not divide either of them into northern and southern provinces. The winds and currents in the Indian Ocean, which Longhurst places in his trades biome, are variable because of the annual monsoon reversal.

The absence, occurrence and timing of **planktonic blooms** are an important part of Longhurst's system. These blooms are best understood by imagining an ocean with stable conditions in which the euphotic layer lies at a constant depth, which is also the depth of the pycnocline. Within this euphotic layer, the light and heat of the sun allow continuous growth of the phytoplankton. However, their total biomass is kept to a constant,

low level by grazing by the **zooplankton** (the tiny animals in the plankton) and by the supply of nutrients, which the phytoplankton normally use up as soon as they are available.

This is more or less the pattern in most of Longhurst's trades biome, which mainly comprises the areas where winds and currents are always from the east and where there is also enough year-round sunlight for the warm euphotic zone to be permanent and of approximately constant depth. Because of the stability of this layer, there is little vertical mixing between the surface waters, in which the phytoplankton exhausts its nutrient supply, and the deeper, more nutrient-rich waters. Consequently, there is little seasonal variation in algal productivity, and therefore no algal bloom. However, the province of the trades biome that lies across the equator in the Atlantic provides an exception because in this province there is an increase in the power of the trade winds in the summer. This leads to an accumulation of surface water in the western part of the basin and a corresponding decrease in the eastern part. This in turn allows deeper, nutrient-rich water to rise upward in this eastern region. As a result, it is now within the euphotic zone, so that its nutrients can fuel a summer algal bloom (Plate 4c). (This phenomenon does not take place within the Pacific because it is too broad for there to be enough time for such a seasonal wind change to have such an effect.)

Longhurst's westerlies biome comprises provinces in which there is an algal bloom in the spring. It includes provinces with three rather different seasonal situations: those in high northern latitudes in the Atlantic, those in similar latitudes in the Pacific, and the rest, which are found in lower latitudes in both hemispheres. These three situations will be described in turn.

In the eastern North Atlantic, phytoplankton productivity is limited by both light and nutrients. The bloom therefore only takes place in the Northern Hemisphere spring, when the amount of sunlight increases and the pycnocline rises towards the surface (Plate 4b). The rate of increase of the phytoplankton is extremely rapid and is unpredictable. As a result, the population of grazing zooplankton has not been able to evolve an annual pattern of increase tied to that of the phytoplankton; their increase therefore lags behind it rather than controlling it. Instead, the increase in the phytoplankton continues until, after a few days, it has exhausted all the available nutrients. The larger phytoplankton such as diatoms then die or lose their buoyancy, and their remains fall downward through the water column to the abyssal levels. Their place in the upper waters is taken by tiny single-celled animals known as flagellates and by blue-green algae known as cyanobacteria (the **picoplankton**). These are responsible for 80–85% of the subsequent primary production. However, because they are too tiny to sink or to be filtered out of the water by zooplankton, they do not form the basis for a classical food web leading to larger animals, such as is found in the Pacific (discussed further in this chapter). This area often has a second bloom in the autumn. This may be because mixing of the waters by storms brings new supplies of nutrients to the surface waters or because many of the zooplanktonic grazers have descended to greater depths to overwinter. This autumn bloom is short-lived, for the failing light intensity soon limits photosynthetic activity, and it declines rapidly when the nutrients in the upper level have been used up. Productivity remains at a very low level throughout the winter.

In contrast with this situation, in the high latitudes of the North Pacific the springtime bloom (Plate 4b) does not use up all the available nitrate. This may be because its timing is more predictable, so that the zooplanktonic grazers that have overwintered at depth, such as the copepod *Neocalanus*, rise to the surface at the appropriate time, ready to graze on the phytoplankton and control their numbers. However, it is also possible that the availability of iron is a limiting factor. Whatever may be the cause, the diatom bloom in the upper waters does not 'crash.' It is therefore available for a longer period of time, into the summer (Plate 4c), as the basis for a food chain upward via zooplankton, fishes and larger crustaceans, and thence to seabirds, seals and whales.

The other provinces of Longhurst's westerlies biome (which together cover over half of the oceans) lie at lower latitudes, so that light is not a limiting factor. Phytoplankton productivity therefore increases during the winter as the progressive deepening of the mixed layer recharges the surface layers with nutrients, and declines in early summer as these nutrients are used up.

In the polar biome it is light, rather than nutrients, that limits algal growth. This growth therefore rapidly increases in the Northern Hemisphere spring (Plate 4b) as the sunlight increases in duration and strength and the ice (if any) melts. The peak of this bloom is near to midsummer (Plate 4c), after which it declines because of grazing by zooplankton. However, a second peak takes place in September (Plate 4d), when the copepod grazers descend to greater depths to overwinter. This secondary peak is less developed in the Antarctic, where these copepods overwinter closer to the surface, among the phytoplankton.

There are some areas where, although the level of nitrates in the surface waters remains high, this does not lead to increased productivity. These regions, known as the high-nutrient/low-chlorophyll regions, comprise the north subpolar Pacific, the eastern tropical Pacific and the Southern Ocean around Antarctica. It has been suggested that this is because productivity is continually limited by grazing zooplankton, or that it is due to a lack of some other nutrient, perhaps iron. Certainly, the addition of iron to samples of seawater from the Southern Ocean in the Antarctic caused a quadrupling of the level of productivity [11].

The UNESCO report [4] comments that Longhurst's provinces do not strictly follow the surface circulation patterns in a number of areas, and that some of his biomes cut across major ocean gyres. It therefore suggests a somewhat different map of the pelagic provinces (Plate 5a), which is likely to form the basis of future research However, Longhurst's work has provided a fascinating account of the pelagic marine ecosystems, and his map is shown here as it is essential in understanding his ecological interpretation.

Invisible Barriers in the Oceans

The final establishment of the Panama link between the Americas about 3.1 million years ago, and its effects on the biogeography of the faunas of the region, is easy to document. But, as we have already seen, there are other examples where it is difficult to recognize the two essential aspects of faunal subdivision – the barriers between the faunal regions and the differences between the faunas themselves. As we have just seen, it is possible to distinguish major water masses by the patterns of movement of the ocean waters and by the differing annual histories of the biota they contain. The patterns of temperature, nutrient availability and phytoplanktonic production seen in Longhurst's provinces are of fundamental importance to the marine organisms that inhabit them. One would therefore expect each population to be physiologically and behaviourally adapted to the conditions in its own water mass. Such a set of adaptations could only have evolved if each population is a separate species, physically and genetically distinct from others and therefore separated from them by obvious barriers. In the terrestrial biotic regions, these barriers are the great mountains, deserts or oceans. It is therefore very surprising to find that the oceanic biomes and provinces appear to be separated by no more than the interface between masses of water moving in different patterns or directions. Surely, one would have thought, these invisible barriers must be so permeable as to permit a great deal of dispersal between one biogeographical unit and another. Perhaps there are no

important taxonomic differences between their biotas, and the differing annual histories that we see are due merely to physical factors that impose these regimes upon biota that basically differ little from one another taxonomically. In the case of terrestrial biota, the biogeographer would seek to confirm or deny such a hypothesis by comparing the taxonomic composition of the different biotas, starting with their morphology.

At first sight, such an endeavour seems to provide little evidence of major taxonomic differences between the marine biomes and provinces – certainly nothing as dramatic as the totally different mammal families of the continents. There are many statements in the literature as to the wide geographical ranges of many marine taxa and the accompanying lack of diversification into separate species. This may be true in some, perhaps many, cases: very many open-sea species do appear to be widespread. For example, the little nektonic, epipelagic arrow-worm *Pterosagitta draco* is found in tropical and subtropical waters of all the oceans, and this is also true of polychaetes (bristleworms) and of the pelagic crustaceans known as euphausids. (These form the 'krill' on which baleen whales, and many other marine animals, feed, and they are by weight the most abundant animal on the planet.) Similarly, arrow-worms that live in the deeper mesopelagic and bathypelagic levels range from the sub-Arctic to the sub-Antarctic because there is little latitudinal change in temperature at these greater depths [12]. This pattern is also found in fishes living there.

However, we should be very cautious in generalizing from these cases. For example, there are not only separate subspecies of the arrow-worm *Sagitta serratodentata* in the Atlantic and in the Pacific, but also a separate species, *Sagitta pacifica*, in the southern Pacific. Furthermore, molecular evidence suggests that the uniform morphological characteristics of some of the widespread species conceal considerable biochemical and physiological diversity, so that in reality they have probably diverged into several geographically defined species. For example, Gibbs [13] comments that the mesopelagic fish *Nominostomias*, which had been thought to contain only eight species, is now known to have over 100. Similarly, there is now molecular evidence suggesting that, though many widespread marine species appear uniform throughout their range, they have in reality diverged into several geographically distinct species. For example, work on the sea urchin *Diadema* [14], whose species range throughout the world, has shown that their morphology is an extremely unreliable guide to their species structure. It has even been stated that the morphological differences are so slight that specimens cannot usually be identified without knowing where they were collected! (This is not a situation that brings joy to the heart of a marine biogeographer!) We are still, almost literally, scratching at the surface in our knowledge and understanding of marine biogeography, for much of our new information merely shows the superficiality of our previous beliefs. It is tempting to see this as merely yet another example of how much less we know about life in the sea than about life on land. But the discovery, using molecular genetic information, that what had been thought to be only nine species of terrestrial earthworm in fact belonged to 14 different species suggests another explanation – that perhaps we have never properly understood the taxonomy of invertebrates at the species level and that the new molecular techniques are merely revealing our ignorance.

The fact that the apparent lack of taxonomic differentiation of marine species is probably illusory is also suggested by the stable behaviour of some species of fish. For example, the populations of the Pacific herring *Clupea harengus pallasi* along the Pacific coast mainly spawn in one of two locations: San Francisco Bay, or Tomales Bay 50 km further north. The juveniles remain in their home bay for about a year and then swim out to the open ocean, where they stay for 2 years before returning to the bay where they themselves spawn for the first time. It would appear to be impossible to be sure whether they have remained 'faithful' to the bay in which they hatched, but studies of their parasites have shown that this is so [15]. Not only do they show differences in the parasites that they contain, but also these differences suggest that the two populations of fish feed in different locations. This is because the intermediate hosts of the parasites of the Tomales Bay population are whales, which lie offshore, while those of the San Francisco Bay population are seals, marine birds and sharks, which are found closer inshore. So, again, the apparent lack of barriers in the oceans may merely be the result of our lack of understanding.

In the long term, it will be interesting to find the extent to which changes in the patterns of connection of the oceans and in the currents that run between and around them (see Figure 5.7) can be detected in faunal differences. Palaeontologists have long recognized the existence of a Tethyan marine fauna, including both vertebrates (fishes and marine reptiles) and invertebrates, that occupied the comparatively narrow seaway between Laurasia and Gondwana during the Cretaceous (cf. Figure 10.3). Similarly, the American oceanographer Brian White [16] believes that one can find parallels between the sequence of appearance of different water masses in the Pacific Ocean, due to plate tectonic changes, and the relationships between the different species of some fishes. It is likely that further work will expand on this, to give the faunas of Longhurst's different provinces the same sort of historical and taxonomic basis that has long been known for terrestrial zoogeographical regions. For example, the limits of distribution of many Pacific epipelagic organisms coincide with one another and seem to be related to the patterns of water masses [17]. At an even larger scale of comparison, the marine fauna of the North Atlantic is much less diverse and has fewer endemic forms than that of the North Pacific. The reason for this may be that the Atlantic is a smaller ocean then the Pacific and therefore provides a smaller area within which evolutionary novelties may appear and disperse to other parts of the ocean. The Australian marine biologists Dave Bellwood and Peter Wainwright have already produced a very valuable review of the relationship between these plate tectonic events and the evolution of reef fish faunas [18]. The question of barriers between shallow-water faunas is discussed in the 'Faunal Breaks within the Shelf Faunas' section.

The Ocean Floor

Patterns of Life on the Deeper Ocean Floor

Apart from the intertidal zone between the high-water mark and the low-water mark, marine biologists recognize four different life zones (Figure 9.2b) for benthic organisms – those that live on or in the floor of the ocean. (It must be emphasized again that changes in conditions and in faunas are always gradual, so that the boundary between two zones is merely a level of more rapid change rather than a level of abrupt change.) The **shelf zone** comprises the continental shelf below the low-water mark down to a depth of 200 m; this will be dealt with later in the chapter. The other levels lie at successively greater depths: the bathyal zone (also known as the slope or **archibenthal zone**), the **abyssal zone** and the **hadal zone**. The hadal zone is easy to define, for it comprises the environment of the deep ocean trenches (see Figure 5.3) at a depth of over 6 km. However, the depth of the boundaries between the shelf zone, bathyal zone and abyssal zone vary according to season, conditions and latitude.

The upper level of the bathyal zone is normally at the edge of the continental shelf, at about 200 m. Its lower level, and therefore the level of transition to the adjacent abyssal zone, varies considerably. Where the surface waters are cold, as in the Arctic region, the bathyal–abyssal transition is similarly at a shallow level, about 400 m, and the bathyal fauna itself extends to within 12 m of the surface. At lower latitudes, closer to the equator, where the water temperatures are in general higher, the bathyal–abyssal transition is at a deeper level, usually at about the 900 m base of the pycnocline. At this level, the water temperature has dropped to 4°C, and below it drops to 1–2.5°C. The Swedish zoologist Sven Ekman [19], who was one of the founders of marine zoogeography, called this change in the depth of the bathyal–abyssal transition the principle of **equatorial submergence**. The pattern of this variation strongly suggests that the depth at which the transition takes place is dependent on temperature.

The seafloor at all these levels is covered by muds and oozes – organic sediments derived from the remains of organisms living in the waters above. As well as fragments of the bodies of larger organisms, most of these remains are of the zooplankton and (especially) phytoplankton, the latter being known as **phytodetritus**, which sticks together to form large flakes known as *marine snow*. At the times of plankton blooms in the waters above, this falls like a blizzard to the seafloor, where it is consumed by the animals that live on or in the sediments there. The intensity of this recycling is shown by the estimate that the entire upper 10 cm of organic sediments in the Santa Catalina Basin off California is eaten and excreted by worms every 70 years.

Turning now to the nature of the faunas in these zones, one can distinguish two quite different environments. At one extreme is the fauna living on the continental shelf. Here, conditions vary in both time and space, but temperatures are higher and there is light, providing the energy basis for a rich ecosystem. At the other extreme is the fauna of the abyssal zone, adapted to lightless, cold waters in which the only source of energy is the rain of phytodetritus from above. In each of these zones there live organisms that are specifically adapted to that environment. The bathyal zone is therefore an intermediate zone within which there is a gradual change with depth, from a mainly shelf-like fauna to a mainly abyss-like fauna. The precise pattern of faunal replacement with depth varies from place to place, according to the physiological tolerances and limitations of the individual species, and according to their interactions with local competitors and predators.

The patterns of change with depth in the faunas of the bathyal and abyssal benthos have been intensively studied off the Atlantic coast of North America. As the continental shelf is the point at which the physical changes in the environment take place most rapidly, it is not surprising to find that this is the region where the rate of faunal change is also most rapid; it then continues at a slower rate with increasing depth.

Another aspect of these faunas is their degree of diversity. This, again, has been most intensively studied in the western North Atlantic, where the faunas show a very clear pattern [20]. The best data so far on the faunal diversity of the continental slope come from the work of Frederick Grassle and Nancy Maciolek [2], who took box-cores totalling 8.69 m² along 176 km of the slope off New Jersey. Not only did these contain nearly 800 species, but also each new core contained additional species. Towing a dredge trawl along the continental slope at 1400 m for an hour produced over 25 000 specimens belonging to 365 species, but the same operation carried out on the abyssal plain at 4800 m produced only 3700 specimens belonging to 196 species, showing a not-unexpected drop in faunal diversity as one moves into the ocean depths. In many groups, the faunal diversity is low on the continental shelf, high at a mid-bathyal depth (2000–3000 m) and low again on the abyssal plain.

The ultimate cause of these changes is probably the gradual change in food and nutrient availability as one moves away from the high-productivity surface waters. The problem has been to identify how this change imposes itself on the community structure of the seabed. Various mechanisms have been suggested, such as changes in the intensity of predation, or competition, arising from differing rates of population growth. An interesting discovery is that a clear correlation exists between faunal diversity and the characteristics of the sediments of the seafloor, especially the diversity of particle size [21]. Because there is no primary production in the lightless deep sea, its economy is based on the organic particles that settle on the bottom, and the fauna that lives on or in the seafloor is dominated by detritivores. Therefore, it may well be that a greater range in particle size provides a greater range of niches for the bathyal benthic fauna. The UNESCO group's suggested pattern of provinces in the lower bathyal zone [4] is shown in Plate 5b.

Whatever the precise pattern (or range of patterns) of relative diversity in the deep-sea benthos, the total diversity on the abyssal plain is very great – even though, because of the low level of food supply, the density of the deep-sea biomass is very low [22]. The ecology of these faunas is so poorly understood that it is difficult to be sure of the reasons for this unexpected faunal richness. It is possible that the lack of barriers to dispersal on the abyssal plain makes it easy for many organisms, which have evolved at different locations on its enormous area, to become widely distributed. The coexistence of the resulting large number of species might be facilitated by another phenomenon. The abyssal benthic environment is often disturbed by the settling of aggregations of plankton or of the remains of larger organisms. It has been suggested that this is so frequent that the community rarely has an opportunity to come to a final balance in which some species become locally extinct due to competition from rival species.

The amount of information on the composition of the fauna of the deep-sea benthos is still very limited and geographically unbalanced. Most of it comes from the western Atlantic, and none is available from central oceanic regions. It is also nearly all derived from soft-bottom communities, little being known of hard-bottom communities.

The international group of experts mentioned earlier in this chapter [4] has therefore had to confine itself to producing maps defining benthic substrate provinces each of which has a particular set of physical characteristics (levels of bathymetry, bottom temperature, salinity, oxygen and organic matter flux). These are provided for the following depth zones: 300–800 m (upper bathyal), 800–2000 and 2000–3500 m (upper and lower portions of the lower bathyal), 3500–6500 m (abyssal) and >6500 m (hadal). The boundaries between these provinces (see Plate 5) approximately correspond to places where oceanographic fronts occur, or where there are known transitions of species or of environmental variables. These provinces are therefore, as the report admits, hypotheses that need to be tested with species distribution data, especially for the lower bathyal provinces. (In this context, it is worrying that research of the distribution of brittle-stars [23] in the Indian, western Pacific and Southern oceans has shown that, rather than being different in each of these oceans as the UNESCO report suggests, they follow broad latitudinal bands across them.) The final classification system of these provinces will need to reflect their taxonomic identity, and to emphasize recognizable communities of species and the changes in the dominant species that determine the structure and function of the ecosystem.

There is even less systematic information on the fauna of the hadal zone, which lies at depths of 6000–11 000 m in the great submarine trenches, most of which are in the western Pacific. As might be expected from its great depth, its fauna is extremely sparse, even more impoverished than that of the abyssal plain. Yet it does also seem to be different – about 68% of the species, 10% of the genera and one family are endemic to the hadal zone. Most of the endemic species have a vertical depth range of less than 1500–2000 m, so there is steady faunal change with increasing depth. These faunas are confined to isolated patches along the deep-sea trenches (Figure 9.6). This has permitted considerable independent evolution of endemic species within each trench fauna, and it is quite possible that their patterns of biogeographical relationship may be similar to those of the hydrothermal vents of the mid-oceanic rises. Unfortunately, it is still too early for marine biologists investigating these patterns to be able to make firm statements as to

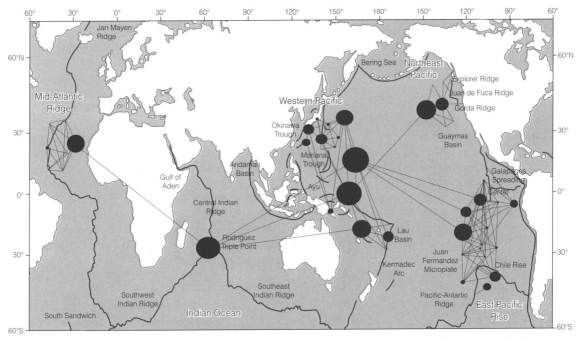

Figure 9.6 The interrelationships of the different hydrothermal vent communities as revealed by network theory. The size of the different circles reflects the relative importance of each province as a connection point between the other provinces. From Moalic [25]. (Reproduced with permission of Oxford Univesity Press.)

what patterns exist and in what variety, or as to how these patterns may vary according to geographical location or seafloor topography.

Similarly, there is still not enough information on possible faunal differences between the faunas of the floors of the different ocean basins as a whole, although it seems extremely likely that these exist. Earlier studies, mainly by Russian marine biologists, have suggested that only 15% of the deep-sea benthic species occur in more than one ocean, and only 4% in all of them. It is also possible that the pattern of mid-ocean ridges, where hot material rises from the depths of the planet to form new seafloor as the tectonic plates separate (see Figure 5.3), may similarly act as more minor barriers to faunal movement, and therefore delimit subsidiary faunal areas within the oceans. But the presence, nature and scale of such possible differences have yet to be established. In continental biogeography, larger scale patterns of faunal change or difference are often found to have been caused by historical events of evolutionary innovation or extinction, followed by dispersal or vicariance arising from plate tectonic events. Smaller scale differences are more likely to result from ecological factors. It will

be interesting to see the extent to which marine biogeographical research reveals similar patterns in the deep sea. Even though the same processes are likely to be operative in the two environments, the very different scale, and probably very different rates, in the oceans may nevertheless lead to significantly different results.

Biogeography of Hydrothermal Vent Faunas

In 1977, marine biologists discovered a dramatically different deep-sea environment containing a fauna that shows a fascinating biogeographical pattern. They are mainly associated with sea-floor spreading at the mid-oceanic ridges, which are at an average depth of 2.5–3.5 km, but also in basins near volcanic island arcs (see Chapter 5). Although the ridges themselves are many hundreds of kilometres wide, they are split by a rift valley only about a kilometre wide, where hot lava is emerging. In some widely scattered areas known as **hydrothermal vents**, each of which covers only a few hundred square metres, the cold seawater penetrates fissures in the surrounding rocks. The water temperature there may reach 400°C (only the enormous pressures at this depth prevent it from turning to steam), and it reacts chemically with the rocks, so that it becomes rich in metals and sulphur. Where this superheated water emerges and is cooled by the surrounding waters of the ocean, these minerals precipitate out of the fluid. Some of them form solid 'chimneys,' which can be many metres high, while others remain as distinct particles in the rising plume of water, which therefore looks like smoke emerging from the chimney.

Accompanying this extraordinary environment is a unique fauna, whose food web is not based on plants that have trapped the sun's energy, but instead on chemosynthetic bacteria that extract energy by the oxidation of the chemicals dissolved in the hot fluids, particularly hydrogen sulphide. Some of these bacteria are consumed by grazing or filtering organisms, while others live symbiotically, rather as photosynthetic algae live in corals (see the 'Coral Reefs' section). They form the food base of a fauna consisting mainly of worms, arthropods and molluscs. The fauna is of low diversity: in the Juan de Fuca section, west of Seattle, there were only 55 species, and 90% of the total number of organisms was contributed by only five species – two gastropod molluscs and three polychaete worms. Such a fauna is typical of other highly disturbed habitats, such as areas being colonized after volcanic eruptions or forest fires. So far, nearly 600 species belonging to 331 genera have been found in vent communities. The harsh conditions in which the vent communities live have stimulated research into their possible relevance to the origins of life on Earth [24].

The vent communities have now been found at many different locations in the oceanic ridge system. Most of these locations are in the tropics and subtropics, where the weather conditions are more favourable for research, but there are indications that there are others at higher latitudes. The faunal composition of the vent faunas varies, providing puzzling and interesting biogeographical problems. Eleven different biogeographical provinces have now been identified. The vent communities of the East Pacific Rise, west of Central and South America, are dominated by giant tube-worms (*Riftia*) up to 2.5 m tall and include also some shrimps. The same community is found 3200 km away in the Juan de Fuca region west of Canada, but there they belong to different genera or species. In fact, 80% of the Juan de Fuca species are endemic, though there are related to those further south. Further away, the vent com-

munities of the North Atlantic ridge are quite different. They lack the tube-worms, and they are dominated by shrimps that belong to the same family as those found in the eastern Pacific. Any connection between the vent communities of these two oceans is most likely to have been east–west via the gap between two Americas before this was closed by the establishment of the Panama Isthmus nearly 3 million years ago. But this poses quite different problems: there has never been any mid-ocean rift system in this area, so any communication between the two vent faunas must have been by long-distance dispersal of larvae (discussed further in this chapter).

Recent research by an international group of marine biologists [25] throws a fascinating light on the interrelationships between the vent communities of the Pacific, Indian and Atlantic oceans and their origins (see Figure 9.6). An analysis using network theory and covering 331 genera shows the West Pacific province in a central position: all the other provinces are connected with it either directly or, in the case of the Atlantic province, indirectly via the Indian Ocean province. This central position of the West Pacific province suggests that it is the oldest of these provinces and the source of the faunas of the others. Molecular results on the faunas suggests that they originated circa 150 million years ago, and the authors also point out that the pattern is similar to that of the early tectonic history of the Pacific plate since the Cretaceous: their maps suggest that this early vent fauna may have evolved on the southwest boundary of that plate. But, if this is so, it raises the question as to what fauna preceded this apparently ancestral vent fauna, and why the former became extinct.

Even more recently, a quite different fauna has been found on the East Scotia Ridge system [26]; it consists of a new species of crab, stalked barnacles, limpets, snails, sea anemones and a predatory seven-armed starfish. It lacks all of the dominant vent species found elsewhere at most mid-oceanic ridges, but shares some elements with communities found at vents in back-arc basins in the west and southwest Pacific, and also with those in the southeast Pacific ridge and the mid-Atlantic ridge.

One of the problems raised by these faunas is the method by which their organisms disperse from vent to vent. In many cases, the degree of similarity between one vent fauna and another is not dependent on the direct distance between them across the abyssal plain. Instead, it depends on the distance between them following a longer path along the pattern of mid-oceanic ridges and faults [27], suggesting that dispersal is directly along these systems. However, some species do have planktonic larvae that may be able to disperse via deep-ocean currents; this may be the reason for the similarities between the faunas of the eastern Pacific ridge and that of the Atlantic ridge. The unusual nature of the East Scotia Ridge fauna may be because it is isolated from faunas to the north by the surface-to-seabed Antarctic Convergence. This major barrier to dispersal that came into existence after the initiation of the Antarctic Circumpolar Current about 37 million years ago, and it is significant that the taxa that are absent from the East Scotia Ridge fauna have planktonic larvae that may be unable to cross the Polar Front. (This Convergence is also important in inhibiting dispersal of benthic faunas; see the 'Transoceanic Links and Barriers between Shelf Faunas' section.)

The Shallow-Sea Environment

Some of the differences between the shallow-sea (or neritic) environment (which includes the waters down to 200 m depth) and the open-sea (or pelagic) environment have already been considered, but some others need to be noted. Even if one subdivides the open ocean into areas corre-

sponding to the movements of the surface waters, each of these is immensely greater than any of the long, narrow individual units of the shallow-sea environment that lie between the coast and the continental edge. Furthermore, each of these shelf seas is heavily influenced by the characteristics of the adjacent land, such as the nature of the coast and the presence of rivers, which may contribute both freshwater and a varying input of sediment. Because of the relative shallowness of the sea, the nature of the seabed also influences conditions through the whole of the overlying water mass, as far as the surface itself; there is no such interaction in the open sea. The shelf seas are therefore also much more heterogeneous than the open ocean.

As a result of these differences, because each shelf fauna may be different from others, and therefore contains its own endemic, locally adapted species, the total number of shelf species is far greater than the total number of pelagic species, which occupy an environment that is far less varied globally. For example, over 970 species of the neritic mysid crustaceans are known, compared with only 86 species of their pelagic relatives, the euphausid crustaceans. Furthermore, there is a fairly sharp distinction between the ranges of the two types of organism, as few of the pelagic species venture into the neritic environment (Figure 9.7). Also, though the shallow seas account for only 8% of the total

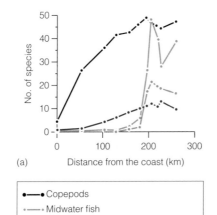

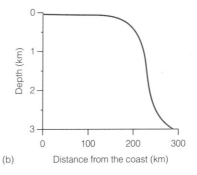

(a) Distance from the coast (km)   (b) Distance from the coast (km)

- ●——● Copepods
- ·——· Midwater fish
- ·——· Decapod crustaceans
- ●——● Euphausid crustaceans

Figure 9.7 The number of species of four different pelagic groups off Florida, showing how the numbers decrease as one moves from the open ocean and towards the coast (a), and often peak at, or close to, the shelf break (shown in (b)). From Angel [6]. (Reproduced with permission of Cambridge University Press.)

area of the oceans, they hold the majority of its diversity of life.

It is not only the free-swimming pelagic organisms of the shallow seas that show this heterogeneity, but also their bottom-dwelling, benthic relatives. They have to rely on a pelagic larval form to ensure that the next generation can be distributed locally, where the presence of the parents suggests that the environment is favourable. However, these larvae may also play a role in more widespread dispersal. This can only be successful if the location that the larva reaches is similarly environmentally favourable, and also does not contain a more competent rival or predator. The potential efficacy of larval dispersal is shown by the fact that species of benthic invertebrate along the western coasts of the Atlantic are more widely distributed if they have planktonic larvae than if they do not (Figure 9.8). Of course, long-lived larvae will need to feed during the days of dispersal, so it is not surprising to find that such larvae are more common in low latitudes, where the phytoplankton season is long, than in high latitudes, in which it is shorter (Figure 9.9).

However, it may be incorrect to assume that the possession of larvae that can potentially live for a longer time indicates that their function is primarily long-distance dispersal [28]. Research using a combination of statistical modelling and observed population genetic structure suggests a surprisingly poor correlation between larval type and the prevalence of long-distance dispersal, and shows that the average scale of dispersal is often very variable for the same taxon and location. In addition, those taxa with larvae that are short-lived or that normally disperse over only a short distance seem to have a disproportionately high rate of long-distance dispersal. Perhaps one role of these long-lived larvae may be merely to give greater reliability to dispersal over small or medium distances, rather than to provide the potential for long-distance dispersal.

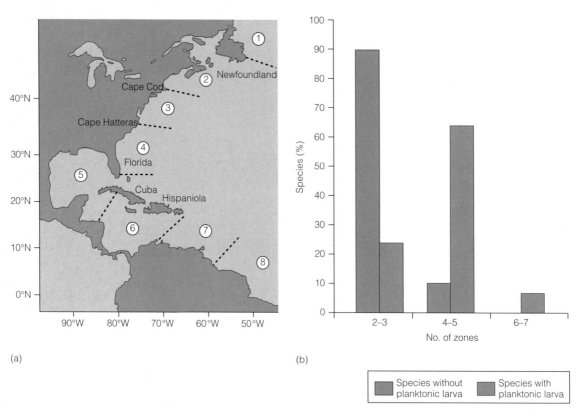

(a) (b)

Figure 9.8 (a) Biogeographical zones down the western coasts of the Atlantic. (b) The number of invertebrate benthic species that occupy these zones, with or without planktonic larvae. Adapted from Scheltema [63].

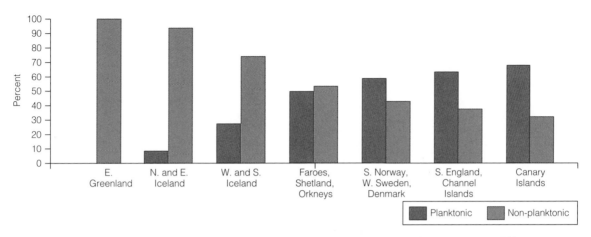

Figure 9.9 The percentage of gastropod species that either have, or do not have, planktonic larvae, at different latitudes. Adapted from Thorson [64].

Bioregions on the Continental Shelves

One result of the system of great oceanic gyres is that the waters that circulate within each of them occupy different bands of latitude and therefore different zones of temperature. It was a Swedish worker, Sven Ekman [19], who first proposed a scheme dividing the waters of the continental shelves into polar, temperate and tropical regions, based on their separation by zoogeographical barriers and on the endemism of their faunas. The American marine zoologist Jack Briggs later [29] developed this scheme further by subdividing the continental shelf regions into tropical, warm-temperate, cold-temperate and polar regions, and by recognizing separate provinces within them.

Over the last few decades, the steadily increasing problems of overuse of the biological resources of the sea have led to a great need for the recognition of ecologically representative systems of protected areas for conservation planning and resource management (see Chapter 14). Trying to find agreed solutions in the coastal and shelf areas is complicated by the fact that these areas lie under the jurisdiction of the nations that they border. In 2007, an international group of 15 experts suggested a system of 12 realms, 62 provinces and 232 ecoregions (Plate 6) to provide a basis for these efforts [30].

The *realms* are very large areas in which the biotas are internally coherent at higher taxonomic levels as a result of their shared and unique evolutionary history. They show high levels of endemism, including some unique families and genera. In the main they follow the scheme mentioned above, following broad latitudinal temperature-based divisions. In the temperate Southern Hemisphere, however, the faunal differences across the wide oceans are so substantial, and the local connections around the individual continents are so great, that it was necessary to adopt separate realms for the waters around South America, Africa and Australasia. In addition to this, as we shall see later in this chapter, the waters of the continental shelves in the interlinked Indian and Pacific oceans contain a huge diversity of life. Though there are gradual changes in the faunas across the whole region, there are no clear subdivisions within it, although the faunas at either end are very different from one another. The group therefore decided to divide this realm into three sub-units, more or less corresponding to a central unit around South-East Asia, flanked by sub-units in the Indian Ocean and the eastern Pacific.

The *provinces* have distinct biotas with some level of endemism, especially at the species level, this resulting from isolation caused by historical and/or geographical factors. The *ecoregions* are areas whose species composition is relatively homogeneous because they contain a small number of distinct local ecosystems resulting from such factors as physical features of the environment and levels or patterns of nutrient input. This is particularly true of those ecoregions that consist of the coastal faunas around isolated oceanic islands, and some of the interesting characteristics of these are considered next.

Coastal Faunas of Islands

Although most coastal faunas lie along the edges of the continents or around islands on the continental shelves, others are to be found around the edges of isolated oceanic islands. Most of these lie around volcanic island-arcs or chains that have resulted from the actions of oceanic trenches or hotspots (see Chapter 5), and most of them lie within the Pacific Ocean. As one might expect, these isolated faunas show a high degree of endemism. Springer [31] studied the patterns of distribution of 179 species of shore fish, belonging to 111 families, from these isolated Pacific islands. He calculated that 20% of them are endemic to the region and that most of these are endemic to a single island. He noted that most of the cartilaginous elasmobranch (shark-like) fishes disappear from the faunal lists as one moves out into the deep Pacific Ocean from the east. This may be because their buoyancy system, based on the presence of a large, oily liver, is less effective than the swim-bladder of bony fish. Perhaps more surprisingly, he found that the number of taxa rapidly diminishes as one moves into the deep ocean from the west, suggesting that these shore fishes find these open wastes of water almost as formidable an obstacle as do land animals and plants. It will be interesting to see whether studies of the shore fishes of the Hawaiian chain show a pattern of relationship similar to that of the insects and plants of the Islands (see Chapter 7), in which the faunas of the younger islands are derived mostly from those of the older ones.

Transoceanic Links and Barriers between Shelf Faunas

A particularly interesting recent paper on benthic faunas of the Southern Ocean [32] by four French biogeographers shows the extent to which our steadily increasing knowledge of past geographies, climates and ecologies, together with new analytical procedures (see Chapter 8), is allowing new insights into biogeographical data. Their studies of echinoids, bivalves and gastropods showed that the benthos of the region was strongly affected by the Antarctic Convergence (which forms the main barrier to northward dispersal) and the Antarctic Circumpolar Current. Their results also suggested past faunal links across the base of the Scotia Peninsula, and also across between East and West Antarctica, which are probably the results of seaways that appeared at the time of the collapse of the West Antarctica ice sheet in the Pleistocene.

A different type of break in the nature of the shallow-water fauna is seen where there is a change in the nature of the bottom sediments. For example, as one travels eastward along the northern coastline of South America, there is an ecological transition at the delta of the Orinoco River in Venezuela. From here eastward, almost to the northeastern corner of Brazil, the bottom of the continental shelf is covered with mud brought down by the great tropical rivers, whose fresh waters also greatly reduce the salinity of the waters of the shallow sea. The coral reefs so characteristic of the Caribbean region to the west are absent, together with their associated fish fauna, which is replaced by such groups as the sea catfishes and croakers.

Another well-defined point of ecologically based faunal change lies at the entrance to the Red Sea, which is separated from the Arabian Sea by a relatively narrow passage, only about 32 km wide, which is also partially blocked by a shallow sill only 125 m deep. As a result, because the waters of the Red Sea evaporate in the dry climate, and there is no significant contribution of fresh water from the land, the Red Sea is unusually saline. There is therefore considerable endemicity in its fauna: corals 25%, crustaceans 33%, cephalopods 50%, echinoderms 15% and fishes 17%.

We have already seen how the invisible barriers within the ocean waters affect the distribution of the pelagic, free-swimming organisms. Inevitably, these barriers also affect the distribution of the benthic organisms that live on the shallow shelves of the continents surrounding these waters. Although currents are potentially capable of carrying the larvae of shelf organisms from one side of an ocean to the other, many larvae live for so short a time that there are comparatively few examples of east–west linkages between their shelf faunas. The eastern Pacific Ocean, because it is so wide (5400 km) and because it contains few islands to provide intermediate havens for shallow-water organisms, is a very effective barrier to many such organisms. Ekman therefore named it the **East Pacific Barrier**. Of the shore fishes that are found either in the Hawaiian Islands or between Mexico and Peru at the eastern end of the Pacific Ocean warm-temperate region, only 6% are found in both. Similarly, Ekman [19] showed that only 2% of the 240 species and 14% of the 11 or 12 genera of echinoderms found in the Indo-West Pacific area had been successful in reaching the west coast of the Americas. (The greater proportion of endemic genera, as compared with endemic species, is because a new taxon first appears as a new species. It is only later that, if enough time has gone by or enough new, related species have appeared, the level of novelty is great enough for us to recognize a new genus. Therefore, genera are older than individual species and will have had a greater length of time in which one of their constituent species may have been able to cross the barrier.)

As always, there are exceptions to this generalization. For example, it has been shown [33] that populations of the sea urchin *Echinothrix diadema* in the eastern Pacific islands of Clipperton Atoll and Isla del Coco are genetically so similar to those of Hawaii that there must have been recent and massive gene flow between these locations. This is despite the fact that it normally takes 100–155 days for water to be carried from one area to another in the northern Equatorial Counter-Current – longer than the duration of the life of the echinoderm larva (50–90 days). However, it may be that larvae were able to make the journey in years when the El Niño regime (see Chapter 12) was strong, when the journey time would have been reduced to 50–81 days. Similar relationships between the populations of crabs and starfish in these regions have also been recorded.

The great antiquity of the East Pacific Barrier has been shown by Richard Grigg and Richard Hey of the University of Hawaii [34], who studied the zoogeographical affinities of fossil and living genera of coral. They found that those of the East Pacific are more closely related to those of the West Atlantic than to those of the West Pacific, even for corals living as long ago as the Cretaceous Period. The fact that the Barrier appears to have been effective in inhibiting dispersal across the Pacific in the Mesozoic is not surprising, for the Americas then

lay much further east. Thus, the gap between their western coastlines and the island arcs of the western Pacific would have been correspondingly wider.

Shelf faunas also provide evidence on the progressive widening of the Atlantic Ocean. Comparison of the shelf faunas on either side of the North Atlantic from the Early Jurassic onward, using the coefficient of faunal similarity, shows a steady decrease in similarity as they are gradually separated by the widening ocean [35]. (This coefficient is calculated as $100C/N$ of each biota, where C is the number of families common to the two regions being compared, while N is the number of families in the smaller fauna.)

The completion of the Panama Isthmus barrier between the western Atlantic and the eastern Pacific about 3 million years ago provides a good example of vicariant evolution in the shelf faunas on either side of the new land connection. Only about a dozen of the approximately 1000 shore fishes of the region still appear to be identical on either side of the barrier, but marine invertebrates seem to have been slower to evolve into new species: 2.3% of the echinoderm species, 6.5% of the porcellanid crab species and 10.8% of the sponge species of the region are found on both the eastern and the western shores of the Isthmus.

The deep waters of the Atlantic form a similar **Mid-Atlantic Barrier** to the dispersal of shelf organisms between the African tropics and the South American tropics. However, as the Atlantic is narrower than the Pacific, this barrier is less effective; thus, in most groups, there is a greater proportion of species that are found on both sides of the ocean. For example, in the shore fishes, there are about 900 species on the western shelf and about 434 species on the eastern; of this total, about 120 species (9%) are common to both faunas [29]. Most of these dispersals appear to have been from South America to Africa, perhaps because the greater richness of the South American shelf fauna, in terms of numbers of both species and individuals, makes it more likely that they will succeed in dispersing. Although the surface South Equatorial Current flows westward, these migrants may have used the deeper Equatorial Under-Current, which runs in the opposite direction.

An example of the appearance of a new link between shelf faunas is the one that took place between the Arctic and the North Pacific about 3.5 million years ago, after submergence of the

Bering Strait. This led to an exchange of cold-water species, in which the majority (125 species) dispersed from the Pacific to the Arctic Basin (in this case, in the direction of the current flow), and only 16 species dispersed in the reverse direction [36]. (But because of the shallowness of the sea across the Bering Strait, it remained an obstacle to the dispersal of deeper living plankton.)

The completion in 1869 of the Suez Canal between the Mediterranean Sea and the Red Sea provided a far more recent example of a link between two marine faunas. It has been named the **Lessepsian exchange**, after the French entrepreneur Ferdinand de Lesseps, who was responsible for its construction. The exchange has been very unbalanced for, although 50 species of fish, 40 species of mollusc and 20 species of crustacean have colonized the Mediterranean from the Red Sea, few if any species have made the reverse journey, which poses an interesting problem as to the cause of this difference [37,38]. Though both seas are small and almost landlocked, there is an important difference between their faunas. That of the Mediterranean is derived from the rather limited fauna of the cold Atlantic Ocean, whereas that of the Red Sea is derived from the rich tropical fauna of the Indian Ocean and is therefore better suited to colonizing the warm, shallow waters of the eastern Mediterranean. Furthermore, the species living at the northern end of the Red Sea already inhabit a shallow, sandy or muddy, and unusually saline environment. They are therefore better suited to surviving in the shallow hypersaline lakes of the Suez Canal through which they have to pass in order to reach the Mediterranean.

Latitudinal Patterns in the Shelf Faunas

Marine organisms show a general latitudinal trend of decreasing diversity as one moves away from the tropics. However, much of that is due to the distribution of coral reefs, which have a faunal diversity unparalleled elsewhere in the shallow seas. Their distribution is centred on the tropics and therefore heavily distorts the underlying pattern. In addition, Crame [39] has shown that Antarctic and sub-Antarctic shelf faunas are far richer (and more ancient) than had previously been thought, further emphasizing that this pattern, and its significance, must now be reconsidered. (For a general discussion of latitudinal patterns of diversity, see Chapter 4.)

Many marine organisms also provide examples of the phenomenon known as **bipolar distribution** (or **antitropical** or **amphitropical distribution**). This term describes a situation in which related organisms are to be found in temperate or polar environments in both the Northern and the Southern Hemispheres, but not in the intervening equatorial region. Whatever may be the reason for terrestrial examples of this pattern, its occurrence in marine faunas has prompted suggestions of specifically marine mechanisms. Charles Darwin suggested that the cooling of the equatorial waters during the Ice Ages had allowed these genera to pass through waters that are now once again too warm for them to inhabit. Brian White [40] has argued that it was instead the warmer temperatures earlier in the Cenozoic that made it impossible for these genera to live in the equatorial waters, so that they now show a relict distribution on either side of this zone. This perspective is supported by Gordon Howes in his useful analysis of the biogeography of the gadoid fishes [41]. Yet another explanation is that of Jack Briggs [42], who links it with his theory that the Indo-West Pacific region is a centre of evolutionary origin; he suggests that bipolar genera have become extinct in the equatorial regions because of competition from genera that have newly evolved there. Perhaps the simplest suggestion is that the organisms living in cool waters on either side of the equator have been able to disperse beneath it via cooler waters at greater depth (cf. the principle of equatorial submergence, discussed later in this chapter) [43]. An example of this was the catching of a notothenioid fish off Greenland [44]. Fishes of this family are normally confined to cold waters in the Southern Hemisphere. This lone example must have travelled at least 10 000 km submerged at depths of 500–1500 m to remain within its normally preferred temperature range. Some species of plankton show precisely this pattern of permanent distribution. For example, the chaetognath *Eukrohnia hamata* is only found in near-surface waters at latitudes greater than 60°, whereas, between those latitudes, where the surface waters are warmer, it is only found at depths of about 1000 m.

In his review of this topic, Crame [45] pointed out that most theories had tacitly assumed two premises: first, that these patterns had originated within the last 5 million years and were possibly related to the climatic changes of the Ice Ages, and, second, that the two now-separate taxa had achieved this pattern as a result of dispersal rather than of vicariance. Concentrating on the distribution of marine molluscs, which have a good fossil record extending over 245 million years, Crame identified three main periods of bipolar distributions. The first was in the Jurassic–Cretaceous and seems to have been caused by vicariance resulting from the breakup of Pangaea. The second was in the Oligocene–Miocene and may have been caused by vicariance resulting from the cooling temperatures at that time, which allowed temperate taxa to spread across the equator, only to become extinct there when temperatures rose again. The third period was during the Pliocene–Pleistocene Ice Ages, which, together with the closing of the Panama Isthmus, caused increased cooling and upwelling along the equatorial divergences in both the Pacific and the Atlantic, allowing dispersal of temperate forms from one hemisphere to the other. Our now steadily increasing knowledge of past patterns of climatic change on both land and sea is likely to lead to further understanding of the problem of bipolar distributions.

Coral Reefs

For many reasons, corals provide fascinating and unique aspects of marine biogeography. One of the most complex and diverse environments on Earth, they are clearly definable in nature, with limits of distribution that are simply explained by fundamental aspects of their biology. They therefore provide a good example of the extent to which marine patterns can be explained when the taxonomic and environmental aspects are simpler than elsewhere in the sea. On the other hand, interpretation of their patterns of diversity raises fundamental problems. Finally, because they are easily recognizable in the fossil record, historical biogeography can contribute more to our understanding of patterns of coral distribution than to most other aspects of marine biogeography. Two books, one by the Australian marine biologist Charlie Veron [46] and the other edited by the American Charles Birkeland [47], have dealt with many aspects of the biology and history of corals (although Veron's theories of coral evolution are controversial [48]).

Coral reefs provide a complex, three-dimensional environment that is home for an immense diversity of marine organisms [49], including 25% of the diversity of life in the oceans, and comprise the greatest diversity of species of vertebrate per square metre known on Earth. To date, 35 000–60 000 different species of reef-dwelling organisms have been described, and this is probably only a fraction of the total number. Between 1950 and 1994, the number of species of fish, molluscs, echinoderms and corals known from the Cocos (Keeling) Islands in the Indian Ocean tripled. Though coral reefs make up only 1% of the area of the oceans, they include 25% of its species, including over 5000 species of fish, over one-third of the total number of species of marine fishes.

Many reef organisms have planktonic larvae, and it has been generally assumed that these would be readily dispersed over distances of hundreds of kilometres, given reasonable longevity and current speed. However, work by a group of American marine ecologists [50] suggests that we may need to reconsider these assumptions. Using molecular techniques, they examined variations in the genetic structure of populations of a shrimp, *Haptosquilla*, taken from 11 reef systems in the East Indies, where the Java Sea and Flores Sea lie between the northern islands of Borneo and Sulawesi and the southern string of Sumatra, Java and smaller islands to the east (cf. Figure 11.9). The planktonic larvae of the shrimp live for 4–6 weeks and, since oceanic currents in the region are strong, it had been supposed that they would easily have been able to disperse across the 600 km that separates these two groups of islands. Surprisingly, however, these studies showed a sharp genetic break between the shrimp populations to the south of the Java–Flores seas and those to the north.

Corals are a type of colonial organism known as **hydrozoans**. The individuals, called *polyps*, resemble tiny sea anemones and feed on zooplankton. The types of coral that form reefs are known as **hermatypic corals**. In these corals, each polyp secretes a hard base composed of calcium carbonate, which is continuous with that of its neighbours, so that they all jointly form the reef. The tissues of the polyps contain a type of algae known as **zooxanthellae**, whose photosynthetic activity provides their energy and nourishment. In return, the algae receive useful nitrogenous waste products from the polyp, the total relationship providing a spectacular example of animal–plant symbiosis.

The biology of the corals limits their distribution to particular combinations of nutrient levels, temperature and light. Corals are found in areas where the nutrient levels are so low that there is too little primary productivity from free-living algae or phytoplankton to provide the basis for a diverse ecosystem. This provides an opportunity for the corals to flourish there because they can rely on the energy from their symbiotic algae rather than relying on gaining it from their environment. Of the other two factors, temperature is more important than light, as is shown by the fact that some corals can grow in deeper water as long as the temperature level is adequate. Coral reefs are only found where there is a minimum sea-surface temperature of at least 18°C, sustained over long periods of time, with a maximum of 30–34°C (Figure 9.10). Most are found between the latitudinal zones where the temperature never drops below 20°C. As a result, relatively diverse coral reef assemblages are found up to about 30° of north and south latitude, with extremes in Japan at 35°N, Bermuda at 32°N and Lord Howe Island at 32°S.

Hermatypic corals also cannot flourish where there is significant sedimentation, for this impedes the light that is vital for their photosynthetic algae. That is the explanation for the gap in the distribution of coral reefs along the tropical northern coastline of South America (Figure 9.11). The westerly winds of the equatorial Atlantic bring heavy rains to the low-lying river basins of tropical South America, which drain back to the sea via the great rivers. The dilution of the seawater by this freshwater, and the immense discharge of

sediment onto the seabed, make this coastal area inimical to the growth of corals. The impact of this phenomenon has been increased by the deforestation of the Amazon Basin, which has resulted in the waters of the great rivers bearing even more sediment and also having an increased concentration of nutriments, while fires in Central America have had a similar effect on corals in the neighbouring part of the Caribbean.

Within these limits, coral reefs are found worldwide. However, it is notable that, in both the Atlantic and Pacific oceans, there are many more corals towards the western ends of the oceans compared to the eastern ends. This has been discussed by the American marine biologist Gustav Paulay [51], who points out that there are several different reasons for this contrast. One major influence is the pattern of ocean currents, for the upwellings of cool, nutrient-rich waters that take place along the eastern margins of the oceans (see Figures 9.3 and 9.4) inhibit coral growth in those regions. In addition, most of the warm, equatorial ocean currents are directed westward; when they reach the edge of the continent, they diverge both northward and southward, bringing warmer waters to higher latitudes. Another contributory factor is that the continental shelves, on which most coral reefs lie, are much narrower along the western margins of Africa and the Americas than along the eastern margin of Asia and in the Caribbean region. However, not all corals grow on the continental shelves; some grow around the margins of volcanic islands. Since these are most common in the western parts of the Pacific Ocean, this provides another increment to the greater reef area of that region. As a result of all of these factors, 85% of the area of coral reef

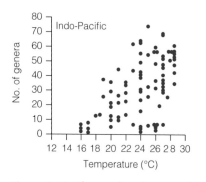
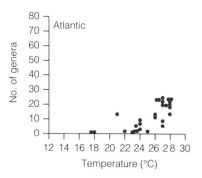

Figure 9.10 The number of genera of coral at different mean annual sea-surface temperatures in the Indo-Pacific and Atlantic oceans. From Rosen [65].

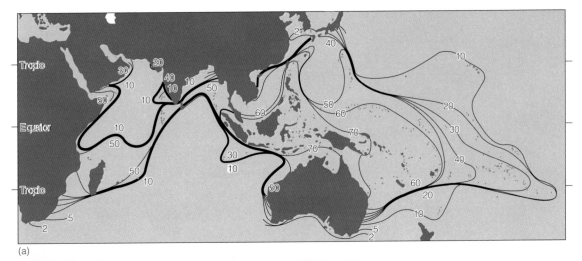

(a)

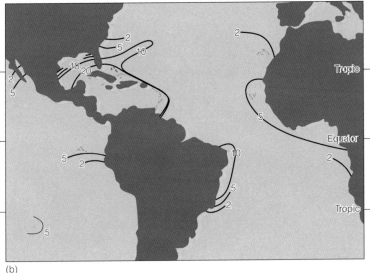

(b)

Figure 9.11 Contours of generic diversity in corals, combining the distribution ranges of all the genera. (a) The Indian Ocean and West Pacific regions. (b) The East Pacific and Atlantic oceans. Adapted from Veron [46].

lies in the Indo-Pacific Ocean and only 15% in the Atlantic Ocean. Similarly, the reefs of the eastern Pacific are only a few metres (or yards) thick, while those of the western Pacific are up to more than a kilometre in thickness.

But the most prominent feature of the diversity of most coral reef organisms is that this decreases both longitudinally and latitudinally (Figures 9.11, 9.12 and 9.13) as one moves away from the group of islands stretching from Sumatra to Papua New Guinea and north to the Philippines, which

make up the Indo-Australian Archipelago (IAA). The IAA area is therefore one of the world's major hotspots of biodiversity, with exceptional levels of endemism. (One locality in the East Indies has been reported to have over 1000 fish species – more than occur in the entire tropical Atlantic.) The taxonomic structure of this hotspot has been analysed as part of a recent wide-ranging review by the marine zoogeographers Dave Bellwood, Willem Renema and Brian Rosen [52], concentrating on information from reef fishes and corals.

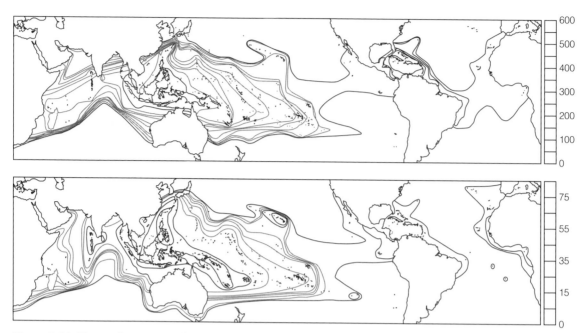

Figure 9.12 The similar patterns of tropical marine species richness found in coral reef fishes (above) and cowrie molluscs (below). From Bellwood and Meyer [66]. (Reproduced with permission of John Wiley & Sons.)

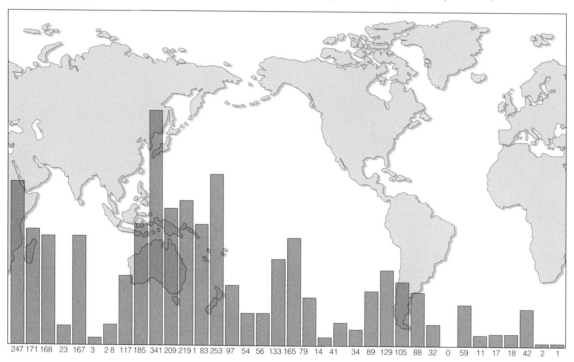

247 171 168 23 167 3   2 8 117 185 341 209 219 1 83 253 97   54 56 133 165 79   14 41   34 89 129 105 88 32   0   59 11   17   18 42   2   1

Figure 9.13 Longitudinal gradients in fish species richness. The columns represent the total number of fish species (from a sample of 799 species) that occur in each 10°-wide band of longitude. Note how the diversity increases in the latitudes that include the West Indies and Caribbean, where there are many coral reefs. Adapted from McAllister *et al.* [67].

Though most of these organisms have very large ranges, those in the IAA hotspot are instead species with a medium to large range.

As in the case of terrestrial gradients of diversity (see Chapter 4), earlier explanations of the high level of diversity in the IAA focussed on possible ecological and evolutionary factors, of which there are four main models: centre of origin, centre of overlap, centre of accumulation and centre of survival.

The centre of origin model suggests that speciation rates are unusually high in the IAA but that the various species have different abilities to disperse outward from it, so causing the gradient in diversity as one moves outward from its centre. Various factors have been invoked as explanations of this high speciation rate, including a high, consistent rate of solar energy input and an abundant area of reef that therefore provides many habitats within which speciation can take place.

Changes in sea level, subdividing and reuniting the reefs and thereby enabling frequent vicariant evolution, are another possible factor. This latter phenomenon is also part of the centre of overlap model: the subsequent enlargement of the ranges of the new species will lead to the overlap of their ranges with those of other species, and so increase local biodiversity. This model also suggests that the IAA has benefited from being between the Indian and Pacific oceans, and has therefore received new immigrant species from both.

The centre of accumulation model suggests two other possible reasons for the high number of species in the IAA, in addition to its higher speciation rates. The first is that, because equatorial currents flow from east to west across the Pacific Ocean, species that may have arisen on its many islands will tend to extend their ranges towards the IAA [53]. Second, because of its large area and high energy input, species in the IAA may have larger population sizes and therefore be less likely to become extinct, so that it is also a centre of survival. Another version of the centre of accumulation model suggests that the high number of species is because entire different faunas have arrived in the IAA, carried on moving continents (Australia and South-East Asia) and on islands, island arcs and terranes carried into the area by plate tectonic movements.

As in the case of terrestrial gradients of diversity, the mid-domain effect (MDE) has been suggested as a possible factor in the gradients shown in the IAA. This effect results from the statistical fact that, if species ranges are randomly placed within an enclosed domain (i.e. area) such as the IAA, the resultant pattern of species richness forms a peak in the middle of the domain – exactly where it is in the IAA, just northeast of New Guinea. The Australian marine biologist Dave Bellwood and his colleagues [54] have analysed the proportions of the variation in reef fish and coral species diversity that are explained by different aspects of the MDE and of two environmental effects (reef area and solar energy input). As shown in Figure 9.14, they found that in the case of both corals and fishes

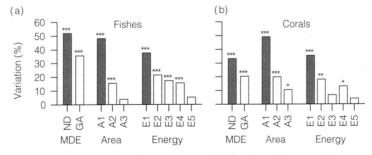

Figure 9.14 Proportion of variation in reef fish and coral species richness at 67 sites, explained by the mid-domain effect (MDE) and eight environmental variables that relate to energy or habitat area. ND, distance of site from the mid-domain relative to domain size; GA, great arc distance of site from the mid-domain; A1, reef area; A2, area of substratum 0–30 m deep; A3, area of substratum 30–200 m deep; E1, mean annual sea-surface temperature; E2, solar irradiance range; E3, sea-surface temperature range; E4, mean solar irradiance; E5, productivity; *** P, 0.001; ** P, 0.01; * P, 0.05. Dark bars indicate variables selected for additional analyses. From Bellwood *et al.* [54].

the most important of these aspects were distance from the centre of the domain (ND), the area of reef (A1) and mean sea-surface temperature (SST, E1). A model combining the roles of these three factors in explaining species richness was then calculated. In fish, ND was the most important single factor, explaining 51% of the variation, followed by reef area and then SST (explaining, respectively, an additional 16% and 2%, to make a total of 69%). In corals, reef area was the most important single factor, explaining 51% of the variation, followed by distance from the centre of the domain, and then SST (explaining, respectively, an additional 7% and 3%, to make a total of 61%). The authors comment that the similarity of the extent to which this analysis explains the patterns in two groups that differ markedly in their mobility, source of energy and manner of reproduction suggests that the results are likely to apply to many other tropical marine organisms. This suggestion has recently been supported by other Australian workers [55].

It would be a truism to point out that a large part of our analyses of the patterns and distributions of reef faunas inevitably relies on an adequate understanding of their genetics and taxonomy. However, research carried out by the American marine biologists Christopher Meyer, Jonathan Geller and Gustav Paulay [56] demonstrates the complete unreliability of our 'knowledge' of these crucial questions. They studied the genetics and distribution of the gastropod mollusc *Astralium rhodosteum* over 40 islands or locations over the 11 000 km distance from Thailand to eastern Polynesia. They found that this apparent species is in reality a species complex that, despite the fact that its larva is non-feeding and short-lived, and so apparently having only limited powers of dispersal, has developed endemic clades on every Pacific archipelago they studied. It comprises two major clades and at least 30 geographically isolated minor clades. One of these minor clades extends for over 750 km, but some others are separated from one another by less than 100 km. Though its powers of distribution allow the mollusc to colonize every island, at the same time the distribution of each clade is sufficiently limited for each one to have been able to evolve into a separate evolutionary unit, in strikingly allopatric fashion. In addition, colonization of an island by one clade seems to create barriers to its subsequent colonization by another. These clades

are also almost indistinguishable from one another morphologically, having virtually identical shells – so the true biogeography of this 'species' could not have been evaluated without genetic studies.

Meyer, Geller and Paulay comment that the data from this study do not support the idea that any of the current theories of the origins of the diversity of the IAA provide the unique key to its biogeography. Instead, this diversity is the product of multiple processes in space and time, including such ecological factors as the differences between oceanic and shelf environments. Of course, it also has to be recognized that a fundamental reason for the great diversity of reef life in this region is its geological history. Volcanic activity and sea-level changes have led to fragmentation of the larger islands each time sea levels rose, providing many opportunities for the appearance of new endemic taxa by vicariance. The distribution of endemic species has been widely used in the evaluation of the different models of the causes of the IAA hotspot, it being assumed that areas with a high proportion of endemic species are areas where new species are appearing. But, as Bellwood, Renema and Rosen [57] point out, most endemics are found peripheral to the IAA, rather than centrally, and are not particularly young. Recent data suggest that the ages of endemic species are quite variable. Also, many species are thought to be endemic to a limited area merely because, after they have first been described there, little has been done to establish how widely they are, in fact, distributed.

But it now seems possible that answer may lie, not in any of the supposed 'inherent' variations in rates of speciation, but in a far more common phenomenon in historical biogeography – evolutionary innovation [58]. Palaeontological studies show that the nature of the coral reefs and of their fauna changed profoundly shortly after the Cretaceous–Tertiary (K/T) boundary 65 million years ago. This was the time of the appearance of the scleractinian corals, which in turn provided the opportunity for the appearance and radiation of most of the groups that dominate the reefs today, including, most importantly, the herbivorous fish. The herbivorous fish maintain a regime of intense grazing on the algae, which creates open areas for fast-growing corals, while the structure of the reef provides shelter for the fish. This took place 42–39 million years ago in the Tethyan–Arabian hotspot

(Figure 9.15), where the approach of Africa to Europe had led to an increase in shallow-water

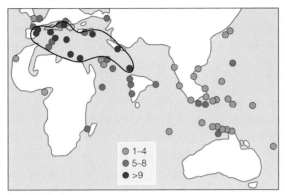

Late Middle Eocene

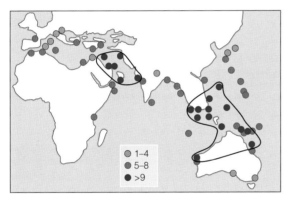

Early Miocene

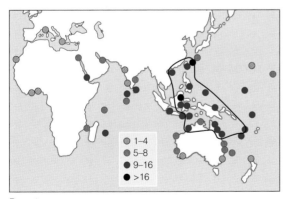

Recent

Figure 9.15 The locations of reef hotspots at three different periods of time. From Bellwood *et al.* [57]. (Reproduced with permission of Cambridge University Press.)

habitats [59]. By the Early Miocene 23–16 million years ago, the diversity in the western Tethys had diminished as the closer approach of Africa to Europe had now eliminated most of the shallow seas. The coral fauna of the Eastern Atlantic suffered badly from extinction after the closure of the Mediterranean seaway, losing some 85% of its coral genera, but long-distance dispersal has led to some exchange of taxa with the Caribbean region. At this same time, the centre of the Tethyan hotspot moved to Arabia, still in the earlier stage of approach between Africa and Asia. The northward movement of Australia had by this time led to its collision with Pacific island-arcs and the southeast margin of Asia, providing shallow-water habitats for a new hotspot in the IAA. This, with its enormous area in the tropical environment that encouraged diversity, became the only major tropical hotspot, and it appears to be where modern coral reefs first appeared. More specialized feeding types of fish (foraminifera feeders, cleaners, detritus feeders, etc.) appeared, and the increasing complexity of the reef ecosystem encouraged increased rates of diversification allied to reduced liability to extinction. (This major radiation of a group in a limited, particularly favourable location is reminiscent of the radiation of the herbivorous bovids in East Africa; see Chapter 11.)

David Bellwood, together with Peter Cowman of Yale University, has recently taken this analysis of reef fish phylogenies further [60], comparing their patterns of origination during the Cenozoic (Figure 9.16). These showed that, from the Eocene onward, the Atlantic and East Pacific regions became increasingly isolated from the rest. The IAA changed its role from being primarily a centre of accumulation in the Paleocene–Eocene, to being a centre of survival from the Oligocene onward. In the Miocene, it became primarily a centre of origination, but also an increasing source of export of some of the resulting lineages into the adjacent regions, especially in the Pliocene. This export increased the diversity in the Indian Ocean and in the West Pacific, but the East Pacific Barrier prevented this spreading also to the East Pacific. According to this analysis, the IAA hotspot of diversity is largely the result of the origination of new lineages in this favourable location over the last 33 million years.

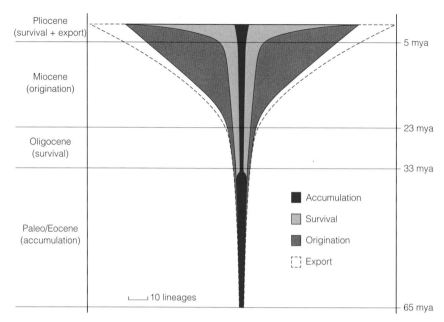

Figure 9.16 Diagram illustrating the changing role of the Indo-Australian Archipelago hotspot in the origins of Indo-Pacific reef fish diversity as inferred from the diversity of labrid fishes. From Cowman and Bellwood [60]. (Reproduced with permission of John Wiley & Sons.)

All in all, the history of investigations and theories of this problem provides a fascinating story of the progress of marine biogeography over the last 20 years, and underlines the importance of not neglecting the historical component when trying to understand contemporary problems.

The next three chapters in the book are concerned with the analysis and history of the biogeography of the continents. This commences with the patterns found some 200 million years ago, and continues through the succeeding millions of years of change in continental patterns, faunas, floras and climates.

Summary

1 The composition, structure and ecology of marine biota are still poorly understood, and their environment differs greatly from that of terrestrial organisms. Nevertheless, it is becoming clear that marine biogeography is basically similar to that of the land, although the boundaries between the units are more gradual and are not fixed in their locations. On the other hand, marine organisms do not seem to show the same degree of correlation between morphology and species definition, which makes the study of their biogeography far more difficult than that of terrestrial species.

2 The major subdivision is between the open-sea environment and the shallow-sea environment.

3 The open-sea environment is divided vertically into zones based on light, temperature and nutrient availability. The surface waters are divided into biomes and provinces, which are related to the rotating patterns of ocean circulation and the patterns of productivity.

4 The units in the shallow-sea environment are much smaller than those in the open-sea environment because they are dependent on the characteristics of the local land, rivers, sediments, seabed, tides and ocean currents.

5 Coral reefs provide the most diverse environments within the seas and the clearest examples of gradients of marine diversity.

Further Reading

Gage JD, Taylor PA. *Deep Sea Biology*. Cambridge: Cambridge University Press, 1991.

Nybakken JW. *Marine Biology. An Ecological Approach*, 4th ed. Menlo Park, CA: Addison Wesley Longman, 1997.

Ormond RFG, Gage JD, Angel MV (eds.). *Marine Biodiversity. Patterns and Processes*. Cambridge: Cambridge University Press, 1997.

Vierros M, Cresswell I, Escobar Briones E, Rice J, Ardron J (eds.). *Global Open Oceans and Deep Seabed (GOODS) Biogeographic Classification*. Intergovernmental Oceanographic Commission (IOC) Technical Series 84. Paris: UNESCO, 2009.

References

1. Vrijenhoek RC. Cryptic species, phenotypic plasticity, and complex life histories: assessing deep-sea faunal diversity with molecular markers. *Deep-Sea Research II* 1970; 11: 1713–1723.

2. Grassle JF, Maciolek NJ. Deep-sea species richness: regional and local diversity estimates from quantitative bottom samples. *American Naturalist* 1992; 139: 313–341.

3. Grant WS, Leslie RW. Inter-ocean dispersal is an important mechanism in the zoogeography of hakes (Pisces: *Merluccius* spp.). *Journal of Biogeography* 2001; 28: 699–721.

4. Vierros M, Cresswell I, Escobar Briones E, Rice J, Ardron J (eds.). *Global Open Oceans and Deep Seabed (GOODS) Biogeographic Classification*. Intergovernmental Oceanographic Commission (IOC) Technical Series 84. Paris: UNESCO, 2009.

5. Angel MV. Spatial distribution of marine organisms: patterns and processes. In: Edwards PJR, May NR, WebbNR (eds.), *Large Scale Ecology and Conservation Biology*. British Ecological Society Symposium no. 35. Oxford: Blackwell Science, 1994: 59–109.

6. Angel MV. Pelagic biodiversity. In: OrmondRFG, Gage JD, Angel MV (eds.), *Marine Biodiversity Patterns and Processes*. Cambridge, Cambridge University Press, 1997: 35–68.

7. Stuiver M, Quay PD, Ostlund HG. Abyssal water carbon-14 distribution and ages of the world's oceans. *Science* 1983; 139: 572–576.

8. Tyrrell T. The relative influences of nitrogen and phosphorus on oceanic primary production. *Nature* 1999; 400: 525–531.

9. Rutherford S, D'Hondt S, Prell W. Environmental controls on the geographic distribution of zooplankton diversity. *Nature* 1999; 400: 749–753.

10. Longhurst A. *Ecological Geography of the Sea*. 2nd ed. London: Academic Press, 2006.

11. Martin JH, Fitzwater SE, Gordon RM. Iron deficiency limits phytoplankton growth in Antarctic waters. *Global Biogeochemical Cycles* 1990; 4: 5–12.

12. Pierrot-Bults AC. Biological diversity in oceanic macrozooplankton: more than merely counting species. In: Ormond RFG, Gage JD, Angel MV (eds.), *Marine Biodiversity Patterns and Processes*. Cambridge: Cambridge University Press, 1997: 69–93.

13. Gibbs RH. The stomioid fish genus *Eustomias* and the oceanic species concept. In: UNESCO, *Pelagic Biogeography*. UNESCO Technical Papers in Marine Science no. 49. Paris: UNESCO, 1985: 98–103.

14. Lessios HA, Kessing BD, Pearse JS. Population structure and speciation in tropical seas: global phylogeny of the sea urchin *Diadema*. *Evolution* 2001; 55: 955–975.

15. Moser M, Hsieh J. Biological tags for stock separation in Pacific Herring *Clupea harengus pallasi* in California. *Journal of Parasitology* 1992; 78: 54–60.

16. White BN. Vicariance biogeography of the open-ocean Pacific. *Progress in Oceanography* 1994; 34: 257–282.

17. McGowan JA. The biogeography of pelagic ecosystems. In: UNESCO, *Pelagic Biogeography*. UNESCO Technical Papers in Marine Science no. 49. Paris: UNESCO, 1985: 191–200.

18. Bellwood DR, Wainwright PC. History and biogeography of fishes on coral reefs. In: SalePF (ed.), *Coral Reef Fishes – Dynamics and Diversity in a Complex Ecosystem*. San Diego: Academic Press, 2002: 3–15.

19. Ekman S. *Zoogeography of the Sea*. London: Sidgwick & Jackson, 1958.

20. Rex MA, Etter RJ, Stuart CT. Large-scale patterns of species diversity in the deep-sea benthos. In: Ormond RFG, Gage JD, AngelMV (eds.), *Marine Biodiversity Patterns and Processes*. Cambridge: Cambridge University Press, 1997: 94–121.

21. Etter RJ, Grassle JF. Patterns of species diversity in the deep sea as a function of sediment particle size diversity. *Nature* 1992; 360: 576–578.

22. Grassle JF. Deep sea benthic diversity. *Bioscience* 1991; 41: 464–469.

23. O'Hara TD, Rowden AA, Bax NJ. A Southern Hemisphere bathyal fauna is distributed in latitudinal bands. *Current Biology* 2011; 21: 226–230.

24. Martin W, Baross J, Kelley D, Russell MJ. Hydrothermal vents and the origin of life. *Nature Reviews Microbiology* 2008; 6: 805–814.

25. Moalic Y, Desbruyères D, Duarte CM, Rozenfeld AF, Bachraty C, Arnaud-Haond S. Biogeography revisited with network theory: retracing the history of hydrothermal vent communities. *Systematic Biology* 2012; 61: 127–137.

26. Rogers AD, Tyler PA, Connelly DP, *et al.* The discovery of new deep-sea hydrothermal vent communities in the Southern Ocean and implications for biogeography. *PLoS Biology* 2012; 10 (1): 1–17, e10011234.

27. Tunnicliffe V, Fowler MR. Influence of sea-floor spreading on the global hydrothermal vent fauna. *Nature* 1996; 379: 531–533.

28. Kinlan BP, Gaines SD, Lester SE. Propagule dispersal and the scales of marine community processes. *Diversity and Distributions* 2005; 11: 139–148.

29. Briggs JC. *Marine Zoogeography*. New York: McGraw-Hill, 1974.

30. Spalding MD, Fox HE, Allen GR, *et al.* Marine ecoregions of the world: a bioregionalization of coastal and shelf areas. *Bioscience* 2007; 57: 573–583.

31. Springer VG. Pacific plate biogeography, with special reference to shore fishes. *Smithsonian Contributions to Zoology* 1982; 367: 1–182.

32. Pierrat B, Saucède T, Brayard A, David B. Comparative biogeography of echinoids, bivalves and gastropods from the Southern Ocean. *Journal of Biogeography* 2013; 40: 1374–1385.

33. Lessios HA, Kessing BD, Robertson DR. Massive gene flow across the world's most potent marine barrier. *Proceedings of the Royal Society London B* 1998; 265: 583–588.

34. Grigg R, Hey R. Paleoceanography of the tropical Eastern Pacific Ocean. *Science* 1992; 255: 172–178.

35. Fallaw WC. Trans-North Atlantic similarity among Mesozoic and Cenozoic invertebrates correlated with widening of the ocean basin. *Geology* 1979; 7: 398–400.

36. Vermeij GJ. Anatomy of an invasion: the trans-Arctic exchange. *Paleobiology* 1991; 17: 281–307.

37. Edwards AJ. Zoogeography of Red Sea fishes. In: Williams AS, Head SM (eds.), *Key Environments: Red Sea*. Oxford; Pergamon Press, 1987: 279–286.

38. Golani D. The sandy shore of the Red Sea – launching pad for the Lessepsian (Suez Canal) migrant fish colonizers of the eastern Mediterranean. *Journal of Biogeography* 1993; 20: 579–585.

39. Crame JA. An evolutionary framework for the polar regions. *Journal of Biogeography* 1997; 24: 1–9.

40. White BN. The isthmian link, antitropicality and American biogeography: distributional history of the Atherinopsidae (Pisces; Atherinidae). *Systematic Zoology* 1986; 35: 176–194.

41. Howes GJ. Biogeography of gadoid fishes. *Journal of Biogeography* 1991; 18: 595–622.

42. Briggs JC. Antitropical distribution and evolution in the Indo-West Pacific Ocean. *Systematic Zoology* 1987; 36: 237–247.

43. Boltovskoy D. The sedimentary record of pelagic biogeography. *Progress in Oceanography* 1994; 34: 135–160.

44. Møller PR, Nielsen JG, Fossen I. Patagonian toothfish found off Greenland. *Nature, London* 2003; 421: 599.

45. Crame JA. Bipolar molluscs and their evolutionary implications. *Journal of Biogeography* 1993; 20: 145–161.

46. Veron JEN. *Corals in Space and Time*. Sydney: University of New South Wales Press, 1995.

47. Birkeland C (ed.). *Life and Death of Coral Reefs*. New York: Chapman & Hall, 1997.

48. Paulay G. Circulating theories of coral biogeography. *Journal of Biogeography* 1996; 23: 279–282.

49. Kohn AJ. Why are coral reef communities so diverse? In: Ormond RFG, Gage JD, Angel MV (eds.), *Marine Biodiversity Patterns and Processes*. Cambridge: Cambridge University Press, 1997: 201–215.

50. Barber PH, Palumbi SR, Erdmann MV, Moosa MK. A marine Wallace's line? *Nature, London* 2000; 406: 692–693.

51. Paulay G. Diversity and distribution of reef organisms. In: Birkeland C (ed.), *Life and Death of Coral Reefs*. New York: Chapman & Hall, 1997: 298–353.

52. Bellwood DR, Renema W, Rosen BR. Biodiversity hotspots, evolution and coral reef biogeography: a review. In: Gower DJ, Johnson K, Richardson J, Rosen B, Rüber L, WilliamsS (eds.), *Biotic Evolution and Environmental Change in Southeast Asia. Spec. Vol. Systematics Assoc.* Cambridge: Cambridge University Press, 2012: 216–245.

53. Jokiel P, Martinelli FJ. The vortex model of coral reef biogeography. *Journal of Biogeography* 1992; 19: 449–458.

54. Bellwood DR, Hughes TP, Connolly SR, Tanner J. Environmental and geometric constraints on Indo-Pacific coral reef diversity. *Ecology Letters* 2005; 8: 643–651.

55. Mellin C, Bradshaw CJA, Meekan MG, Caley MJ. Environmental and spatial predictors of species richness and abundance in coral reef fishes. *Global Ecology and Biogeography* 2010; 19: 212–222.

56. Meyer CP, Geller JB, Paulay G. Fine scale endemism on coral reefs: archipelagic differentiation in turbinid gastropods. *Evolution* 2005; 59: 113–125.

57. Bellwood DR, Renema W, Rosen B. Biodiversity hotspots, evolution and coral reef biogeography: a review. In: Gower DJ, Johnson K, Richardson J, Rosen B, RüberL, Williams S (eds.), *Biotic Evolution and Environmental Change in Southeast Asia. Spec. Vol. Syst. Assoc.* Cambridge: Cambridge University Press, 2012: 216–246.

58. Cowman PF, Bellwood DR. Coral reefs as drivers of cladogenesis: expanding coral reefs, cryptic extinction events, and the development of biodiversity hotspots. *Journal of Evolutionary Biology* 2010; 24: 2543–2562.

59. Renema W, Bellwood DR, Braga JC, *et al.* Hopping hotspots: global shifts in marine biodiversity. *Science* 2008; 321: 654–657.

60. Cowman PF, Bellwood DR. The historical biogeography of coral reef fishes: global patterns of origination and dispersal. *Journal of Biogeography* 2012; 40: 209–224.

61. Levitus S, Conkright ME, Reid JL, Najjar RG, Mantyla A. Distribution of nitrate, phosphate and silicate in the world oceans. *Progress in Oceanography* 1993; 31: 245–274.

62. Berger WH. Global maps of ocean productivity. In: Berger WH, Smetacek VS, Wefer G (eds.), *Productivity of the Ocean: Past and Present*. London: Wiley, 1989: 429–455.

63. Scheltema RS. Planktonic and non-planktonic development among prosobranch gastropods and its relationships to the geographic ranges of species. In: Rylands JS, Tyler PA (eds.), *Reproduction, Genetics and Distribution of Marine Organisms*. 23rd European Marine Biology Symposium, Fredensborg, 1989. Fredensborg, Denmark: Olsen & Olsen, 1989.

64. Thorson G. Reproductive and larval ecology of marine bottom invertebrates. *Biological Review* 1950; 25: 1–45.

65. Rosen BR. Reef coral biogeography and climate through the late Cainozoic: just islands in the sun or a critical pattern of islands? *Special Issue, Geological Journal* 1984; 11: 201–262.

66. Bellwood DR, Meyer CP. Searching for heat in a marine hotspot. *Journal of Biogeography* 2009; 36: 569–576.

67. McAllister DE, Schueler FW, Roberts CM, Hawkins JP. Mapping and GIS analysis of the global distribution of coral reef fishes on an equal-area grid. In: Miller R (ed.), *Advances in Mapping the Diversity of Nature*. London: Chapman & Hall, 1994: 155–175.

Patterns in the Past

Chapter 10

*The past is a foreign country; they do things dif-
ferently there.*

(L.P. Hartley, *The Go-between*)

This chapter explains how the very different
geographies, climates, faunas and floras of our
planet gradually changed, over the last 400 million years, into those that we see today. At first
there was a pattern of separate continents, which
later united into a single Pangaea landmass, followed by renewed fragmentation and some collisions. The early history of mammals and flowering plants is described, together with the changing
climates and floras from the Mid-Cretaceous up to
the beginning of global cooling at the end of the
Miocene.

The ecological approach that was explained in
Chapters 2–4 can explain some of the aspects of
the distributions of the different groups of animals
and plants within the different continents. But
these groups are also quite differently distributed
among the continents, and biogeographers also
want to understand how this has come about. This
historical approach to continental biogeography is
the subject of Chapters 11 and 12.

Although very few groups have precisely the
same pattern of geographical distribution, there are
some zones that mark the limits of distribution of
many groups. This is because these zones are barrier regions, where conditions are so inhospitable
to most organisms that few of them can live there.
For terrestrial animals, any stretch of sea or ocean
proves to be a barrier of this kind – except for flying
animals, whose distribution is for this reason obviously wider than that of solely terrestrial forms.

Extremes of temperature, such as exist in deserts
or in high mountains, constitute similar (though
less effective) barriers to the spread of plants and
animals.

These three types of barrier – oceans, mountain chains and large deserts – therefore provide
the major discontinuities in the patterns of the
spread of organisms around the world. Oceans
completely surround Australia. They also virtually
isolate South America and North America from
each other and completely separate them from
other continents. Seas, and the extensive deserts
of North Africa and the Middle East, effectively
isolate Africa from Eurasia. India and South-East
Asia are similarly isolated from the rest of Asia by
the vast, high Tibetan Plateau, of which the Himalayas are the southern fringe, together with the
Asian deserts that lie to the north. Each of these
land areas, together with any nearby islands to
which its fauna or flora has been able to spread, is
therefore comparatively isolated. It is not surprising to find that today's patterns of distribution of
both the faunas (faunal provinces or zoogeographical regions) and the floras (floral regions) largely
reflect this pattern of geographical barriers.

Before the detailed composition of these faunal
provinces and floral regions can be understood
fully, it is first necessary to follow the ways in
which today's patterns of geography, climate and
distribution of life came into existence. From what
has been discussed in this volume, it is clear that
the differences between the faunas and floras of
different areas might be due to a number of factors. First, any new group of organisms will appear
first in one particular area. If it competes with

Biogeography: An Ecological and Evolutionary Approach, Ninth Edition. Edited by C. Barry Cox, Peter D. Moore, Richard J. Ladle.
© 2016 John Wiley & Sons, Ltd. Published 2016 by John Wiley & Sons, Ltd.

another, previously established group in that area, the expansion in the range of distribution of the new group may be accompanied by contraction in that of the old. However, once it has spread to the limits of its province or region, whether or not it is able to spread into the next will depend initially on whether it is able to surmount the geographical barrier between them, or adapt to the different climatic conditions to be found there. (Even if it is able to cross to the next province or region, however, it may be unable to establish itself because of the presence there of another group that is better adapted to that particular environment.) Of course, changes in the climate or geographical pattern could lead to changes in the patterns of distribution of life. For example, gradual climatic changes, affecting the whole world, could cause the gradual north- or southward migrations of floras and faunas because these extended into newly favourable areas and died out where the climate was no longer hospitable. Similarly, the possibilities of migration between different areas could change if vital links between them became broken by the appearance of new barriers, or if new links appeared.

Early Land Life on the Moving Continents

Our understanding of the movements of the continents and of the timing of the different episodes of continental fragmentation or union is now fairly detailed. A 1993 understanding of this, showing also the distribution of the epicontinental seas, is shown in Figure 10.1, while an up-to-date set of maps showing the movements of the continents and mini-continents is shown in Plate 7. The distribution of fossil organisms correlates very well with the varying patterns of land. The earliest time at which there is enough evidence to discern patterns of life is the Early Devonian, about 380 million years ago (Plate 7a and Figure 10.1), by which time separate floras and/or fish faunas can be distinguished in the northerly placed Siberian continent, the equatorially placed Euramerican continent, Kazakhstan, northern Africa and Australia [1]. Early amphibians are found in near-equatorial positions in both Euramerica and Australia, where the climate would have encouraged a rich growth both of plants and of the invertebrates that the earliest

terrestrial vertebrates would have fed upon. The fossil record suggests that all the early amphibian and reptile groups evolved in the continent called Euramerica, where they were largely confined to a warm, humid equatorial zone, bordered by dry subtropical belts, until that continent collided with the great supercontinent made up of all of today's southern continents known as Gondwana in the Late Carboniferous [2].

The great expansion of the land plants began in the Devonian, and this itself may have been one of the causes of a marked change in the world's climate. Today, vegetated surfaces decrease the **albedo** (or reflectivity) of an area by 10–15%, while plants recycle much of the rainfall (up to 50% in the Amazon Basin). Theory suggests that the photosynthetic activities of the plants would have reduced the carbon dioxide content of the atmosphere, which would have led to an 'ice-house' effect (the reverse of the greenhouse effect of higher temperatures, which results from *increased* carbon dioxide), and an accompanying increase in the levels of oxygen produced by the plants. There is botanical, geological and experimental evidence [3–5] that this took place, and the enormous build-up of the world's vegetation at this time may have been the cause of the next significant event, which was the onset of global cooling.

This cooling began in the middle of the Carboniferous Period and led to the appearance of ice sheets around the South Pole, similar to those of Antarctica today. Through this period of time, the whole of Gondwana was rotating clockwise and moving towards Euramerica, with which it collided in the Late Carboniferous (Plate 7b and Figure 10.1). As Gondwana moved across the South Pole, the glaciated area moved across its surface. Although the whole South Polar area must have been icy cold, the proportion that was actually glaciated probably varied according to the position of the Pole itself. When this was near the edge of the supercontinent, the adjacent ocean would have provided enough moisture to create the heavy snowfalls that formed extensive continental ice sheets. But when the Pole lay farther inland, away from the ocean, it is possible that the inland areas would have been a polar desert rather than being glaciated.

The low temperatures around the South Pole caused the latitudinal ranges of the Carboniferous floras to be compressed towards the equator. In the equatorial region, there was a great swampy

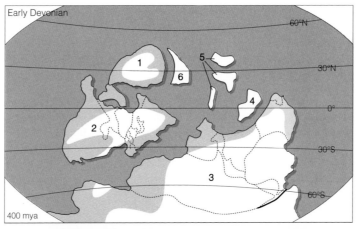

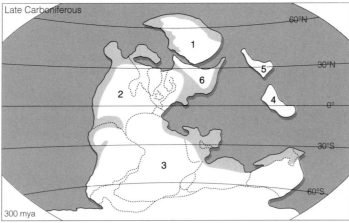

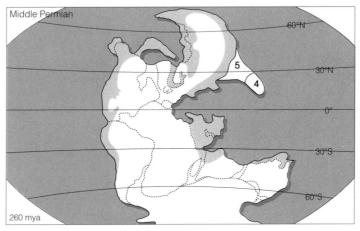

Figure 10.1 World geography at three different stages in the past: Tripel–Winkel projection. Dark tint indicates ocean; epicontinental seas (light shading) after Smith *et al.* [38]. Dotted lines indicate modern coastlines. Continental positions after Metcalfe [39]. (1) Siberia; (2) Euramerica; (3) Gondwana; (4) southern China; (5) northern China; (6) Kazakhstan.

tropical rainforest, rather like that of the Amazon Basin today. This lay across Euramerica and was fed by rains from the warm westward equatorial ocean current that would have washed the eastern shore of that continent. Surprisingly, the distribution of different types of rock laid down at this time shows no sign of the presence of monsoonal, dry conditions in the interior of the great supercontinent, and the climatic pattern seems to have been essentially latitudinal [6]. The equatorial wet belt was bordered to the south by a subtropical desert that stretched across northern South America and northern Africa. Beyond this lay a warm temperate zone across central South America and Africa, to the south of which lay the cold glacial or desert lands around the South Pole. Another desert belt covered northern North America and northeastern Europe, but Siberia (still a separate island continent) lay in a warm temperate zone farther to the north.

The absence of dormant buds and annual growth rings in the fossil remains of the equatorial coal-swamp flora indicates that it grew in an unvarying, seasonless climate. The flora was dominated by great trees belonging to several quite distinct groups (Figure 10.2). *Lepidodendron*, 40 m tall, and *Sigillaria*, 30 m tall, were enormous types of lycopod (related to the tiny living club-moss, *Lycopodium*). Equally-tall *Cordaites* was a member of the group from which the conifer trees evolved,

and *Calamites*, up to 15 m high, was a sphenopsid related to the living horsetail, *Equisetum*. Tree ferns such as *Psaronius* grew up to 10 m high, and seed ferns such as *Neuropteris* were among the most common smaller plants living around these great trees. In the eastern United States, and in parts of Britain and central Europe, the land covered by this swamp forest was gradually sinking. As it sank, the basins that formed became filled with the accumulated remains of these ancient trees. Compressed by the overlying sediments, dried and hardened, the plant remains have become the coal deposits of these regions. Far to the south of the equatorial coal-swamp flora, the lands around the growing south polar ice sheets bore a different flora, lacking many of the northern trees and with fewer ferns and seed ferns.

After the Carboniferous gave way to the Permian, the coal swamps of southern Euramerica disappeared, and deserts lay in their place by the Late Permian. This was partly because these regions had moved northward, away from the equator, and partly because the mountain ranges of northern Africa and eastern North America had extended and risen higher, blocking the moist winds from the ocean that lay to the east.

In the Middle Permian, four main floras can be distinguished [7]. In the north, the still-separate continents of Siberia and Kazakhstan bore the cold

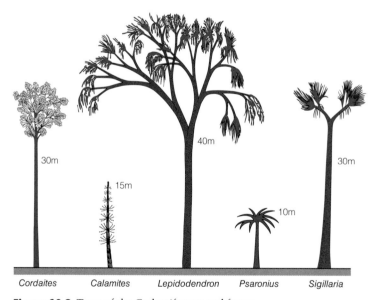

Figure 10.2 Trees of the Carboniferous coal forest.

temperate 'Angara' flora, with *Cordaites*-like conifers, and also herbaceous horsetail plants, ferns and seed ferns. The flora is richest towards the eastern coast and becomes less diverse towards the colder north of Siberia. The second flora is the rich, varied, ever-wet tropical 'Cathaysian' rainforest flora, which is found in the still-separate land mass of China in the Pacific. This flora was made up of sphenopsid and *Lepidodendron* trees, *Gigantopteris* lianas and many types of seed fern; conifers and *Cordaites* were rare. The old coal swamps of southern Euramerica had disappeared, and the rest of Euramerica bore a tropical summerwet 'Euramerican' flora, with ginkgoes, conifers and *Cordaites*. Subtropical deserts lay to the north and south of this, across northern Euramerica, northern South America and north-central Africa. The Gondwana flora, characterized by the seed fern *Glossopteris*, occupied the whole of that supercontinent in a cold-temperate environment.

In the Late Permian (Plate 7c and Figure 10.1), the equatorial belt narrowed and the subtropics expanded, while the polar ice caps disappeared. The world became warmer and drier, causing changes in both the plants and animals as they adapted to these new environmental conditions, with new groups evolving and spreading widely through the world, while many of the original groups became extinct. As one might expect, land vertebrates did not reach Siberia or China until after those landmasses had joined the world supercontinent in the Mid to Late Permian. Rich faunas of Late Permian fossil reptiles have been found in mid-latitude regions of southern South America and Africa that, according to climatic modelling experiments, would have had annual temperature changes of 40–50°C – similar to those of Central Asia today, and not very congenial to reptiles [8]. However, many of the rivers drained internally into great lakes, whose waters would have had a cooling, stabilizing effect on the climate.

One World – for a While

The coalescence of the different continental fragments to form Pangaea led to climatic changes. In general, the world became steadily warmer and drier during the Triassic. The disappearance of the oceans that had previously separated the continents, and the formation of the enormous supercontinent of Gondwana, had left vast tracts of land far from the oceans and the moist winds that originated there. Furthermore, the new, lofty mountain ranges that still marked the regions where Euramerica, Siberia and China had collided provided physical and climatic barriers to the dispersal of their floras and faunas.

In the floras, we can identify a general evolutionary change in which older types of tree, such as those belonging to the lycopods, sphenopsids and *Calamites*, disappeared. They were replaced either by the radiation of existing types of tree, such as caytonias and ginkgoes, or by new groups such as cycads, bennettitaleans and conifers. This floral change was complete by the end of the Triassic Period. However, there were also local variations: for example, in the Gondwana flora, the Early Triassic *Glossopteris* was replaced by the seed fern *Dicroidium*.

In the Jurassic and Early Cretaceous, floras seem to have gradually become more similar to one another, approaching the modern pattern in which there are gradual latitudinal changes governed by climate, the patterns of dominance of different groups changing from lower latitudes to higher latitudes [9]. There was a broad equatorial-band summerwet tropical biome, covering southern North America and northern South America and Africa, to the north and south of which lay in turn a winterwet Mediterranean-like band and a warm-temperate and cool-temperate band with decreasing diversity. The flora of the warm-temperate band included cycads, bennettitaleans and large-leaved conifers such as podocarps and *Araucaria*, which would have provided suitable food for the herbivorous dinosaurs, including the long-necked sauropods, which include the largest land vertebrates we know, weighing up to 17 metric tonnes.

These Mesozoic floras extended to high (about 70°) northern and southern latitudes, to areas that, though clearly warm, must have had seasonal very brief periods of daylight. Oxygen-isotope measurements of the composition of fossil shells of marine Cretaceous plankton show that the intermediate to deep waters of these oceans were 1.5°C warmer than those of today. On land, the presence and spread of plants, dinosaurs and early mammals through high-latitude routes such as the Bering region and Antarctica support these observations.

To examine the biogeographical history of the vertebrate land animals (amphibians and reptiles) of Pangaea, one must return to the Permian and Triassic. By the middle of the Permian, these animals appear to have been able to disperse through regions of different climate, for Pangaea soon came to contain a fairly uniform fauna, with little sign of distinct faunal regions [10]. Great changes took place in this worldwide fauna during the Triassic [11]. The bulk of the Permian faunas had been made up of synapsid or 'mammal-like' reptiles (so named because they include the ancestors of the mammals) and other older types of reptile, but these disappeared during the Early and Middle Triassic. They were at first replaced by a radiation of early reptiles known as archosaurs. However, these in turn were soon replaced (in the Late Triassic) by their own descendants, the dinosaurs, which came to dominate the world throughout the Jurassic and Cretaceous.

As we have seen (Chapter 5), Pangaea became progressively more fragmented during this period of time, partly because of the break-up of the land masses and partly because rising sea levels also subdivided them. The region that we now know as Europe was often merely an archipelago of separate islands.

The fact that, to begin with, the whole of the land in the Northern Hemisphere was interconnected allowed the Triassic and Early Jurassic dinosaurs to range throughout the whole northern area; they were also able to spread into Gondwana [2]. As a result, most of the dinosaur groups that evolved in the Jurassic and Early Cretaceous (ostrich dinosaurs, dome-headed dinosaurs, dromaeosaurs, primitive duck-billed dinosaurs and larger carnivorous tyrannosauroids) were able to disperse throughout the undivided Northern Hemisphere (Plate 7d and Figure 10.3). There, they were able to replace older groups such as the ceratosauroids, allosauroids and titanosaurs. But because Gondwana was now separate, they were unable to reach that southern landmass, and the older groups were therefore able to survive there. The extent to which the groups that evolved later were able to spread probably differed according to the time at which each appeared [12,13]. Those that had evolved earlier, before the subdivisions of the land, must have been able to spread more widely than those that appeared later. So, for example, the primitive

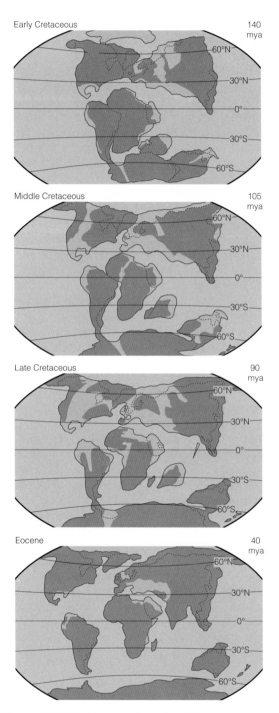

Figure 10.3 World geography at four different stages in the past: Tripel-Winkel projection; outlines of epicontinental seas (light shading) after Smith *et al.* [38]. Dotted lines indicate modern coastlines. Continental positions after Cambridge Palaeomap Services [40].

subfamily of duck-billed dinosaurs, the hadrosaurines, evolved by the Lower Cretaceous and was able to spread throughout the Northern Hemisphere. Three new subfamilies (the saurolophines, cteniosaurines and lambeosaurines), evolved in the Upper Cretaceous. However, this took place after the Mid-Continental Seaway subdivided that land mass into Euramerica and Asiamerica (Figure 10.3), and they were therefore unable to spread out of their homeland in Asiamerica [2].

Our knowledge of the dinosaurs of the Southern Hemisphere is more limited than that of the Northern Hemisphere (though much is currently being discovered both there and in China). We know that sauropods and more advanced duck-billed dinosaurs were present in the Late Cretaceous of South America. It is possible that their ancestors had reached South America earlier in the Cretaceous, from Eurasia via Africa before Africa split from South America, as has been suggested for early placental mammals. Though it seems likely that a chain of volcanic islands lay between North and South America during the Cretaceous (cf. Figure 11.13), it seems unlikely that animals the size of dinosaurs could have dispersed via this route.

It might have been hoped that we could now identify differing dinosaur faunas in the different various land masses in the Southern Hemisphere, as we can for mammals in the Cenozoic. However, there are several difficulties in doing this. In order to distinguish separate faunas, we need to be able to identify a variety of groups in each area, so that we can distinguish different faunas by the unique presence or absence of particular taxa. Unfortunately, both the patchy nature of the fossil record and the rapidity (in geological terms) with which the pattern of geography changed make this impossible. Another difficulty is that we are still uncertain about the interrelationships of many of the taxa, so that we cannot trace the patterns of diversification of the groups.

However, the subdivision of Pangaea had some fundamental biogeographical consequences, for evolution could now take place independently in each new continent to produce new, unique groups, thus leading to greater global diversity. This can be contrasted to the reduction in global diversity that took place when previously separated continents became connected, as when North and South America became linked, leading to the extinction of many groups of South American mammals (see 'The Great American Interchange' in Chapter 11). It is also interesting that the Late Cretaceous dinosaur fauna of one of the islands that made up what we now call Europe shows the phenomenon of island dwarfing that we have seen in other, living groups (see Chapter 7) [14].

In addition to this, the biogeography of the dinosaurs has other things to teach us. Recent research by the American biogeographers Chris Noto and Ari Grossman [15] has revealed some interesting aspects of dinosaur ecology. In more arid biomes, there are more large-bodied groups, with large herbivorous sauropods, better able to cope with the lower resource density and quality of the vegetation, together with large theropods that preyed upon them. On the other hand, there are few ground-foraging herbivores and small carnivores, as there is little ground cover in which the herbivores could hide. In semi-arid or seasonally wet biomes, where there is more ground vegetation, the small groups are more common.

Finally, the biogeography of dinosaurs also suggests that some aspects of the world today that we assume as unvarying may be less constant than we think. Today there is a latitudinal gradient in which biodiversity is controlled by climate, peaking in the tropics and declining towards the poles. Work by a group of British and American palaeontologists [16] suggests that dinosaur diversity is instead correlated with the distribution of land area, and is therefore highest in temperate palaeolatitudes, where there was more land than closer to the poles. This may have been the result of a Mesozoic climatic gradient weaker than today's, which may have weakened its control of biodiversity, a situation that may have continued until after the Eocene–Oligocene change in world climate (see later in this chapter).

The climate and biogeography of the whole world were transformed by the great meteor that struck just off the coast of northern Yucatan in northern Mexico 64.5 million years ago. The meteor, probably moving at more than 50 000 km/h, was at least 15 km in diameter and caused a crater 200–300 km in diameter and 30 km deep. The impact caused a fireball many hundreds of kilometres across, and burning hot winds must have swept around the world. This, together with the red-hot ejecta from the crater, would have started forest fires in many areas, and the thickness of the resulting soot layer that is still visible in the geological record suggests that 90% of the world's forests may have burned. The impact also threw 1000–4000 km^3 of heated limestone rock into the atmosphere. The combination of this, plus aerosols and the smoke from the forest fires, would have caused an initial darkening of the world's skies, cutting out 80% of the sun's warmth and leading to a temperature drop of about 10°C for about 6 months. This would have been followed by a 'greenhouse effect' temperature rise of 3–10°C, caused by the carbon dioxide and sulphur dioxide released by the heated carbonate rock, which would have lasted several tens of years. All of this would have had profound and complicated effects on the world's weather [17].

The most obvious biological result of these changes was the total extinction of the dinosaurs, but even the marine biota was profoundly affected, for studies of changes in marine microfossils suggest that the temperature of the sea dropped by 7°C. This led to major dislocations in the food chains of the oceans, with the extinction of many groups of marine plankton and therefore also of groups higher up the food chain, such as the great marine reptiles (plesiosaurs and ichthyosaurs) and the ammonites. The event also annihilated the North American forests, with the extinction of up to 75% of their flora, but the changes to the world's floras were far less extensive and largely temporary. As described in Chapter 6, many plants are polyploid, which provides them with an extra set of genes that are available for modification and adaptation when the environment changes. The Belgian biologist Kevin Vanneste and his colleagues wondered whether this ability might have been important in the greatly changed end-Cretaceous world. They studied the genomes of plants belonging to 41 different major angiosperm groups. Of these, 24 were polyploid, and molecular-clock analysis showed that half of these had arisen at the time of the meteor impact or soon after, suggesting that their polyploidy may have given them a useful additional flexibility in adapting to the new conditions.

All these changes in the world's biota were recognized long ago, and led to the establishment of the geological boundary between the Cretaceous and the Tertiary (or Cenozoic). It is therefore known as the Cretaceous–Tertiary (or K/T) Boundary Event.

Biogeography of the Earliest Mammals

One of the most exciting developments in historical biogeography over the last 10 years has been the gradually increasing understanding of the relationships in time between three phenomena: the diversification of the placental mammals, the disappearance of the dinosaurs, and the patterns of break-up of the supercontinental land masses by plate tectonics.

The first, most primitive mammals appeared in the Triassic Period, not long after the first dinosaurs, but they almost certainly laid eggs, as do the living monotremes (the platypus and spiny anteater of Australia). The modern mammals are divided into two major groups, the marsupials and the placentals. In the **marsupial** type of mammal, the young leave the uterus at a very early stage and complete their development in the mother's pouch, whereas in the **placental** mammals the whole period of embryonic development takes place in the uterus.

The families of mammal are each distinguished by easily recognizable features of their skeletons and teeth, and these readily fossilize. It is therefore easy to trace the history of the appearance and diversification of these families, and of their dispersal around the world, directly from the fossil record. To take, for example, the history of the placentals, we can see in the fossil record that their diversity gradually diminishes as we pass back in time through the early Cenozoic and into the Late Cretaceous (Figure 10.4).

Until fairly recently, this was the only source of information on their history of evolution and divergence. However, the development of molecular methods has given us a mass of new evidence, and shown us that their divergences were far earlier than the fossil record had suggested. But, in order to understand this properly, it is necessary first to appreciate the difference between what are called the *stem group* and the *crown group* during the evolution of a clade.

To take the marsupials as an example, we can be confident that any feature that we find in all, or a great majority of, the living marsupial orders was also present in their common ancestor; otherwise, we should have to put forward the unlikely hypothesis that these features had evolved independently

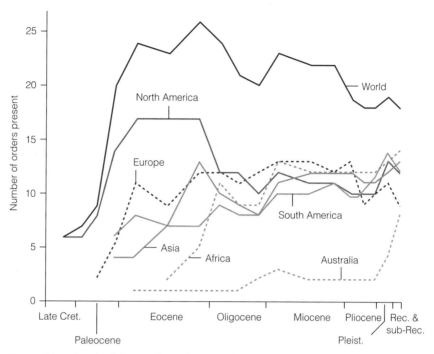

Figure 10.4 Graph of the numbers of mammalian orders through time for the world as a whole and for each continent. Pleist, Pleistocene; Rec, Recent. Adapted from Lillegraven [41].

in each order. These features together make up the character complex that led to the success of the clade. Their common ancestor, plus all of its descendants, is known as the **crown group**; the molecular methods enable us to date the time of origin of this crown group, indicated as M in Figure 10.5. Other, related orders that lived earlier in time may well have had some of these marsupial features, but we cannot be sure which, for it is unlikely that all of them evolved simultaneously. Instead, these features would have evolved as individual characters, or character complexes, over millions of years. These earlier orders comprise the **stem group** of the marsupials.

The placental mammals have a similar history, but their crown group (P in Figure 10.5) is characterized by, among other features, the possession of the specializations that enable the embryo to develop within the mother's uterus, instead of being expelled earlier to avoid rejection by the mother's autoimmune systems that would have treated them as foreign tissue, as in the marsupials. The fossil record allows us to ascertain how charac-

teristics of the skeleton and teeth of the placentals gradually evolved during their earlier history. But this does not give us information on other features, such as those of their physiology or reproduction. So, though we can be sure that the common ancestor of all the placentals had the placental method of reproduction, we cannot tell how and when this evolved from the marsupial method. We therefore cannot be sure that any fossil mammal that is not part of the crown group of placentals had the placental method of reproduction; it might still have had the marsupial method. The evolutionary histories of the stem groups of the marsupials and placentals converge back in time to the common ancestor of all the modern mammals (A in Figure 10.5), which we can identify from such features as its dentition, but which must also have had the marsupial method of development. It lived about 176 million years ago, in the Jurassic. Though other fossil mammals are known, which lived during the Jurassic or Cretaceous, they were probably egg layers like (as mentioned earlier) their modern Australian descendants, the platypus and spiny anteater.

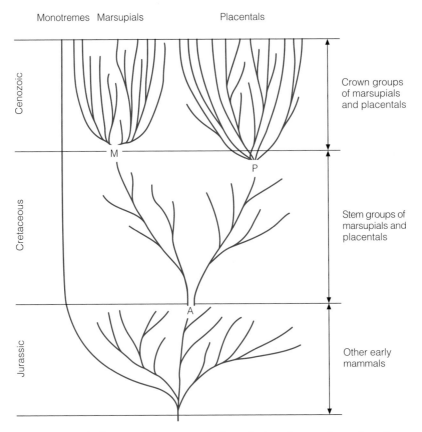

Monotremes Marsupials   Placentals

Cenozoic

Cretaceous

Jurassic

M

P

A

Crown groups
of marsupials
and placentals

Stem groups of
marsupials and
placentals

Other early
mammals

Figure 10.5 The history of the mammals through time. See text for explanation.

As already noted, molecular methods have given us a new, more extensive and detailed source of information on the times of evolution and differentiation of the mammals (see Hedges and Kumar (2009) in Further Reading). It has shown that the dates of divergence of the groups took place much earlier than the fossil record had suggested (Figure 10.6). So, for example, the divergence between the marsupials and the placentals took place about 176 million years ago (mya), in the Early-Mid Jurassic; the crown group of the living marsupials originated about 78 mya, in the Late Cretaceous; and that of the living placentals about 105 mya, in the Early Cretaceous. This leads to the surprising conclusion that the early ancestors of the living orders of marsupial and placental were living for millions of years alongside the dinosaurs, rather than only being able to evolve after their disappearance.

The solution to this problem may lie in trying to imagine what life was like for these early members of the crown groups of both marsupials and placentals. They existed almost throughout the world, as isolated small populations in the varied ecosystems of the Cretaceous. In any of these populations, random mutations may have led to the evolution of adaptations, for example to carnivorous life by having sharper teeth and more powerful jaw muscles. But they were unable to capitalize fully on the potential of these adaptations by becoming larger and more aggressive until after the sudden disappearance of the dinosaurs. So the Cretaceous world was full of populations of little mammals, showing little obvious difference from one another, that were potentially lions, horses or rabbits, but were as yet unable to fulfil their potential. However, molecular studies of the relationships between today's lions, horses and rabbits

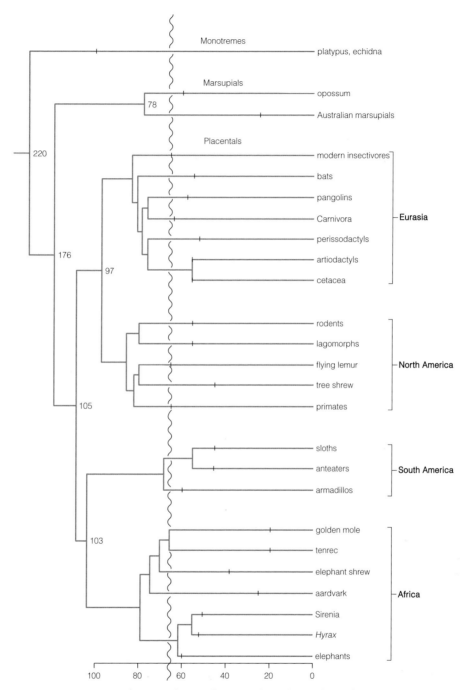

Figure 10.6 Timetree of mammalian evolution and its relationship to biogeography. The cladistic diagram shows the relationships between the different groups. They are joined together at the times suggested by molecular studies; the short vertical bar on each lineage indicates the age of the oldest known fossil of that group. NB: The timescale is only linear back to 100 million years. The vertical wavy line at 65 million years indicates the date of the K/T boundary. Based on data in references [18–20,42–44].

would place their origins at the points at which the little Early Cretaceous populations diverged from one another, rather than at the beginning of the Cenozoic when each was at last able to diversify. This scenario is supported by the molecular work, which suggests that the evolution of larger size, and of the individual features that allowed eventual specialization into distinct ecological roles, did not take place until after the K/T extinction event [18].

The biogeographical history of the early mammals, too, poses some interesting problems. The earliest fossil records of both marsupials and placentals are from the Early Cretaceous of Asia. In Eurasia, the placentals were dominant throughout the Cretaceous. In North America, however, the marsupials radiated in the Late Cretaceous; placentals were absent from that continent for most of that period of time and did not attain appreciable diversity there until the last 10 million years of the Cretaceous. By the Eocene, 55 million years ago, marsupials had arrived in both Europe and Africa, but later became extinct in both the Northern Hemisphere and Africa. Neither marsupials nor placentals were able to reach India or Madagascar in the Cretaceous, for those two areas of land, still joined together, had separated from the rest of Gondwana in the Jurassic.

There are serious gaps in our understanding of the biogeographical history of mammals in South America, Antarctica and Australia. (Antarctica had warm climates at this time, with forests extending close to the location of the current South Pole.) Both marsupials and placentals are known from the Late Cretaceous of South America, and are also found in western Antarctica (in the Antarctic Peninsula, which stretches up to South America), which was still connected to it at that time. But we have no fossil record of the mammals of the much larger land mass of eastern Antarctica, nor of those of the Early Tertiary of Australia. As a result, in many cases we do not know when the ancestors of the later groups that we recognize evolved or diversified, nor do we know their patterns of dispersal. Genetic work [19] suggests that there were two dispersals of South American marsupials across Antarctica to Australia: the peramelids (bandicoots) are related to the South American *Caenolestes*, while the remainder of the Australian forms (now known as

the Eometatheria) are related to the little South American *Dromiciops*. (Alternatively, perhaps *Dromiciops* was the result of a dispersal in the opposite direction.)

The marsupials of South America and Australia were later separated by the extinction of mammals in Antarctica, caused by its glaciation. When we eventually find diverse Early Tertiary marsupials in Antarctica, it will be interesting to see whether the continent contains two different faunas (e.g. South American forms in western Antarctica, and Australian forms in eastern Antarctica), separated by a geographical barrier that the early marsupials had to cross and that we at present cannot identify. Such a barrier might also have been responsible for the absence of placentals from Australia: there is no trace of them alongside the varied Australian marsupials that we know from the Oligocene onward. (Because it is the placentals, rather than the marsupials, that have usually succeeded whenever the two groups have been in competition in other parts of the world, most biogeographers have tended to assume that placentals are absent from Australia in the Tertiary because they never reached that continent. But note the success of the South American marsupials, discussed later in this chapter.)

The wealth of data on the times of divergence of different clades of mammal has revealed another set of interesting possibilities. For example, they suggest that the date of divergence between the early placentals of South America (known as the Xenarthra) and those of Africa (known as the Afrotheria) was about 103 mya. This date is almost identical to that of the final separation of the two continents by plate tectonics. This has led to the suggestion by an American group led by Derek Wildman [20] that the ancestors of the two faunas were present in West Gondwana before that separation, and that the distinction between the two groups was an example of vicariant evolution caused by that event. Unfortunately, our knowledge of the mammals of these two continents only begins in the Cenozoic, 65 mya.

The African placental fauna is surprising in its lack of ecological diversity. That of the northern continental mass seems to have had a much greater ecological potential, for it radiated into a very diverse range of families that fed on leaves, fruit, seeds, nuts and invertebrates, and evolved into carnivores and hoofed herbivores (and bats and whales). Hardly any of these diverse ecological opportunities seem to have been taken by the early placentals of Africa, as can be seen by merely noting those that were occupied by the families of African origin shown in Figure 10.6. The vacant niches in Africa were only filled later, by the arrival of other families from Europe about 50 mya (see Chapter 11). Though Figure 10.6 gives the impression that the placental fauna of South America was similarly limited in its ecological potential, this is because it only shows orders that are alive today. In fact, the early Cenozoic mammal fauna of South America included members of two groups: first, that of over 20 families of hoofed herbivorous placental whose ancestors had arrived there from North America; and, second, that of marsupials that had evolved from the arboreal opossums and carnivorous borhyaenids. All of these varied groups, except the opossums, became extinct in the Pliocene (see Chapter 11). (An even greater puzzle is the presence of a 50-million-year-old South American anteater in Germany, far away in time and space from where one would have expected it. But science would cease to be intriguing if we knew all the answers!)

Early History of the Flowering Plants

Molecular evidence places the origin of the flowering plants in the Middle Jurassic [21] (Figure 10.7), but, as is usual, this is far earlier than the first fossils, which date from the Early Cretaceous (see Willis and McElwain (2014) in Further Reading), when they also underwent a major radiation and dispersal into higher latitudes world-wide at a time of high global temperatures. It is still uncertain whether the earliest angiosperms were trees, shrubs or herbs [22]. Their rise was paralleled by a corresponding reduction in the numbers and variety of mosses, club mosses, horsetails, ferns and cycads, but there was less change in the overall diversity of conifers (Figure 10.8). Even at the end of the Cretaceous, although angiosperms formed 60–80% of the low-latitude floras, they comprised only 30–50% of those in high latitudes. They appeared most abundantly in a warm-temperate zone that in the Northern Hemisphere covered northern North America, southern Greenland, Europe, Russia and northern China, and coastal Antarctica and parts of southern South America and Australia in the Southern Hemisphere.

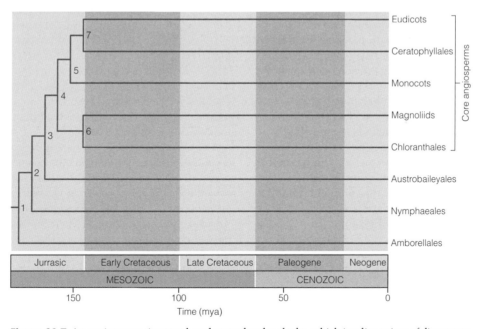

Figure 10.7 An angiosperm timetree based on molecular clocks, which implies a time of divergence in the Jurassic. From Magellon [21]. (Reproduced by permission of Oxford University Press.)

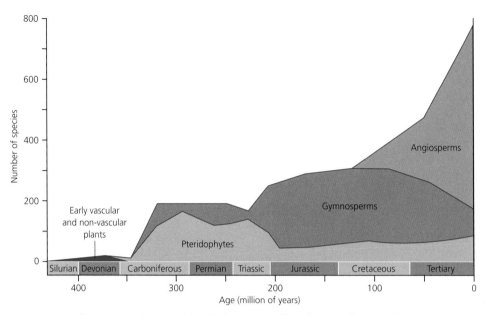

Figure 10.8 The number of species of each plant group through time, showing the great increase in the number of angiosperm species from the beginning of the Cretaceous. From Niklas [45]. (Reproduced with permission of John Wiley & Sons.)

As noted in Chapter 3, botanists analysing the floras of the world today can distinguish different biomes (e.g. forest, woodland, prairie/savanna, etc.), each of which bears a community of plants that are similar in their structure and appearance, and their own preferred climate. Many can be recognized in more than one continent, though the precise families of plant that are found in them may be quite different, so that the constitution of each biome may vary from place to place. For example, the five major rainforest biomes of today (in South America, Africa, Madagascar, South-East Asia and New Guinea) are quite distinct from one another in their detailed taxonomic composition. The fossil record shows not only that different combinations of ecologically compatible species can exist, but also that there were some biomes quite unlike any found today, such as the Eocene flora of the London Clay or the Oligocene semi-deciduous tropical dry forest of southeastern North America (see later in this chapter).

We have seen here how easy it is for palaeozoologists to understand the history of the mammalian families, but, unfortunately, the fossil record is not very helpful in revealing the history of the

appearance and diversification of the flowering plants. This is because, though it contains their leaves, fruits, seeds and pollen, the fossil record hardly ever provides us with a complete plant. Further information can be gained from the climate preferred by the living representatives of the families of plant found in each flora [23]. However, it would be unwise to assume that living species could never have existed beyond the range of environmental conditions within which they are found today. They may in the past have lived in environments from which they later became excluded by competitive interaction with other, newly evolved types. In addition, the fossil data are often biased towards near-coastal areas of deposition, whose climates are usually milder and less seasonal than those of more inland areas. Their fossilized leaves provide an additional source of information after the flowering plants became common and diverse, in the middle Cretaceous. In areas of high mean annual temperature and rainfall, the leaves have 'entire' margins, not subdivided into lobes or teeth; they are large and leathery, and are often heart-shaped, with tapering, pointed tips, and a joint at the base of the leaf. These features are less common in floras from areas of low mean annual temperature and rainfall. In the attempt to recreate the environment in which the plants lived, it is also possible to use some features of the rocks in which they are found, for some of these contain minerals such as coal, evaporites, bauxite (which indicates a hot, wet, seasonal climate) or desert sandstones, which give an indication of the climate in which the plants had lived.

Reconstructing Early Biomes

In attempting to reconstruct the floras of the past, palaeobotanists can use all of the evidence listed in this chapter, as has been done by Ziegler [24]. The palaeobotanists Kathy Willis and Jenny McElwain have used the results to reconstruct up to seven global biomes for past periods of earth history (see Further Reading); of course, these palaeobiomes are inevitably less detailed than the 12–14 biomes that are recognized today. Also, the Australian palaeobotanist Bob Morley [25] has written a useful integration of global climate changes and the distribution of tropical rainforests. Both these sources will be used in this section. Morley also suggests whereabouts in the world some of the angiosperm families may have originated, though, as he notes, our information on this is heavily biased by the fact that we know much more about the palaeobotany of Europe and North America than about the rest of the world.

Changes in Climate and Plant Distributions: Late Cretaceous–Middle Eocene

During the Late Cretaceous, the Earth's climate changed, becoming cooler and more seasonal (Figure 10.9). The Cretaceous cooling is clearly shown in a series of floras, ranging over 30 million years, from about 70°N in Alaska [26]. The earliest contains the remains of a forest dominated by ferns and by gymnosperms such as cycads, ginkgoes and conifers. The nearest living relatives of this flora are found in forests at moderate heights in warm-temperate areas at about 25–30°N. By the time of the last of these middle Cretaceous Alaskan floras, the flora had changed in two ways. First, the angiosperms had by this time diversified to such an extent that they dominated the flora. Second, this forest was similar to that found today at a latitude of 35–40°N – much farther north than the living relatives of the earlier flora. By this time, too, latitudinal differences in floras have become apparent, with higher latitudes bearing larger leaved deciduous vegetation, suggesting moist mesothermal forests, while lower latitudes have thicker, smaller leaved evergreen vegetation, suggesting subhumid megathermal (preferring mean annual temperatures above 40°C) forests, which were probably mainly single-story forests. By the end of the Cretaceous, other Alaskan floras show that the mean temperature had dropped by about 5°C and that the diversity of the flowering plants had dropped very greatly [27].

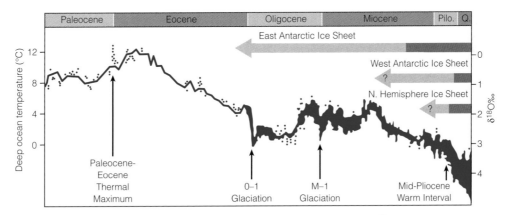

Figure 10.9 Changes in Cenozoic temperature, based on deep-sea benthic $\partial^{18}O$ isotopic records. From Solomon *et al.* [46].

The steady cooling in the world's climate that had taken place during the Cretaceous was reversed at the end of that period (Figure 10.9). It is possible that the immense climatic change caused by the meteorite strike at that time may have played an important role in that cooling, though changes in the positions of the continents, with consequent changes in ocean currents, may also have been involved. The period of time from the beginning of the Paleocene until the middle of the Eocene (66–50 mya) was one of the warmest in the world's history. This culminated in what is called the Paleocene–Eocene Thermal Maximum (PETM), when global temperatures rose by about 5°C in less than 10 000 years, to about 31–34°C. This may have been caused by a sudden increase in the CO_2 content of the atmosphere, perhaps due to a release of methane from the Atlantic Ocean. The enormous gyres (see Chapter 9) of the huge Pacific Ocean would have carried warm equatorial waters to high northern and southern latitudes, so that there were no polar ice caps, and the temperature gradient between the equator and the poles was therefore much lower than it is today.

Morley states that the first appearance of angiosperms with the leaf types typical of megathermal forests was at middle latitudes in the Middle Eocene, and that the first evidence for the presence of closed-canopy, tropical, megathermal, everwet rainforest was not until the Late Cretaceous in West Africa. The presence of large seeds and fruit shows that this was also the first known multistorey rainforest. In the Paleocene, this everwet tropical forest spread to a huge area, including most of South America, Africa, South-East Asia and southern and western North America, to latitudes of 40°N and S (Figure 10.10a). It may be significant that this was after the disappearance of the dinosaurs, which had browsed upon these forests, and it was accompanied by a radiation of fruit-eating mammals. This rainforest was dominated by angiosperms, which there diversified and modernized, including many that are still found in tropical environments today, and with an abundance of evergreen families and diverse palms. Its conifers included araucarians and podocarps; ginkgoes were present but rare.

To the north and south of this huge rainforest lay a unique subtropical summerwet 'paratropical' rainforest, up to 50–60°N and S, covering northern Europe, Russia and eastern North America to the north, and Argentina and southern Australia in the south. It contained a mixture of angiosperms found in today's tropical and temperate areas, with lianas, climbers and palms, and covered an area that in the Cretaceous had been occupied by a winterwet biome with some Mediterranean-like characteristics. These forests reached their maximum extent at the Paleocene–Eocene boundary (Figure 10.10b).

These two types of rainforest covered most of the world and may have had major importance in changing world climates. The leaves of angiosperms have large veins and can therefore transpire water rapidly. Boyce and Lee [28] have suggested that the resulting increased amount of moisture in the atmosphere may have led to increased rainfall in the tropics, with cooler, wetter and more aseasonal climates, which would have strongly altered tropical climates and the global hydrological cycle. Palms (Arecaceae) were also common and diverse in these forests, and Morley [25] therefore refers to the equatorial area as the Palmae Province (Figure 10.10a).

The enormous extent of these megathermal rainforests would have pushed the warm-temperate and cool-temperate biomes farther towards the poles. So is not surprising to find that taxa that are today restricted to particular latitudes were, in this warm Eocene world, found much farther from the equator, or existed over a much wider band of latitudes. For example, the diverse fauna of Ellesmere Island in the Canadian Arctic (81°N) included mammals, snakes, lizards and tortoises [29,30]. The forests of such high northern latitudes were unlike any modern forests, for they included elements of more southern temperate broad-leaved deciduous forests, as well as characteristically northern needle-leaved conifers such as pine, larch and spruce.

The appearance of the de Geer route between North America and Europe (see Chapter 11) facilitated the exchange of some of the mesotherms and mammals between the two continents. However, after the ending of the land bridges between the two continents in the Early Eocene, these two floras and faunas diverged. That of Europe is well known from the fossils of the London Clay, then at a latitude of about 45°N; these fossils contain the seeds and fruits of 350 species of plant, belonging to over 150 genera [31]. The flora includes magnolias, vines, dogwood, laurel, bay and cinnamon, as well as the palm trees *Nipa* and *Sabal* and the conifer *Sequoia*,

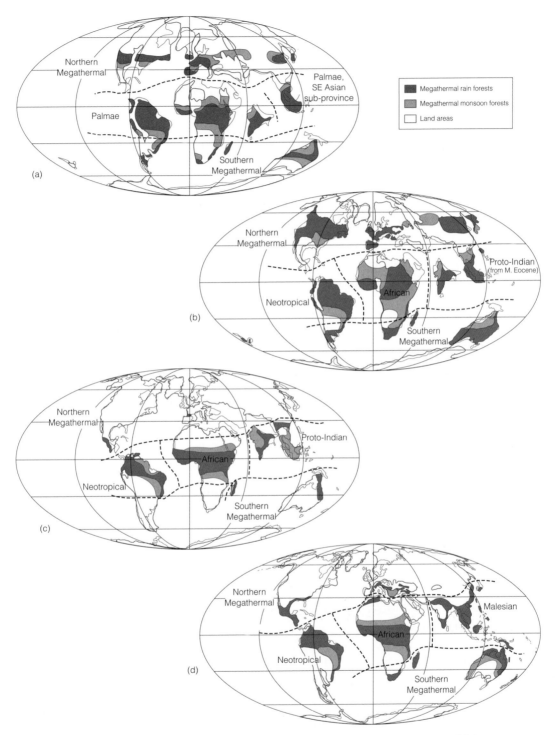

Figure 10.10 Distribution of closed-canopy megathermal rainforests in (a) the Early Paleocene; (b) the Late Paleocene/Early Eocene; (c) the Oligocene; (d) the Middle Miocene. Dotted lines are the boundaries of floristic provinces. From Morley [25].

with mangrove swamps along the coasts; there are no growth rings in this flora. Though its closest analogue today would be a subtropical rainforest, its composition is not identical to that of any single modern flora. Some of these plants are now found in the tropics, especially in South-East Asia, while others live in temperate conditions like those of east-central China today. The accompanying fauna, which includes crocodiles and turtles, is similar to that found today in the tropics.

In the Southern Hemisphere, the warm/cool temperate biome included proteas and the southern beech tree, *Nothofagus* (Fagaceae), plus podocarp and araucarian conifers and ferns; this flora is very like that of New Zealand today. The PETM would also have facilitated the exchange of plants and mammals between South America and Australia across Antarctica.

Changes in Climate and Plant Distributions: Middle Eocene–Oligocene

A steady cooling of the world's climate took place between the Eocene and the Late Oligocene, with a rapid temperature fall at the end of the Eocene. This led to the beginning of the Antarctic ice sheet,

with a consequent fall in sea levels, withdrawal of epicontinental seas and increased aridity. One of the most important factors in this was probably the plate tectonic changes that took place in the Southern Hemisphere.

As long as Australia and Antarctica were parts of a single weather system, some of the warmth that Australia received was circulated by wind systems southward to its southern neighbour. But the separation of both South America and Australia from Antarctica, completed in the Late Eocene, permitted the appearance of the Antarctic Circumpolar Current of cooler water and associated westerly winds (Figure 10.11). Now isolated from the warmth of Australia, Antarctica cooled and continental ice sheets started to form there. There would also have been less evaporation from the now cooler seas around Australia, reducing the rainfall on that continent, with a consequent increase in its arid, desert areas.

Most of the rainforests of southern South America and South Africa disappeared as the climates of these regions cooled. However, the floras of South America and Australia still contained, as they do today, descendants of the old Late Cretaceous floras, with such families as the proteas, myrtles,

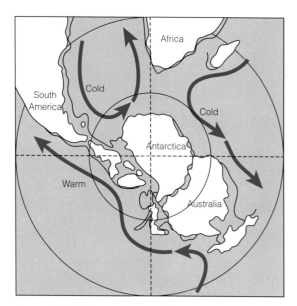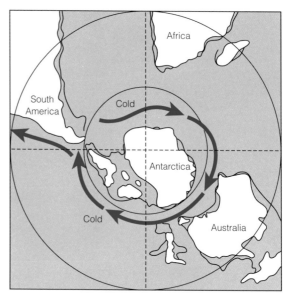

Figure 10.11 The separation of Australia from Antarctica in the Late Eocene, causing the development of a seaway around Antarctica, which enabled the development of an ocean circulation system around that continent. Adapted from Condie and Sloan [47].

Nothofagus and the 'southern conifers' such as *Araucaria*, *Podocarpus* and *Dacrydium*. These forests also covered at least the periphery of Antarctica, but the inland flora of that continent is unknown.

The global cooling caused a great reduction in the tropical everwet biome, and the accompanying movements equatorward of the subtropical summerwet biome that lay to the north and south. This was particularly marked in the Northern Hemisphere. Elements of its megathermal forests therefore dispersed southward into South-East Asia, where they are still present. However, those of North America and Europe were unable to disperse southward, due to the ocean gap between North and South America and the presence of the Mediterranean Sea, but those of northern Australia were able to survive there because the area had moved northward.

The summerwet biome now included semi-desert areas of woody savanna (the first appearance of this biome) in central North America, southern Asia (not including India) and parts of South America. This was the environment for the first significant areas of grassland with its associated grazing mammals. Grasses (Poaceae) evolved at about the time of the K/T meteorite strike, probably in South America or Africa, and the first grassland habitats had appeared by 40 mya. Grasses are distinguished by the fact that growth takes place, not at the apices of the leaves, but instead where the plant emerges from the ground. They are therefore resistant to wildfires and also to grazing animals such as horses, camels and artiodactyls that evolved in the Early Oligocene to take advantage of this new source of food [32]. With their long, hoofed legs, they could run swiftly over the grasslands, and they had teeth adapted to the heavy wear caused by the high silica content of the grasses.

Other biomes characterized today by their arid-adapted components, such as tundra and subtropical desert, also began to appear and spread at this period of time, with a corresponding decrease in the world's forest [33]. This provided an evolutionary opportunity for the appearance of new types of non-woody flowering plant, and therefore of herb-dominated flowering plant communities. As a result, the plant biomes started to become more like those of today, both in their systematics and in their structural composition. For example, the warm-temperate deciduous forests of the low-lands of North America and the coniferous forests of its highlands were similar to those of today. But there were also exceptions: for example, the tropical rainforest of southeastern North America was replaced by a semi-deciduous tropical dry forest unlike anything that can be seen in the world today. In Eurasia, this climatic change is accompanied by a marked change in the mammal faunas, to smaller types such as rodents and rabbits, and the old forest types become extinct [34].

By the Mid-Late Eocene, the floras of tropical South America and West Africa were becoming progressively more different as the climate cooled and the two continents were moving farther apart, their floras becoming recognizable as the beginnings of today's South American (or 'Neotropical') and African floras. Many of the taxa that these two floras had shared in the Late Cretaceous became rarer, but the remainder make up what has been called a common 'amphi-Atlantic' element. It was also at this time that the Indian flora, with its dipterocarps, dispersed into South-East Asia after the two continents collided. Though some of the rainforest elements of Asia were able to disperse southward into southern China, the more southern regions of India and South-East Asia were affected by changing patterns of atmospheric circulation caused by the rise of the Himalayas; their rainforests were mainly replaced by more seasonal, monsoonal vegetation. Because Australia had drifted northward into warmer latitudes, a few of its megatherms survived along its northeast coast.

Changes in Climate and Plant Distributions: Oligocene–Late Miocene

During this time, from 23 to 5.3 mya, there was at first a short period of global warming that permitted a brief poleward expansion of the megathermal forests. This climatic change was most pronounced in South-East Asia, where elements of the South-East Asian rainforest flora spread into northern India. The grasslands of that region disappeared, leading to the appearance of the modern 'Malesian' flora there. Morley suggests that this local change may have been due to the beginning of the collision of the Australian and Asian plates, which may have resulted in the warm moist air from the Pacific dropping its rains on South-East Asia rather than farther west. This, then, may have been the beginning of the modern climate of that region, in which the El Niño phase of climatic oscillation triggers the Sunda rainforest trees to fruit at the same time, so reducing total seed predation (see Chapter 12) [35].

The highest global temperatures were at the beginning of the Middle Miocene (the Miocene Climatic Optimum). But then, despite an increase in temperature in equatorial regions, there was a steady decrease in global temperatures that led to an increase in Antarctic continental ice sheets and the beginning of ice in the Arctic (Figure 10.9). The combination of this and the higher equatorial temperatures led to a steep temperature gradient between the poles and the equator. At the same time, the withdrawal of water from the oceans to create the polar ice sheets led to lower sea levels and the disappearance of the Turgai Seaway. This in turn caused a reduction in atmospheric moisture, and increased aridity in continental interiors, leading to the beginning of the Saharan and Middle Eastern deserts. Because of the lower sea levels, a land bridge appeared between northwest Africa and Spain, closing the Mediterranean to Atlantic water and leading to its eventual drying out (see Chapter 11).

The tropical everwet biome increased slightly in its latitudinal extent during this period of time, extending through northern South America and central Africa and on through South-East Asia to northern Australia. This period also saw the beginning of the great Amazonian rainforest [33], with a highly diversified and multistratal structure. The flora of this biome included abundant evergreen angiosperm trees, with lianas, vines, palms, *Araucaria* (the 'monkeypuzzle tree') and podocarps. To its north and south lay bands of paratropical summerwet forest with, as in the Oligocene, areas of open woody savanna with increasing grasslands. A new feature was the reappearance of the Cretaceous winterwet biome, found in all the areas that today have Mediterranean-like sclerophyll vegetation.

The climatic deterioration that began in the Early Oligocene reduced the variety of plants that could disperse through the Bering link (see Chapter 11), although the North Pacific was bordered by continuous broad-leaved deciduous forests until at least the Middle Miocene. It led to the expansion of the Northern Hemisphere family Pinaceae – pine, fir, spruce and larch. It was at one time thought that the climatic cooling caused a wholesale southward movement, through the whole of the Northern Hemisphere, of an 'Arcto-Tertiary' flora. This was thought to have evolved in the Arctic during the Cretaceous and to have survived until today, little changed, in southeastern North America and east-central Asia. However, this concept has not been supported by later knowledge of the floral history of the Alaskan region. Instead, the Northern Hemisphere angiosperm floras appear to have adapted to the Late Cenozoic climatic change in three ways: by the adaptation of some genera to changed, cooler climates; by the restriction of the range of some genera and their replacement by other already existing genera that preferred a cooler climate; and by the evolution of new genera that preferred those cooler climates. (As noted in this chapter, we cannot assume that the structure and composition of biomes of the past are similar to those of today.) Open savanna with a high proportion of grasses and herbs appeared at this time.

By the Late Miocene, all of today's 12–14 biomes are recognizable. Much of the Late Cenozoic microthermal vegetation of the Northern Hemisphere appears to have evolved there from ancestors within the same area. For example, a recent molecular phylogenetic study of 10 angiosperm lineages found in both California and the Mediterranean [36] shows that their times of divergence into separate lineages in the two regions was at this time (perhaps during the Miocene Thermal Maximum). Their pattern of distribution is therefore the result of independent adaptations of these plants to the environment of the newly established Mediterranean climate, rather than being due to vicariance after the completion of the North Atlantic much earlier in time, in the Eocene.

Since there was little exchange of plants between North America and Eurasia during this time, these two floras steadily diverged. It has also been suggested that they shared a common Madro-Tertiary dry flora, exchanged via a low- to mid-latitude dry corridor. Though later work on the times of divergence of members of the flora in the two continents, and on the distribution of fossil relatives, does not support the concept, there is as yet no convincing alternative explanation [37].

Changes in Climate and Plant Distributions: Later Middle Miocene–Pliocene

This was a period of steady cooling in the global climate (Figure 10.9). As a result, the megathermal rainforests withdrew to the tropics (except in Northern Australia), while grasslands and deserts increased across the lower to mid-latitudes. At this time, much of the Indian rainforest was replaced by deciduous forests, grasslands, savannas and savanna woodlands. This flora was the beginning of the modern Indo-Pacific floral region. When, in the Pliocene, plants were able to disperse from North America into Central and South America over the new Panama land bridge, the survivors of the old northern megatherms were now to be found only there and in South-East Asia, forming an 'amphi-Pacific' element in these two floras. In West Africa, cycles of wet and dry climate led to depauperation of the rainforest flora during the dry cycles.

We have now followed the biogeographical history of the terrestrial flora, and some of its fauna, from their beginnings in the Devonian to after the great Cretaceous–Tertiary extinction event 65 million years ago, and the climatic and floral history into the Cenozoic. In the next chapter, we turn to consider the Cenozoic biogeographical history of the mammals, and the changes in flowering plant biogeography later in the Cenozoic.

Summary

1 The patterns of distribution of animals and plants are controlled mainly by the patterns of geography – by the positions of oceans, shallow epicontinental seas, mountains and deserts. Because these were different in the past, mainly because of the movement of continents due to plate tectonics, the patterns of distribution of life in the past were different also.

2 The problems involved in trying to estimate the date of appearance of the ancestors of any living group are explained, and the discrepancies between estimates derived from the fossil record and those derived from molecular clocks are discussed.

3 The contrasting patterns of success of the marsupial and the placental mammals in the different continents seem to have developed because placentals did not colonize the South America–Antarctica–Australia chain of continents until after the marsupials of those continents had radiated into a great variety of families and these continents had started to separate from one another.

4 Climate changes have also had a great effect on these patterns. In the Late Cretaceous and Early Cenozoic, the world's climate was warmer than it is today. Many organisms were then able to spread via high-latitude routes that are now closed both by climatic changes and by the separation of continents. The turning point came with the great cooling at the end of the Eocene, after which warmth-loving plants and their ecosystems either became more restricted in their distribution or were replaced by those adapted to cooler climates.

5 It is now possible to follow in considerable detail the way in which the changing patterns of geography and climate led to changes in the floras of the different continents and to their diversification. Increasing aridity led to the replacement of forests by woodland and grasslands. These in their turn caused changes in the nature of the mammal faunas that lived in them.

Further Reading

Cronin TM. *Paleoclimates.* New York: Columbia University Press, 2010.

Culver SJ, Rawson PS (eds.). *Biotic Response to Global Change: The Last 145 Million Years.* London: Natural History Museum, 2000.

Graham A. *Late Cretaceous and Cenozoic History of North American Vegetation.* New York: Oxford University Press, 1999.

Hedges SB, Dudley J, Kumar S. *Time Tree: A Public Knowledge-Base of Divergence Times among Organisms.* Bioinformatics 2006; 27: 2971–2978.

Hedges SB, Kumar S. (eds.). *The Timetree of Life.* New York: Oxford University Press, 2009. (Also available online.)

Morley RJ. Cretaceous and Tertiary climate change and the past distribution of megathermal rainforests. In: Bush M, Flenley J, Gosling W. (eds.), *Tropical Rainforest Responses to Climatic Change.* Berlin: Springer, 2011: 1–34.

Willis KJ, McElwain JC. *The Evolution of Plants.* Oxford: Oxford University Press, 2014.

References

1. Edwards D. Constraints on Silurian and Early Devonian phytogeographic analysis based on megafossils. In: McKerrow WS, Scotese CR (eds.), *Palaeozoic Palaeogeography and Biogeography*. Geological Society Memoir no. 12. London: Geological Society of London, 1990: 233–242.

2. Cox CB. Vertebrate palaeodistributional patterns and continental drift. *Journal of Biogeography* 1974; 1: 75–94.

3. Mora CL, Driese SG, Colarusso LA. Middle to Late Paleozoic atmospheric CO_2 levels from soil carbonate and organic matter. *Science* 1996; 271: 1105–1107.

4. Milner AR. Biogeography of Palaeozoic tetrapods. In: Long JA (ed.), *Vertebrate Biostratigraphy and Biogeography*. London: Bellhaven, 1993: 324–353.

5. Lenton TM, Crouch M, Johnson M, Pires N, Dolan L. First plants cooled the Ordovician. *Nature Geoscience* 2012; 5: 86–89.

6. Wnuk C. The development of floristic provinciality during the Middle and Late Paleozoic. *Review of Palaeobotany and Palynology* 1996; 90: 6–40.

7. Ziegler AM. Phytogeographic patterns and continental configurations during the Permian Period. In: McKerrow WS, Scotese CR (eds.), *Palaeozoic Palaeogeography and Biogeography*. Geological Society Memoir no. 12. London: Geological Society of London, 1990: 367–379.

8. Rees PM, Gibbs MT, Ziegler AM, Kutzbach JE, Behling PJ. Permian climates: evaluating model predictions using global paleobotanical data. *Geology* 1999; 27: 891–894.

9. Batten DJ. Palynology, climate and the development of Late Cretaceous floral provinces in the Northern Hemisphere: a review. In: Brenchley P (ed.), *Fossils and Climate*. Geological Journal, Special Issue no. 11. London: Wiley, 1984: 127–164.

10. Cox CB. Triassic tetrapods. In: Hallam A (ed.), *Atlas of Palaeobiogeography*. Amsterdam: Elsevier, 1973: 213–223.

11. Cox CB. Changes in terrestrial vertebrate faunas during the Mesozoic. In: Harland WB (ed.), *The Fossil Record*. London; Geological Society, 1967: 71–89.

12. Upchurch P, Hunn CA, Norman DB. An analysis of dinosaurian biogeography: evidence for the existence of vicariance and dispersal patterns caused by geological events. *Proceedings of the Royal Society B* 2002; 269: 613–621.

13. Serrano PC. The evolution of dinosaurs. *Science* 1999; 284: 2137–2147.

14. Stein K, Csiki Z, Rogers KC, Weishampel DB, Redelstorff R, Carballido JL, Sander PM. Small body size and extreme cortical bone remodeling indicate phyletic dwarfism in *Magyarosaurus dacus* (Sauropoda: Titanosauria). *Proceedings of the National Academy of Science USA* 2010; 107: 9258–9263.

15. Noto CR, Grossman A. Broad-scale patterns of Late Jurassic palaeoecology. *PLoS One* 2010; 5: e12553.

16. Mannion L, Benson RBJ, Upchurch P, Butler RJ, Carrano MT, Barrett PM. A temperate palaeodiversity peak in Mesozoic dinosaurs and evidence for Late Cretaceous geographical partitioning. *Global Ecology and Biogeography* 2012; 21: 898–908.

17. Crowley TJ, North GR. Palaeoclimatology. Oxford: Oxford University Press, 1991.

18. Murphy WJ, Eizirik E. Placental mammals (Eutheria). In: Hedges SB, Kumar S (eds.), *The Timetree of Life*. Oxford: Oxford University Press, 2009: 19–25.

19. Asher RJ, Horovitz I, Sanchez-Villagra MR. First combined cladistic analysis of marsupial mammal interrelationships. *Molecular Phylogenetics and Evolution* 2004; 33: 240–250.

20. Wildman DE, Uddin M, Opazo JC, *et al.* Genomics, biogeography, and the diversification of placental mammals. *Proceedings of the National Academy of Science USA* 2007; 104: 14395–14400.

21. Magallon S. Flowering plants (Magnoliophyta) In: Hedges SB, Kumar S (eds.), *The Timetree of Life*. Oxford: Oxford University Press, 2009.

22. Soltis DE, Bell CD, Kim S, Soltis PS. Origin and early evolution of angiosperms. *Annals of the New York Academy of Sciences* 2008: 1133: 3–25.

23. Collinson ME. Cenozoic evolution of modern plant communities and vegetation In: Culver SJ, Rawson PS (eds.), *Biotic Response to Global Change: The Last 145 Million Years*. London: Natural History Museum, 2000.

24. Ziegler AM. Phytogeographic patterns and continental configurations during the Permian period. In: McKerrow WS, Scotese CS (eds.), *Palaeozoology, Palaeogeography and Biogeography*. London: Geological Society, 1990: 363–379.

25. Morley RJ. Cretaceous and Tertiary climate change and the past distribution of megathermal rainforests. In: Bush M, Flenley J, Gosling W (eds.), *Tropical Rainforest Responses to Climatic Change*. Berlin: Springer, 2011: 1–34.

26. Smiley CJ. Cretaceous floras from Kuk River area, Alaska: stratigraphic and climatic interpretations. *Bulletin of the Geological Society of America* 1966; 77: 1–14.

27. Parrish JT, Spicer RA. Late Cretaceous vegetation: a near-polar temperature curve. *Geology* 1988; 16: 22–25.

28. Boyce CK, Lee J-E. An exceptional role for flowering plant physiology in the expansion of tropical rainforests and biodiversity. *Proceedings of the Royal Society B* 2010; 485: 1–7.

29. McKenna MC. Eocene paleolatitude, climate, and mammals of Ellesmere Island. *Palaeogeography Palaeoclimatology Palaeoecology* 1980; 30: 349–362.

30. Estes R, Hutchinson JH. Eocene lower vertebrates from Ellesmere Island, Canadian Arctic Archipelago. *Palaeogeography Palaeoclimatology Palaeoecology* 1980; 30: 325–347.

31. Collinson ME. *Fossil Plants of the London Clay*. London: Palaeontological Association, 1983.

32. Stromberg CAE. Evolution of grasses and grassland ecosystems. *Annual Review of Earth and Planetary Sciences*. 2011; 39: 517–544.

33. Hoorn C. An environmental reconstruction of the palaeo-Amazon River system (Middle-Late Miocene, NW Amazonia). *Palaeogeography, Palaeoclimatology, Palaeoecology* 1994; 112: 187–238.

34. Meng J, McKenna MC. Faunal turnovers of Paleogene mammals from the Mongolian Plateau. *Nature* 1998; 394: 364–367.

35. Ashton P, Givnish T, Appanah S. Staggered flowering in the Dipterocarpaceae; new insights into floral induction and the evolution of mast fruiting in the aseasonal tropics. *American Naturalist* 1988; 132: 44–66.

36. Vargas P. Testing the biogeographic congruence of floras using molecular phylogenetics: snapdragons and the Madro-Tethyan flora. *Journal of Biogeography* 2014; 42: 932–943.

37. Graham A. *Late Cretaceous and Cenozoic History of North American Vegetation*. New York: Oxford University Press, 1999.

38. Smith AG, Smith DG, Funnell BM. *Atlas of Mesozoic and Cenozoic Coastlines*. Cambridge: Cambridge University Press, 1994.

39. Metcalfe I. Palaeozoic and Mesozoic geological evolution of the SE Asian region. In: Hall R, Holloway JD (eds.), *Biogeography and Geological Evolution of SE Asia*. Leiden: Backhuys, 1998: 25–41.

40. Cambridge Paleomap Services. *AtlaS Version 3.3*. Cambridge: Cambridge Paleomap Services, 1993.

41. Lillegraven JA. Ordinal and familial diversity of Cenozoic mammals. *Taxon* 1972; 21: 261–274.

42. Cifelli RL, Davis BM. Marsupial origins. *Science* 2003; 302: 1899–2000.

43. Springer MS, Murphy WJ, Eizirik E, O'Brien SJ. Molecular evidence for major placental clades. In Rose KD, Archibald JD (eds.), *The Rise of the Placental Mammals*. Baltimore: Johns Hopkins University Press, 2005.

44. Bininda-Edmonds ORP, Cardillo M, Jones KE, *et al.* The delayed rise of present-day mammals. *Nature* 2007; 446: 507–512.

45. Niklas KJ. The influence of Paleozoic ovule and cupule morphologies on wind pollination. *Evolution* 1983; 37 (5): 968–986.

46. Solomon S, Qin D, Manning M, *et al.* (eds.) *Climate Change 2007. The Physical Science Basis. Working Group I Contribution to the Fourth Assessment Report of the Intergovernmental Panel on Climate Change*. Cambridge: Cambridge University Press.

47. Condie K, Sloan R. *Origin and Evolution of Earth; Principles of Historical Geology*. Upper Saddle River, NJ: Pearson Education, 1998.

Setting the Scene for Today

In Chapter 10, we saw how earlier forms of life were distributed over the very different geographies that then patterned the face of our planet, and how the major groups that we see today (the flowering plants and mammals) came into existence and occupied the world. In this chapter, we follow the histories of these groups as they diversified and spread through the still-changing pattern of continents and climates to occupy the regions we recognize today.

The Biogeographical Regions Today

The currently accepted systems of biogeographical regions have their roots in the 19th century, when our growing knowledge of the world allowed biologists to realize that its surface could be divided into separate areas that differed in their endemic animals and plants. Naturally enough, these divisions were based on the distribution of dominant, easily visible groups. Thus, Candolle in 1820, followed by Engler in 1879, used the patterns of distribution of flowering plants as the basis for a system of floral regions, while Sclater in 1858, working on birds, and Wallace in 1860–1876, working on mammals, defined the system of zoogeographical regions (see Chapter 1). Apart from some minor modifications, these 19th-century interpretations survived almost unchanged until the end of the 20th century.

However, we now know how these faunal and floral regions gradually developed in the past, and in particular how their interconnections were differ-

ent in the past. Sometimes this was due to changes in the positions of the continents, for example as Australia moved northward away from Antarctica and closer to South-East Asia. More often, it was caused by changes in the world's climates, which allowed animals and plants to spread and disperse via high-latitude routes that are now too cold to allow this, such as the Bering land bridge between Alaska and Siberia, and the varied connections between North America and Eurasia in the early Cenozoic (see later in this chapter). (It is important to note that the identification of these regions is based on the history of their faunas and floras. Each of the regions will today contain areas of different climate, within which evolutionary change in the region's biota will lead to the appearance of clades adapted to those conditions, to form a local biome. The pattern and content of these biomes are therefore the results of local ecology over the last few million years, and must not be confused with those of the historically based biogeographical regions [1], whose origins go much further back into geological time.)

This knowledge has made it possible to revise the 19th-century interpretations [2], and also provided an opportunity to replace the old Classical names (e.g. Palaeotropical and Nearctic) with the names of the continents themselves; Figure 11.1 shows the result of this. That of the mammals follows Wallace's system, except that the boundaries of his Oriental and Australian regions in the East Indies follow the edges of the continental shelves of the two regions, rather than the two regions meeting along a line within the East Indies pattern

Biogeography: An Ecological and Evolutionary Approach, Ninth Edition. Edited by C. Barry Cox, Peter D. Moore, Richard J. Ladle.
© 2016 John Wiley & Sons, Ltd. Published 2016 by John Wiley & Sons, Ltd.

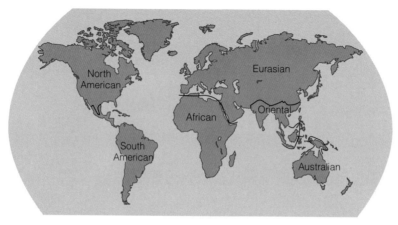

Figure 11.1 Biogeographical regions of a group that does not disperse readily – the mammals.

of islands (see Box 11.1). It also draws the northern boundary of the African region at the northern edge of that continent rather than along the southern edge of the Sahara. That of the flowering plants (Figure 11.2) has been similarly updated, especially to take account of the extent to which the ability of angiosperms to disperse across ocean barriers has made it possible for elements of the flora of South-East Asia to disperse across the Pacific, and that of southern South America to disperse eastward through the islands to the north of Antarctica,

taking advantage of the strong west wind that blows around that continent.

Though we can draw sharp, simple lines between most of the different biogeographical regions on today's maps, this is merely the result of coincidences of current geography and recent climatic changes. As we shall see, the Ice Ages of the last couple of million years progressively reduced the northward extent of the range of the more warmth-loving flowering plants and mammals. As they were pushed southward, however,

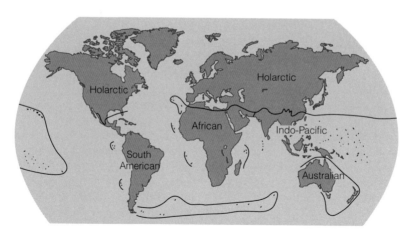

Figure 11.2 Biogeographical regions of a group that disperses readily – the flowering plants.

they found their way blocked in North America by the narrow filter of the Panama Isthmus, and blocked in Eurasia by the combination of the Mediterranean Sea, the deserts of North Africa, the Middle East and southern Asia, and the Himalayan Mountain Range. As a result, many of these families became extinct north of those barriers. When the climate recovered during the interglacial periods, those same barriers prevented them from returning northward. So, instead of there being broad zones of gradual transition between neighbouring faunal or floral regions, they are now separated by areas within which climate or geography forms a barrier between very different biotas.

The Basis of Mammal Biogeography

The biogeography of mammals is uniquely well understood and documented, for several reasons. Until the dinosaurs suddenly became extinct, the mammals had been confined to the roles of small animals living mainly on invertebrates, fruit and nuts. Now they were suddenly able to increase in size and diversity and occupy the roles previously taken by the dinosaurs, and to become in their turn the dominant group of animals, common and easily identified. But, because they had already spread throughout the world, the nature of their diversification was different and unique on each continent. The differences between these faunas were perpetuated by the fact that, with the exception of the bats, mammals are unable to cross wide stretches of water, such as those that surround the major continents. As a result, taxa that evolve in a particular continent are normally confined to it, as endemic taxa. Furthermore, there are relatively few (about 120) families of non-marine mammal, and, because their remains are readily identifiable in the fossil record, we have a reliable record of their origins and interrelationships. We can therefore trace the histories of these continent-based mammal faunas through time, relate their histories to that of the movements of the continents, and understand how the dispersal of a fauna or group from one continent to another was from time to time permitted by new continental connections.

Such analyses can also explain the histories of individual taxa. One of the best documented groups is the superfamily Equoidea [3], whose modern representatives include the horses, zebras and asses (Figure 11.3). Their early evolution was in Euramerica, and in particular in North America after that continent became separate. The genus *Hipparion* spread across the Bering region to Eurasia in the Late Miocene and thence to Africa when that continent became connected to Eurasia (see Figure 11.8). The genus *Equus* appeared in the Pliocene and gave rise to three distinct lineages: the asses, the zebrids and, from the zebrids, the true horses. Four different species of zebrid are still found in Africa, and the quagga, related to the true horse, only became extinct there during the 19th century. Both *Hippidion* and true horses reached South America after the Panama Isthmus formed; but, surprisingly, all the equids of the New World became extinct during the Pleistocene.

As a result of all the factors mentioned here, the biogeographical history of the mammals is so well documented and understood that it is likely to provide an enduring scenario that will be used in any attempt to explain the histories of origin and dispersal of other groups, both plant and animal. For this scheme is *not* a general scheme of zoogeographical regions, but merely a scheme for mammals. Other groups will show different patterns, and some of these are being shown by the new analytical techniques that enable useful comparisons between the distributions of other groups and that of mammals. So far, however, these relate only to the distribution of these groups today, for the fossil history of most other groups is not known in enough detail to provide a parallel to that of mammals, though a start has been made on some groups, such as dung beetles [4] and birds. We shall now follow the analyses and history of the mammal groups and the development of their biogeographical regions as we see them today, before turning to that of the dominant group of plants today, the flowering plants or angiosperms.

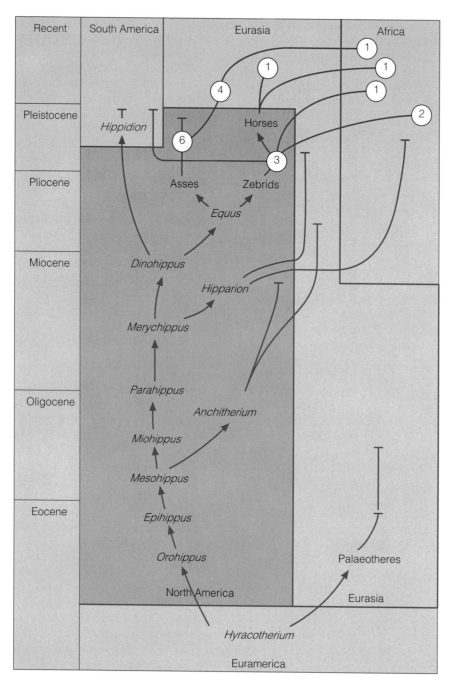

Figure 11.3 The biogeographical history of horses (Equidae), showing how most of their diversification took place in North America (tinted in dark blue), followed by later dispersal to other continents. The circled figures show the number of genera of horse, ass or zebra that existed in each continent at particular times. Adapted from MacFadden [3].

Patterns of Distribution Today, I: The Mammals

As mentioned in this chapter, the possibilities for mammals to disperse between the different continents varied in the past due to changes in the positions of the continents, in the climate or owing to the rise or withdrawal of shallow seas that from time to time subdivided the continents. Figure 11.4 summarizes the times and directions of dispersal that took place. It is useful to compare this figure with Figure 11.7, which shows

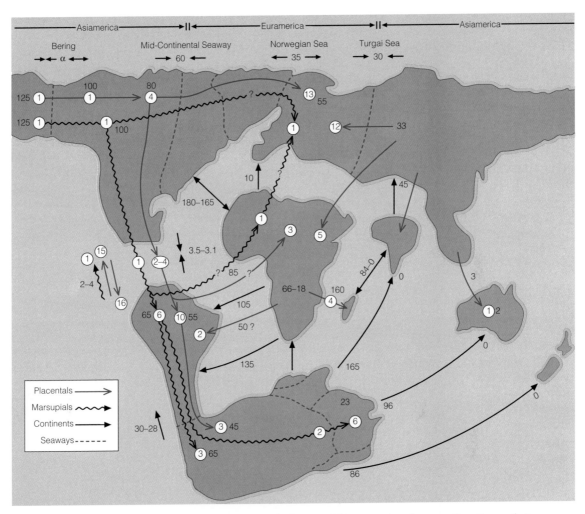

Figure 11.4 Diagram to illustrate the main events in the historical biogeography of mammalian dispersal. Arrows indicate the directions of dispersal of mammalian families or of continents. Dotted lines indicate the positions of the Bering, Mid-Continental, Labrador, Greenland and Turgai seas, with arrows indicating whether the sea opened or closed at the time indicated by the adjacent figure (in millions of years ago). (The alpha symbol at the Bering Sea indicates that it opened and closed a number of times.) The position of the Labrador Sea, west of Greenland, is shown as a dotted line, but no further information is given as it is irrelevant to mammalian dispersal. Figures in circles indicate the number of mammal families in the area in question, the adjoining figure indicating the relevant time in millions of years ago. No attempt has been made to give the total numbers of families in the different continents at any period later than 55 mya.

Gondwana's pattern of progressive break-up in the form of a dated cladogram.

Because both South America and Africa were largely isolated from other continents in the Late Cretaceous and Early Cenozoic, each developed a characteristic mammalian fauna. In the Late Cenozoic, India and South-East Asia had a fauna similar to that of Africa. But their different climatic histories, and the appearance of desert and mountains separating the two areas, led to the divergence of their mammalian faunas. The Australian region, with its unique marsupials, forms another distinct fauna. The two landmasses of the Northern Hemisphere have mammalian faunas that differ somewhat from one another, although both are similar in having been greatly impoverished by the climatic change of the Pleistocene Ice Ages.

The distribution pattern of terrestrial mammals in the Late Cenozoic Miocene–Pliocene Epochs is shown in Table 11.1. The final pattern found today is slightly different from this, because elephants became extinct in the Northern Hemisphere during the Pleistocene, and because edentates and marsupials dispersed to North America via the Panama land bridge. The final total of orders for each region in Table 11.1 takes account of these changes. The last line of Table 11.1 also shows the total number of terrestrial families of mammal in each region. These figures therefore exclude whales, sirenians (dugongs and manatees), pinnipeds (seals etc.) and bats, as well as humans and the mammals they took with them in their travels (such as the dingo and rabbit in Australia).

The individual families within the orders of mammal show considerable variations in their success at dispersal. A few have been extremely successful. Nine families have dispersed to all the regions except the Australian: soricids (shrews), sciurids (squirrels, chipmunks and marmots), cricetids (hamsters, lemmings, voles and field mice), leporids (hares and rabbits), cervids (deer), ursids (bears), canids (dogs), felids (cats) and mustelids (weasels, badgers, skunks etc.). In addition, the bovids (cattle, sheep, impala, eland etc.) have

Table 11.1 Distribution of terrestrial mammals during the Late Cenozoic (Miocene–Pliocene).

	Africa	Oriental	Eurasia	North America	South America	Australia
Rodents	×	×	×	×	×	×
Insectivores, carnivores, lagomorphs	×	×	×	×	×	
Perissodactyls, artiodactyls, elephants	×	×	×	×	×	
Primates	×	×	×		×	
Pangolins	×	×				
Conies, elephant-shrews, aardvarks	×					
Edentates					×	
Marsupials					×	×
Monotremes						×
Total number of orders today*	12	9	7	8	9	3
Total number of terrestrial families today*	44	31	29	23	32	11

*The final totals of orders and families also take account of the Quaternary extinctions and dispersals (see text in this chapter).

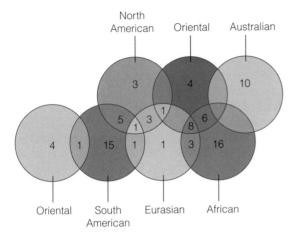

Figure 11.5 Venn diagram showing the interrelationships of the families of terrestrial mammals of their six zoogeographical regions today, excluding the 11 'wandering' families.

Table 11.2 Degree of endemicity of the families of terrestrial mammal (number of endemic families × 100 ÷ total number of families).

Region	Endemicity
Australia	$10 \times 100 \div 11 = 91\%$
South America	$15 \times 100 \div 32 = 47\%$
Africa	$16 \times 100 \div 44 = 36\%$
Holarctic	$7 \times 100 \div 37 = 19\%$
North America	$3 \times 100 \div 23 = 13\%$
Oriental	$4 \times 100 \div 31 = 13\%$
Eurasia	$1 \times 100 \div 30 = 3\%$

dispersed to all the regions except South America and Australia, and the murids (typical rats and mice) have dispersed everywhere except North and South America. This group of 11 families can conveniently be called the 'wanderers.' Their inclusion in any analysis of the patterns of distribution of the living families of terrestrial mammal tends to blur the underlying patterns of relationship of these zoogeographical regions. These 'wanderers' have therefore been excluded from Figure 11.5, which shows the distribution of the remaining 79 families. The Oriental region is shown twice, so that the single family shared with South America (the relict distribution of camelids) can be shown.

As can be seen in Figure 11.5, the majority of all the families of terrestrial mammal (51 out of 90, or 57%) are endemic to one region or another. The degree of endemicity of the mammals in each of the different regions is calculated in Table 11.2. (Rodents, the most successful of all the orders of mammal, contribute 19 of these endemic families: two North American, one Eurasian, ten South American and six African.) It is clear from Figure 11.3 and Table 11.2 that the degree of distinctiveness of the six zoogeographical regions of today, if judged by the endemicity of their mammals, varies greatly. These figures are the result of three main factors: isolation, climate and ecological diversity.

The results of the long isolation of the Australian and South American regions are obvious from the distribution of the families in the Miocene, when the other four regions were all interconnected (Figure 11.6). Comparison of Figures 11.5 and 11.6 also shows how many South American families became extinct after the Pliocene connection with North America (as discussed further in this chapter). The North American region was connected to Eurasia via the high-latitude Bering region, and many mammal groups were able to disperse across this region during those warmer times. Nevertheless, these groups did not include the tropical and subtropical groups that were then found throughout Africa and

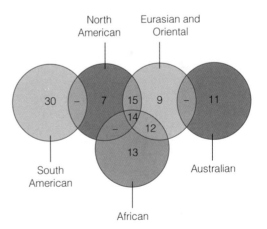

Figure 11.6 Venn diagram showing the interrelationships of the families of terrestrial mammals of the continents at the end of the Miocene, excluding the 11 'wandering' families.

southern Eurasia – including parts of Eurasia well north of the present limits of the Oriental region, which was therefore not recognizable as a separate zoogeographical region at that time. The Pleistocene glaciations of the Northern Hemisphere caused the extinction of many of the mammal faunas of both North America and Eurasia. These two regions therefore contain few mammal families, and these families are also still found mostly in the adjoining southern regions. As a result, North America and Eurasia contain few endemic mammal families. Tropical and subtropical Old World mammal families are therefore now found only in the Oriental and African regions. Without the three families found only in Madagascar, the degree of endemicity of the mammals of the African region would be slightly lower ($13 \times 100 \div 41 = 32\%$), but it would still be significantly higher than that of the Oriental region. Interestingly, this difference is also found in birds: 13 families are endemic to the African region, but only one is endemic to the Oriental region. These differences are probably mainly the result of the greater latitudinal spread of Africa, which extends from the equator northward through the Sahara Desert to approximately 30°N, and southward to 35°S, and therefore includes a wider range of environments than India, which only extends between 14°N and 35°N. But the greater area of Africa may also have played a part in this, providing more space for the evolution of novel groups.

Eisenberg [5] has made an interesting analysis of the extent to which the different ecological niches have been filled in the different mammalian biogeographical regions. Because East Africa has an extensive area of grasslands and savanna, 21% of the mammal genera of Africa are browsers or grazers, compared with only 9% of those in South America. Similarly, because great areas of South America are covered by rainforests, 22% of the mammal genera of northern South America are fruit-eating or omnivorous, compared with only 11% in southern Africa. For the same reason, northern South America also has a high proportion of mammals (bats) that feed on insects, either in the air or on leaves. Surprisingly, only two genera of South-East Asian bat feed on insects living on leaves, compared with nine in South America.

Comparison of data of this kind provokes interesting questions. Australia and northern South America contain similar proportions of arboreal genera, although Australia today has a much smaller area of forests. Eisenberg suggests that this may be an inheritance from the past, when Australia was more heavily forested (see Chapter 10), so that many early marsupials were arboreal. Similarly, he speculates on the possible reasons for the fact that there is a comparatively small number of fruit-eaters or omnivores in Australia. This may be because there are fewer fruit trees in Australia, or because parrot-type birds there have been more successful in that niche than have the mammals. It is also possible that, because the Australian fruit trees live in a more seasonal climate, their crops may be more seasonal and therefore less reliable as a year-round source of food. This may similarly be the reason that there are many fewer bats in Australia.

In a similar analysis, the ecologists Cris Cristoffer and Carlos Peres [6] compared the composition of the forests and its herbivores in the Old World tropics with those of the tropics of South America. They point out that the Old World forests have fewer lianas and a higher proportion of large and robust trees. The size of these trees allows them to withstand the foraging of the large herbivores, such as elephants and rhinos, which consume or damage smaller trees, breaking off branches or pushing the tree over. They noted that the Old World also has a larger area of savanna, and speculated that the savanna herbivores could have provided a source for the evolution of the large herbivores of the Old World forests. These have a much greater diversity and biomass of large terrestrial herbivores than those of South America, which instead has a greater number of larger arboreal herbivores and a greater number of smaller herbivores, both vertebrate and invertebrate.

Patterns of Distribution Today, II: The Flowering Plants

Compared to the mammals, there are many more families of flowering plant (about 450 vs about 100); the group originated and diversified much earlier (Early Cretaceous vs Late Cretaceous–Early Cenozoic), and it is far more effective at dispersal. As we have seen in Chapter 10, it is also far more difficult to trace the histories of the flowering plant families through time as we can those of the mammals. It is therefore at present not possible to compose brief summaries of their distribution today or of their times and directions of dispersal in the same way that it is possible to do for mammals (cf. Figure 11.4).

It is only on the largest scale that one can see a simple pattern of relationship between geography and the distribution of flowering plants. As early as the 19th century, the German botanist Adolf Engler noticed that the islands and continents of the southernmost part of the world contained elements of a single flora, which he called the Ancient Ocean Flora (see Chapter 1). We now know that the floral similarities between these areas are because they all contain survivors of the Late Cretaceous cool-temperate flora of Gondwana (see Chapter 10). Elements of this flora have survived in southernmost South America and in western Tasmania, and in scattered locations further north, such as New Caledonia and the mountains of New Guinea. Furthermore, as we shall see later in this chapter, the flora of the Australian region, too, is derived mainly from this original Gondwana flora, although it has become much altered in adapting to the steadily increasing aridity of that continent.

The difficulties involved in trying to analyse the biogeographical history of individual angiosperm families with a Gondwana pattern of distribution are well exemplified by the work of a group of Australian biogeographers, studying the classic southern family Winteraceae [7]. This family of woody evergreen shrubs or small trees, which dates from the Early Cretaceous, includes nine genera and about 130 species, found mainly in the South-West Pacific, Australia and New Guinea, but also occurs in South America and Madagascar. The results of the group's molecular analysis shows a complex pattern of events, involving long-distance dispersal both to existing continents (South America and Australia) and to parts of the now partly submerged continent of Zealandia (cf. Figure 11.11) (New Caledonia, New

Zealand and the volcanic Lord Howe Island), while the Madagascar genus provides a single example of vicariance.

History of Today's Biogeographical Regions

The Cenozoic history of each of the main regions, as well as that of the large islands of New Zealand and Madagascar, will now be considered in turn. As we shall see, that of the southern continents is complex, as these are all the result of the break-up and dispersal of Gondwana, whereas that of the northern continents is mainly the result of the appearance and disappearance of land or sea barriers within North America and Eurasia.

As can be seen from Figures 10.3 and 11.7 and Plate 7d, e, the break-up of the great southern supercontinent of Gondwana commenced 175 million years ago (mya), when a seaway appeared between India–Madagascar and the eastern coastline of Africa. This seaway gradually extended clockwise around Africa, to form in turn the southern South Atlantic about 135 mya, and the northern South Atlantic about 105 mya. India–Madagascar parted from Antarctica about 132 mya, while the split between India and Madagascar took place 90–85 mya. After this split, India moved rapidly northward; its northeastern corner collided with an island arc about 57 mya, before finally colliding with Tibet in southern Asia about 35 mya, near the Eocene/Oligocene boundary (Plate 7h), or even later [8]. Further west, the comparatively small northern movement of South America led to its connection with Northern America about 3 mya, which led to a complex interchange between their faunas and floras. Further east, New Zealand separated from Australia 84 mya, followed by the small continental fragment we call New Caledonia 80–65 mya. Australia itself gradually separated from Antarctica 52–35 mya and moved far northward to its present position near South-East Asia, allowing a complex interchange between the biotas of those two areas.

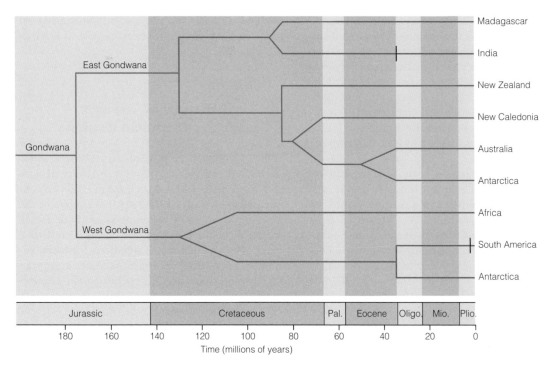

Figure 11.7 Diagram to show the progressive fragmentation of Gondwana, showing the dates at which the different events took place. Sloping lines of separation indicate the uncertainty over the exact times of certain separations. Cross-bars on the lines indicate the times of collision of Africa and India with the southern margin of Eurasia, and of the establishment of the Panama Isthmus linking South America to North America. Antarctica is part of East Gondwana, but it is also shown in the West Gondwana group in order to show its time of separation from South America. For details, see text in this chapter.

The Old World Tropics: Africa, India and South-East Asia

Two of these continents, India and Africa, were originally part of Gondwana (see Figures 10.1 and 10.3). Africa became united with the northern continents during the Mid-Miocene, 16–10 mya, but it had never lain far from southern Eurasia, and it was probably from Africa that tropical flowering plants dispersed to the tropics of southern Asia. Many elements of the tropical biota doubtless became widespread throughout the region and, before the Late Cenozoic cooling of the Northern Hemisphere, would also have ranged northward into higher latitudes of Eurasia. However, that cooling, together with the spread of seas and deserts in the Middle East, led to a new division of the Old World tropics into a western division, made up of Africa alone, and an eastern Oriental section made up of India and South-East Asia.

It is not yet possible to trace in full the separate contributions of Africa, India and South-East Asia to the final Old World tropical biota. As always, the fossil record of mammals is easier to interpret and can then be used as a guide for the probable biogeographical histories of the angiosperm floras. However, because they have better powers of dispersal, a greater variety of flowering plants than of mammals were able to disperse from Eurasia into Africa.

Africa

Africa contains the best known and most distinctive of the Old World tropical mammal fauna. As explained in Chapter 10, molecular studies of the interrelationships of the different orders of placentals suggest that the common ancestor of a group of African placentals entered the continent in the Late Cretaceous. This group is made up of the elephants, hyracoids (conies), aquatic sirenians (sea cows), elephant shrews, aardvarks, Cape golden moles, insectivorous tenrecs and extinct embrithopods; they have been placed together in a superorder, Afrotheria, that is endemic to Africa [9].

Much of the continent was covered by rainforests of the tropical everwet and subtropical summerwet biomes (see Figure 10.10) for much of the early Cenozoic, until the Late Miocene.

Though sea separated Africa from Eurasia (Figure 11.8a) in the Late Cretaceous and early Cenozoic, some placentals managed to enter Africa from the north. An invasion by early primates and creodonts (early carnivorous mammals) may have occurred near the Paleocene–Eocene boundary 56 mya [10], for such an event is needed to explain the more diverse fauna found in the Late Eocene to Early Oligocene of northern Africa. Apart from members of the Afrotheria, this fauna include creodonts, rodents, the cloven-hoofed artiodactyls

and the earliest members of the anthropoid primate line (that later evolved into apes and human beings). Though their only representatives today are the rabbit-like conies, in the Oligocene (and perhaps earlier) the hyracoids were the dominant small and medium-sized terrestrial herbivores of the continent.

After this early arrival, the placental fauna of Africa did not receive any further additions until much later in the Tertiary, about 19 mya in the Miocene, when a firm land connection with Eurasia took place in what is now the Middle East (Figure 11.8b). This closure of the seaway between the two continents may have been the cause of the climatic deterioration that took place in central Europe at that time. Elephants, creodonts, primates and cricetid rodents passed across the new land bridge from Africa to Asia, while carnivorans, suids (pigs) and bovids (cattle, antelope, etc.) entered Africa, causing the extinction of many of the earlier types of African mammal.

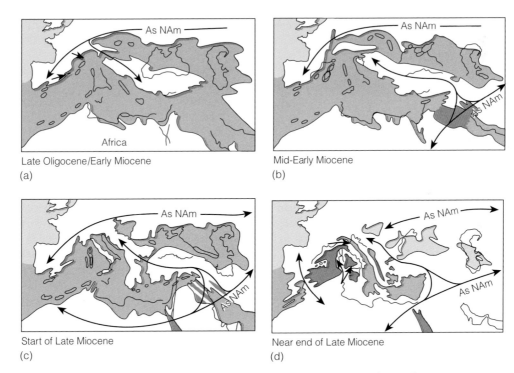

Figure 11.8 Reconstructions of the Mediterranean area at different times during the Cenozoic. Light tint, sea; dark tint, evaporitic deposits laid down as the seas dried up. Arrows show directions of mammal dispersals. As, Asia; NAm, North America. Some present-day geographical outlines have been added to aid recognition and location. Adapted from Steininger *et al.* [54].

Although the seaway reopened later in the Miocene, so that the climate through central Europe became warmer and moister, it was not enough of a barrier to prevent the dispersal to Eurasia of new types of primate, elephant and suid that had evolved in Africa. The seaway was finally broken near the beginning of the Late Miocene, 12 mya, by rising mountains in Arabia, Turkey and the Middle East. It was at this time that the early horse *Hipparion*, which had evolved in North America, appeared in both Eurasia and Africa (Figure 11.3), while rhinoceros, hyenas and sabre-toothed cats dispersed from Africa to Eurasia (Figure 11.8c). A final, dramatic event was the closure of the western connection between the Mediterranean and the Atlantic near the end of the Late Miocene, 6 mya. It was caused by a worldwide fall in sea levels (caused by an increase in the polar ice caps), as well as by the rise of mountains in both Spain and northwest Africa. This particularly affected the western part of the Mediterranean Sea, as there are no major rivers to provide water for this region – unlike the eastern part, which receives inflows from western Asia via the Black Sea, and from Africa via the River Nile. As a result, over the next 2 million years, the western Mediterranean from time to time dried up completely, leaving an immense plain 3000 m below sea level, covered with a thick deposit of rock salt (Figure 11.8d).

The results of the Miocene interchange of mammals between Africa and tropical Eurasia can still be seen today, for a number of groups are found exclusively in these two areas. However, the two areas were sufficiently isolated from one another for separate evolution to take place in each, so that different genera of these groups are found in each, except that the scaly anteater (*Manis*) is found in both areas. For example, the African rhinoceros, elephant and porcupine all belong to genera different from those found in the Oriental region. Similarly, the lemurs of Madagascar and the chimpanzees and gorillas of Africa are not found in the Oriental region, where these groups are represented by the lorises and by the orangutan and gibbon. However, some of the groups that today are found only in these two areas once had a wider distribution. For example, the elephants originated in Africa in the Eocene, migrated into Eurasia in the Early Miocene and migrated back later in the

Miocene [11]. The ancestors of the mammoths and modern Asian elephant probably evolved in Africa 4 to 5 mya and migrated back into Eurasia 2 million years later. Although they spread through much of the world, the mammoths became extinct everywhere during the Ice Ages. The mammal faunas of Africa and Eurasia also became more isolated from one another by the development of the Red Sea in the Miocene and by the extension of deserts in the Middle East.

Africa was uplifted from the Mid-Tertiary, so that today only 15% of its area is low enough to be within the altitudinal range of rainforest. The climate also became cooler and dryer in the Late Miocene. The Sahara Desert, whose 9.4 million km^2 area is greater than that of Australia and dominates the map of Africa today, began to appear at this time, about 5 mya. East Africa, too, became cooler and dryer and became covered by woodland and bushland with a ground cover of herbs and grass. This led to a great radiation of browsing and, especially, grazing mammals, particularly the bovids (cattle, sheep, antelope, gnu, impala etc.); there are 76 species of bovid in Africa, compared with only 37 species in Asia. Together with the giraffes, warthogs and zebras, they form huge herds in East Africa and are commonly thought of as the 'typical' fauna of Africa. But in reality these are latecomers to the African scene; their ancestors are not known in Africa until the Middle Miocene. Our own genus, *Homo*, also seems to have evolved in this environment (see Chapter 13), and human activities may have been responsible for the characteristic flora of the seasonally arid, overgrazed and burned grasslands of Africa [12]. The drying of East Africa also reduced the eastward extent of the African rainforest, which now became restricted to West Africa and the Congo Basin, and led to extinctions in its biota. This decrease in its area is probably one of the reasons why the African tropical flora is much less diverse than that of South America and South-East Asia. The Pleistocene glaciations of the Northern Hemisphere were reflected by dry, cool periods in Africa, which led to fluctuating subdivision and reunion of the rainforests.

The closure of the Panama Isthmus 3 mya, caused by changes in North Atlantic ocean currents, led to drier climates in Europe and the consequent development of the modern Mediterranean climate with its hot, dry summers and cool, wet winters, and the expansion of sclerophyll evergreen woodlands in the region. The drier European environment received many Asian mammals that inhabited steppe and savanna conditions, and it was still warm enough for some mammals that we now view as typically African (such as lions, hyenas, giraffids, hippopotamuses, rhinoceroses and macaques) to be found also in southern Europe [10]. The fauna of northern Africa was therefore similar to that of the rest of Africa during most of the Tertiary. However, in the early Holocene, about 10 000 years ago, the climate of the region became much wetter, so that humid zones invaded many of the presently desert areas of Africa, the Middle East and northwest India. Saharan cave paintings of this date show a fauna typical of savanna grasslands, including giraffe, hippopotamuses, elephants and bovids (see also Chapter 12). However, aridity set in again about 5000 years ago, and the cave paintings of this date show cattle, indicating a pastoral way of life, but then later change to images of horses and camels as the climate became even more severe.

At the margins of Africa, two areas contain biota that merit special consideration: the flora of the Cape region of southern Africa, and the fauna of Madagascar.

The Cape Flora

It has been customary for plant biogeographers to recognize the flora of the Cape region of southern Africa as a separate floral kingdom, thus placing it at the same level of importance as the floras of each of the major continents of the world, or of the whole of the temperate Northern Hemisphere [2]. However, the Cape region is one of several regions that have a Mediterranean-type climate, the others being California, coastal Chile, southwestern Australia and the Mediterranean Basin itself [13]. The Cape flora is therefore merely one of several floras that have resulted from similar ecological and evolutionary histories, leading to high levels of endemicity, and should be regarded as a plant province or region, rather than a kingdom. These five regions occupy less than 5% of the Earth's surface, yet contain around 48 250 species of flowering plant (almost 20% of the world total), as well as exceptionally high numbers of rare and locally endemic plants. Before the beginning of global cooling and drying in the Pliocene, all five regions were covered by subtropical forest, but now have a mixture of floras that include some relicts of the former forest, plus sclerophyllous shrublands and woodlands, with drought- and fire-adapted lineages predominating.

India and Madagascar

The history and biological connections of these two land masses have been described by the geologists Ian Ali and Jonathan Aitchison [8]. As can be seen from Plate 7, they were originally linked together and only became separate from the rest of Gondwana in the Early Cretaceous, 132 mya. Their biota would have been like that of the rest of Gondwana at that time, and would have included dinosaurs as well as early types of Mesozoic mammal of a more primitive evolutionary grade than the marsupials and placentals (as these primitive types were widespread in the Mesozoic world). India and Madagascar became separate from one another in the Late Cretaceous, 90–85 mya.

Madagascar

With its area of 587 000 km^2, Madagascar is second only to New Guinea in size. Its plate tectonic history is complex [8] (Plate 7d–g). Madagascar reached its present position relative to Africa by 121 mya; it separated from India about 88 mya and so became an island, while India continued its journey toward Asia. This long period of isolation explains the high degree of endemicity in the biota of Madagascar: about 96% of its 4200 species of trees and shrubs are endemic, as are 9700 species of plant and 770 species of vertebrate. The eastern part of the island still contains a large area of rainforest, which is species-rich, probably because it did not suffer as badly from prolonged drying during the glacial periods as did Africa. The biota of Madagascar has been comprehensively described [14].

Although the Cretaceous biota of Madagascar must have been similar to that of India, we have no fossil evidence of that, and recent molecular evidence suggests that the flowering plants of Madagascar are more likely to have arrived by dispersal from Africa [15], probably mostly by wind dispersal.

Since it was never narrower than today (380 km wide at its narrowest), the Mozambique Channel has always been wide enough to provide a formidable barrier for terrestrial African mammals. Nearly all of Madagascar's native terrestrial mammal species belong to only four groups: the Insectivora (tenrecs), Primates (lemurs), Carnivora (fossas) and Rodentia. In each of these groups, the Madagascan species all belong to a single lineage. (The only other terrestrial mammals that reached the islands naturally are a pygmy hippopotamus, only 200 cm long, that became extinct during the Pleistocene, and a river hog.) The remaining three groups of Madagascan mammals represent only a small part of the African mammal fauna, and molecular techniques suggest that the dates at which these different groups of mammal diverged from their non-Madagascan ancestors, after arrival in the island, varied from 60 to 18 mya [16].

Jonathan Ali and Matthew Huber [17] have described how the possibilities of the arrival of mammals in Madagascar must have been strongly affected by the pattern of ocean currents in the Indian Ocean. Today, strong currents in the Mozambique Channel sweep any rafts south or north, making dispersal from Africa impossible. This instead favours dispersals from Madagascar to Africa, and explains the discovery, based on molecular work, that chameleons appear to have evolved in Madagascar and spread from there to Africa and other oceanic islands [18]. In the Eocene, however, Africa and Australia were more than 10° south of their present positions, and the resulting strong westerly currents in their latitude would have impacted directly onto Madagascar. This would explain the fact that the coastal lizards *Cryptoblepharus* of Madagascar seem to have dispersed there from Indonesia or Australia [19], and also the possibility that the rodents of Madagascar may have reached the island from Asia, not Africa, and went on to invade the African mainland [20]. This would also have caused a strong anticlockwise gyre in the Mozambique Channel, which would have directed flow along the coast of Africa eastwards, favouring dispersal of African mammals to Madagascar.

The presence of two lineages of the freshwater cichlid fishes in Madagascar provides another complication. Molecular studies suggest that these are most closely related to the cichlids of India rather than to those of Africa [21]. However, the time of divergence between the cichlids of the two areas, 29–25 million years ago, is far too late for them to be the result of vicariant evolution.

Because its immigrants came from only a few groups of mammals and birds, the successful colonists were able to occupy niches that would not normally be available to them – a phenomenon that is often found in such island biotas (see Chapter 7). Some of the now-extinct types of lemur included forms similar to the tree sloths and ground sloths of South America, and others similar to gorillas of Africa and to the koala of Australia. The Madagascar fossas, which belong to the mongoose family of the Carnivora, included something very like a lion, while the birds include the massive extinct herbivorous flightless elephant bird, as well as radiations of endemic passerine birds – the shrike-like vangas and the warblers [15].

It is obvious from all the above that molecular work is continually providing both new solutions and new problems in interpreting the biogeography of Madagascar. It is also a good example of the extent to which historical biogeographers today have to take into account many different changing aspects of the geography of the Cenozoic world. It will be fascinating to see how these studies progress in the future.

India

After its separation from Madagascar, India was isolated from direct contact with other land masses during its long journey northward to collide with southern Asia that, as described earlier in this chapter, probably took place less than 35 mya. However, Ali and Aitchison [22] believe that oceanic islands and now-submerged oceanic plateaux could have provided stepping stones for organisms to disperse between it and other parts of East Gondwana in the Late Cretaceous and early Paleocene. A good example of this is the discovery that there were once islands on the 5000 km long Ninetyeast Ridge, so-named because it approximately parallels that longitude today. Though they are now eroded away, in the Paleocene to Late Oligocene these islands bore fossil floras of a distinctively East Gondwanan character, with tropical forests and palms, even though they were over 1000 km away from Australia [23] This is another example of the powers of dispersal of plants, but, since the island chain would have affected ocean currents in the ocean to the east of Madagascar, it also affects our interpretation of the relevance of these to the fauna of that island, as we have already seen.

As would be expected, India's Cretaceous flora includes such southern groups as *Nothofagus*, Proteaceae and Podocarpaceae. However, it also includes elements of northern plant groups (see Willis and McElwain (2014) in Further Reading), as well as northern types of lizard, snake and frog. India would have received progressively more flowering plants from Asia, as it approached and eventually collided with that continent: there is clear evidence of typically Indian or Gondwanan types of pollen in South-East Asia after the collision [24]. These new angiosperms replaced much of the former flora of South-East Asia, and those that survived are still present in the flora that is found in both regions today. This forms the Indo-Malesian floral kingdom and has also extended eastward far into the islands of the Pacific Ocean. After the Mid-Tertiary global cooling, the rainforests of South-East Asia and China also received many frost-sensitive plants from the mid-latitudes of Asia. A similar process led to the appearance of northern frost-sensitive plants in northern South America after the completion of the Panama Isthmus. The resulting similarity between elements in the forests of South America and China, sometimes referred to as *amphi-Pacific elements*, are thus the result of the southward migration of these plants at either end of the Northern Hemisphere, rather than being evidence for some 'Pacific Ocean baseline,' as suggested by panbiogeographers (see Chapter 1).

Unfortunately, the mammal fauna of India in the Cretaceous and Paleocene is unknown, but it must have been quickly colonized by the placental mammals of Asia after its collision with that continent. The later history of exchange of mammals between India–Eurasia and Africa has been described earlier in this chapter. About 3 mya, a similar increase in aridity, related to the uplift of the Himalayan Mountains to new heights, led to the northern parts of the Indian subcontinent becoming much drier and also led to an increase in the numbers of grazing mammals such as horses, antelope and camels, and of elephants.

A complex pattern of islands lies on the continental shelves of South-East Asia (the Sunda shelf) and that of New Guinea–Australia (the Sahul shelf), and these have been colonized from both these areas (Box 11.1 and Figure 11.9).

Between two worlds – Wallacea

Concept Box 11.1

The series of islands between the mainlands of Asia and Australia contain a transition between the Asian flowering plants and placental mammals, and the Australian flowering plants and marsupial mammals. To plant biogeographers, the whole area is the Malaysian province of the Indo-Pacific floral kingdom, which extends eastward to include New Guinea and most of the islands of the Pacific Ocean. Zoologists, on the other hand, found that New Guinea contained marsupials, but very few placentals, and a predominantly Australian bird fauna. In the 19th century, the zoogeographer Alfred Russel Wallace had suggested a line of faunal demarcation, later called Wallace's Line, which separated the predominantly Asian bird fauna from the more eastern, predominantly Australian bird fauna. This line, which runs close to the Asian continental shelf, has in the past therefore been recognized as the boundary between the Oriental and the Australian zoogeographical regions. However, the area between the Asian and the Australian continental shelves contains relatively few mammals of any kind or origin. Later zoogeographers proposed six different variants of Wallace's Line [55]. The debate on where to draw a line diverted attention from the real interest of the area, which is the extent to which animals or plants have been able to enter or cross this pattern of islands from either direction. It is therefore best to draw the boundary of the Oriental and Australian faunal regions at the continental shelves, as is done in the other faunal regions, and to exclude the intervening area, which some earlier biogeographers have named **Wallacea** (Figure 11.9).

The biogeographical problems of Wallacea are not as straightforward as a simple inspection of the modern map might suggest. One example is provided by the island of Sulawesi, which has a varied mammal fauna and complex geological origin. The western part of the island was part of Borneo until it drifted away in the Eocene; it may have been a major source of Asian plants for the islands to the east, and for New Guinea. In contrast, the eastern parts of Sulawesi were originally fragments of Australasia and only joined the rest of the island in the Early Miocene. The Makassar Strait between Borneo and Sulawesi is 104 km wide today. Sulawesi has a varied mammal fauna, including bats, rats, shrews, tarsiers, monkeys, porcupines, squirrels, civets, pigs, deer and a fossil pygmy elephant. Many of these are endemic and so were not introduced by humans; they probably

crossed the Makassar Strait during the Pleistocene, when lower sea levels would have reduced its width to only 40 km. Although some types of mammal found in Sulawesi are found in islands further to the east, most were probably taken by humans; only the bats and rats appear to have made these additional travels unaided. Sulawesi also contains two species of the marsupial phalangers. If this is the result of natural dispersal, it is the only island in Wallacea where there is a natural overlap between Asian and Australian mammals. But phalangers are also found on several other islands, and it is quite possible that they have been taken by islanders as pets.

In general, the biotas of Wallacea are less diverse than those to the east or west, because of the difficulties of colonization across a series of ocean straits, and because of the inherent vulnerabilities of island biotas (see Chapter 7).

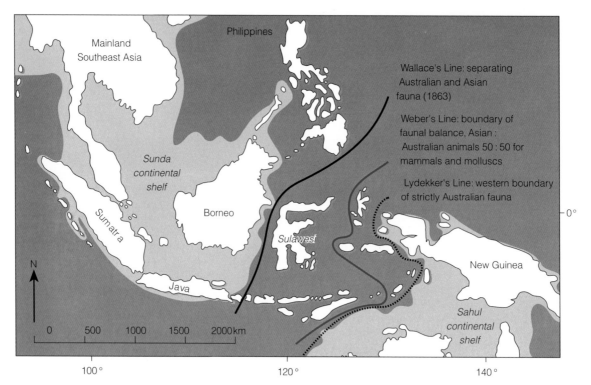

Figure 11.9 Map of the East Indies. The continental shelves are shown in light blue, and the deeper ocean in dark blue. Three of the 'Lines' of faunal division are shown and explained. 'Wallacea' is the area that lies between Wallace's Line and Lydekker's Line. Adapted from Moss and Wilson [56].

Australia

The history, characteristics and biogeographical affinities of the biota of Australia are the most unusual and interesting of any in the world, and their explanation requires a very rewarding understanding of the interplay between continental movement, climatic change and biotic dispersal [24–26].

By the Middle and Late Cretaceous, Gondwana had lost both India and Africa and retained only a narrow connection with South America, so that Antarctica–Australia was the most substantial remnant of the old supercontinent. Its biota included marsupial mammals, araucarian and podocarp conifers, ginkgoes, bennettitaleans, cycads and some of the earliest flowering plants (including early members of the *Nothofagus* and *Protea* families). They were living in far-south latitudes which, though ice-free, had the almost sunless winters and almost continuously light summers seen in the Antarctic today. The Cretaceous and early Tertiary flora is known almost exclusively from southern Australia, where there is evidence of low-diversity, moist temperate rainforests dominated by gymnosperms, unlike any vegetation known today. During the mid to late Eocene, these forests were replaced by more diverse forests dominated by *Nothofagus*. Comparison with later times, when there is evidence from other Australian environments, suggests that there were dense evergreen rainforests in more northern regions. In central Australia, there were floras like that of today's sclerophyll and monsoonal tropics biomes, dominated by proteas rather than *Nothofagus*, and with a greater variety of angiosperms.

Australia and Antarctica started to split apart about 52 mya, but at first they separated quite slowly (Plate 7f, g). They therefore remained parts of a single weather system, so that some of the warmth that Australia received was circulated by wind systems southward to its Antarctic neighbour. As a result, and also because at this time atmospheric CO_2 levels and global temperatures were quite high, there was no Antarctic ice cap. There was an extensive warm sea inlet between the two continents, so the climate of this region must have been warm with high rainfall. By 50 mya, there is evidence of major meso-megathermal gymnosperm-dominated rainforest there, similar to today's aseasonal-wet biome, with *Nothofagus*, and coastal mangroves. The old cool-temperate *Nothofagus* forests were now mainly confined to the southern part of Australia, and *Ginkgo* became extinct at about this time.

This, then, was the genial environment of Australia in the early Cenozoic, when its marsupials were undergoing their great radiation, whose diversity paralleled that of the absent placentals. It was also the time and environment in which the songbirds (oscines) started their diversification, prior to dispersing to Asia in the Eocene, where their descendants started the worldwide radiation that today comprises nearly half of all living species of bird.

Australia only started to move rapidly northward 46 mya, in the Eocene. By the Early Oligocene, around 35 mya, it had separated sufficiently far from Antarctica for the deep-water Antarctic Circumpolar Current of cooler water and associated westerly winds to become established (see Figure 10.11). This was probably a major factor in the marked worldwide cooling that took place at this time. Now isolated from the warmth of Australia, Antarctica cooled and its ice sheets started to form. There would also have been less evaporation from the now-cooler seas around Australia, reducing the rainfall on that continent, with a consequent increase in its arid, desert areas. This was exacerbated by the fact that Australia's northward movement brought it into the 30°S high-pressure zone of low rainfall. For all these reasons, Australia became increasingly arid from the Late Oligocene and into the Miocene.

The continent's only significant mountain chain is the Great Dividing Range, which formed about 280 mya. Since then, there have been no new mountains whose erosion might have provided new sediments and minerals for the vast flat expanses of the rest of the continent. Their soils therefore have about half the levels of nitrates and phosphates of equivalent soils elsewhere. Furthermore, because these mountains run along the eastern side of the continent, the rains dropped by the prevailing east winds fall there, so that the lands to the interior to the west have become a vast semi-arid zone. In addition, Australia's annual rainfall is also extremely variable, because it is highly affected by the El Niño events in the Pacific Ocean. At times when the rainfall is very low, about every 10–20 years, the Australian flora is subject to devastating wildfires. These are particularly fierce because many of the plants, in trying to protect themselves from herbivores, have developed tough resinous prickly foliage, which is highly flammable.

Because of all these factors, the productivity of the vegetation of Australia is very low and very variable, which in turn leads to wide fluctuations in the populations of Australian herbivores. As Flannery [27] points out, this is why the continent has very few large carnivores, as they are inevitably less numerous than their prey, and are therefore very vulnerable to extinction. Flannery notes that Australia has an unusually high number of reptilian predators, such as pythons and varanid lizards, and that in the Pleistocene Period there were giant members of these groups, as well as a large land crocodile. He suggests that the reptiles, being cold-blooded and therefore not needing to consume as much food as their warm-blooded mammalian competitors, may have been better able to survive periods of starvation and so retain a higher, safer level of population. Milewski and Diamond [28] have suggested that the lack of micronutrients containing iodine, cobalt and selenium in Australia's impoverished soils may similarly have affected the evolution of large herbivores in the continent, for these may be of limited size, fecundity and intelligence compared with those of other continents.

Today, a major characteristic of Australia is its 'sclerophyll' flora, 45% of whose genera are endemic to the continent. **Sclerophyll** (or **scleromorph**) plants grow slowly, readily stop growing and have a small total leaf area, made up of small, broad, evergreen, leathery leaves. It was long thought that this flora had evolved in response to the aridity of the continent. However, plants with these characteristics appeared in the Australian record far earlier than this, 60–55 mya, and seem to have evolved from the rainforest flora, for all the larger families with sclerophyll types are also found in the rainforests. It is now clear that the sclerophyll habit started as an adaptation to Australia's low soil nutrients, and only later turned out to be adaptive also to increasing aridity. The most spectacularly successful sclerophyll forms are the gum-tree genus *Eucalyptus* (Myrtaceae), which includes over 500 species, and the family Proteaceae. The sclerophyll vegetation is low in

nutrients and high in toxic biochemicals to deter the herbivore. The effects of this in depressing the population density of herbivores are shown by the brush opossum, *Trichosurus vulpecula*, whose density in the very different vegetation of New Zealand is five to six times greater than its density in its native Australia [29].

This increasing aridity was accompanied by the reduced dominance of plants characteristic of the aseasonal-wet biome (*Nothofagus* and conifers) and the increasing dominance of *Eucalyptus*. Over this same period of time, burning became an important aspect of Australian ecology. All of these changes led to the appearance of open grasslands and savanna in central Australia, containing radiations of the eucalypts, casuarinas, pea-flowered legumes and *Acacia* (Fabaceae). By the end of the Pliocene, the *Eucalyptus* forests had changed to become dry, rather than wet, sclerophyll forests, and much of the continent had been taken over by the huge Eremean biome with its open, arid shrublands, woodlands and grasslands, into which such herbs as brassicas and chenopods had arrived from elsewhere (Figure 11.10).

This plate tectonic and climatic history has therefore combined to make Australia the driest of all continents; two-thirds of it have an annual rainfall of less than 500 mm, and one-third has less than 250 mm. This also explains why, although both the rainforest and the sclerophyll floras once covered much greater areas of Australia, they are today found only in a relict distribution, in scattered peripheral areas.

The mammals of Australia had to adapt to these climatic and vegetational changes. This fauna is made up of marsupials alone (plus the few egg-laying monotremes), and they have radiated into a great variety of forms, occupying the niches that placentals have filled everywhere else in the world. The marsupial equivalents of rats, mice, squirrels, jerboas, moles, badgers, anteaters, rabbits, cats, wolves and bears all look very like their placental counterparts – only the kangaroo looks quite unlike its placental equivalent, the horse. Unfortunately, their fossil record in Australia does not begin until the Mid-Miocene. Molecular results suggest that the (surprisingly diverse) arboreal possum families radiated in the Eocene, when the Nothofagus forests appeared. The early marsupial herbivores were mostly browsers, the terrestrial

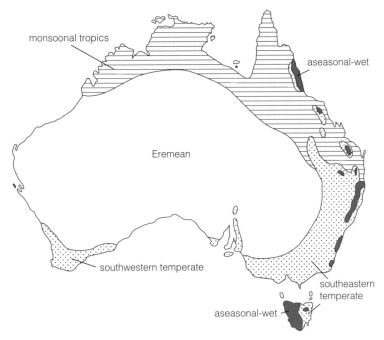

Figure 11.10 Map of Australian biomes. Southeastern temperate: sclerophyll eucalypt forest, woodlands and heath, seasonally dry. Southwestern temperate: sclerophyll eucalypt forest, woodlands and heath, Mediterranean climate. Monsoonal tropics: savannah, mostly sclerophyl eucalypt and acacia, seasonally dry. Eremean: arid shrubland, low woodland and grassland. Aseasonal wet: year-round high rainfall, tropical to temperate or subalpine, closed-canopy rainforest on volcanic soils, to heath on poor soils. From Crisp *et al.* [57].

grazers such as the diverse kangaroos and wallabies only appearing after the forests began to break up in the Oligocene, and after the Eremean grassland ecosystem appeared in the Pliocene.

Although Australia had parted from its original Gondwana relationships, its northward movement into the Pacific eventually brought its northern edge close to a great oceanic trench, where old ocean-crust material was sinking downward. This lighter material therefore came to underlie the northern edge of the Australian continent, which started to rise from the sea to form New Guinea at the Oligocene–Miocene boundary, 25 mya, and reached its present position in the Late Miocene, 6–8 mya. New Guinea is the largest island in the world (nearly 800,000 km²). Its mountains provide a high, cool environment that today is the wettest area on Earth, although it is close to the driest continent, and it was was colonized by

the Australian rainforest flora, including *Nothofagus*. However, the surrounding lowlands of New Guinea were colonized by a mixture of Asian and Australian plants. A recent fossil-calibrated molecular analysis of dispersals of the Asian plant *Trichosanthes* [30] shows several histories that differ according to the time at which the dispersal took place. The two earliest occurred before the rise of New Guinea and were therefore originally limited to Australia; one of these remained there, while the other later colonized New Guinea. The other lineages placed after that island had grown significantly, and so are found both there and in Australia itself. These lineages underwent major diversification either in the wet highlands of New Guinea or in the monsoonal tropics of northern Australia, depending on the environment in which they had originated and to which they were therefore already adapted.

Placental mammals, too, started to spread eastward from Asia, but, apart from humans and the aerial bats, only the rats spread naturally as far as Australia, where their 50 species now form 50% of the diversity of the Australian land mammal fauna. Human beings probably arrived 60 000 to 40 000 years ago and brought the domestic dog (the ancestor of the dingo) about 3500 years ago.

New Caledonia

Geological studies show that New Caledonia is the smallest (about 1600 km²) surviving fragment of Gondwana (Figure 11.11), having rifted away from Australia in the Late Cretaceous 80–65 mya and reaching its present position 1500 km east of it 50 mya. It is a hotspot for diversity: 77% of its flowering plant species are endemic to the island, as are 98% of the species of the flowering plant family Sapotaceae. The position and history of New Caledonia make its biota an ideal test case for the two current theories about the origins of such isolated floras or faunas, which completely conflict with one another.

In view of the large amount of research that shows the frequency of long-distance dispersal in such scenarios, most biogeographers today believe that this was the mechanism of arrival of the ancestors of the New Caledonian taxa. Geological studies show that the island was submerged for long periods in the Paleocene–Eocene, so that the island was not available for colonization until about 37 mya. The New Caledonian species must therefore have diverged from their ancestors after this date.

Panbiogeographers (see Chapter 1), however, reject long-distance dispersal. They therefore have to theorize that the Cretaceous ancestors of the present biota were widespread across the Pacific islands, including on supposedly formerly exposed large plateaux such as the Ontong Java plateau [31]. These plateaux are then presumed to have fragmented into separate isolated areas of land, on each of which new species arose by vicariance, and 7–8 of these areas amalgamated to form the island of New Caledonia. A major methodological problem with their approach is that no explanation is given as to precisely how this situation arose. For example, how did this original biota come to be widespread across these areas within the Pacific

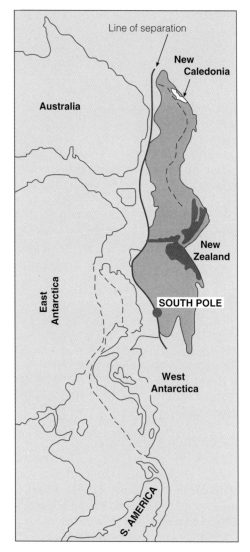

Figure 11.11 The original position of Zealandia (shaded) close to the South Pole, before it broke away from Gondwana. From McDowell [34].

Ocean? And how far is this compatible with our now-detailed geological knowledge of the nature, environment and age of the different components of all these terranes and islands, many of which originated as volcanic island arcs or as rocks that originated deep within the Earth?

However, we now have methods that allow us to evaluate other requirements of the panbiogeographical approach. If their theory is correct, the separation of the New Caledonian species from one another must have taken place in the Cretaceous, and their ancestral area must have been the Pacific, across which all these patches of land lay scattered. Using nuclear ribosomal DNA (nrDNA) molecular analysis to estimate phylogenetic divergence times and ancestral areas using BEAST (see Chapter 8), biogeographers in Sweden and France [32] have investigated the history of the New Caledonian species of the angiosperm family Sapotaceae. They found that these resulted from nine separate colonizations of the island between 33 and 4.2 mya, dates that are about 40 million years after the rifting of New Caledonia from Australia. New Guinea and Australia were the most important sources of the arrivals, those from Pacific islands arriving only more recently, 27–24 mya. These results are clearly totally incompatible with the panbiogeographers' theories.

New Zealand

New Zealand is unique. Its total area, 270 000 km^2, makes it far larger than any other Pacific oceanic island, and it is therefore not so vulnerable to the fluctuations of population and diversity that characterize them (see Chapter 8). It is also unlike other Pacific oceanic islands in being a fragment of old Gondwana, rather than being entirely volcanic, and has suffered from a series of changes in area. When it first separated from Gondwana in the Late Cretaceous, about 82 mya, it was part of a minicontinent known as Zealandia [33,34], almost half the size of Australia, and stretching as far as New Caledonia to the northwest and Chatham Island to the southeast (Figure 11.11). This small landmass drifted northeastward to its present position, separated from Australia by the Tasman Sea, 1400 km wide. Most of it sank below sea level 30–25 mya, during the Oligocene, and it is possible that all of it was submerged at this time. More recently, the New Zealand biota was greatly affected by the Ice Age climatic cooling, which caused extensive glaciation there and must have led to many extinctions.

New Zealand's isolated position and history have made the interpretation of its flora and fauna an interesting battlefield. On one side are the panbiogeography (see Chapter 1) proponents of vicariance, for whom patterns of scattered, disjunct distribution are the result of the subsequent fragmentation of an originally single homeland, so that the members of the present biota of New Zealand are merely the descendants of that original biota. On the other side are the proponents of dispersal, for whom these patterns instead resulted from the organisms crossing the intervening barriers after they formed.

Molecular studies that provide the dates of divergence of the New Zealand lineages from their closest external relatives have shown that many of New Zealand's organisms (especially the plants) reached there in comparatively recent times, and therefore by dispersal. The origin of the islands' vascular plant flora has been discussed by Mike Pole of the University of Tasmania [35]. The nature of the earliest flora, from the Late Cretaceous, reflects New Zealand's cold, high-latitude position at that time, while the excellent record from the Miocene (by which date the islands were already extremely isolated) is of floras characteristic of rainforest or sclerophyll forest–woodland. This Miocene flora must have been the result of dispersal from Australia, which also then had a similar climate and flora but, like the earlier flora, is quite unlike the flora of New Zealand today. Its Miocene flora appears to have become largely extinct following the climatic cooling at the Pliocene–Pleistocene boundary. Its modern flora was likely the result of often multiple, long-distance dispersal from the present-day flora of Australia, aided by the strong west winds that blow around Antarctica, followed by major radiations of the successful immigrants within Australia. All of this led Pole to suggest that New Zealand had been totally submerged during the Oligocene, all of its present biota being the result of dispersal.

However, more recent work has led to a more balanced view. A group of mainly New Zealand scientists has recently reported the results of an analysis of the mitochondrial genome of the endemic New Zealand frog *Leiopelma* [36]. The divergence times between the two species are well over 65 mya, far earlier than the suggested total drowning of the islands. Other elements of the fauna similarly seem likely to be parts of an ancient and diverse fauna that are unlikely to have been able to arrive by overseas dispersal. This includes earthworms, the velvet worm *Peripatus*, arachnids, the unusual endemic lizard *Sphenodon* (whose ancestors were in the island in the Miocene) and the flightless birds the moa and the tinamou.

This may also be true of the only fossil mammal known from New Zealand; found in Miocene rocks, it is probably a relict of the early egg-laying mammals (see Figures 10.5 and 10.6), as are the monotremes of Australia. The balance of the evidence now seems to favour the view that, though much of the islands may have been submerged during the Oligocene, a varied though limited fauna survived and was joined by plants, most of which originated in Australia.

Not all of New Zealand's immigrants, however, arrived from the west. Molecular phylogenetic results suggest that a few plants have dispersed in a westerly direction, against the prevailing currents of wind and water, to Australia and New Guinea. These may be the result of dispersal by oceanic birds, which fly for long distances. An even more surprising result suggests that the little creeping flowering plant *Tetrachondra* spread to New Zealand from South America. However, this may not have been directly across the thousands of miles of ocean that now separate the two areas of land; fossil wood of *Nothofagus* has been found in Pliocene deposits (5–2 million years old) in Antarctica, so this continent could have provided an intermediate stepping stone.

As a result of all these factors, the New Zealand flora is highly endemic at the species level (86% of its less than 2500 species are endemic), but with no endemic families, perhaps because as yet there has not been sufficient time for their evolution to have taken place. The flora also shows especially high endemicity in such plants as ferns and orchids and the tree *Metrosideros*, all of which are characteristic of floras of distant islands, as is the fact that some of the herbs have evolved into endemic trees (see Chapter 7).

The West Indies

Though superficially similar in their geographical position, poised between two continents (Figure 11.12), the islands of the Caribbean do not present the same fascinating problems as the islands of Wallacea because they do not lie between continents with totally different faunas and floras. Instead, they have posed a different, interesting problem: are their faunas and floras primarily the result of vicariance, due to evolution on islands that arose by the break-up of a larger 'proto-Antillean' landmass, or are they the result of dispersal from the neighbouring continents? This question, and the geological history of the region, have been reviewed by the American biogeographer Blair Hedges [37] (Box 11.2).

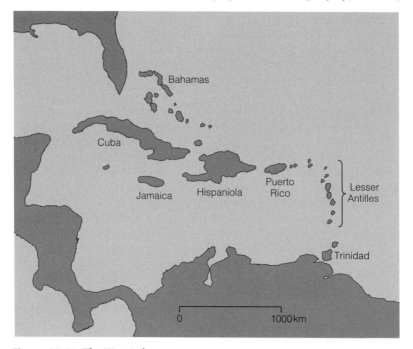

Figure 11.12 The West Indies.

Box
11.2

Guest Author

Molecular clocks, flotsam and Caribbean islands

Dr S. Blair Hedges, Center for Biodiversity, Temple University, Pennsylvania

The islands of the Caribbean have provided a classic test of the two major mechanisms of historical biogeography: vicariance and dispersal. Formed in the Mid-Cretaceous (about 100 mya), they have had a long and complex geologic history that included an early connection between North and South America (proto-Antilles) and a catastrophic asteroid impact (about 66 mya). During the Cenozoic, some large islands (Greater Antilles) broke apart and fused, a stable carbonate platform (Bahamas Bank) kept up with sea-level changes and a chain of volcanic islands (Lesser Antilles) migrated slowly from west to east. Soon after the theory of plate tectonics became accepted, it was recognized that the current biota of the islands may be the fragmented (vicariant) remnant of a once-continuous proto-Antillean biota. For the last three decades, a debate has ensued over the importance of vicariance versus dispersal in the origin of the biota.

The answer has not come easily, nor has it been agreed upon by everyone. Nonetheless, most research suggests that the entire living biota of the Caribbean islands arrived by dispersal and not through the geologic break-up of an ancient landmass. Initially, it was thought that the key information to answer the question would come from the phylogenetic relationships of organisms. In part, this thinking emerged from the popularity – in the 1980s – of the field of cladistics (and vicariance biogeography) which emphasizes the relationships of organisms over most other types of data. Undoubtedly, relationships are important, but the problem with this line of thinking is that the branching order of species might match the geologic break-up of land areas, but the timing could be very different. So it was soon realized that data on the times of divergence of organisms from their closest relatives on the mainland were critical data. The fossil record in this region is poor, but molecular clocks provided those data.

Molecular clocks (see Chapter 8) need calibration against some external events, such as well-dated fossils, or geological events, such as the time of emergence of an island above sea level – for this is the earliest time at which it could be occupied by terrestrial organisms. For the Caribbean islands, it turned out that relationships were not important in answering the basic question of vicariance versus dispersal. This is because nearly all of the times of divergence measured by molecular clocks, for many different groups of terrestrial vertebrates, have been too young to have resulted from a Late Cretaceous vicariance event. Instead, the times were scattered throughout the Cenozoic, almost randomly, and in accord with a mechanism that relies on chance events such as dispersal. However, relationships were useful in determining the source area of dispersal. For most terrestrial vertebrates that cannot fly or otherwise disperse over water on their own powers, their closest relatives are in South America, while a majority of the birds, bats and freshwater fishes in the Caribbean islands appear to have come from North and Central America.

Other diverse evidence, too, supports a dispersal origin for the Caribbean biota. Most important among them is the taxonomic composition of the endemic groups. There are some enormous adaptive radiations, often with species filling niches different from those in the same genus on the mainland. For example, some of the smallest and largest species of major groups (e.g. cycads, swallowtail butterflies, frogs, lizards and snakes) occur on islands in the Caribbean. Yet at the same time, many major groups are absent, such as salamanders and caecilian amphibians, marsupials, rabbits, armadillos and carnivorous placental mammals. The fossil record, including that of the 15–20 million-year-old Dominican amber, which includes the remains of insects, frogs, lizards and small mammals, shows a similar taxonomic composition. This is best interpreted as a strong filter effect, whereby a few colonists survive long-distance dispersal and then radiate into a diversity of unoccupied ecological niches. This same evidence also argues against an origin for the biota by way of a Mid-Cenozoic (about 34 mya) land bridge from South America, which has also been suggested. Such a land bridge would not have acted as a strong filter and would have allowed many other groups to enter the archipelago that we do not in fact find there.

Ocean currents and geographical proximity best explain the source areas of the island colonists identified in molecular phylogenies. Water flows almost unidirectionally from east and southeast to west and northwest in the Caribbean – and this was true even prior to the uplift of the Isthmus of Panama. As a result, flotsam ejected and carried down from the rivers in northern and northeastern South America will end up in the Caribbean, if it continues to float. For example, even though Cuba is much closer to North America than to South America, it is much easier for a lizard to arrive in Cuba by floating on vegetation from South America; this is reflected in the composition of the Cuban lizard fauna. But for organisms that can fly or swim, the geographically closer areas are the more likely sources, and the common air current direction in the Caribbean – northeast to southwest – might even assist dispersers flying from North America.

Two lineages of island vertebrates that show old (Cretaceous) times of divergence from their closest relatives on the mainland, using molecular clocks, have been debated as possible examples of proto-Antillean vicariance. These are the giant shrews (solenodons) of Cuba and Hispaniola and the night lizards (Xantusiidae) of Cuba. While an ancient origin cannot be ruled out, both groups are biogeographical relics, for their mainland fossil record demonstrates a wider distribution in the past. This raises the possibility – not normally considered for other groups – that they diverged more recently from close relatives on the

mainland that are now extinct and hence inaccessible to molecular clocks. Some geologists also are uncertain about whether there was any continuously emergent land in the Caribbean before the late Eocene (about 37 mya), which would have been necessary for maintaining such lineages. Moreover, it is not clear how these organisms might have survived the end-Cretaceous asteroid impact, which occurred a short distance away. The origin of these two groups will likely continue to be debated.

Now we know that flotsam was critical for the origin of the Caribbean terrestrial biota, but surprisingly little is known about this mode of dispersal across ocean waters. How abundant are floating islands? How long do they stay afloat, and how far do they travel? What organisms do they typically carry? There are many anecdotal accounts of floating islands but almost no scientific studies. Analysis of satellite imagery, GPS tracking and taxonomic surveys of floating islands might answer some of these questions. Whatever the details, we can be certain that long-distance dispersal by flotsam did occur and that fragile animals – such as small frogs – successfully colonized Caribbean islands millions of years ago after riding the ocean waves for weeks on a jumble of logs.

What is now the Caribbean region began as merely a gap between North America and South America (Figure 11.13a), and therefore also between the Atlantic plate to the east and the Pacific plate to the west. The Caribbean plate between the two was therefore bounded by ocean trenches (see Chapter 5), where old ocean crust was disappearing into the Earth, causing volcanic activity and the appearance of a chain of volcanic islands. The Americas moved westward past this whole system, leaving it in a progressively more and more eastern position relative to these continents. The island chain to the east, where the Atlantic plate seafloor is still being consumed, today forms the Lesser Antillean island chain and trench system (Figures 11.12 and 11.13). The islands in the chain to the west gradually enlarged and coalesced to form the Panama Isthmus, linking North and South America

(Figure 11.14). (The Scotia Sea today is a replica of this situation, for it is formed from a small portion of the South Pacific plate that extended between the Antarctic Peninsula and South America as the latter moved westward.)

There was also, in the Early Cenozoic, a northward component of movement as South America approached North America, resulting in the appearance of the Greater Antilles (the larger islands of Cuba, Jamaica, Hispaniola and Puerto Rico) along the northern margin of the Caribbean plate (Figure 11.13b). Although Hispaniola appears to have been formed by the fusion of two smaller islands, the remaining Caribbean islands remained separate units that individually appeared or disappeared. This makes it likely that the organisms found there must have arrived by overwater dispersal followed by independent evolution, rather than

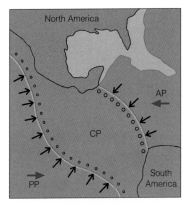

(a) Late Cretaceous 150 mya

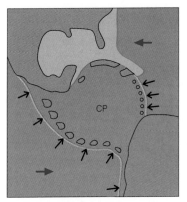

(b) Eocene 50 mya

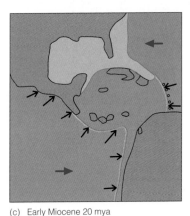

(c) Early Miocene 20 mya

Figure 11.13 The formation of the West Indies. Dark grey tint indicates dry land; light blue tint indicates shallow water; dark blue tint indicates deep water. Filled arrows indicate plate motion: AP, Atlantic Plate; CP, Caribbean Plate; PP, Pacific Plate. Smaller arrows indicate where ocean crust is being consumed, leading to the appearance of volcanic islands. Adapted from Huggett [58].

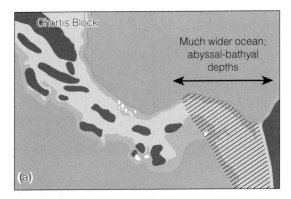

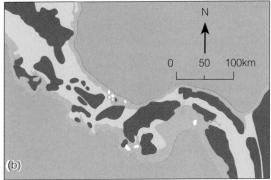

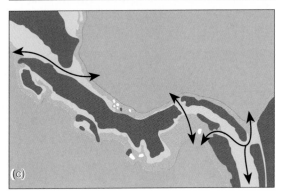

Figure 11.14 The formation of the Panama Isthmus. Dark grey tint indicates dry land; shallow marine sediments are indicated by light grey; deep ocean sediments are indicated by oblique parallel lines. (a) Middle Miocene, 16 to 15 mya; (b) Late Miocene, 7 to 6 mya; (c) Late Pliocene, approximately 3 mya. From Coates and Abando [59].

by vicariance. As Hedges points out, the nature of the faunas of the Caribbean islands strongly supports this interpretation. The ecological diversity of the islands is largely based on the radiation of a comparatively small number of higher taxa, some of which contain a very large number of species. This suggests that a small number of immigrants, finding themselves in environments lacking their normal competitors, were able to diversify opportunistically into the vacant ecological niches. Today, we recognize over 1300 native species of freshwater or terrestrial vertebrate in the islands, of which 75% are endemic (Table 11.3).

The results of these radiations are impressive. The eleutherodactylid family of frogs has radiated into 161 species, all of which are endemic to the islands. The anguid, iguanid, gekkonid and teiid families of lizard have produced 338 species, all but three of which are endemic. The boid, dipsadid, leptotyphlopid, tropidophiid and typhlopid families of snake have given rise to 129 species, of which all but two are endemic. However, it must be remembered that these species are distributed over the 29 islands of the Lesser Antilles, plus the Bahamas and the four larger islands of the Greater Antilles, and that most species are restricted to one island, and often to a small area within that island. There are also interesting examples of parallel evolution, most clearly seen in the lizards of the genus *Anolis* that live on trees on the Greater Antilles. Ecologists have distinguished at least six

morphological types of this lizard, differing in characteristics of body size, limb proportions and so on, each of which occupies a different part of the tree (crown; twigs; upper, middle or lower trunk; on the ground or on bushes). Cuba and Hispaniola contain all six types, Puerto Rico contains five and Jamaica four, but each type is represented by one or more independently evolved species on each island, so that there is a total of 128 species of *Anolis* on the four islands.

The West Indies islands contain 672 species of amphibian and reptile, which belong to about 75 different evolutionary lineages. Of these, 49 are related to South American taxa, and only four to North American taxa (the origins of the remaining 22 lineages are Africa, Central America or uncertain). This pattern is probably the result of dispersal within a Caribbean area in which the patterns of both sea currents and wind currents (sometimes of hurricane strength) run from east to west. It is also significant that the great rivers of northern South America, such as the Orinoco, open to the east of the Caribbean; others opened to the south of the area earlier in the Cenozoic, and all may have acted as sources for flotsam that may have borne amphibians and reptiles. In contrast, most of the 425 lineages of birds of the Caribbean region are related to those of North America, probably because the islands are closer to that continent, and also were used as overwintering refuges. Finally, the Caribbean islands were also colonized

Table 11.3 The taxonomic diversity of native West Indian land vertebrates. From Hedges [37].

Group	Orders	Families*	Genera			Species		
			Total	Endemic	% Endemic	Total	Endemic	% Endemic
Freshwater fishes	6	9	14	6	43	74	71	96
Amphibians	1	4	6	1	17	173	171	99
Reptiles	3	19	50	9	18	499	478	96
Birds	15	49	204	38	19	425	150	35
Mammals:								
Bats	1	7	32	8	25	58	29	50
Other†	4	9	36	33	92	90	90	100
Totals	30	97	342	95	28	1319	989	75

* Includes one endemic family of birds (Todidae) and four of mammals (Capromyidae, Heptaxodontidae, Nesophontidae and Solenodontidae).

† Edentates, insectivores, primates and rodents.

by some South American mammals, the cavio-morph rodents (four lineages), ground sloths (two lineages) and New World monkeys (one lineage), which evolved into endemic genera and species.

Hedges points out that the geological and ecological picture suggesting that these land vertebrates arrived in the West Indian islands by dispersal, rather than by vicariance, is supported by the times of divergence of these lineages. Molecular methods studying the amino-acid sequence divergence of the protein serum albumin (see Chapter 6), which has a remarkably constant rate of evolution, and also DNA-sequence 'clocks,' suggest that these lineages diverged at a variety of times during the mid-Tertiary (Figure 11.15). The geological evidence on the palaeogeography of the Caribbean indicates that its pattern of large and small islands during that time was similar to that of today, though not identical. This evidence therefore does not support suggestions that there had been immigration to the region over a land bridge that later broke up into independent islands. Such a pattern would instead have resulted in a clustering of divergence times around a few dates corresponding to the times of break-up.

In addition to these biogeographical relationships between the West Indies and other parts of the New World, two reptiles (the amphisbaenian *Cadea* and the gecko *Tarentola*) appear to have arrived in the West Indies from the Mediterranean region about 40 mya.

The West Indies islands thus provide a fascinating picture in which geological, faunal and molecular studies seem to be converging on a convincing story of dispersal into and through a region of complex history. It is also an interesting contrast to the East Indies or Wallacea, where the history of the neighbouring continents has been quite different.

South America

The history of the biota of South America has been dominated by the effects of the great earth-engine of plate tectonics. Some of these effects were caused by the westward drift of the continent, which led to mountain building and consequent climatic changes, but most were the result of changes in its relationship with North America. As noted in the description of the origin of the Caribbean islands in the 'The West Indies' section, there had been islands in that region ever since the Americas started to move westward in the Early Cretaceous. This may have been the route by which some North American dinosaurs dispersed southward (see Chapter 10) in the Mid-Cretaceous, accompanied by some early types of marsupial and placental mammal. But this connection has always been tenuous, sometimes permitting passage and sometimes becoming broken. This led to cycles of immigration, isolation and evolution, and final new immigration and extinction.

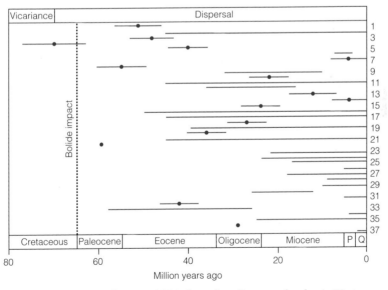

Figure 11.15 Times of origin of 37 independent lineages of endemic West Indian amphibians and reptiles. From Hedges [37].

The Early Cenozoic

The mammal fauna of South America today is characterized by a few marsupials (opossums) and a diversity of edentate placental mammals known as the Xenarthra (sloths, armadillos and South American anteaters) that are hardly known elsewhere. Otherwise, it does not seem particularly unlike other mammal faunas, for it includes members of most of the placental orders (see Table 11.1). However, in the Late Cretaceous, South America had a mammal fauna of strange marsupials and of ungulate placentals quite unlike anything known elsewhere in the world. By the early Cenozoic, the South American placentals had diversified into 22 families of unusual herbivorous ungulate, the edentates had diversified into the three groups mentioned above and the marsupials had evolved into the opossums and a group of carnivores called the borhyaenids. As we shall see, it was only when, much later, South America became joined to its northern neighbour that a great faunal change led to the extinction of most of this early endemic mammal fauna.

The northern part of the continent in which this fauna lived was, as we saw in Chapter 10, largely covered by megathermal rainforests of the Equatorial Belt or Palmae Province (cf. Figure 10.10). The Terminal Eocene Cooling Event does not seem to have led to any major change to drier environments here, as happened in the Far East, so that the present Amazon rainforest is the direct descendant of this early megathermal forest. In the Paleocene, there was a narrow band of similar forest in the Southern Megathermal Province. This expanded southward during the Eocene thermal maximum, but disappeared during the Terminal Eocene Cooling Event. It was replaced by microthermal forest containing the southern beech tree *Nothofagus* and other characteristic Southern Hemisphere elements, such as members of the Proteaceae, Restionaceae and Gunneraceae, and conifers such as *Araucaria*, *Podocarpus* and *Dacrydium* (all of which survived through the Cenozoic both in southern South America and in Australia).

Two other endemic South American mammal groups are the New World monkeys and the caviomorph rodents (which include the guinea pig and the capybara), whose closest relatives live in Africa. The earliest fossils of these two groups are from the Early Oligocene, 37 mya, a time when South America had already drifted a considerable distance away from Africa. However, recent molecular work [38] on the date of divergence of these groups from their African relatives suggests that they arrived much earlier, about 50 mya, when the ocean gap between the two continents was significantly smaller. As today, the ocean currents in the equatorial South Atlantic would have been westward, aiding such a transoceanic rafting.

Ever since it started to separate from Africa in the Middle Cretaceous, South America had been moving westward toward an oceanic trench that lay in the eastern part of the South Pacific. Along this trench, old eastern Pacific seafloor was being drawn back into the earth (see Figure 5.2). When South America eventually reached the trench, this tectonic movement caused volcanic and earthquake activity that led to the appearance of the Andean mountain chain. Today, nearly 8000 km long, this is the longest continental mountain chain in the world and rises to heights of up to nearly 7000 m. Mountain formation began in the south about 40 mya, the southern Andes reaching a height of less than 1000 m in the Early Cenozoic, and gradually spread northward, the most northeastern Andes being the last to appear, about 11 mya. The rise of these mountains caused a reversal in the direction of flow of northwestern rivers such as the Amazon and the Orinoco, and led to vicariant evolution in the biota of that region.

The rain shadows caused by the Miocene uplift of the Andes led to the replacement of much of the former forests by woodlands and grasslands, and the associated radiation of the endemic South American placental herbivores. This is also the time at which cool-temperate plants first appear in northern South America, having dispersed from the north. Grasslands as a major element in the flora first appeared in South America about 30 mya (earlier than in North America, where this did not happen until 25–20 mya). The Early Oligocene fauna of South America is the oldest in the world that is dominated by grazing herbivores.

The Late Cenozoic–Pleistocene

In addition to their westward movement, North and South America also moved roughly 300 km closer together from the Middle Eocene to the Middle

Miocene. From the Middle Miocene onward, the water level along the westward margin of the Caribbean plate became steadily shallower. Volcanic islands formed and enlarged to create an increasingly complete link between the two continents (Figure 11.14). The first biogeographical evidence of this link came in the Late Miocene and Early Pliocene, when two families from each continent crossed this island chain to reach the other continent. The Panama Isthmus became a complete land bridge for the first time about 3 mya, in the Middle Pliocene.

The Pliocene witnessed another geological event that had a major impact on the biota of South America. This was the great Pliocene uplift of the Andes, which doubled their height from 2000 to 4000 m. As a result, the extensive intermingling of the faunas of North and South America that took place after the final completion of the Panama land bridge also took place at a time of profound ecological change in the South American continent. The fascinating interplay of geology and biogeography has been analysed in rewarding detail, particularly by the American palaeontologists Larry Marshall and David Webb [39–41], who have called it the Great American Interchange.

Before the Interchange, each continent had 26 families of land mammal, and about 16 families from each dispersed to the other continent. Of the North American mammals, 29 genera dispersed southward, mainly in the Late Pliocene–Early Pleistocene, about 2.5 mya, plus a few about 1 million years later. At the same time, nine genera of South American mammal dispersed northward; about 1.5 mya, they were followed northward by opossums, porcupines and armadillos (Figure 11.16).

The fauna that was exchanged between the two continents was one that was adapted to a savanna, open-country environment, suggesting that this was the environment of the connecting Panama Isthmus when the exchange took place. That inference is supported by the fact that several types of birds and xerophilous shrubs that are typical of savannas are now found both north and south of the Isthmus.

Although similar proportions of the two faunas (9–10% of the genera) emigrated, the North American emigrants were far more successful than those from South America (Figure 11.17). In South America, 85 (50%) of the living land mammal genera are descended from South American immigrants,

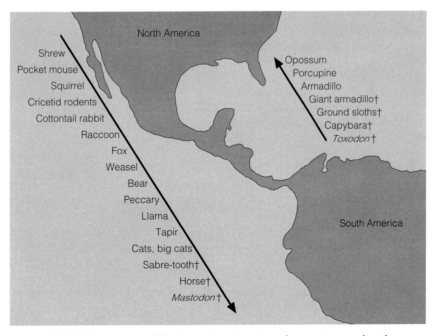

Figure 11.16 The migrants of the Great American Interchange. A cross after the name indicates that it later became extinct in its new continent.

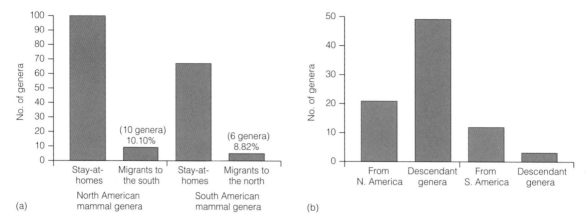

Figure 11.17 (a) The numbers of Pliocene genera of North and South American mammals that stayed at home versus the numbers that immigrated to the other continent. (b) Relative success of the immigrants during the Pleistocene. The blue bars show the number of genera that had arrived from the other continent; the grey bars show the number of genera that resulted from that immigration. Adapted from Marshall *et al.* [60].

while the corresponding figure for North America is only 29 (21%). Easily the most successful immigrants were the cricetid rodents, which diversified into 45 genera in South America. But several other families, such as canids, horses, camelids (which evolved into llamas) and peccaries, contributed to the success of North American families there. The most notable extinction in South America was that of all 13 genera of endemic ungulate, which were unable to cope with the new North American carnivores, or with competition from the immigrant North American ungulates. It has been suggested that the North American mammals in general may have had greater success because they were the survivors of many millions of years of competition between mammal groups within the whole of the Northern Hemisphere, while the South American mammals had been protected in their isolated continent. It is also true that the North American ungulates had proportionately larger brains than their South American counterparts. However, statistical analysis suggests that size was an important factor: the types that became extinct were the larger ones, and, because the South American mammals were larger than the North American ones, more of them became extinct [42]. The South African palaeontologist Elizabeth Vrba of South Africa, working at Yale University, has offered a more sophisticated suggestion, based on the ecological history of the region [43]. She points out that the mammals of southern North America contained many more

lineages adapted to the open, arid savanna environment that was spreading in the region, as the result of a general cooling of the climate. This alone gave them an advantage over the South American mammals, which were mainly adapted to a closed forest environment – which, as a result of this climatic change, was becoming fragmented and changing to savanna. So the South American mammals were suffering from a dual, climate-induced change in their environment that worked in favour of their North American competitors.

A final phase in the transformation of the South American land mammal fauna came during the Pleistocene. The climatic change that in the Northern Hemisphere caused the Ice Ages and the associated biotic changes (see Chapter 12) also caused a number of extinctions in South America. These included the last of the giant ground sloths, the giant armadillo-like glyptodonts and the herbivorous South American ungulate *Toxodon* that had made a fleeting appearance in southern North America. In addition, the mastodont elephants and the horses, which had emigrated from North America, died out both there and in South America – although both survived in the Old World.

The tapirs and llamas that had originally dispersed from North America became extinct there but survived in South America. As a result, these two groups now show a disjunct, relict distribution in which their surviving relatives (tapirs and camels) are found in Asia. Perhaps it is not surprising that few of the South American forms that had dispersed to North America when it was warm were able to survive there in the colder environment: only the opossum, armadillo and porcupine survived.

The final result of the ecological and biogeographical changes is thus a South American mammal fauna that shows little trace of its original inheritance from the Early Cenozoic. The characteristic flowering plant families today include the xerophytic Cactaceae, the Bromeliaceae (the latter include the pineapple and are often xerophytic), the Tropaeolaceae (including the garden nasturtium) and the Caricaceae (pawpaw or papaya tree).

The Rainforests of South America

The adjoining basins of the Amazon and Orinoco rivers (which together are about the size of the continental United States) contain approximately half of today's rainforests, the highest number of genera of palms, the richest of the world's vertebrate biotas and more species of fish than in the whole of the North Atlantic Ocean. (South America was not as badly affected by the Terminal Eocene cooling as were the other tropical regions, and it has been suggested that this may have been one of the reasons why it was able to retain a richer flora.) Birds and bats are particularly diverse in such forest regions, because their ability to fly allows them to take full advantage of the three-dimensional nature of that environment, with its many ecological niches.

The causes of the biogeographical patterns to be found in Amazonia have been much debated. It is now clear that, as mentioned in this chapter, some of these are the result of the Andean orogeny, while others were caused by the climatic changes of the Pleistocene. Research on quite different topics shows that the climatic history of the Amazon Basin [44] may have been dominated by rhythms in the strength of the low-level atmospheric jet stream, rather than showing the same pattern and timescale as the Ice Ages of the Northern Hemisphere – though the latter may have affected the climates of the Andes themselves.

Many of the more recent studies show how necessary it is to use modern methods of analysis and, as always in biogeographical work, to ensure that the taxonomy of the group being investigated is correct. This is most clearly seen in recent work on the parrot genus *Pionopsitta* [45], where the introduction of cladistic and molecular studies showed the errors in earlier classifications, and therefore also of the biogeographical patterns that they suggested. The revised taxonomy reveals the presence of two separate clades, east and west of the northern Andes, whose Pliocene dates of divergence suggest that they resulted from vicariance caused by that geological event. Other work, on curassow birds and howler monkeys, shows a similar pattern of relatedness to the Andean orogenies. The comparative phylogeography of 11 monophyletic groups of small mammals living in the forests of Amazonia, in the separate forest of the Atlantic coast and in the intervening area was investigated by Leonora Costa [46]. She found that the times of diversification of 11 different lineages of rodent and opossum varied greatly, but many of them were far older than the Pleistocene.

As noted in Chapter 10, the Amazon rainforest seems to have appeared in the Miocene, at the time of the Climatic Optimum. There has been much debate on the reasons for the immense biotic diversity of the biome, beginning when the American zoologist Jurgen Haffer [47] suggested that the biogeography of many Amazon forest birds showed two features. First, there are a few (about six) areas that each contain clusters of endemic species with rather restricted and similar ranges; these centres of endemism together contain about 150 species of birds, which make up 25% of the forest bird fauna of the Amazon Basin. Second, Haffer also found evidence, between these centres, of zones where the related species from different centres of endemism hybridized. Haffer hypothesized that, during the Pleistocene glacial periods of the Northern Hemisphere, when the climate became drier and cooler, the Amazon rainforest became restricted to separate areas, or 'refugia', surrounded by areas of savanna (Figure 11.18). Many bird species had been able to survive in these refugia, now separate from their nearest relatives in the other areas of surviving rainforest. That situation had permitted their independent evolution towards becoming separate species. However, when the forests expanded again during the wet interglacial periods and the grasslands shrank in their turn, the bird populations had not yet become fully separate species. As a result, when the related forms met again, they were still

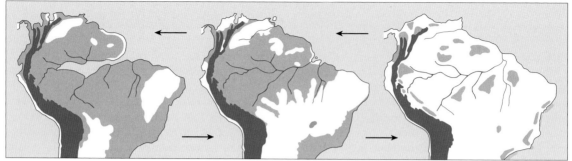

Figure 11.18 Maps of northern South America showing the suggested extent of the lowland rainforests (light blue) during the different phases of the glacial cycle. Dark blue areas represent land above 1000 m, and white represents grassland. The coastlines have been adjusted to reflect the changes in sea levels, but the pattern of the Amazon Basin drainage is left unaltered in all three maps, to facilitate comparison. Adapted from Lynch [61].

able to hybridize. Haffer referred to this phenomenon as a 'species pump'.

Haffer's views have been challenged by other workers, such as the British ecologist Paul Colinvaux [48], who points out that pollen data from the time of the last glacial maximum show little difference from the floral composition of the biome today. He interprets this as proving that Haffer's refugia never existed. However, the pollen samples come from only limited areas, and in any case are restricted to the last 62 000 years, and so provide a poor basis for such an extrapolation. In contrast, a recent paper by three biologists working in Tennessee [49] has used molecular methods based on mtDNA sequence comparisons to identify the times at which the members of 131 sister species of 35 different genera of Amazonian butterflies diverged from one another. They found that 72% of these speciation events took place within the last 2.6 million years, which is consistent with Haffer's hypothesis. Similar results have been obtained by other workers on rainforest trees, toucans and monkeys. Recently, results from trees in the similar African rainforests [50] have also provided support for Haffer's suggestion.

Long ago, Alfred Wallace, one of the founders of zoogeography (see Chapter 1), noted that the range boundaries of Amazonian birds seemed to coincide with the pattern of the rivers. It is fascinating that recent research [51] shows that closely related taxa of some birds are to be found on either side of rivers throughout Amazonia. Molecular investigation of the timing of their diversification demonstrates

that this was closely linked to vicariance events caused by episodes of Andean orogeny that led to changes in the river patterns due to drainage capture, supporting Wallace's suggestion.

All in all, this varied work highlights the need for such research to take full account of work in other areas, to be careful of generalizing beyond the scope of the data discovered and to be wary of the extent to which a previously unsuspected variety of factors may interact in producing biogeographical patterns. Though this may be especially true of the Amazon region, it would be sensible to assume that this lesson also applies to other areas and continents.

The Northern Hemisphere: Holarctic Mammals and Boreal Plants

In contrast to the complexity of the geological and geographical history of the Southern Hemisphere, with large areas of land splitting far apart from one another, moving through significant bands of latitude and climate and colliding with one another, that of the Northern Hemisphere has been comparatively uniform. Although shallow epicontinental seas and the developing North Atlantic have from time to time subdivided the land areas of North America and Eurasia into different patterns (Figure 11.19), the two continents have never been far apart, so that dispersal between them has usually

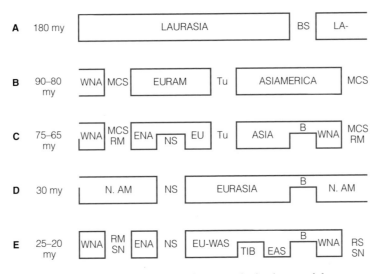

Figure 11.19 The interrelationships between the land areas of the Northern Hemisphere over the last 180 million years. (a) Early/Mid-Jurassic, 180 mya (cf. Plate 7d). A single landmass, Laurasia, extends around nearly the whole of the Northern Hemisphere; the Bering Sea (BS) between Asia and Alaska is the only break in the near-circle of land. (b) Early/Late Cretaceous, 90–80 mya (cf. Plate 7e). The Mid-Continental Seaway (MCS) and the Turgai Sea (Tu) have divided this into two landmasses, Euramerica (EURAM) and Asiamerica. (c) End of the Late Cretaceous, 75–65 mya (cf. Plate 7f). The Mid-Continental Seaway (MCS) or the Rocky Mountains (RM) divide North America into two parts. A western part (WNA) is linked to Asia by the Bering region (B), where the former Bering Sea has dried up to create a land bridge. The eastern part (ENA) is linked to Europe. The Turgai Sea is still in place, but the expanding Norwegian Sea (NS) now separates the southern parts of Europe from North America–Greenland. (d) Early Oligocene, 30 mya (cf. Plate 7h). Now the continents are only divided by the completed Norwegian Sea, and their northern parts are linked by the Bering connection. (e) Late Miocene, 26–20 mya (cf. Plate 7i). The Rocky Mountains (RM) and Sierra Nevada (SN) subdivide North America, while the Tibetan Plateau (TIB) and adjacent Gobi Desert separate southern China and South-East Asia (EAS) from Europe and Western Asia (EU-WAS).

been fairly easy. Furthermore, all these areas lay to the east or west of one another, at almost identical latitudes, and their faunas and floras were therefore similar in their nature and adaptations. Consequently, the biogeographical results of plate tectonic movements in the Northern Hemisphere were far less complex and dramatic than those in the Southern Hemisphere [52]. In addition, their faunas and floras have in the recent past all suffered from the severe climatic effects of the Ice Ages. As a result, they are the two great regions of temperate-

and cold-adapted biota. Although the two continents are usually distinguished as being separate 'North American' and 'Eurasian' zoogeographical regions, they are sometimes considered as a single 'Holarctic' or 'Boreal' region.

The controls on the dispersal of animals and plants in these northern continents have therefore been changes in sea level or climate, and the effects of these in four different crucial areas (Figure 11.20) have recently been described by the Greek biogeographer Leonidas Brikiatis [53].

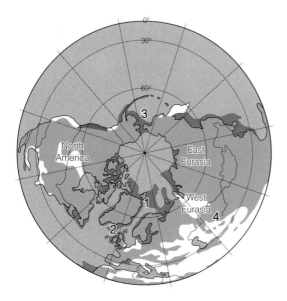

Figure 11.20 Simplified palaeogeographical reconstruction of the high northern latitudes during the Paleocene, to show the key dispersal locations in the Paleocene and earliest Eocene. (1) de Geer route; (2) Thulean route; (3) Beringia; (4) Turgai Sea. See text for details. From Brikiatis [52]. (Reproduced with permission of John Wiley & Sons.)

In Figure 11.20a, the de Geer route (1) connected northeastern Greenland to Scandinavia in the Latest Cretaceous–Early Paleocene (71–63 mya); the Thulean route (2) connected northern North America to Europe via a ridge of land that is now submerged; and the route of land across the Bering Sea between Siberia and Alaska (3) was dry during parts of the Paleocene. The Turgai Seaway (4) ran from north to south near the Ural Mountains. Much of Europe was, until about 30 mya, covered by a shallow epicontinental sea, so that the area was an archipelago of islands of varying size, rather like the East Indies today. Because of their position in fairly high northern latitudes, some of these routes acted as a filter whose intensity varied according to the climate.

In addition to these, the Mid-Continental Seaway and Rocky Mountains, with their rain shadow to the east, in turn formed a barrier between western and eastern North America from the Cretaceous until the Early Oligocene, 30 mya, by which time erosion had levelled the early Rocky Mountains to a plain, so that east–west dispersal within North America was once again possible (cf. Figure 11.19). However, biogeographical relations within the Northern Hemisphere also changed at this time because of the withdrawal of the seas from Eurasia. This allowed some mammal groups to enter Europe from Asia about 30 mya. From then on, the high-latitude Bering link was the only route between Siberia and Alaska. The disappearance of the Turgai Sea brought a drier climate to Central Asia. As a result, a cool-temperate deciduous flora evolved there and later replaced the mixed-mesophytic forests of Europe, while its megathermal elements found refuge in the rainforests of South-East Asia.

In the Late Oligocene, 25 mya, renewed mountain building in North America caused a new rain shadow with cooler and drier climates over the Great Plains, leading to the development of the grassland biome there. This probably caused a great deal of vicariance in the biota on either side of the mountains. The forests with their mesophytic elements became restricted to the western side of the mountains and to the eastern seaboard, and eventually disappeared during the Pleistocene glaciations.

The biogeographical importance of the climate of the Bering region can be seen in its influence on faunal exchange between North America and Eurasia. When the climate became cool, as in the Early Oligocene, few mammals crossed. When it improved again a little later in the Oligocene, a number of Asian mammals dispersed to North America. Some of these had evolved within Eurasia, while others had dispersed to that continent from Africa. The final climatic deterioration in the Bering region began in Miocene times and may have been related to the increase in Antarctic glaciation at that time. From then on, most of the mammals that dispersed were large forms and, even more significantly, types that are tolerant of cooler temperatures; such warmth-loving forms as apes and giraffes could not reach North America. This climatically based exclusion became progressively more restrictive, until in the Pleistocene only such hardy forms as the mammoth, bison, mountain sheep, mountain goat, musk ox and human beings were able to cross. The final break between Siberia and Alaska took place 13 000–14 000 years ago.

Despite the long history of intermittent connection between the North American and Eurasian regions, each has a few endemic groups, while other groups did reach both regions, but later became extinct in one of them. Pronghorn antelopes, pocket gophers, pocket mice and sewellels (the last three groups are all rodents) are unknown in the Eurasian region, whereas hedgehogs, wild pigs and murid rodents (typical mice and rats) are absent from the North American region. The domestic pig has been introduced to North America by human beings, as have mice and rats at various times. Horses crossed the Bering connection to Eurasia, but then became extinct in their North American homeland, probably because of competition from the immigrant bison. Horses were therefore unknown to the American Indians until they were introduced by the Spanish conquistadores in the 16th century.

Although the Ice Ages did not begin until the end of the Pliocene, the steadily cooling climates had already exerted a great influence on the floras of the northern continents. For example, there was a considerable change in the European flora during the Pliocene. The modern flora of the region contains over 60% of its Late Pliocene flora, but only 10% of that of the Early Pliocene. The intervening 3 million years had therefore seen a rapid change in the flora of Europe, as it adapted to climatic changes that became greatly exaggerated as the Pleistocene Ice Ages commenced. These Ice Ages stripped North America and Eurasia of virtually all tropical and subtropical animals and plants. This happened so recently that the faunas and floras have as yet had no time to develop any new, characteristic groups. Since they also have no old relict groups, such as the marsupials, it is the poverty and the hardiness of their faunas that distinguish them from those of other regions. Many groups of animals are absent altogether and, of the groups that are present, only the hardier

members have been able to survive. Even these become progressively fewer toward the colder, Arctic latitudes. In North America there is, in addition, a similar thinning out of the fauna in the higher, colder zones of the Rocky Mountains; this is a general feature of the fauna and flora of high mountains.

The Eurasian fauna was almost completely isolated from the warmer lands to the south by the Himalayas and by the deserts of North Africa and southern Asia, and has therefore received hardly any infiltrators to add variety. The situation was eventually rather different in the Western Hemisphere. During the Early Cenozoic there had been hardly any exchange of animals between North and South America, presumably because there was still a wide ocean gap between the two continents. After the Panama Isthmus was completed at the end of the Cenozoic (see Figure 11.14), many North American mammals dispersed to South America. However, North America was successfully colonized by only three types of South American mammal (opossums, armadillos and anteaters), along with a number of birds such as hummingbirds, mockingbirds and New World vultures.

For some reason, the situation was reversed for plants. Instead of surviving in the Panama Isthmus, few of the North American megathermal plants survived the Late Cretaceous climatic cooling. The bulk of the lowland vegetation of Central America is, instead, of South American origin, perhaps because that great tropical region had produced an enormous variety of tropical plants. The mesothermal plants of North America were more successful at colonizing South America, presumably using the cooler mountainous spine of Central America as their route from the Rockies to the Andes.

This chapter completes the review of the history of the appearance, evolution and development of the faunas and floras of the different continents until 2 million years ago, which were mainly gradual and long term. In contrast, their subsequent history was rapidly and fundamentally changed by the sudden climatic somersaults of the Ice Ages, which are described in the next chapter.

1 Today, the major elements in the distribution patterns of living mammals and flowering plants are the individual continents, each showing significant differences from the others. However, this is the result of comparatively recent tectonic and climatic events, such as the Ice Ages, which greatly reduced the variety of animals and plants in the northern continents. Barriers of sea, mountain and desert prevented them from returning northward when the climate later improved. As a result, the tropical and subtropical biotas of Africa, India, South-East Asia and South America are very different from the impoverished biotas of Eurasia and North America.
2 Australia separated early from the rest of Gondwana and was isolated until only about 10 mya, when New Guinea at its northern edge became close to South-East Asia. Australia has very few native placental mammals but a great variety of marsupials. Because of the steadily increasing aridity of Australia, its original 'Antarctic' flora, descended from the Cretaceous Southern Gondwanan cool-temperate flora, was progressively transformed into a sclerophyll flora, which was in turn replaced by grassland and savanna. The Antarctic flora therefore now has a relict distribution, from Patagonia to New Zealand and the mountains of New Guinea.
3 The West Indies provide a fascinating story of dispersal from neighbouring South and North America into and through a region of complex geological history.
4 South America too was isolated until only a few million years ago, when the Panama Isthmus was completed. Most of its older fauna of unusual herbivorous placental mammals and carnivorous marsupials then became extinct due to competition from immigrant North American mammals. The rich tropical flora of South America was more successful and colonized the lowlands of Central America.
5 The floras of North America and Eurasia each adapted to the climatic changes of the Late Cenozoic and to the Ice Ages of the Pleistocene, both by evolutionary change and by changes in their patterns of distribution.

Summary

Further Reading

Culver SJ, Rawson PS (eds.). *Biotic Response to Global Change: The Last 145 Million Years*. London: Natural History Museum, 2000.

Goldblatt P (ed.). *Biological Relationships between Africa and South America*. New Haven, CT: Yale University Press, 1993.

Hall R, Holloway JD (eds.). *Biogeography and Geological Evolution of SE Asia*. Leiden: Backhuys, 1998.

Pennington RT, Cronk QCB, Richardson JA (eds.). Plant phylogeny and the origin of major biomes; a discussion meeting. *Philosophical Transactions of the Royal Society B* 2004; 359: 1453–1656.

Willis KJ, McElwain JC. *The Evolution of Plants*. Oxford: Oxford University Press, 2014.

References

1. Cox CB. Underpinning global biogeographical schemes with quantitative data. *Journal of Biogeography* 2010; 37: 2027–2028.

2. Cox CB. The biogeographic regions reconsidered. *Journal of Biogeography* 2001; 28: 511–523.

3. MacFadden BJ. *Horses*. Cambridge: Cambridge University Press, 1992.

4. Davis ALV, Scholtz CH, Philips TK. Historical biogeography of scarabaeine dung beetles. *Journal of Biogeography* 2002; 29: 1217–1256.

5. Eisenberg JF. *The Mammalian Radiations. An Analysis of Trends in Evolution, Adaptation and Behavior*. Chicago: University of Chicago Press, 1981.

6. Cristoffer C, Peres CA. Elephants versus butterflies: the ecological role of large herbivores in the evolutionary history of two tropical worlds. *Journal of Biogeography* 2003; 30: 1357–1380.

7. Thomas N, Bruhl JJ, Ford A, Weston PH. Molecular dating of Winteraceae reveals a complex biogeographical history involving both ancient Gondwana vicariance and long-distance dispersal. *Journal of Biogeography* 2014; 41: 894–404.

8. Aitchison JC, Ali JR, Davis AM. When and where did India and Asia collide? *Journal of Geophysical Research* 2007; 112: B05423.

9. Springer MS *et al.* Endemic African mammals shake the phylogenetic tree. *Nature, London* 1997; 388: 61–64.

10. Gheerbrant E. On the early biogeographical history of the African placentals. *Historical Biology* 1990; 4: 107–116.

11. Kalb JE. Fossil elephantids, Awash paleolake basins, and the Afar triple junction, Ethiopia. *Palaeogeograpy, Palaeoclimatology Palaeoecology* 1995; 114: 357–368.

12. Retallak GJ. Middle Miocene fossil plants from Fort Ternan (Kenya) and evolution of African grasslands. *Paleobiology* 1992; 18: 383–400.

13. Cowling R, Rundel PW, Lamont BB, Arroyo MK, Arianoutsou M. Plant diversity in Mediterranean-climate regions. *Trends in Ecology & Evolution* 1996; 11: 362–366.

14. Goldman SM, Benstead JP (eds.). *The Natural History of Madagascar*. Chicago: University of Chicago Press, 2003: 1130–1134.

15. Renner SS. Multiple Miocene Melastomataceae dispersal between Madagascar, Africa and India. *Philosophical Transactions of the Royal Society B* 2004; 359: 1485–1494.

16. Poux C *et al.* Asynchronous colonization of Madagascar by the four endemic clades of primates, tenrecs, carnivores, and rodents as inferred from nuclear genes. *Systematic Biology* 2005; 54: 719–730.

17. Ali JR, Huber M. Mammalian biodiversity on Madagascar controlled by ocean currents. *Nature* 2012; 463: 653–655.

18. Raxworthy CJ, Forstner MRJ, Nussbaum RJ. Chameleon radiation by oceanic dispersal. *Nature, London* 2002; 415: 784–787.

19. Rocha S, Carretero MA, Vences M, Glaw F, Harris DJ. Deciphering patterns of transoceanic dispersal: the evolutionary origin and biogeography of coastal lizards (*Cryptoblepharus*) in the Western Indian Ocean region. *Journal of Biogeography* 2006; 33: 13–22.

20. Jansa SA, Goodman SM, Tucker PK. Molecular phylogeny and biogeography of the native rodents of Madagascar (Muridae: Nesomyinae); a test of the single-origin hypothesis. *Cladistics* 1999; 15: 253–270.

21. Vences M *et al.* Reconciling fossils and molecules: Cenozoic divergence of cichlid fishes and the biogeography of Madagascar. *Journal of Biogeography* 2001: 28; 1091–1099.

22. Ali JR, Aitchison JC. Gondwana to Asia: plate tectonics, paleogeography and the biological connectivity of the Indian sub-continent from the Middle Jurassic through the latest Eocene (166–35 Ma). *Earth-Science Reviews* 2008; 88: 145–166.

23. Renner SS. Biogeographic insights from a short-lived Palaeocene island in the Ninetyeast Ridge. *Journal of Biogeography* 2010; 37: 1177–1178

24. Morley RJ. *Origin and Evolution of Tropical Rainforests*. New York: Wiley, 1999.

25. Hill RS. Origins of the southeastern Australian vegetation. *Philosophical Transactions of the Royal Society of London* 2004; B359: 1537–1549.

26. Crisp M, Cook L, Steane D. Radiation of the Australian flora: what can comparisons of molecular phylogenies across multiple taxa tell us about the evolution of diversity in present-day communities? *Philosophical Transactions of the Royal Society of London* 2004; B359: 1551–1571.

27. Flannery T. The mystery of the Meganesian meat-eaters. *Australian Natural History* 1991; 23: 722–729.

28. Milewski AV, Diamond RE. Why are very large herbivores absent from Australia? A new theory of micronutrients. *Journal of Biogeography* 2000; 27: 957–978.

29. Tyndale-Biscoe CH. Ecology of small marsupials. In: Stoddart DM (ed.), *Ecology of Small Mammals*. London: Chapman & Hall, 1979: 342–379.

30. de Boer HJ, Steffen K, Cooper WE. Sunda to Sahul dispersals in *Trichosanthes* (Cucurbitaceae): a dated phylogeny reveals five independent dispersal events to Australasia. *Journal of Biogeography* 2015; 42: 519–531.

31. Heads M. Biogeographical affinities of the New Caledonia biota: a puzzle with 24 pieces. *Journal of Biogeography* 2010; 37: 1179–1201.

32. Swenson U, Nylinder S, Munzinger J. Sapotaceae biogeography supports New Caledonia being an old Darwinian island. *Journal of Biogeography* 2014; 34: 797–809.

33. Trewick SA, Paterson AM, Campbell HJ. *Hello New Zealand. Journal of Biogeography* 2007; 34: 1–6.

34. McDowell RM. Process and pattern in the biogeography of New Zealand – a global microcosm? *Journal of Biogeography* 2008; 35: 197–212.

35. Pole MS. The New Zealand flora – entirely long-distance dispersal? *Journal of Biogeography* 1994; 21: 625–655.

36. Carr LM *et al.* Analyses of the mitochondrial genome of *Leiopelma hochstetteri* argues against the full drowning of New Zealand. *Journal of Biogeography* 2015; 42: 1066–1076.

37. Hedges SB, Paleogeography of the Antilles and origin of West Indian terrestrial vertebrates. *Annals of the Missouri Botanical Garden* 2006; 93: 231–244.

38. Rowe DL, Dunn KA, Adkins RM, Honeycutt RL. Molecular clocks keep dispersal hypotheses afloat: evidence for trans-Atlantic rafting by rodents. *Journal of Biogeography* 2010; 37: 305–324.

39. Marshall LG. The Great American Interchange – an invasion-induced crisis for South American mammals. In: Nitecki MH (ed.), *Third Spring Systematic Symposium: Crises in Ecological and Evolutionary Time*. New York: Academic Press, 1981: 133–229.

40. Webb SD, Marshall LG. Historical biogeography of Recent South American land mammals. In MA Mares, HG Genoways (eds.), *Mammalian Biogeography in South America*. Spec Publ Series 6, Pymatuning Lab. of Ecology, University of Pittsburgh, 1982: 39–52.

41. Webb SD. Late Cenozoic mammal dispersals between the Americas. In: Stehli FG, Webb SD (eds.), *The Great American Biotic Interchange*. New York: Plenum, 1985: 357–386.

42. Lessa EP, Fariña RA. Reassessment of extinction patterns among the Late Pleistocene mammals of South America. *Palaeontology* 1996; 39: 651–659.

43. Vrba ES. Mammals as a key to evolutionary theory. *Journal of Mammalology* 1992; 73: 1–28.

44. Bush MB. Of orogeny, precipitation, precession and parrots. *Journal of Biogeography* 2005; 32: 1301–1302.

45. Ribas CC, Gaban-Lima R, Miyaki CY, Cracraft J. Historical biogeography and diversification within the Neotropical parrot genus *Pionopsitta* (Aves: Psittacidae). *Journal of Biogeography* 2005; 32: 1409–1427.

46. Costa LP. The historical bridge between the Amazon and the Atlantic forest of Brazil: a study of molecular phylogeography with small mammals. *Journal of Biogeography* 2003; 30: 71–86.

47. Haffer J. Speciation of Amazonian forest birds. *Science* 1969; 165: 131–137.

48. Colinvaux PA, Irion G, Räsänen ME, Bush MB, Nunes de Mello JAS. A paradigm to be discarded: geological and paleoecological data falsify the Haffer and Prance refuge hypothesis of Amazonian speciation. *Amazoniana* 2001; 16: 609–646.

49. Garzón-Orduña IJ, Benetti-Longhini JE, Brower AVZ. Timing the diversification of the Amazonian biota: butterfly divergences are consistent with Pleistocene refugia. *Journal of Biogeography* 2014; 41: 1631–1638.

50. Duminil J *et al.* Late Pleistocene molecular dating of past population fragmentation and demographic changes in African rain forest tree species supports the forest refuge hypothesis. *Journal of Biogeography* 2015; 42 (8): 1443–1454.

51. Fernandes AM, Wink M, Sardelli CH, Aleixo A. Multiple speciation across the Andes and throughout Amazonia: the case of the spot-backed antbird species complex (*Hylophylax naevius/Hylophylax naevioides*). *Journal of Biogeography* 2014; 41: 1094–1104.

52. Sanmartin I, Ronquist F. Southern Hemisphere biogeography inferred by event-based models: plant versus animal patterns. *Systematic Biology* 2004; 53: 216–243.

53. Brikiatis L. The De Geer, Thulean and Beringia routes: key concepts for understanding early Cenozoic biogeography. *Journal of Biogeography* 2014; 41: 1036–1054.

54. Steininger FF, Rabeder G, Rogl F. Land mammal distribution in the Mediterranean Neogene: a consequence of geokinematic and climatic events.In:Stanley DJ, Wezel FC (eds.), *Geological Evolution of the Mediterranean Basin*. New York: Springer, 1985: 559–571.

55. Simpson GG. Too many lines; the limits of the Oriental and Australian zoogeographic regions. *Proceedings of the American Philosophical Society* 1977; 121: 107–120.

56. Moss SJ, Wilson MEJ. Biogeographic implications of the Tertiary palaeogeographic evolution of Sulawesi and Borneo. In: Hall R, Holloway JD (eds.), *Biogeography and Geological Evolution of SE Asia*. Leiden: Backhuys, 1998: 133–163.

57. Crisp M, Cook L, Steane D. Radiation of the Australian flora: what can comparisons of molecular phylogenies across multiple taxa tell us about the evolution of diversity in present-day communities? *Philosophical Transactions of the Royal Society of London* 2004; B359: 1551–1571.

58. Huggett RJ. *Fundamentals of Biogeography*. London: Routledge, 1998.

59. Coates AG, Obando J. The geologic evolution of the Central American Isthmus. In: Jackson JBC, Budd AF, Coates AG (eds.), *Evolution and Environment in Tropical America*. Chicago: University of Chicago Press, 1996: 21–56.

60. Marshall LG, Webb SD, Sepkoski JJ, Raup DM. Mammalian evolution and the Great American Interchange. *Science* 1982; 215: 1351–1357.

61. Lynch JD. Refugia. In: Myers AA, Giller PS (eds.), *Analytical Biogeography*. London: Chapman & Hall, 1988: 311–342.

Ice and Change

The preceding chapters have demonstrated that an understanding of the biogeography of the modern world requires a knowledge of past events. Most of the changes considered so far relate to the distant past and the processes of changing continental arrangements and the evolution of the major groups of living organisms. But the current distribution of plants and animals has been greatly affected by relatively recent events in the Earth's history, especially those of the last 2 million years when extensive ice periodically covered many parts of the Earth's surface. The world's major ice caps were already in existence around 42 million years ago, having formed in Antarctica during the Eocene [11]. Global temperatures dropped very rapidly at the end of the Eocene [22], and this appears to correspond with the first formation of an ice cap over Greenland in the Northern Hemisphere [33].

Many landform features in the temperate areas of the world show that major, geologically rapid changes in climate have taken place since the Pliocene. The general cooling of world climate that started early in the Tertiary continued into the Quaternary; the boundary between the two is placed at about 2 million years ago, but difficulties in definition as well as in dating techniques and geological correlation leave this date open to some doubt. The definition of the boundary comes from Italian marine sediments, where the appearance of fossils of cold-water organisms (certain foraminifera and molluscs) suggests a fairly sudden cooling of the climate, which has been dated at 1.8 million years. Similar evidence of cooling has been found in sediments from the Netherlands, and this is believed to mark the end of the final stage of the Pliocene (locally termed the Reuverian) and the first stage of the Pleistocene (the Pretiglian) [4]. Evidence from sediment cores in the North Atlantic indicates that debris was being carried into deep water by ice rafts as long ago as 2.4 million years, and many geologists feel that this would be a more appropriate date for the Pliocene–Pleistocene boundary. This is now the most widely accepted date for the boundary, but evidence from Norway suggests that Scandinavian glaciers extended down to sea level as long as 5.5 million years ago [5]; the point in time when we mark the opening of the Quaternary is therefore inevitably disputed.

At various stages during the Pleistocene, ice covered Canada and parts of the United States, northern Europe and Asia. In addition, independent centres of **glaciation** were formed in lower latitude mountains, such as the Alps, Himalayas, Andes, East African mountains and in New Zealand. At the height of its development, about 80% of glacial ice lay in the Northern Hemisphere. A number of present-day geological features show the effects of such glaciations. One of the most conspicuous of these is the glacial drift deposit, boulder clay or **till** covering large areas and sometimes extending to great depths. This is usually a clay-rich material containing quantities of rounded and scarred boulders and pebbles, and geologists consider it to be the detritus deposited during the melting and retreat of a glacier. The most important feature of this till, and the one by which it may be distinguished from other geological deposits, is that its constituents are

Biogeography: An Ecological and Evolutionary Approach, Ninth Edition. Edited by C. Barry Cox, Peter D. Moore, Richard J. Ladle.
© 2016 John Wiley & Sons, Ltd. Published 2016 by John Wiley & Sons, Ltd.

completely mixed; the finest clay and small pebbles are found together with large boulders. Often, the rocks found in such deposits originated many hundreds of miles away and were carried there by the slow-moving glaciers. Fossils are rare, but occasional sandy pockets have been found that contain mollusc shells of an Arctic type. Some enclosed bands of peat or freshwater sediments within these tills provide evidence of the warmer intervals. They often show that there were phases of locally increased plant productivity, and they may contain fossils indicative of warmer climates.

Many of the valleys of hilly, glaciated areas have distinctive, smoothly rounded profiles because they were scoured into that shape by the abrasive pressure of the moving ice. In places, the ice movement has left deep scratches on the rocks over which it has passed, and tributary valleys may end abruptly, high up a main valley side, because the ice has removed the lower ends of the tributary valleys. Such landscape features provide the geomorphologist with evidence of past glaciation. Understanding the significance of the shapes of hills and valleys depended on a geological principle put forward by the Scottish geologist James Hutton (1726–1797), who expounded the idea of **uniformitarianism**. In essence, this states that present-day conditions can be used as a key for understanding past processes and that there was no need to interpret geology in the light of supposed global catastrophes, such as the biblical Flood of Noah. Using uniformitarianism as his basic premise, the Swiss geologist Louis Agassiz (1807–1873) saw how glaciers could affect landforms and proposed that areas now clear of ice had once been covered by great depths of ice in an ice age. The validity of this proposal soon became evident following further research in Europe and North America.

Immediately outside the areas of past glaciation were regions of treeless tundra that experienced **periglacial** conditions (Figure 12.1). The present-day equivalents of these areas are very cold, and their soils are constantly disturbed by the action of frost. When water freezes in the soil, it expands, raising the surface of the ground into a series of domes and ridges. Stones within the soil lose heat rapidly when the temperature falls, and the water freezing around them has the effect of forcing them to the surface, where they often become arranged in **stone stripes** and **polygons**. Similar patterns are produced

Figure 12.1 Contorted birch tree on the tundra in northern Lapland, marking the northern limit of tree growth. Lying beyond the permanent ice, these regions experience periglacial conditions.

by **ice wedges** that form in ground subjected to very low temperatures. Some of these patterns, which are so evident in present-day areas of periglacial climate, can be found in parts of the world that are now much warmer. For example, they have been discovered in eastern parts of the United States and in Western Europe as a result of air photographic survey. Such 'fossil' periglacial features show that, as the glaciers expanded, so the periglacial zones were pushed before them towards the equator.

Climatic Wiggles

The Pleistocene Epoch has not been one long cold spell. Careful examination of tills and the orientation of stones embedded within them soon showed that several advances of ice had taken place during the Pleistocene, often moving in different directions. Occasional layers of organic material were sometimes discovered trapped between tills and other deposits, and these have provided fossil evidence of warm periods alternating with the cold. Where terrestrial sequences of deposits are reasonably complete and undisturbed, as in the eastern part of England (East Anglia) and parts of the Netherlands, it has been possible to construct schemes to describe these alternations of warm and cold episodes, to name them and to determine their relationships in time. But in many parts of the world, this has not proved at all easy, and the correlation of events, as reflected in terrestrial deposits, between different areas has often been speculative and unsatisfactory, mainly because of the difficulty experienced in obtaining secure dates for the deposits. At one time, for example, geologists considered that there were four episodes of ice advance in Europe, defined mainly by sequences of tills in the Alps. The four glaciations were named Günz, Mindel, Riss and Würm. This is now regarded as a gross oversimplification of the true situation, as there have been far more climatic fluctuations in the Pleistocene than this simple model suggests.

Because of the difficulties experienced in climatic reconstruction using land-based evidence, attention has turned to the seas, where marine sediments provide a more complete and uninterrupted sequence. The retrieval of long, deep ocean cores of sediment has provided an opportunity to follow the rise and fall of various members of plankton communities in the past, particularly those, like the **foraminifera**, which, though tiny, have robust outer cases that survive the long process of sedimentation to the ocean floor and there accumulate as fossil assemblages. Some members of the foraminifera, like some species of *Globigerina* and *Globorotalia*, are sensitive to ocean temperature, so their relative abundance in the fossil record provides evidence of past climates.

An even more powerful tool for reconstructing long-term climatic changes has been the use of **oxygen isotopes** retained in the fossil material of the sediments. 'Normal' oxygen (^{16}O) is far more abundant than the heavier form of oxygen (^{18}O).

For example, the heavy form comprises only about 0.2% of the oxygen incorporated into the structure of water (H_2O). Water evaporates from the sea, but those molecules containing ^{18}O condense from a vapour form rather more readily than their lighter counterparts, so this heavy form tends to return rapidly to the oceans. Water containing ^{16}O, on the other hand, remains in the atmosphere as vapour for longer and is more likely to fall eventually over the ice caps and to become incorporated into these as ice. Under cold conditions, the volume of global ice increases, and this (since it is formed largely from precipitation) tends to lock up more of the ^{16}O, leaving the oceans richer in ^{18}O. So the ratio of ^{18}O–^{16}O left in the oceans increases during periods of cold. This ratio is then reflected in the skeletons of foraminifera and other planktonic organisms and is deposited in the ocean beds. Analysis of the oxygen isotope ratios in ocean sediments thus provides a long and continuous record of changing water temperatures going back millions of years [6]. It has even been possible to use these methods for the analysis of oxygen isotope ratios in inland areas, as in the gradual deposition of calcite in the Devil's Hole fault in Nevada [7].

As a result of such oxygen isotope studies of a series of cores in the Caribbean and Atlantic oceans, Cesare Emiliani of the University of Miami, Florida, was able to construct **palaeotemperature** curves for the ocean surface waters [8]. A summary curve for the past 3 million years is shown in Figure 12.2, and it is quite obvious from

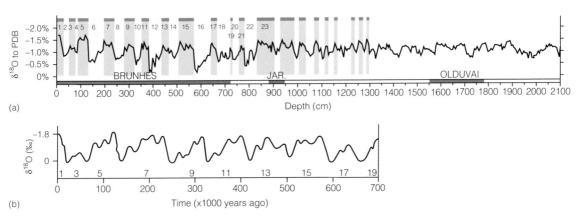

Figure 12.2 Oxygen isotope curve covering the past 3 million years. The timescale reads from right to left. Peaks represent periods of relative warmth, and troughs correspond to cold episodes. From Emiliani [8].

this diagram that the climatic changes in this latter part of the Pleistocene have been numerous and complex [9]. Translating oxygen isotope ratios into mean annual temperature fluctuations requires a number of assumptions about global temperature and ice volumes, but it is generally agreed that the difference in mean annual temperature between a glacial maximum (the troughs in this diagram) and a period of interglacial warmth (the peaks) is about 8–10°C.

Examination of the changing isotope ratios in Figure 12.2, which have been assembled from numerous studies of deep-sea sediments and ice cores, shows that the fluctuations in temperature become stronger as time proceeds (note that the time sequence runs from right to left). Gentle wanderings around a slowly falling mean value gradually become more pronounced as greater extremes of temperature are experienced. Sharper changes also become more apparent, with the temperature changing both radically and rapidly. One can also see that the course of change, although showing a general wave-like form, is not simple or regular, but has many minor wiggle patterns imposed on it. This latter point implies that we should not expect the terrestrial record in currently temperate lands such as North America or central Europe to show a simple alternation of temperate conditions with arctic ones, but a much more varied pattern in which cold and warmth alternate in varying degrees and intermediate conditions are often found.

Interglacials and Interstadials

The fluctuations in global temperature represented in the ocean cores are reflected in the terrestrial geological sequence by glacial and interglacial deposits. A warm episode (usually represented by an organic, peaty deposit) that is sandwiched between two glacial events (often represented by tills), and that achieved sufficient warmth for a long enough duration for temperate vegetation to establish itself, is termed an **interglacial**. The sequence of events demonstrated in the fossil material of such an interglacial shows a progressive change from high arctic conditions (virtually no life) through subarctic (tundra vegetation) to boreal (birch and pine forest) to temperate

(deciduous forest) and then back through boreal to arctic conditions once more. If the warm event is of only short duration, or if the temperatures attained are not sufficiently high, then the vegetation changes may only reach a boreal stage of development. In this case, it is termed an **interstadial**. We are currently living in the most recent interglacial (termed by geologists the **Holocene**).

Interglacials are often times of increased biological productivity (except in the more arid parts of the world), and they are often represented in temperate geological sequences as bands of organic material (Figure 12.3). This material usually contains the fossil remains of the plants, animals and

Figure 12.3 Organic sediments from an interglacial (dark band), resting on gravel of a former raised beach and covered by deposits laid down under periglacial conditions. A cliff exposure at West Angle, Pembrokeshire, Wales. From Stevenson and Moore [11].

microbes that existed at or near the site during its formation, and it is this evidence, often stratified in a time sequence, that allows us to reconstruct past conditions and habitats. One of the most valuable sources of fossil evidence for this purpose have been the **pollen grains** from plants that are preserved in the sediment and that reflect the vegetation of the area at that time. Pollen grains are sculptured in such a way that they are often recognizable with a considerable degree of precision. They are also produced in large numbers (especially those of wind-pollinated plants) and are widely dispersed. Finally, they are preserved very effectively in waterlogged sediments, such as peat and lake deposits. Therefore, the analysis of pollen grain assemblages, termed **palynology**, can provide much information about vegetation and hence about climate and other environmental factors [10].

Cores of sediment from lakes, peat bogs or even ice are extracted intact, and then samples are taken at different depths. Pollen and spores are preserved in these samples and can be concentrated by removing or dissolving the matrix material. Eventually, these microfossils are dense enough to be counted with the aid of a light microscope, and the proportions of different types can be determined.

Fossil pollen data are usually presented in the form of a pollen diagram, and Figure 12.4 shows a pollen diagram belonging to the last (so-called Ipswichian) interglacial in Britain. The vertical axis is the depth of deposit, which is directly related to the age of the samples, so the diagram should be read from the bottom up. The sequence begins with boreal trees, birch and pine, which are then replaced by deciduous trees such as elm, oak, alder, maple and hazel [11]. Later in the sequence these trees go into decline, to be replaced by hornbeam, and then pine and birch once more. The details of such a sequence will obviously vary from one locality to another. Figure 12.5 shows the pollen diagram from the same interglacial further east in Europe, in Poland, where it is called the Eemian interglacial. (It is conventional to name interglacials and other geological episodes with local names that can later be used in correlation between different areas.) In the Polish Eemian interglacial, lime, spruce and fir play a more important role than in equivalent western sites. The vegetation sequence thus varies considerably between regions and also between different interglacials, but all such sequences show a consistent pattern, for they all pass through a predictable series of developmental stages. Often these are shown on such

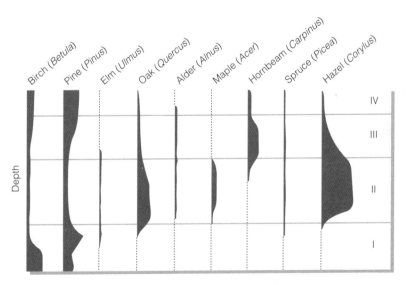

Figure 12.4 Pollen diagram from sediments of the last (Ipswichian) interglacial in eastern Britain. Only tree taxa are shown. The depth axis is related to age, with the oldest sediments at the base.

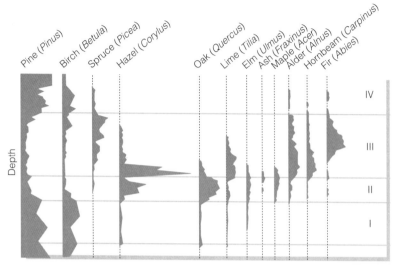

Figure 12.5 Pollen diagram from sediments of the last (Eemian; equivalent to the British Ipswichian) interglacial in Poland. Note the greater importance of lime, spruce and fir than is the case in the British Ipswichian.

diagrams by dividing them into four zones (usually labelled in Roman numerals, I–IV): pre-temperate, early temperate, late-temperate and post-temperate, respectively. Pollen diagrams from our present interglacial suggest that we are well advanced into the late-temperate stage.

The last interglacial is the easiest to identify stratigraphically because it occurred in the very recent past. Dating it precisely is difficult because the radiocarbon method becomes less accurate as one proceeds back in time, and dating methods relying on argon isotopes are dependent on the presence of volcanic material [12]. Nevertheless, it is believed that the last interglacial began about 130 000 years ago (although a well-dated site from Nevada suggests commencement at 147 000 years ago, and this has caused considerable controversy among geologists and palaeoclimatologists [13]). It ended some 11 500 years ago with the beginning of the last major glaciation. This means that we can identify the terrestrial record of the last interglacial with the final dip in the oxygen isotope curve seen in Figure 12.2. Earlier interglacials are much more problematic; they are usually given local names when they are first described, and attempts are later made to correlate them, often on the basis of the fossils they contain. Some possible correla-

tions are shown in Figure 12.6, but any correlation of terrestrial deposits in the absence of firm dating can only be tentative, especially in the case of older glacials and interglacials.

Biological Changes in the Pleistocene

With the expansion of the ice sheets in high latitudes, the global pattern of vegetation was considerably disturbed. Many areas now occupied by temperate deciduous forests either were glaciated or bore tundra vegetation. For example, most of the north European plain probably had no deciduous oak forest during the glacial advances. The situation in Europe was made more complex by the additional centres of glaciation in the Alps and the Pyrenees. These would have resulted in the zone compression and often the ultimate local extinction of species of warmth-demanding plants and animals during the glacial peaks. Figure 12.7 shows the broad vegetation types that occupied Europe during interglacial and glacial times. During the interglacials, tundra species became restricted in distribution due to their inability to cope with high summer temperatures and their

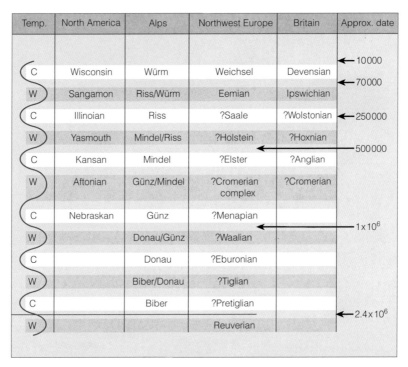

Temp.	North America	Alps	Northwest Europe	Britain	Approx. date
C	Wisconsin	Würm	Weichsel	Devensian	← 10000
W	Sangamon	Riss/Würm	Eemian	Ipswichian	← 70000
C	Illinoian	Riss	?Saale	?Wolstonian	← 250000
W	Yasmouth	Mindel/Riss	?Holstein	?Hoxnian	500000
C	Kansan	Mindel	?Elster	?Anglian	←
W	Aftonian	Günz/Mindel	?Cromerian complex	?Cromerian	
C	Nebraskan	Günz	?Menapian		← 1×10⁶
W		Donau/Günz	?Waalian		
C		Donau	?Eburonian		
W		Biber/Donau	?Tiglian		
C		Biber	?Pretiglian		← 2.4×10⁶
W			Reuverian		

Figure 12.6 Conventional correlations assumed between local glacial and interglacial events in the Pleistocene. Local complexities render such correlations tentative, particularly in the earlier stages. C, cold; W, warm.

failure to compete with more robust, productive species. High-altitude sites and disturbed areas sometimes served as **refugia** within which groups of such species may have survived in isolated localities. Similarly, during glacials particularly favourable sites that were sheltered, south-facing or oceanic and relatively frost-free may have acted as refugia for warmth-demanding species through the time of cold. In Europe, for example, deciduous forest species are thought to have survived glacial episodes in what is now Spain and Portugal [14], Italy, Greece, the Danube Delta, Turkey and around the Black and Caspian seas [15]. This hypothesis is supported by a study of intraspecific (within-species) genetic diversity of 22 common European tree species led by French scientist Rémy Petit [16]. As predicted, tree populations from Mediterranean areas that contained refugia had high genetic diversity, especially for species with low dispersal. Interestingly, the highest genetic diversity was found at intermediate latitudes, probably as a result of the subsequent mixing of lineages that had been isolated in different refugia. This general pattern of a reduction in genetic (allelic) diversity from southern to northern Europe and species subdivision is seen in many species (reviewed in Ref. [17]). These patterns are driven by the rapid northward expansion of many species after the last Ice Age and variations in the topography of southern refugia allowing populations to diverge through several ice ages. The DNA evidence suggests that some species have been diverging in refugial regions for only a few ice ages at most, whilst other species are more genetically distinct indicating a much more ancient separation [17]. But it would be a mistake to think that temperate European areas were solely colonized from southern refugia at the end of the last glacial maximum. There is increasing evidence, both palynological and genetic, that northern refugia also existed – the most prominent of which was in the area around the Carpathians and was characterized by deciduous and coniferous woodland, mainly on southern facing slopes [18].

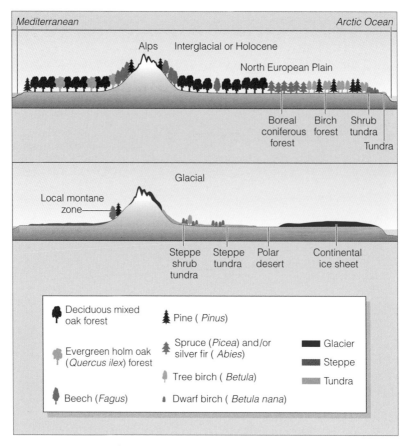

Figure 12.7 A simplified diagrammatic representation of the vegetation belts of Europe during glacial and interglacial time. The profiles run south–north from the Mediterranean Basin to the Arctic Ocean. During glacial times, the temperate forest became confined to refuge areas mainly south of the Alps.

Animal species with good dispersal mechanisms were able to shift their populations to cope with the new conditions more easily than plants. Plants were dependent on the movement of their fruits and seeds to spread into new areas, and meanwhile old populations died out as conditions became less favourable. The fact that each species faced its own peculiar problems, in terms of both climatic requirements and dispersal capacity, means that species must have become shuffled into new assemblies (or *communities*; see Chapter 4) during times of change. Figure 12.7 expresses vegetation as life forms (with examples), so the picture is one of major biome shifts under different climatic regimes, but the constitution of these biomes changed according to the patterns of movement of different tree species. The same is true of the animal components of the biomes. An extensive study [19] of the ranges of North American mammals during Late Quaternary times has shown that individual species shifted at different times, in different directions and at different rates during the climatic swings. Only during the last few thousand years have the familiar modern groupings of plants and animals assembled themselves, so the idea that species assemblages moved as intact 'communities' must be abandoned in the light of these findings.

Extinctions occurred during this process of successive climatic changes and distributional shifts.

In Europe, many of the warm-preferring genera and species so abundant in the Tertiary Epoch were lost to the flora. The hemlock (*Tsuga*) and the tulip tree (*Liriodendron*) were lost in this way, but both survived in North America, where the generally north–south orientation of the major mountain chains (the Rockies and the Appalachians) allowed the southward migration of sensitive species during the glacials, and their survival in what is now Central America. The east–west orientation of the Alps and Pyrenees in Europe permitted no such easy escape to the south. The wing nut tree (*Pterocarya*) was also extinguished in Europe, but it has survived in Asia, the Caucasus and Iran. Studies relating to the histories of many tree species during the latter part of the Pleistocene have greatly assisted in explaining the present-day distribution of trees and the composition of forests [20].

Climate-related extinctions also occurred in animal taxa, especially bigger species. It has been estimated that 65% of mammal genera weighing over 44 kg went extinct sometime between 50 000 and 3000 years before present (BP). The cause and the possible role of humans in these extinctions has been hotly debated (see Chapter 13), but it is becoming increasingly clear that changes in climate played a major role in many of these disappearances. Spanish biogeographer David Nogués-Bravo and colleagues recently demonstrated that continents with greater magnitudes of climate changes during the Late Quaternary had more megafauna extinc-

tions than continents that experienced less (with the notable exception of South America) [21]. In another study, Nogués-Bravo shows that climate may have had a role in one of the most iconic extinctions of the Holocene, that of the woolly mammoth (*Mammuthus primigenius*) [22]. Using bioclimatic envelope modelling, they show that suitable climate conditions for the mammoth rapidly diminished between the Late Pleistocene and the Holocene, with the last remaining suitable areas being mainly restricted to Arctic Siberia – the site of the latest records of mammoths in continental Asia. This massive climate-induced range contraction almost certainly made the mammoth more vulnerable to other factors, including the growing populations of human hunter-gatherers.

The Last Glacial

The most recent glacial stage lasted from approximately 115 000 years ago until about 10 000 years ago, so our current experience of a warm Earth is rather unusual as far as recent geological history is concerned. However, even a glacial is not a time of uniform cold, and there have been numerous warmer interruptions to the prevailing cold climate. Many interstadials are recorded within the last glacial. Figure 12.8 shows a detailed oxygen isotope curve of an ice core taken from the Greenland Ice Cap [23]. The instability of the temperature during the glacial is evident, and it is

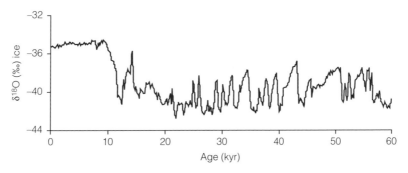

Figure 12.8 Oxygen isotope curve from an ice core extracted from the Greenland Ice Cap. (Less negative values of ^{18}O in ice indicate higher temperatures.) The sequence covers a period of 60 000 years, approximately the latter half of the last glacial event. The instability of the temperature, as reflected by the isotope curve, is evident, the cold period being frequently interrupted by short episodes of higher temperature. From Dansgaard *et al.* [23].

clear that there were many short-lived warm episodes all the way through the glacial. The cycles of alternating warmth and cold during the last glacial have been termed **Dansgaard–Oeschger cycles**, after the scientists who first described them. Each cold event lasts about 1000–2000 years, separated by intervals of approximately 7000 years, and the cold phase is usually accompanied by massive discharges of icebergs into the North Atlantic Ocean, first discovered by Hartmut Heinrich in 1988 and now called **Heinrich events**. These icebergs carried debris, such as sand and limestone fragments, out into the Atlantic where they eventually melted and released their loads of detritus, which joined the marine sediments and left a record of the frequency of Heinrich events. The sediments also record a lowered abundance of plankton fossils, such as foraminifera, during Heinrich events, suggesting reduced oceanic productivity. In addition, there is evidence of major changes in ocean salinity at the time of their occurrence, and this has led some oceanographers to propose that the entire circulation pattern of the Atlantic Ocean changed during

the Dansgaard–Oeschger cycles [24], an idea that is discussed in more detail in this chapter.

The plant and animal life on land also responded to the climatic fluctuations during the last glacial. Continuous sediment records that contain fossil pollen and allow us to trace terrestrial vegetation back through the last glacial into the preceding interglacial are relatively uncommon. Many possible sites were actually scoured by glaciers and so have lost their record, but there are some deep lake sites, often associated with old volcanic craters, where a complete record is available. One such profile that takes the vegetation record back about 125 000 years is shown in Figure 12.9. This is a pollen diagram from Carp Lake, situated on the eastern side of the Cascade Mountains in the Pacific Northwest of America [25]. It lies just outside the limit of the great Laurentide ice sheet that covered much of the northern part of North America during the final Wisconsin glacial. It currently lies at the boundary of two biomes, the sagebrush (*Artemisia*) steppe and the montane pine forest (*Pinus ponderosa*),

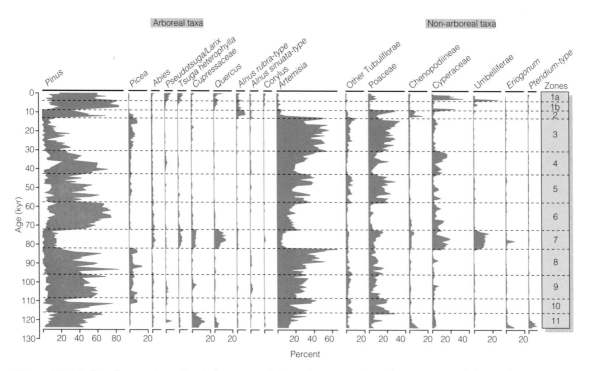

Figure 12.9 Pollen diagram from Carp Lake, a crater lake in the eastern Cascade Mountains of the northwestern United States. The sediments cover approximately 125 000 years and display the erratic response of vegetation to climatic instability during the last (Wisconsin) glacial. From Whitlock and Bartlein [25].

and its closeness to this boundary (called an **eco-tone**) means that the vegetation will have been particularly sensitive to climatic swings in the past. The basal part of the diagram (Zone 11) shows the open pine and oak (*Quercus*) forest of the last interglacial, with some indications that conditions were then warmer and drier than at present. There follows, for the bulk of the diagram, alternating peaks of *Artemisia* and pine, showing the constant alteration of the vegetation over this sensitive ecotone as the climate warmed and cooled. At one stage (Zone 7, about 85 000–74 000 years ago), there was an episode of such warmth that an open mixed forest including oak established itself. It is likely that this inter-

lude was warmer and wetter during the summers than at present. Cooler conditions then returned until the glacial ended and the current vegetation established itself in the last 10 000 years (Zone 1).

This remarkable record demonstrates that the temperature fluctuations recorded in the ice-core profile of oxygen isotopes (see Figure 12.8) were reflected in the response of vegetation and that the instability of the glacial climate caused constant adjustments in the ranges of species and in the boundaries of biomes.

Even in the tropics, such as northern Australia, vegetation changed markedly in response to climatic fluctuations over the past 120 000 years, as discussed in Box 12.1 and Figure 12.10.

The Australian tropics in times of cold

Concept Box 12.1

In the tropics, the effect of ice was not experienced directly except on the very high mountains, but the climate was generally colder, and changes in vegetation reflected in pollen diagrams from tropical regions suggest that there were important variations in precipitation. In Queensland, Australia, for example, Peter Kershaw [26] has analysed the sediments of a volcanic crater lake from a rainforest area, and the results are shown in Figure 12.10. Dating at this site is difficult, but the total span of the diagram is thought to cover about 120 000 years, extending back into the closing stages of the last interglacial, a similar time coverage to that of the Carp Lake diagram from the Pacific Northwest of North America (Figure 12.9). During the last interglacial (basal zone E3), this Australian site was occupied by many of the rainforest tree genera that currently occupy the area, but at the time when the final glacial began in high latitudes, the rainforest was replaced by forest of simpler structure in which the palm-like trees of the genus *Cordyline* played an important part. There was then a brief reversion (dated at about 86 000–79 000 years ago, very close to the warm spell in the North American diagram) when rainforest became re-established. But then much more arid conditions set in, and the forest became increasingly dominated by gymnosperm trees, such as the monkey puzzle (*Araucaria*). Between about 26 000 and 10 000 years ago, the climate became very dry, and the former forests took on a sclerophyllous form, being dominated by *Casuarina*, a tree currently associated with hot dry conditions. At the end of the 'glacial', however, there

was a very rapid change in vegetation, with the invasion of rainforest trees once again. The period of maximum aridity at this site, 26 000 to 10 000 years ago, encompasses the period of maximum extent of the Northern Hemisphere glaciers at about 22 000 years ago.

In the Nullabor Plain in south-central Australia, studies on fossil vertebrates show that a high proportion of these organisms became extinct during the Pleistocene [27], but climate on its own is unlikely to account for this loss in biodiversity. An increasing incidence of bushfires is a more likely immediate cause of the losses, and the impact of humans (who arrived in Australia around 40 000 years ago) cannot be excluded as a contributory factor: 90% of the megafauna became extinct soon after the first archaeological evidence for human colonization of the continent. Interestingly, a very similar pattern is observed on the neighbouring island of Tasmania (which was connected to the mainland when sea levels were lower). Although most of the Tasmanian megafauna went extinct between 43 000 and 40 000 years ago, before the arrival of humans on the island, recent analysis of fossil remains and their associated sediments suggests that at least some species hung on until at least 41 000 years ago and therefore overlapped with humans [28]. The fact that this latter period, which was not associated with significant regional climatic or environmental change, occurred between 43 000 and 37 000 years ago suggests that humans probably played a major role in the demise of these relict populations.

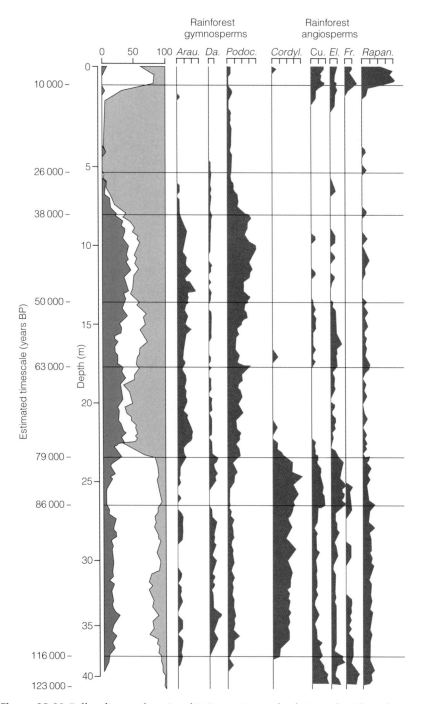

Figure 12.10 Pollen diagram from Lynch's Crater, Queensland, Australia. The column on the left shows the percentage of rainforest gymnosperms (grey), rainforest angiosperms (white) and sclerophyll taxa (blue). The frequencies of pollen of all taxa are shown as percentages of the dry land plant pollen total; each division represents 10% of the pollen sum. Abbreviations for taxa: Arau., *Araucaria*; Da., *Dacrydium*; Podoc., *Podocarpus*; Cordyl., *Cordyline*; Cu., *Cunoniaceae*; El., *Elaeocarpus*; Fr., *Freycinettia*; Rapan., *Rapanea*; Casuar., *Casuarina*; Euc., *Eucalyptus*. Adapted from Kershaw [26].

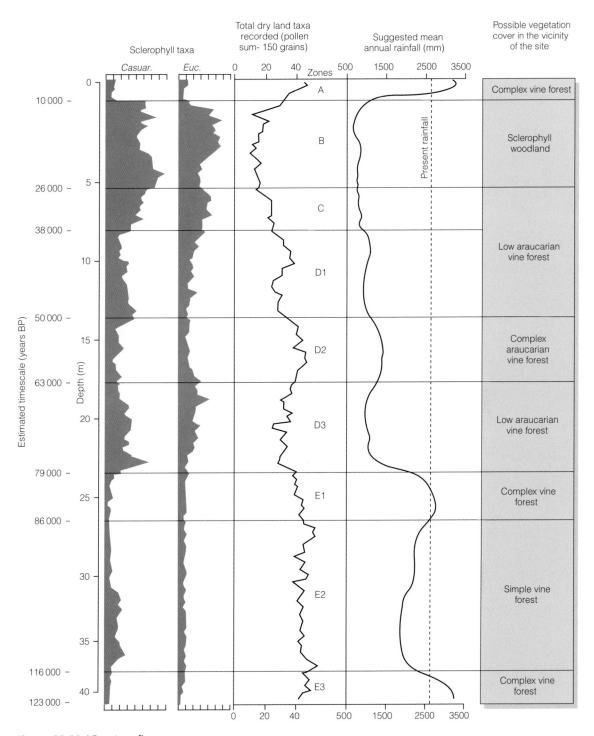

Figure 12.10 (*Continued*)

Tropical aridity seems to have been widespread when the higher latitudes were experiencing their glacial episodes, as is confirmed by evidence from tropical Africa, India and South America. Much of the humid tropical forest of the Zaire Basin was probably replaced by dry grassland and savanna during the glacial, although some forest may well have survived in riverine and lakeside situations. The temperature records from tropical locations, such as East Africa, follow closely the trends observed in other parts of the globe. Pollen analyses from lake sediments along the Uganda–Zaire border [29] have shown that rainforest was replaced by dry scrub and grassland to the east of the Western Rift Valley during the height of the last glacial episode. Any refugial fragments must have been situated further west, in the lowlands of the Congo Basin.

Figure 12.11 shows a proposed reconstruction of the approximate areas of rainforest that existed at the time of maximum glacial extent in the high latitudes (about 22 000 years ago). From this, it can be seen that rainforest was much fragmented as a result of the glacial drought in tropical latitudes [30]. Many areas currently occupied by rainforest were greatly modified at this time. This is an important point to bear in mind when considering the high species diversity of the rainforests (see Chapter 2). Most rainforests have not enjoyed a long and uninterrupted history but have been disrupted by global climatic changes, especially cold and drought. Even some of the regions (like the Uganda–Zaire borders), which are now hotspots of biodiversity, bore a very different vegetation during the last glacial. However, there is still considerable debate about the vegetation of the last glacial maximum. At one time it was believed that savanna woodlands occupied much of the region, but current opinion [31] favours dry seasonal forests with an admixture of montane, cold-tolerant tree species. During the glacials the temperate forests were forced to occupy new areas at lower latitudes, but the tropical rainforests had nowhere to which they could retreat, so they became fragmented and dismembered as their composition was drastically altered (see Chapter 11). Some ecologists believe that this fragmentation may even have added to their diversity by permitting

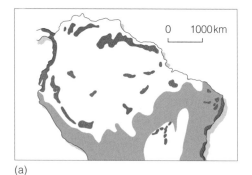

(a)

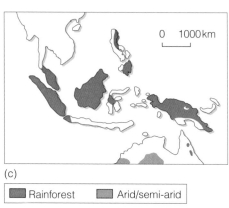

(b)

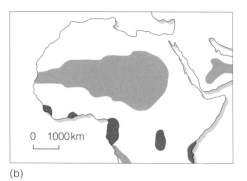

(c)

■ Rainforest   ■ Arid/semi-arid

Figure 12.11 Possible distribution of regions of humid rainforest and arid–semi-arid areas 22 000 years ago, when the glaciation of the higher latitudes was at its maximum. From Tallis [30].

the isolation of populations and the development of new species [32].

Figure 12.11 also shows possible areas of drought at the time of maximum glacial extent, and the likely global pattern of dry regions, as evidenced by sand deposits, is shown in Figure 12.12, where it is

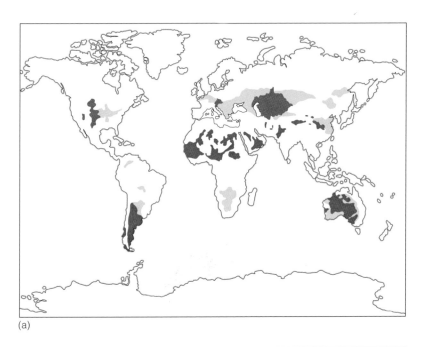

(a)

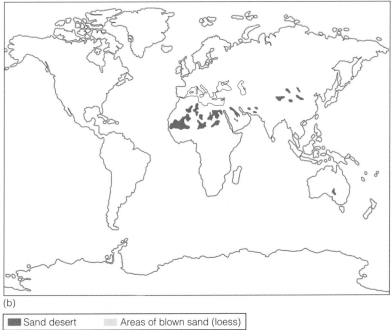

(b)

| ■ Sand desert | ▨ Areas of blown sand (loess) |

Figure 12.12 (a) Proposed distribution of sand deserts at the height of the last glacial compared with (b) their modern distribution. Mobile sand (loess) was a feature of many areas during the glacial episode. From Wells [33].

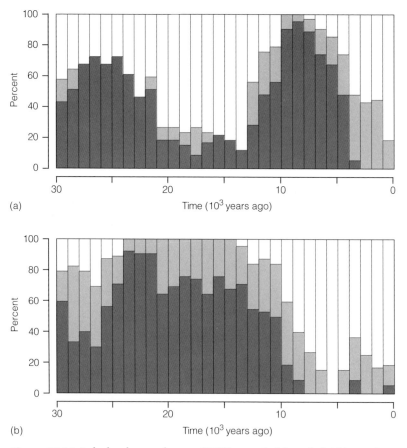

Figure 12.13 Lake levels over the past 30 000 years in (a) tropical Africa and (b) the western United States. The bars represent the proportion of lakes studied with high (dark shading), intermediate (light shading) or low (white) lake levels. From Tallis [30].

compared to modern desert distribution [33]. The global aridity of the last stages of the last glacial is very apparent. Blown sand, termed **loess**, from these deserts is found fossilized in many parts of the eastern United States, in central Europe and in southern Australia, where fossil vertebrates have confirmed the arid nature of much of the Pleistocene [27]. Lake-level studies have similarly shown that the period between 20 000 and 15 000 years ago was particularly dry in some regions. For example, the lake levels from tropical Africa were at a low point during this time period, as shown by the data summarized in Figure 12.13.

It would be misleading, however, to give the impression that drought prevailed throughout the Earth at this time. Just as at the present day some

areas are wetter than others, so it was in glacial times, and the western part of the United States enjoyed a time of wet climate and high lake levels between 10 000 and 25 000 years ago. Such wet periods are called **pluvials**, and many parts of the world experienced pluvials at various times in Quaternary history. In subtropical Africa, for example, 55 000 and 90 000 years ago seem to have been times of pluvial climate.

Evidence for the existence of pluvial lakes in western North America is provided not only by the geological deposits but also by the present-day distribution of certain freshwater animals. In western Nevada, there are many large lake basins that are now nearly dry, but in the remaining waterholes there live species (or, more accurately, a *species*

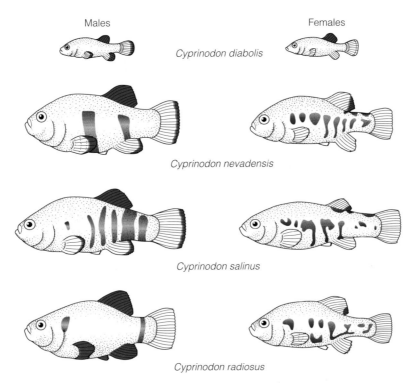

Males   Females

Cyprinodon diabolis

Cyprinodon nevadensis

Cyprinodon salinus

Cyprinodon radiosus

Figure 12.14 The four 'species' of desert pupfish that have evolved in the streams and thermal springs of the Death Valley region. Males are bright iridescent blue; females are greenish. (Their different markings are shown by the blue tint.) Adapted from Brown [84].

complex) of the desert pupfish (*Cyprinodon*) (Figure 12.14). The isolated populations have gradually evolved into what are considered at least three different species (one now extinct), each adapted to its own specific environment, rather like the Hawaiian honey-creepers (see Chapter 8). In many respects, the wet sites in which these fish live can be regarded as evolutionary 'islands' separated from each other by unfavourable terrain. The species have probably been isolated from one another since the last pluvial at the beginning of the present interglacial, whereas populations within each species may still be in partial contact during periods of flooding. The isolation of populations during the Pleistocene and Holocene has resulted in remarkable levels of genetic divergence [34], normally only seen in some marine fish.

Although pluvial conditions prevailed in western North America during the glacial maximum, this was not generally the case elsewhere. Glacial conditions in the high latitudes were associated with colder and drier conditions over much of the Earth. Figure 12.15 shows the modelled reconstruction of glacial conditions at different latitudes [35]. Of particular note in this diagram is the greater reduction in temperature apparent in the high latitudes of the Northern Hemisphere, and also the increased wind stress in the mid-latitudes of the north during the glacial. Fossil evidence from various parts of the world dating from the last glacial confirm that the glacial stages were times of severe disruption to the entire biosphere. This fact, coupled with the likelihood that we are still locked into the oscillating climatic system that has operated over the past 2 million years, makes it imperative that we understand the mechanisms that have generated the glacial–interglacial cycle.

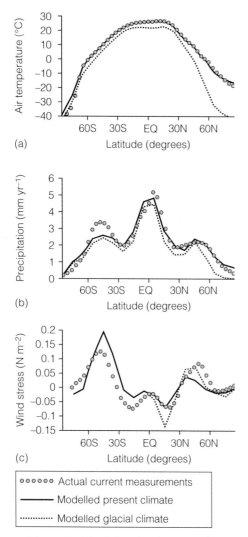

(a)

(b)

(c)

○○○○○○ Actual current measurements

—— Modelled present climate

·········· Modelled glacial climate

Figure 12.15 Models of glacial and current climatic conditions (annual means) at different latitudes compared with actual modern measurements. Note that the colder, drier, windier conditions during the glacial are especially pronounced in the Northern Hemisphere. From Ganopolski *et al.* [35].

Causes of Glaciation

Ice Ages are relatively rare events in the Earth's 4.6 billion years of history. Although the polar regions receive less energy from the sun than the equatorial regions, they have been supplied with warmth by a free circulation of ocean currents through most of the world's history. Only occasionally do land masses pass over the poles, or form obstructions to the movement of waters into the high latitudes, which can result in the formation of polar ice caps. Table 12.1 shows the approximate timing of glacial episodes during Earth's history.

The table shows that ice ages tend to occur approximately every 150 million years, although there was not one during the Jurassic Period, 150 million years ago, probably because there was no major continental mass over the South Pole at that time. During the most recent Ice Age, the movement of Antarctica into its position over the South Pole led to the development of a Southern Polar Ice Cap perhaps as long as 42 million years ago. Falling global temperature then led to the development of the Greenland Ice Cap around 34 million years ago. Rearrangements of the Northern Hemisphere land masses have subsequently resulted in the isolation of the Arctic Ocean so that it received relatively little influence from warm-water currents, leading to its becoming frozen some 3 to 5 million years ago. The presence of two massive polar ice caps increased the albedo, or reflectivity, of the Earth, for whereas the Earth as a whole reflects about 40% of the energy that falls on it, the ice caps reflect about 80%. The formation of two such ice caps therefore significantly reduced the amount of energy retained by the Earth. The scene was set for the development of an ice age.

It is still necessary, however, to explain why the recent Ice Age has not been a period of uniform cold, but has consisted of a sequence of alternating warm and cold episodes. A proposal to explain this pattern was put forward in the 1930s by the Yugoslav physicist Milutin Milankovich and has become widely accepted by climatologists. Milankovich constructed a model based on the fact that the Earth's orbit around the sun is elliptical and that the shape of the ellipse changes in space in a regular fashion, from more circular to strongly elliptical. When the orbit is fairly circular there will be a more regular input of energy to the Earth through the year, whereas when it is strongly elliptical the contrast between winter and summer energy supply will be much more pronounced. It takes about 96 000 years to complete

Table 12.1 The occurrence of Ice Ages in the history of the Earth.

Name	When?	Putative cause
Huronian Glaciation	2.4 to 2.1 billion years ago	Volcanic activity
Cryogenian Ice Ages	850 to 630 million years ago	Loss of CO_2 from the atmosphere due to recently evolved multicellular organisms sinking to the sea bed
Andeab-Saharan Ice Age	460 to 430 million years ago	Triggered by volcanic activity that deposited new silicate rocks, which draw CO_2 out of the air as they erode
Karoo Ice Age	360 to 260 million years ago	Falling CO_2 as a result of the expansion of land plants. As plants spread over the planet, they absorbed CO_2 from the atmosphere and released oxygen
Antarctica freezes	14 million years ago	Falling CO_2 caused by the rise and subsequent erosion of the Himalayas. The weathering sucked CO_2 out of the atmosphere and reduced the greenhouse effect
Quaternary glaciations Older Dryas Younger Dryas	2.58 million to 12 000 years ago 14 700 to 13 400 years ago 12 800 to 11 500 years ago	Triggered by a fall in atmospheric CO_2 due to continued weathering of the Himalayas. Timing of the glacials and interglacials driven by periodic changes in Earth's orbit and amplified by changes in greenhouse gas levels (see the main text)

a cycle of this change in orbital shape, as shown in Figure 12.16.

A second source of variation is produced by the tilt of the Earth's axis relative to the sun, which again affects the impact of seasonal changes, with a cycle duration of about 42 000 years. The third consideration is a wobble of the Earth's axis around its basic tilt angle, which shows a cycle of about 21 000 years. All of these cycles affect the strength of solar intensity that is received by the Earth, and the pattern of climatic change should, according to Milankovich, be a predictable consequence of summing the effects of these three **Milankovich cycles**.

Geophysicists working on the chemistry of ocean-floor sediments have expended much effort in seeking evidence for cyclic periodicity in past ocean temperatures and in checking the cycles found against those proposed by Milankovich. In 1976, Jim Hays, John Imbrie and Nick Shackleton [36] were able to confirm that all three levels of Milankovich cycles could be detected in the marine sediments. The Milankovich pattern has now been found in sediments dating back 8 million years, but one important change has been detected in their effects. Whereas 8 million years ago the effects of the 96 000-year cycle were weak, in the last 2 million years it is this cycle that has been very strong

and that has dominated the glacial–interglacial sequence. Additional factors must have amplified this particular cycle in recent times.

One possible explanation for the current exaggeration of the effects of the 96 000-year cycle is that ice masses themselves are responsible. The ice grows slowly and decays relatively quickly, which can itself modify the global climate. Computer models have been constructed that take account of this effect, and they produce a better fit to observed data than the Milankovich cycles on their own.

Detailed testing of the Milankovitch theory is, of course, dependent on well-dated records, and at present the dating of climatic fluctuations in the Pleistocene is crude. The difference in the dating of the beginning of the last interglacial, for example, varies by over 17 000 years, depending on the materials used. Until dating is more firmly established, full confirmation of the correlation between orbital cycles and climate cannot be achieved.

Solar forcing, in which variations in astronomical conditions determine global climate, as reflected in the Milankovitch cycles, appears to be a major element underlying the observed pattern of glacials and interglacials in the past 2.4 million years. There are complicating factors, however, many of which have yet to be fully

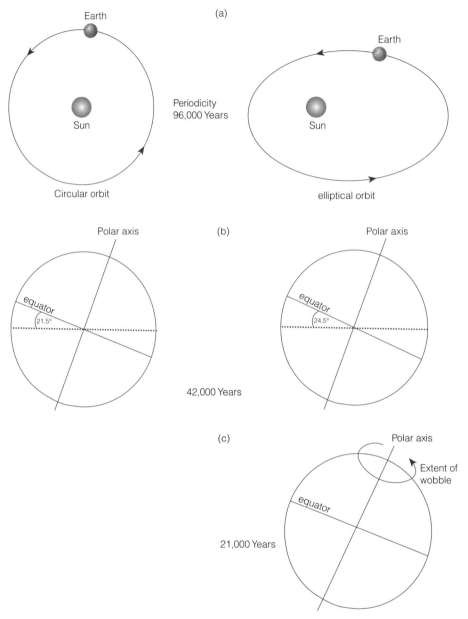

Figure 12.16 The three variations in orbital and axial conditions that affect the solar radiation arriving at the surface of the Earth, as proposed by Milankovitch. (a) The eccentric nature of the Earth's orbit around the sun, which has a periodicity of about 96 000 years. (b) The variation of the tilt of the Earth on its axis, having a periodicity of about 42 000 years. (c) The wobble of the Earth on its axis, which affects the season of tilt and has a periodicity of about 21 000 years.

investigated and explained. Over the past 25 years, it has become possible to investigate the composition of the Earth's atmosphere during the glacial–interglacial cycles. The technique depends on the chemical analysis of small bubbles of air trapped within the ice of ice sheets and glaciers during the past and retained in a 'fossil' form until brought to the surface by modern coring. The extracted ice is carefully sealed from contamination by modern air and is crushed to force out the gases trapped within it at different levels. It is assumed that the gas thus extracted is a true reflection of the contemporaneous air enclosed by falling snowflakes and sealed in the ice over the course of hundreds of thousands of years.

Figure 12.17 shows the results of an analysis of ice from the Vostok research station in eastern Antarctica [37]. The ice core is 3300 m in depth (top scale) and covers the last 400 000 years of the Earth's history (bottom scale). Superimposed on these scales is the projected solar energy input, the **insolation** (curve e), the oxygen isotope record (curve d), the inferred temperature (curve b) and two of the gases found within the bubbles that have significance in climatic fluctuations, namely,

carbon dioxide (curve a) and methane (curve c). The first point to notice about this diagram is that the insolation curve ties in very acceptably with the oxygen isotope curve and therefore with the temperature curve. Major peaks in temperature, of which there are five including the current, correspond well with peaks in insolation. But the two atmospheric gases, carbon dioxide and methane, also follow closely the inferred temperature curve, both rising in concentration during the interglacials and falling during the glacials. Significantly, both of these compounds are **greenhouse gases**; both have a high capacity for energy absorbance in the infrared region of the spectrum and so are able to act as thermal insulators for the Earth.

We must try to explain the correspondence of these gases with global temperature change and

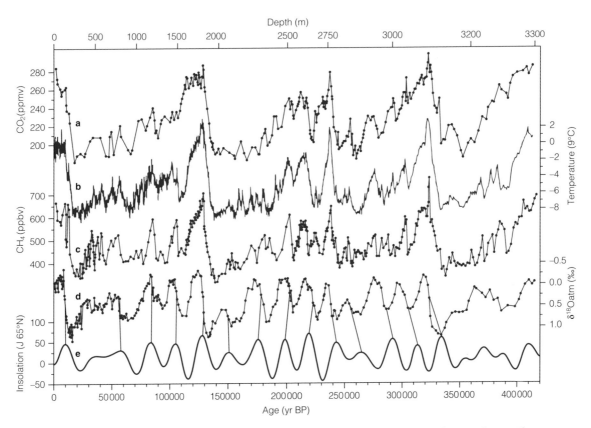

Figure 12.17 Analyses of gas bubbles in an ice core taken from the Antarctic Ice Cap at the Vostok site. The gases extracted are considered fossil samples of the ambient atmosphere of the past. The concentrations of carbon dioxide (curve a) and methane (curve c) are shown, together with a reconstruction of temperature change (curve b) derived from oxygen isotope studies (curve d). At the bottom of the diagram (curve e) is a proposed curve for insolation based on Milankovitch cycle predictions. From Petit *et al.* [37].

also assess their influence on such changes, since they are capable of altering the energy exchange system of the planet. One could explain the rise in these gases during times of warmth by considering the additional respiratory activity of organisms, particularly microbes, during the interglacials. Respiration of all organisms produces carbon dioxide, and microbial decay is stimulated by warmth, especially in the cooler regions of the Earth where glacial conditions may have held back decomposition. The weathering of rocks, such as limestone, in soils would also increase during times of warmth, resulting in additional carbon dioxide production. Methane is produced in a variety of ways: by the activities of termites, by generation within the rumens of large herbivores and by the incomplete decomposition of organic matter in wetlands, among others. But on the reverse side, warmer conditions could result in additional plant productivity and biomass accumulation, which would lead to more accumulation of organic matter at the expense of atmospheric carbon dioxide. Changing levels of carbon dioxide in the atmosphere may even have resulted in structural changes in plants (Box 12.2). The role of the oceans is also likely to be important here. At present, it is estimated that about a third of the carbon dioxide entering the atmosphere each year ends up in the oceans, largely because many of the microscopic planktonic organisms (especially the coccolithophorids and the foraminifera) construct shells of calcium carbonate that sink to the ocean floors after their death. The carbonate they use is ultimately derived

from carbon dioxide from the atmosphere dissolving in the surface waters and then being taken up by the plankton. The question is, was this oceanic sink for carbon greater during the interglacials, resulting in the extraction of carbon from the atmosphere? If oceanic circulation changes during a glacial result in upwelling and the conveyance of deep, nutrient-rich waters to the surface, then planktonic growth would be stimulated and a biological pump would be brought into action, taking carbon out of the atmosphere and depositing it in the ocean sediments [38]. It is also possible that changes in deep-sea currents could have flushed carbon-rich waters from the abyssal depths at the commencement of interglacials, thus increasing greenhouse gases in the atmosphere and enhancing global warming. Evidence from the North Pacific seems to favour this argument [39].

Whatever the precise mechanism at work, the outcome is a close correspondence between atmospheric carbon dioxide (and methane) and atmospheric temperature. The raised levels of these gases during the interglacials act as a positive feedback, enhancing the higher temperatures because of their thermal, greenhouse properties in the atmosphere.

Wallace S. Broecker of Columbia University has emphasized the importance of the circulation of the Earth's ocean currents, coupled with those of the atmosphere, to explain the rapidity with which some of the climatic shifts in the Pleistocene have come about [42]. His theory does not contradict the ideas of Milankovitch, but supplements them. Much of the warmth of the North Atlantic is brought into the

Carbon dioxide and plant porosity

Concept

Box 12.2

One consequence of fluctuations in atmospheric carbon dioxide is that the anatomy of some plants may have altered in response to the changes. Since plants take up the gas through the pores (stomata) in their leaves, they may have needed fewer stomata under conditions of high carbon dioxide concentration. Ian Woodward of Sheffield University, England, examined herbarium specimens from museums and has shown that plants have reduced their stomatal density in parallel with carbon dioxide increase in the last century [40]. Evidently, it has proved advantageous for the plant to cut down on the pore density, and hence reduce water loss by transpiration, in a situation where carbon dioxide is easier to obtain. Jenny McElwain of University College Dublin [41] has followed up these modern herbarium studies by looking at stomatal densities in more ancient plant materials. She examined fossil plant epidermises and found a close link with supposed carbon dioxide levels in the distant past. Thus, the technique may be sufficiently robust to use for long-term atmospheric reconstructions.

region by highly saline waters heading northward at intermediate depths of about 800 m. This current effectively redistributes the tropical warmth into the high latitudes, as seen in Chapter 3 (Figure 3.10). Broecker estimates that this input of energy into the North Atlantic is equivalent to 30% of the annual solar energy input to the area. Fossil evidence, however, indicates that this oceanic conveyor belt became switched off during the glacial episodes, and the reduction in the heat transfer to high latitudes led to the development of the extended ice sheets. If this were so, it would explain the severe and relatively sudden drop in temperature in the high latitudes of the Northern Hemisphere seen in Figure 12.17. Smaller changes in the thermohaline circulation patterns of the North Atlantic could account for the climatic oscillations experienced during the last glaciation, the Dansgaard–Oeschger cycles [24].

One further suggestion is that massive volcanic eruptions may precede and initiate the process of glaciation [43]. There is certainly some evidence for a correlation between glacial advances and periods of volcanic activity during the last 42 000 years in New Zealand, Japan and South America. Volcanic eruptions produce large quantities of dust, which are thrown high into the atmosphere. This has the effect of reducing the amount of solar energy arriving at the Earth's surface, and dust particles also serve as nuclei on which condensation of water droplets occurs, thus increasing precipitation. Both of these consequences would favour glacial development. Attempts to correlate volcanic ash content with evidence of climatic changes in ocean sediments, however, have not met with much success. There has, however, been an increase in the general frequency of volcanic activity during the last 2 million years, which correlates with a time of overall global cold conditions.

The Current Interglacial: A False Start

Following the maximum extent of the most recent glaciation at about 22 000 years ago, the climate began to warm, as shown in the oxygen isotope record in Figure 12.8, and the great ice sheets started to recede. Every indication pointed to the beginning of a new interglacial. But the increasing warmth was quite suddenly interrupted by a return to extremely cold conditions, forming a cycle reminiscent of the Dansgaard–Oeschger cycles of the earlier part of the last glacial.

The instability of climate around 14 000 to 10 000 years ago was first noted by geologists in Denmark, who found that the sediments of lakes dating from the transition between the glacial and the current interglacial exhibited some unusual features. The inorganic clays, typical of sediments formed in lakes surrounded by arctic tundra vegetation, where soils are easily eroded and the vegetation has a low organic productivity, were replaced by increasingly organic sediments as the development of local vegetation stabilized the mineral soils, and the aquatic productivity of the lakes led to an increasing organic content in the sediments. But this process, reflecting the increased warmth of the climate, was evidently interrupted, for the sediments then reverted to heavy clay, often with angular fragments of rock denoting a return to severe periglacial or, locally, even glacial climatic conditions. Above this layer, the organic sediments reappeared, which is interpreted as a sign of increased plant productivity associated with climatic warming. This time, the increased temperature led to the development of our present interglacial. The evidence for the climatic interruption proved to be a consistent feature of sediments of this age throughout northwestern Europe.

The cold episode that caused the deposition of these sediments was severe enough to lead to the regrowth of many glaciers. Geomorphologists have shown that the glaciers of Scandinavia and Scotland extended considerably during this episode, while small glaciers began to form on north-facing slopes in more southerly mountains, from Wales to the Pyrenees. The event was called the **Younger Dryas**, because the fossil leaves of the arctic plant the mountain avens (*Dryas octopetala*) were found in abundance in the clay layers during the original Danish studies (Figure 12.18). Radiocarbon dating of the Younger Dryas from a range of sites indicates that it took place between 12 700 and 11 500 solar (calendar) years ago [44]. The precise dating of this event is in some doubt because radiocarbon dating lacks precision during this stage in Earth's history; the calibration curve for converting radiocarbon years into solar years, based on tree growth rings, has a flat plateau at this time.

Although stratigraphic and fossil evidence for the Younger Dryas cold phase was abundant in

Figure 12.18 Mountain avens (*Dryas octopetala*), an arctic–alpine plant, remains of which were found in late-glacial sediments from Denmark and which has lent its name to the cold event called the Younger Dryas stadial.

northwestern Europe, it was more difficult to discern in the sediments of the European Alps, especially on the southern side. On the western side of the Atlantic its influence, though apparent on the eastern seaboard of North America, is difficult to detect when one moves inland. There is some evidence for cooling at this time in the northern Pacific, as far south as California [45], and even in southern Chile [46], but again the episode is less strongly recorded than in the North Atlantic region. Although evidence for climatic cooling at this time has been collected from around the world, the evidence suggests that this event was centred on the North Atlantic. In addition, oceanic information from fossils in the sediments shows that a cold front of polar water did indeed extend as far south as Spain and Portugal during the Younger Dryas stadial.

One possible explanation of how this climatic reversion, centred in the North Atlantic, may have come about was offered by Claes Rooth of the University of Miami and was later supported by the data of Wallace Broecker [47] and his co-workers. They suggested that the change resulted from an alteration in the pattern of discharge of meltwater from the great Laurentide glacier of North America. As this ice mass, covering the whole of eastern Canada, melted, its waters initially flowed south to the Gulf of Mexico. During the Younger Dryas stadial, however, it is proposed that the meltwater was re-routed via the St Lawrence into the North Atlantic (Figure 12.19). Not only would this bring large volumes of cold water into the North Atlantic, but the freshwater would also dilute the saline waters on the oceanic conveyor belt (see Chapter 4) and could disrupt the global movement of this conveyor. Such modification would explain why the Younger Dryas was most strongly felt in the regions around the North Atlantic, where the warm influence of the conveyor has its greatest impact. But if meltwater flow and salinity changes were responsible for switching off the conveyor, we would expect the onset of the Younger Dryas to be accompanied by a rapid rise in sea level. The work of R. G. Fairbanks [48] suggests that ice sheets were not melting rapidly at the time of the onset of the Younger Dryas, so the meltwater hypothesis is not supported. Subsequent work on the growth of deep-sea corals in the North Atlantic also confirm these findings [49], and the Broecker model looks less likely to be the complete explanation for these climatic changes. The cold episode certainly took place however, and the record of marine sediments confirms that oceanic circulation changed [50], but the relationship between climate change, oceanic circulation, terrestrial hydrology and the composition of the atmosphere still needs to be clarified [51].

Perhaps the most interesting and important aspect of the Younger Dryas is that it demonstrates how quickly the Earth's climate can shift from interglacial to glacial mode and back again. The brief warm spell before the Younger Dryas, called the **Allerød interstadial**, was sufficiently hot to bring beetles with Mediterranean affinities as far north in western Europe as the middle of England. Yet within a few centuries, glacial conditions once more prevailed in the area. Perhaps even more remarkable is the rapidity of change at the end of the Younger Dryas. Evidence suggests that within a period of 50 years, the mean annual temperature rose by 7.8°C [44], and some have suggested an even more rapid transition. Close study of this episode may help biogeographers to understand the impact of rapid climate change on the distribution patterns of living organisms, and may provide a basis for projection of the possible impact of current climate changes.

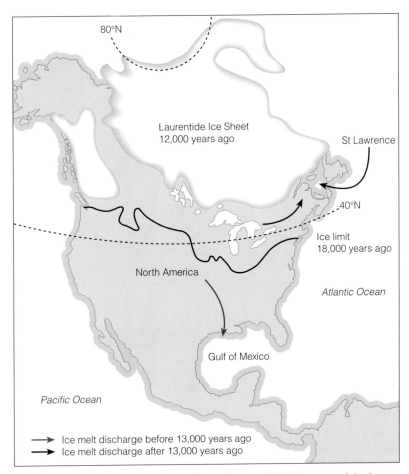

80°N

Laurentide Ice Sheet
12,000 years ago

St Lawrence

40°N

Ice limit
18,000 years ago

North America

Atlantic Ocean

Gulf of Mexico

Pacific Ocean

→ Ice melt discharge before 13,000 years ago
→ Ice melt discharge after 13,000 years ago

Figure 12.19 Extent of the Laurentide Ice Sheet in the final stages of the last Ice Age. At its greatest extent the ice sheet terminated south of the Great Lakes, so that 18 000 years ago its meltwater outflow is likely to have been south into the Gulf of Mexico. Once it had retreated north of the St Lawrence River, about 12 000 years ago, the meltwater became diverted into the North Atlantic.

Forests on the Move

After this faltering start to the current interglacial, the generally raised temperature (recorded in the oxygen isotope ratios of the ocean sediments and in the ice accumulations of the ice caps) was consistently maintained. The oxygen isotope curves indicate that there was less climatic instability during the warmer interglacial than had been the case during the glacial. The vegetation and animal life of the Earth had to adjust to a new set of conditions, but at least the increased warmth was fairly consistent for several thousand years. The pollen grains preserved in the accumulating lake sediments laid down since the end of

the Younger Dryas (the Holocene) provide detailed records of the arrival and expansion of tree populations as species distribution patterns changed and forests were reconstituted. The pollen diagrams from Minnesota shown in Figure 12.20 [52] illustrate a typical forest progression from the Northern Hemisphere temperate region.

In part, the sequence of trees (*Picea, Betula, Quercus* and *Ulmus*) reflects the general climatic tolerances of these tree genera responding as the climate warmed, but it is also a function of their relative speed of migration and the distances they had to cover to invade land exposed by the retreating glaciers. Birch (*Betula* spp.), for example, has light, airborne fruits that can travel considerable

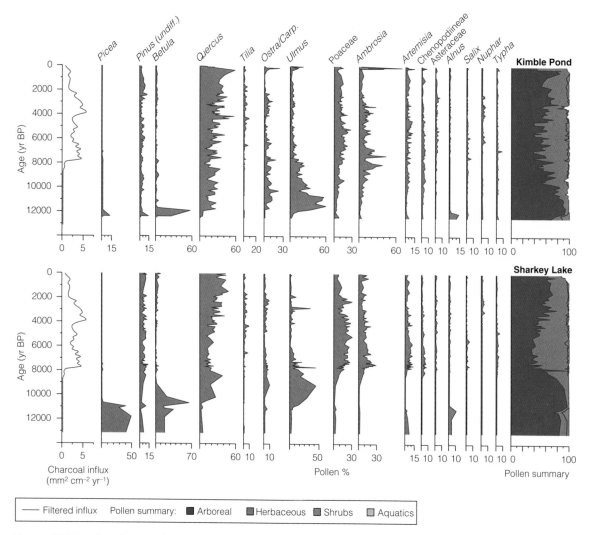

Figure 12.20 Pollen diagrams from two lake sites in Minnesota. Quantities of charcoal within the sediments are also shown. These sites lie in the transition region between deciduous forest and prairie biomes. From 8000 years ago trees decline, and grasses and sagebrush elements increase, together with charcoal derived from fire. From Camill *et al.* [52].

distances: it is also able to produce fruits when only a few years old, which permits a rapid expansion of its range. Add to this the fact that it is reasonably cold-tolerant and may have survived quite close to the glacial limits, and it is only to be expected that it should be one of the first trees to appear in any abundance in the postglacial pollen record. One must also remember that birch produces large quantities of pollen, so it could be over-represented in the pollen record. Many factors must be considered before a pollen sequence can be interpreted in terms of changing climate. Some elegant techniques

have been employed to try to translate pollen densities in sediments into estimates of the population densities. In this way, one can trace the population expansion of trees as they invaded new areas [53].

Overall, it is evident from the oxygen isotope data that there was a warming of the climate, reaching its maximum during the early stages of the interglacial, and cooling over the past 5000 years. The most warmth-demanding trees, however, such as lime (*Tilia* spp.) in western Europe, were relatively late arrivals, possibly in part because of their slow rates of spread. Vegetation changes in the later parts of the Holocene are difficult to interpret because the general climatic cooling was associated with varying patterns of precipitation and aridity in different parts of the world. In some areas, such as western Europe, the intensification of human agriculture during the later part of the Holocene was associated with forest clearance, and this obscures any climatic signal in the pollen record. The diagrams from Minnesota show increasing quantities of grass (Poaceae) and ragweed (*Ambrosia*) pollen accompanied by an input of charcoal, derived from fire in the surrounding vegetation, to the lake sediments over the past 8000 years. These two lake sites lie on the borderline between two major biomes—temperate forest and prairie grassland—so it is possible that the increased fire frequency is associated with climatic conditions, the warmer, drier climate favouring the spread of prairie grasslands at the expense of deciduous forest. Around 4000 years ago, this trend was reversed. Charcoal levels began to fall, and oak

began to increase in its abundance at the expense of the prairie herbs as cooler and more humid conditions took over once more, reducing fire abundance and allowing the development of open oak woodland. Not until very recent times, in the upper layers of sediment at Kimble Pond, do we see an impact on the vegetation that clearly has its origin in human activity. The sudden expansion of ragweed indicates the arrival of settlers from Europe.

Large numbers of pollen diagrams covering the current interglacial are now available from all over the world, and many of them are firmly dated by means of radiocarbon. This has made it possible to study the movement of individual species and genera of plants by constructing pollen maps for particular periods in the past. In this way, one can follow the spread of trees, for example, and observe the routes along which they have dispersed and the way in which they have reassembled themselves into reconstituted forests. Figure 12.21 shows

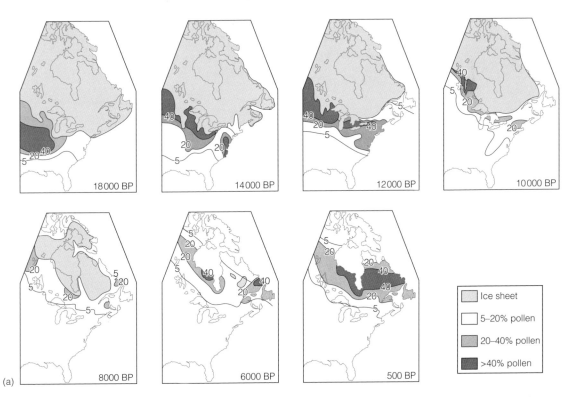

Figure 12.21 Spread of selected trees through North America since the last glacial maximum. Dates are in radiocarbon years before present (BP); they have not been corrected to solar years. Contours ('isopolls') join sites of equal pollen representation in lake sediments from the appropriate time period. (a) Spruce (*Picea*). Overleaf: (b) pine (*Pinus*); (c) oak (*Quercus*). From Jacobson *et al.* [54]. (*Continued on next page.*)

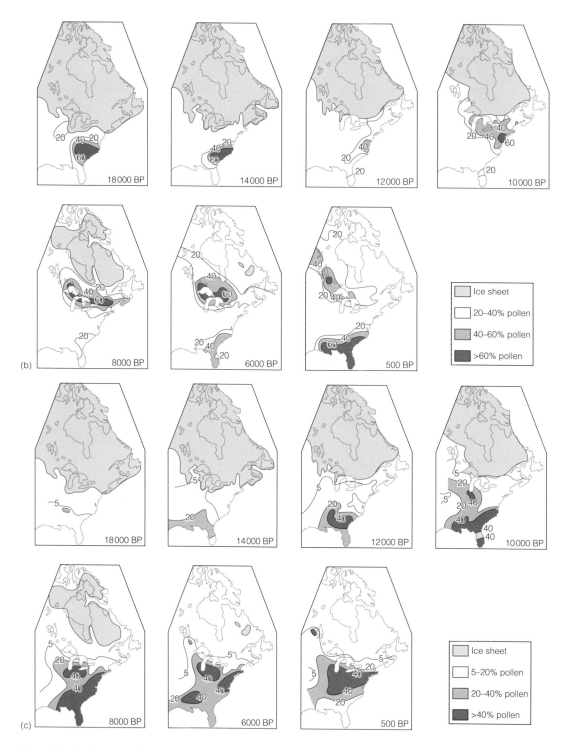

Figure 12.21 (*Continued*)

some examples of this type of work taken from the data analyses of George Jacobson, Tom Webb II and Eric Grimm on the recolonization of North America by trees after the last glaciation [54].

Just three tree taxa have been selected from their extensive studies to illustrate the patterns of tree movement in the last 18 000 years. Spruce survived the glacial maximum in the Midwest and along a broad front immediately to the south of the Laurentide Ice Sheet. It followed close on the retreating ice front, eventually settling in Canada. Pollen levels suggest that it has increased in density over the past 6000–8000 years, just as it did in the later stages of many earlier interglacials. One problem with pollen maps is that it is rarely possible to differentiate between the different species of a genus, and this is the case with spruce. It is not possible, therefore, to make allowance for the different ecological requirements of the various species when interpreting pollen maps. Pine is even more difficult to interpret than spruce because there are many native species in North America, all with very different requirements, and they cannot be effectively separated on the basis of their pollen grains. Other fossil material, such as their needles, however, indicates that both southern and northern pine types survived the glaciation in the southeastern United States, mainly on the Atlantic coastal plain. The two groups subsequently separated, one group heading north to invade areas left bare by retreating ice, and the other group becoming firmly established in Florida. The oaks had found refuge from the glaciation in Florida, where they had achieved a marked dominance by 8000 radiocarbon years (about 9000 solar years) ago. Subsequently, they have declined in Florida and have moved into their current stronghold to the south of the Great Lakes area.

The technique of mapping tree movements has provided some valuable information about the rapidity with which different species can respond to climate change, which may well prove useful knowledge when we are concerned with predicting future responses to our currently changing climate. The rates of spread in response to climate change, even of large-seeded species such as the oaks, are surprisingly rapid. In Europe, for example, the rate of spread of the oaks reached 500 m per year in the early part of the current interglacial. Rates of 300 m per year are common for many tree and shrub species [55].

The Dry Lands

The application of pollen analysis to the reconstruction of vegetation history in the Holocene is not confined to the cool temperate regions. Where lakes exist and sediments can be sampled, it has proved a very valuable tool in understanding the history of the world's dry lands. Pollen diagrams from Syria show that oak forest was advancing into the dry, steppic vegetation that had persisted during the glacial maximum. In fact, many of those parts of the world currently occupied by desert or semi-arid scrub suffered a similarly dry climate during the glacial maximum, but the close of the glaciation brought renewed rains to many of these dry areas. Studies of lakes in the vicinity of the Sahara suggest that conditions became more humid in a series of stages, commencing around 14 000 years ago, but there was a short period of aridity during the Younger Dryas. This is also supported by studies of the discharge rate of the Congo River and isotopic ratios in fossil plant waxes from this time [56]. In northeastern Nigeria, swamp forest vegetation occupied the hollows between what are now the dunes of modern savanna grasslands [52]. The early part of the Holocene thus provided a time of wetness for many currently desert areas, extending from Africa through Arabia to India. Lakes existed in the middle of the very arid Rajasthan Desert of northwest India, and the surge of freshwater down the Nile created stratified waters in the eastern Mediterranean, the low-density freshwater lying over the top of the high-density saltwater. As a result, the lower layers became depleted in oxygen, and black anoxic sediments, called **sapropels**, were deposited [57,58].

The climatic wetness in the African tropics at this time permitted the northward extension of savanna and rainforest into formerly dry areas, and Figure 12.22 summarizes data from many pollen diagrams taken from sites in the southern Sahara [59]. The age axis on this diagram is expressed in radiocarbon years: at 10 000 radiocarbon years, these are approximately 1000 years younger than the true solar dates. The expansion of the more

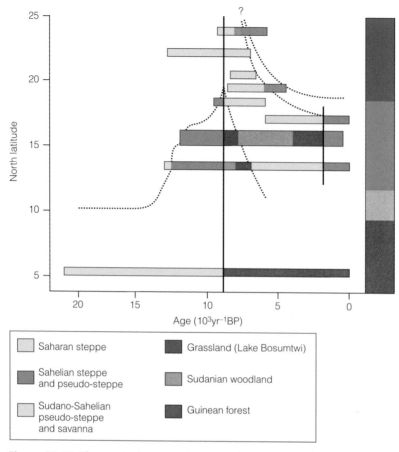

Figure 12.22 Changes in the vegetation in northern tropical Africa over the past 20 000 years. The present latitudinal zonation of vegetation is shown on the right of the diagram and it can be seen that this zonation was displaced about 58°N during the period 9000 to 7000 years ago, in the early stages of the current interglacial. From Lézine [59]. (Reproduced with permission of Elsevier.)

Legend:
- Saharan steppe
- Sahelian steppe and pseudo-steppe
- Sudano-Sahelian pseudo-steppe and savanna
- Grassland (Lake Bosumtwi)
- Sudanian woodland
- Guinean forest

humid biomes in the early Holocene, around 9000 radiocarbon years ago, can be seen on this diagram, followed by their contraction when aridity set in once more about 5000 years ago.

The Sahara Desert is quite rich in ancient rock paintings, some dating back over 8000 years. The older ones depict big-game animals now associated with the savanna grasslands, confirming the evidence of the pollen. Between 7500 and 4500 years ago, the painters of the pictures were evidently pastoralists who depicted their cattle on rocks in locations where cattle could certainly not graze today. After that time, the pictures of cows were replaced by camels and horses as the arid climate became more severe, and the activities and the domesticated animals of the local peoples altered accordingly.

Many of the great deserts of the world were evidently initiated by climatic changes in the latter half of the Holocene. The involvement of increased human activity during this time, however, has obviously complicated the picture, and some researchers believe that human exploitation of the limited arid land resources, even in prehistoric times, may have contributed to the development of deserts, as in the case of the Rajasthan Desert, where the increasing aridity coincided with the cultural development of the Indus Valley civilization [60]. It is very difficult, however, to determine the role of humankind in the development of desert spread under such circumstances.

Changing Sea Levels

Interglacial warm episodes are times of changing sea levels, and the current warm period is no exception. The melting of glacial ice has released considerable quantities of water into the oceans, and this, combined with the increase in water volume as its temperature rises, has resulted in a **eustatic** rise in sea level relative to the land. This may have amounted to as much as 100 m in places. On the other hand, the loss of ice caps over those land masses that acted as centres of glaciation relieved the Earth's crust of a weight burden, resulting in an **isostatic** upwarping of the land surface with respect to sea level. The relative importance of these eustatic and isostatic processes varied from one place to another, depending on how great a load of ice an area had borne. In western Europe, the result was a general rise in the level of the southern North Sea and the English Channel with respect to the local land surface. In this way Britain, which was a peninsula of the European mainland during the glaciation, gradually became an island. Prior to this, the rivers Rhine, Thames, Somme and Seine had all converged to flow west into the Atlantic, between the hilly areas that were later to become Cornwall and Brittany [61]. Evidence from submerged peat beds in the Netherlands suggests a rapid rise in sea level between about 10 000 and 6000 years ago, which has subsequently slowed down gradually. By this latter date, England's links with continental Europe had been severed. At this time, many plants with slow migration rates had still not crossed into Britain and were thus excluded from the British flora. The separation of Ireland from Britain occurred earlier because there are deeper channels between these land masses, and many species native to Britain have not established themselves as far west as Ireland. As a result, plants such as the lime tree (*Tilia cordata*) and herb paris (*Paris quadrifolia*) are not found growing wild in Ireland. Other plants, however, were more successful and invaded Ireland along the western seaboard of Europe before sea levels fragmented this route (Box 12.3).

Some mammals also failed to make the crossing into Ireland, and it is difficult to explain why they failed when closely related species succeeded. For example, the pygmy shrew (*Sorex minutus*) reached Ireland, yet the common shrew (*Sorex araneus*) did not (Figure 12.23). Perhaps this is related to their habitat preferences, for the pygmy shrew is found

---

The legacy of Lusitania

There is one group of Western European plants that has proved of great interest to plant geographers and that succeeded in reaching Ireland before the rising sea level separated that country from the rest of the British Isles; this is known as the Lusitanian flora. Lusitania was the name for a province of the Roman Empire consisting of Portugal and part of Spain, and, as its name suggests, the Lusitanian flora has affinities with that of the Iberian Peninsula. Some of the plants, such as the strawberry tree (*Arbutus unedo*) (see Chapter 3) and giant butterwort (*Pinguicula grandiflora*), are not found growing wild in mainland Britain. Others, such as the Cornish heath (*Erica vagans*) and the pale butterwort (*Pinguicula lusitanica*), are found in southwestern England as well as in Ireland. It therefore seems likely that these plants spread from Spain and Portugal up the Atlantic seaboard of Europe in early postglacial times but were subsequently cut off by the rising sea levels.

Recent research is finally beginning to shed some light on this classic biogeographical conundrum. Gemma E. Beatty and Jim Provan used a combination of palaeodistribution modelling and phylogeographical genetic analyses to reconstruct the distribution and spread of the Lusitanian plant species *Daboecia cantabrica* during and after the Last Glacial Maximum (LGM) [62]. Their data indicate that *D. cantabrica* survived the LGM in two separate southern refugia in western Galicia and off the coast of western France. Spain was recolonized from both refugia, whilst Ireland was probably only recolonized from the Biscay refugium. These findings strongly suggest that smaller, more northerly refugia in Ireland did not exist – although this cannot be completely ruled out.

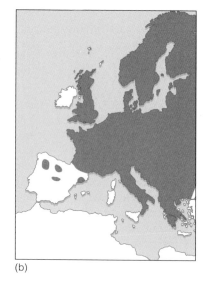

(a)   (b)

Figure 12.23 Distribution maps of two mammal species in Europe: (a) pygmy shrew (*Sorex minutus*); (b) common shrew (*S. araneus*). The pygmy shrew reached Ireland, but the common shrew failed to do so.

on moorlands and might have survived better than the common shrew if the conditions of the land bridge were wet, peaty and acidic. The stoat (*Mustella ermines*) also reached Ireland, but the weasel (*Mustella nivalis*) did not. Arrival in this case may have been due to sheer chance. A single pregnant female arriving on a raft of floating vegetation could have been sufficient to populate the island. Discoveries in some archaeological sites in Ireland of small mammals, such as the wood mouse (*Apodemus sylvaticus*), raise the possibility that some plants and animals could have been carried over the water by prehistoric humans. The arrival of some large mammals on isolated islands could also be an outcome of human transport, even in pre-agricultural times. Mesolithic (Middle Stone Age) people in the British Isles, for example, may have been responsible for carrying the red deer (*Cervus elaphus*) to Ireland and to other offshore islands such as Shetland. The red deer or elk was the major prey animal of these people, who did not have truly domesticated animals and plants, and the transport of young animals would have presented few difficulties even in small, primitive boats.

Rising sea levels during the present interglacial also were responsible for the severing of land connections in many other parts of the world. For example, Siberia and Alaska were connected across

what are now the Bering Straits, which in places are only 80 km across and 50 m deep. This high-latitude land bridge would have been a suitable dispersal route only for arctic species, but it is believed to have been the route by which humans entered the North American continent (see Chapter 13).

A Time of Warmth

The period of maximum warmth during the present interglacial lasted from early in the interglacial until about 5500 years ago. At this time, warmth-demanding species extended farther north than they do at present. For example, the hazel (*Corylus avellana*) was found considerably further north in Sweden and Finland than it is today. In North America, fossils of the plains spadefoot toad (*Scaphiopus bombifrons*) have been found 100 km north of its current distribution limits. These examples indicate that conditions have become cooler since that time. The remains of tree stumps, buried beneath peat deposits at high altitude on mountains and far north of the tree line in the Canadian Arctic, also bear witness to more favourable conditions in former times. Things are not always what they seem, however, and one has to remember the possible involvement of humans in the clearance of forests and the modification of habitats. Humans may have played an important part, for example, in the forest clearance that led to the formation of many of the so-called blanket mires of western Europe [63]. By clearing the trees, they created a new set of hydrological conditions, and the saturated soils began to accumulate peat. Continued burning and grazing by prehistoric pastoralists ensured that the forests were not able to reinvade. But the lack of trees does not mean that the climate is no longer suitable for their growth. When grazing animals are removed from these bogs, the woodland is often able to regenerate. In the case of deserts, it is also often difficult to appreciate the extent to which people and their domestic grazers currently limit the distribution of plants and animals.

Climate and fire also interact with one another, as in the prairie region of central North America. Measurements of fire frequency in the past history of the prairies show that the time between fires increases when the climate is cooler, so the composition of vegetation may be determined by climate but in an indirect way [64]. But the use of fire as a management tool by Native Americans also complicates the history and ecological impact of fire [65]. In the Mediterranean region of Europe, the extent of oak woodland was once greater than it is today, but here also the activities of humans have had a strong influence for many thousands of years, and late Holocene climatic influences on the vegetation are therefore difficult to discern.

The spread of forest in temperate areas during the early Holocene, caused by the increasing warmth, created heavily shaded, unsuitable conditions for many of the plants that had previously been widespread at the close of the glacial stage. Some of these, the arctic–alpine species, are also physiologically unsuited to warmer conditions. Many such plants, for example the mountain avens (*D. octopetala*), grow poorly when the summer temperatures are high (above 23°C for Britain and 27°C for Scandinavia). The climatic changes that occurred during the Holocene therefore proved harmful to such species, and many of them became restricted to higher altitudes, especially in the lower latitudes. Other plant species are more tolerant of high temperatures but are incapable of survival under dense shade. Low-latitude, low-altitude habitats that became covered by forests may have become unsuitable for the continued growth of these species, and many of them

also became restricted to mountains, where competition from shade-casting trees and shrubs did not occur to the same extent. It is likely, however, that some clearings and open areas were created in the forest by storms, catastrophes and the impact of large grazing animals. Some ecologists feel that the temperate woodlands were relatively open in structure, forming a kind of wood pasture during the mid-Holocene as a result of the pressures imposed by large grazing animals, but the evidence from pollen analysis and other sources does not support this view [66]. Lowland environments that for some reason bore no forest must have provided some suitable places of refuge, however. Coastal dunes, river cliffs, habitats disturbed by periodic flooding, and steep slopes all supplied sufficiently unstable conditions to hold back forest development and allow the survival of some of these open-habitat organisms, whether herbs, insects or molluscs.

An outcome of these processes of vegetation change was the development of relict distribution patterns (see Chapter 4). Sometimes, the separation of species into scattered populations, even though it has lasted only about 10 000 years, has encouraged genetic divergence, as in the case of the sea plantain (*Plantago maritima*) in Europe. This species has survived in both alpine and coastal habitats, but the different selective pressures of the two environments have resulted in physiological divergence between the two races. The coastal race is able to cope with high salinity but tends to be more frost-sensitive than the montane race. The development of molecular techniques for studying the genetic composition of organisms has led to extensive documentation of populations becoming fragmented by the changes of the Pleistocene and Holocene, resulting in the formation of distinct races. The white-tailed eagle (*Haliaeetus albicilla*), for example, shows a separation into two genetic groups, one from the west of its range, in Europe, and the other from the far east of its range, in Japan and eastern Asia [67]. Environmental changes thus fuel the pace of evolution.

Some species, limited by competitive inadequacies rather than by climatic factors, and favoured by physiological or behavioural adaptability, have taken advantage of the disturbed conditions provided by human settlements and agriculture. These plants, which fared so poorly during forested times in the temperate latitudes, have become latter-day weeds, pests and opportunists. Thus, climatic change and human habitat disturbance have interacted to provide different histories for each species.

Climatic Cooling

Climatic reconstruction and modelling can be based on many sources of evidence. Pollen analysis of lake sediments provides some useful data, but results of this kind, though giving information on past vegetation changes, are not very precise in supplying data about climate. Vegetation tends to respond rather slowly to climate change and is also subject to other factors, especially the influence of human activities. For example, pollen stratigraphic evidence from the Amazon Basin in eastern Brazil has indicated a change from closed forest to open savanna around 5000 years ago [68]. This could be associated with climatic change, such as drier conditions, and the increased incidence of charcoal in the corresponding sediments could be regarded as supportive of this view. On the other hand, the role of local human populations in burning the forest cannot be discounted, so the evidence is inconclusive regarding climate change. A recent article from Oxford palaeoecologist Professor Kathy Willis tries to distinguish the roles of climate and culture (anthropogenic fire) in shaping the current vegetation of the moist semi-evergreen rainforest of the Congo basin in West Africa [69]. A combination of pollen, microscopic charcoal and geochemical data was used to assess how vegetation dynamics have been affected by climate change, anthropogenic burning and metal smelting. The forest was found to have changed in response to burning and climate change, but not metallurgy. Interestingly, anthropogenic burning had the strongest influence, starting approximately 1000 years ago.

Oxygen isotope curves provide a more precise means of detecting climate change than by using shifts in vegetation and have supplied valuable information on temperature variations. In Greenland, for example, the oxygen isotope curve shows warm conditions between about AD 700 and 1200, followed by colder conditions until the late 19th century. This accords well with historical records for these centuries.

Another dating technique that is proving increasingly valuable and reliable is the measurement of the rate of peat growth in bogs. Peat accumulates in acid bogs because the rate of decomposition fails to keep pace with the rate of litter deposition at the bog surface. The wetness and acidity of the bog vegetation lead to low microbial activity, so that decomposition is slow. Under wetter conditions, the bog mosses grow rapidly and decomposition is curtailed, so the peat formed under such conditions looks fresh and undecomposed and is often light in colour, whereas well-humidified peat tends to be dark. The main problem with using this approach as a climate proxy is that of correlation of horizons between different sites. Local factors, such as microclimatic conditions or patterns of surface-water movement and pool formation, could vary between sites. Radiocarbon dating, particularly by the use of atomic mass spectroscopy, which enables very small samples of peat to be dated, has led to a considerable rise in interest in the use of peat growth rates over wide areas as a means of reconstructing climatic change [70,71]. On the basis of such studies, times of cooler, wetter conditions have been identified from around 5400 years ago. A time of particularly strong bog growth in the first millennium BC (especially around 800 to 400 BC) has been found through much of western Europe.

Studies of tree rings also provide a means of reconstructing past climates. Wood from the trees of temperate climates, and other types of climate with strong seasonal variation where there is a distinct alternation of summer and winter, shows annual growth rings; the width of a ring corresponds to the amount of growth the tree has achieved in that season. Care is needed in interpreting tree rings from dry climates, however, as increased growth may be related to irregular episodes of water supply. Generally, years in which the climate is suboptimal for the tree species under study produce narrower growth rings. The growth of black spruce (*Picea mariana*) in northern Canada, for example [72], shows a series of increasingly narrow annual rings between AD 1500 and 1650, indicating declining growth conditions. In this case, the rings probably reflect lower temperatures because this period of time marks the commencement of a severe stage in the so-called **Little Ice Age**. This was a lengthy spell of generally low temperature between about AD 1350 and 1850 that is apparent over a very wide area in the Northern Hemisphere. The Little Ice Age is thought to have been caused by a combination of low summer insolation in the northern hemisphere, low solar activity and several strong tropical volcanic eruptions [73].

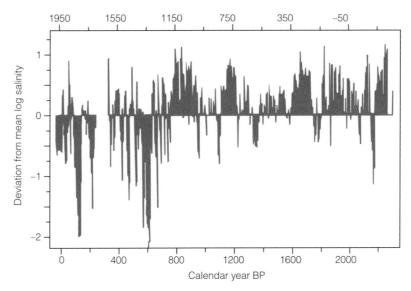

Figure 12.24 Changes in the salinity in the sediments of Moon Lake, North Dakota (Great Plains), determined from analysis of fossil diatoms. Saline episodes, caused by drought and evaporation, are more frequent prior to AD 1200 than they have been since. From Laird *et al.* [74].

Other sources of evidence of climate change are provided by lake levels that are sensitive to the balance of water input, discharge and evaporation. In the northern Great Plains region of North America, drought periods can result in evaporation increase, which in turn leads to a rise in the salinity of the water. Changes in a lake's salinity have many effects on its biota, especially the microscopic planktonic diatoms. Studies of lake salinity changes in North Dakota [74] show a sharp decline in salt content around AD 1150. The climatic change that caused the Little Ice Age in the North Atlantic region resulted in increased precipitation and lower salinity here in the continental region of North America (Figure 12.24).

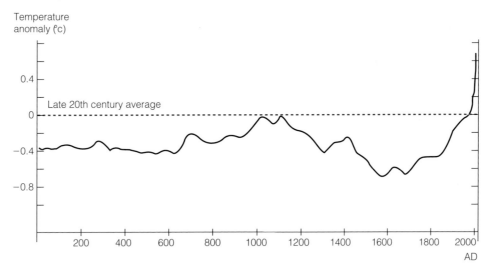

Figure 12.25 Reconstructed annual average temperature curve for the past 2000 years, based on various sources of proxy evidence.

By using a combination of all the available sources of evidence about recent climate change, it has proved possible [75] to construct a curve for the likely mean temperature of the Northern Hemisphere over the past 2000 years (Figure 12.25). There was a peak in warmth around AD 1000 to 1150, often termed the **Medieval Warm Period**, followed by the cold of the Little Ice Age. During the last 200 years, instrumental records have been available for many parts of the world, so recent changes in temperature are well documented.

Recorded History

As soon as human beings appeared on the scene, they often inadvertently began to leave information about climate and its changes. Early records provide clues rather than precise information, such as the ancient rock drawings of hunting scenes discovered in the Sahara, which indicate that its climate was much less arid at the time they were made (early Holocene) than it is today. With the development of writing, more precise information concerning climate was recorded. For example, there are reports of pack ice in the Arctic seas near Iceland in 325 BC, indicating the very low winter temperatures at the time. During the heyday of the Roman Empire, however, climate steadily grew warmer, allowing the growth of such crops as grapes (*Vitis vinifera*) in the British Isles.

In the year AD 1250, historical records suggest that alpine glaciers began to grow, and pack ice advanced in the Arctic seas to its most southerly position for 10 000 years. In 1315 a series of poor summers began in northern Europe, leading to crop failures and famine. During the Little Ice Age, glaciers reached their most advanced positions since the end of the Pleistocene glacial epoch, and tree lines in the Alps were severely depressed. The climate became somewhat warmer after 1700 and especially since 1850. There was a slight cooling after 1940, when Northern Hemisphere winters in particular became colder, but since 1970 average temperatures globally have been rising again.

With more precise climate records from different parts of the world, short-term fluctuations within the longer term trends have become apparent. It is important to understand these patterns and their causes if future climate change is to be predicted.

As will be discussed in Chapter 13, many factors influence the current global climate, including the activities of our own species. But there are also relatively short-term shifts in climate that exhibit a pattern of their own. For example, the energy output from the sun itself varies, as reflected in the number of dark sunspots on its surface. Very few sunspots were present in the 17th century, known by astrophysicists as the Maunder Minimum; calculations suggest a decrease in solar energy output at that time of about 0.4%. This was a time when the Little Ice Age was at its height. There is a general 11-year cycle in sunspot activity that could affect climatic conditions on Earth, but overall change in solar output over the last 150 years can only account for a small proportion of the observed rise in global temperature [44].

Atmosphere and Oceans: Short-Term Climate Change

Many aspects of global climate are influenced by movements of air and water masses, powered by winds and ocean currents. The oceanic conveyor described in Chapter 4 is an example of how ocean currents redistribute the energy received by the Earth from the sun and greatly modify the climate of different regions. Changes in salinity, as we have seen, can strongly influence this global movement of water, resulting in rapid climate change.

Some changes in the movement of the oceans are periodic and cause cycles of climate change, some of which are local and others global. An example that has received much attention from climatologists of late is the **El Niño Southern Oscillation** (ENSO). The west coast of South America periodically experiences particularly warm dry conditions for several months, often beginning around Christmas time, and these have been called El Niño, meaning literally the male child. The event occurs every few years—usually every 3–7 years—but varies in frequency and intensity. The general pattern of air movement in the region is dominated by the Trade Winds, which blow offshore from the west coast of South America and westward over the Pacific Ocean towards South-East Asia (see Figure 3.9), and under normal conditions these easterly winds are strong. The winds drive the surface waters of the Pacific Ocean westward, resulting in the upwelling of deep, cold water along the coast, so that the **thermocline** (the boundary between warm, less dense water and the deeper, denser water) comes close to the sea surface. On the other side of the Pacific in Indonesia, the constant influx of warm tropical water creates a deep thermocline and even has the effect of elevating the sea level in this region by about 60 cm. The strong easterly winds bring warm, moist air that creates heavy rainfall over the islands of South-East Asia, as shown in Figure 12.26.

During El Niño the east winds are weaker, so that the upwelling along the west coast of South America is less pronounced, the thermocline remains deep and the surface waters may rise in temperature by up to 7°C. The upwelling of deep water brings nutrient-rich waters to the surface and creates a highly productive ecosystem, resulting in considerable economic gains for the local people. Thus the reduced upwelling during an El Niño event can lead to decreased fish stocks, declines in the populations of piscivorous mammals, such as the fur seals, and economic hardship for local people. The lighter east winds result in the rain-bearing clouds failing to reach Indonesia, which then experiences drought. Drought in Indonesia can be very severe during periods of a pronounced ENSO cycle, as in 1982–1983,

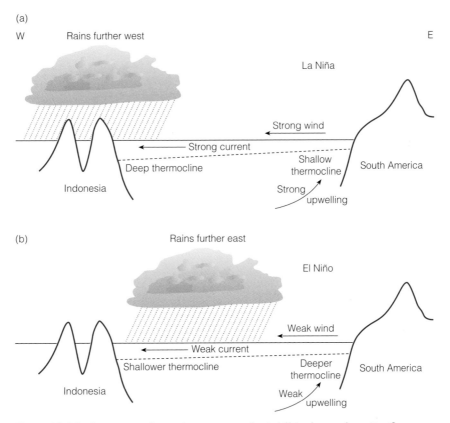

Figure 12.26 The pattern of oceanic currents and rainfall in the southern Pacific Ocean showing two contrasting situations. In (a), sometimes called La Niña, there is a strong airflow from east to west and hence a strong movement of surface water from continental South America towards Indonesia. This creates an upwelling of cold, nutrient-rich water along the South American coast and leads to high rainfall over Indonesia. In (b), called El Niño, the easterly airflow and surface ocean currents are weak, resulting in a poor upwelling of cold water along the South American coast and a shortage of rainfall in Indonesia.

when forest fires devastated large areas of Indonesia, and again in 1997–1998. The effects of El Niño are widespread throughout the world. The 1982–1983 event was associated with droughts in Australia, Central America and east Africa, together with floods in Florida and the Caribbean. The alternative state, when east winds are stronger and upwelling is more pronounced, is known by the feminine term La Niña. Exactly what causes ENSO remains unclear, but its global significance has stimulated a great deal of research into the cycle.

Corals in the oceans grow by depositing layers of calcium carbonate in annual bands, and these can be analysed isotopically to provide a record of past ocean temperature, which is affected by ENSO. Studies of the history of ENSO [76] reveal that its strength has varied considerably over the past thousand years, being particularly prominent in the mid-17th century. The strength of ENSO has also increased in recent times but is not in an exceptionally high state in comparison to the past. There appears to be no simple link between general global climate and the frequency or strength of ENSO, which means that current global warming is unlikely to be a strong influence on the system.

A similar cycle takes place in the Indian Ocean to the west of the Indonesian islands, called the **Indian Ocean Dipole**. Work on corals from this region [77] indicates that the two systems are not linked, as was once thought, but that the Indian Ocean circulation is dominated by the Asian monsoon system, independently of ENSO.

The Future

Biogeographers need to understand the underlying patterns of climate change, including short-term change, in order to predict the future changes in the distribution patterns of organisms on the planet. We are clearly in one of the warm interludes, or interglacials, that have regularly interrupted the generally cold conditions of the Pleistocene. However, what happens next is somewhat uncertain and depends upon unknowables such as the future development of human societies. What is clear is that average global temperatures have been steadily rising since the 1970s (Figure 12.27) and are set to further increase throughout the current century.

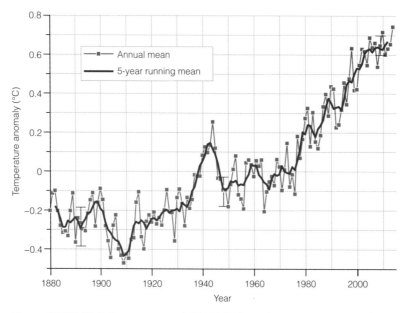

Figure 12.27 Global temperatures 1880–2012, from the Intergovernmental Panel on Climate Change. Data from NASA http://data.giss.nasa.gov/gistemp/graphs_v3/.

The most authoritative analysis of the Earth's future climate has been performed by the Intergovernmental Panel on Climate Change (IPCC), who published their fifth climate change report in 2013 [78]. Their conclusions are both significant and deeply worrying for the short-term future of life on our small blue planet: 'many of the observed changes are unprecedented over decades to millennia. The atmosphere and ocean have warmed, the amounts of snow and ice have diminished, sea level has risen, and the concentrations of greenhouse gases have increased'. It is also a fact that each of the last three decades has been successively warmer than any preceding decade since 1850, and the evidence suggests that, in the Northern Hemisphere, 1983–2012 was probably the warmest 30-year period of the last 1400 years. Some scientists even believe global warming will be sufficient to disrupt the natural progression of ice age cycles with projected increases in atmospheric CO_2 concentration during the 21st century being of sufficient magnitude to either delay or entirely prevent the next ice age [79].

As we will discuss in detail in Chapter 14, the global impacts of climate change on the diversity and distribution of species will be profound and lasting. They are also very difficult to predict, partly because global warming will soon start to produce areas that experience 'novel climates' with no current equivalent. Such novel 21st-century climates will probably lead to novel biological communities and other 'ecological surprises'. Moreover, as extant climates become extinct, they may be followed by species with narrow geographical or climatic distributions [80]. Climate scientists John Williams and Stephen Jackson predict that novel climates will initially be concentrated in tropical and subtropical regions since, under global warming, the warmest areas are necessarily the first to surpass existing climate conditions [81]. Thus, their simulations indicate that novel climates are likely to first develop in lowland Amazonia, the southeastern United States, the African Sahara and Sahel, the eastern Arabian Peninsula, southeast India and China, the Indo Pacific and northern Australia. Many of these areas (e.g. lowland Amazonia) contain enormous levels of known and unknown biodiversity, and there is a strong possibility that the shifting climate could 'push' some areas across ecological tipping points where the current biome is no longer viable [82]. More generally, novel climates represent an enormous challenge for ecological forecasting since the researchers are literally straying into unchartered territory where records of past communities provide only limited insights into what might happen [83].

Geological and meteorological studies of the past 2 million years have thus demonstrated that the climate has been extremely unstable, varying between periods of warmth and those of considerable cold. It is also clear that climate change has occasionally taken place very rapidly and that such changes have had strong effects on the biogeography of the planet. It is against the background of these fluctuating conditions that our own species enters the story, bringing new pressures to bear on the global environment. This is the theme of the next chapter.

Summary

1 The overall downward trend in global temperature observed in the Cenozoic finally resulted in the Quaternary Ice Age of the last 2 million years.

2 Within the Quaternary, glacials have alternated with interglacials in a cyclic pattern, but generally the cold phases have dominated.

3 Biomes have altered their distribution patterns in response to changes, and communities of plants and animals have broken up and reconstituted in new assemblages. Some species have become extinct in the process.

4 Many areas that are now regarded as biodiversity hotspots were greatly altered during cold episodes, including the rainforests of equatorial regions, which are likely to have been reduced in area, changed in their composition and fragmented.

5 Many factors have contributed to the development of the glacial–interglacial cycle, primarily external, astronomic factors, but also including internal changes, such as ocean–atmospheric circulation patterns, feedbacks with living ecosystems, and perhaps volcanism.

6 Climate change is not necessarily a smooth, steady process; it can alter abruptly and vary significantly over a matter of decades, as in the case of the cold Younger Dryas stadial at the opening of

the current warm episode. Critical in such change is the failure of the oceanic conveyor to maintain its global heat transfer.

7 Pollen analysis of lake and peat sediments has permitted detailed reconstruction of the rates of movement of major tree genera and the directions they took during the warming of the climate in the current interglacial.

8 Climatic changes over the past 10 000 years (the Holocene) are well recorded in lake sediments, ice sheets, marine deposits and human historical records. There appears to have been a climatic opti-mum in the first half of the Holocene, after which conditions have become cooler. Global sea-level changes during these times have altered and created barriers to dispersal for species in the course of their spread.

9 Atmosphere–ocean interactions are responsible for some short-term climatic alterations and cycles, such as the El Niño Southern Oscillation (ENSO), which is centred on the southern Pacific Ocean but has climatic impacts throughout the world.

Summary

Further Reading

Beerling DJ, Chaloner WG, Woodward FI (eds.). Vegeta-tion–climate–atmosphere interactions: past, present and future. *Philosophical Transaction of the Royal Society London B* 1998; 353: 1–171.

Delcourt HR, Delcourt PA. *Quaternary Ecology: A Pal-aeoecological Perspective.* London: Chapman & Hall, 1991.

Lowe JJ, Walker MC. *Reconstructing Quaternary Envi-ronments.* London: Longman, 1996.

Moore PD, Chaloner B, Stott P. *Global Environmental Change.* Oxford: Blackwell Science, 1996.

National Research Council. *Abrupt Climate Change: Inevitable Surprises.* Washington DC: National Acad-emy Press, 2002.

Pielou EC. *After the Ice Age: The Return of Life to Glaci-ated North America.* Chicago: University of Chicago Press, 1991.

Tallis JH. *Plant Community History.* London: Chapman & Hall, 1991.

Willis, KJ, Bennett, KD, Walker D (eds.). The evolution-ary legacy of the Ice Ages. *Philosophical Transaction of the Royal Society London B* 2004; 359: 155–303.

References

1. Edgar KM, Wilson PA, Sexton PF, Suganuma Y. No extreme bipolar glaciation during the main Eocene calcite compensation shift. *Nature* 2007; 448 (7156): 908–911.

2. Zanazzi A, Kohn MJ, MacFadden BJ, Terry DO. Large temperature drop across the Eocene–Oligocene transi-tion in central North America. *Nature* 2007; 445 (7128): 639–642.

3. Eldrett JS, Harding IC, Wilson PA, Butler E, Roberts AP. Continental ice in Greenland during the Eocene and Oligocene. *Nature* 2007; 446 (7132): 176–179.

4. West RG. *Pleistocene Geology and Biology*, 2nd ed. London: Longman, 1977.

5. Jansen E, Sjøholm J. Reconstruction of glaciation over the past 6 Myr from ice-borne deposits in the Norwegian Sea. *Nature* 1991; 349 (6310): 600–603.

6. Moore PD, Chaloner B, Stott P. *Global Environmental Change.* Oxford: Blackwell Science, 1996.

7. Winograd IJ, Coplen TB, Szabo BJ, Riggs AC. A 250,000-year climatic record from Great Basin vein cal-cite: Implications for Milankovitch theory. *Science* 1988; 242 (4883): 1275–1280.

8. Emiliani C. Quaternary paleotemperatures and the duration of the high-temperature intervals. *Science* 1972; 178 (4059): 398–401.

9. Zachos JC, Dickens GR, Zeebe RE. An early Cenozoic perspective on greenhouse warming and carbon-cycle dynamics. *Nature* 2008; 451 (7176): 279–283.

10. Moore PD, Webb JA, Collison ME. *Pollen Analysis.* Oxford: Blackwell Scientific Publications, 1991.

11. Stevenson A, Moore P. Pollen analysis of an intergla-cial deposit at West Angle, Dyfed, Wales. *New Phytolo-gist* 1982; 90 (2): 327–337.

12. Lowe JJ, Walker MJ. *Reconstructing Quaternary Environments.* 3rd ed. London: Routledge, 2014.

13. Kerr RA. Second clock supports orbital pacing of the ice ages. *Science* 1997; 276 (5313): 680–681.

14. López de Heredia U, Carrión JS, Jimenez P, Collada C, Gil L. Molecular and palaeoecological evidence for multiple glacial refugia for evergreen oaks on the Iberian Peninsula. *Journal of Biogeography* 2007; 34 (9): 1505–1517.

15. Leroy SA, Arpe K. Glacial refugia for summer-green trees in Europe and south-west Asia as proposed by ECHAM3 time-slice atmospheric model simulations. *Journal of Biogeography* 2007; 34 (12): 2115–2128.

16. Petit RJ, Aguinagalde I, de Beaulieu J-L, *et al.* Glacial refugia: hotspots but not melting pots of genetic diversity. *Science* 2003; 300 (5625): 1563–1565.

17. Hewitt GM. Post-glacial re-colonization of European biota. *Biological Journal of the Linnean Society* 1999; 68 (1–2): 87–112.

18. Provan J, Bennett K. Phylogeographic insights into cryptic glacial refugia. *Trends in Ecology and Evolution* 2008; 23 (10): 564–571.

19. FAUNMAP Working Group. Spatial response of mammals to late Quaternary environmental fluctuations. *Science* 1996; 272 (14): 1601–1606.

20. Svenning JC, Skov F. Ice age legacies in the geographical distribution of tree species richness in Europe. *Global Ecology and Biogeography* 2007; 16 (2): 234–245.

21. Nogués-Bravo D, Ohlemüler R, Batra P, Araújo MB. Climate predictors of late Quaternary extinctions. *Evolution* 2010; 64 (8): 2442–2449.

22. Nogués-Bravo D, Rodríguez J, Hortal J, Batra P, Araújo MB. Climate change, humans, and the extinction of the woolly mammoth. *PLoS Biology* 2008; 6 (4): e79.

23. Dansgaard W, Johnsen SJ, Clausen HB, *et al.* Evidence for general instability of past climate from a 250-kyr ice-core record. *Nature* 1993; 364 (6434): 218–220.

24. Schmidt MW, Vautravers MJ, Spero HJ. Rapid subtropical North Atlantic salinity oscillations across Dansgaard–Oeschger cycles. *Nature* 2006; 443 (7111): 561–564.

25. Whitlock C, Bartlein PJ. Vegetation and climate change in northwest America during the past 125 kyr. *Nature* 1997; 388 (6637): 57–61.

26. Kershaw A. A long continuous pollen sequence from north-eastern Australia. *Nature* 1974; 251: 222–223.

27. Prideaux GJ, Long JA, Ayliffe LK, *et al.* An arid-adapted middle Pleistocene vertebrate fauna from south-central Australia. *Nature* 2007; 445 (7126): 422–425.

28. Turney CS, Flannery TF, Roberts RG, *et al.* Late-surviving megafauna in Tasmania, Australia, implicate human involvement in their extinction. *Proceedings of the National Academy of Sciences* 2008; 105 (34): 12150–12153.

29. Jolly D, Taylor D, Marchant R, Hamilton A, Bonnefille R, Buchet G, Riollet G. Vegetation dynamics in central Africa since 18,000 yr BP: pollen records from the interlacustrine highlands of Burundi, Rwanda and western Uganda. *Journal of Biogeography* 1997; 24 (4): 492–512.

30. Tallis JH. *Plant Community History: Long-Term Changes in Plant Distribution and Diversity.* London: Chapman & Hall, 1991.

31. Colinvaux PA. *Amazon Expeditions: My Quest for the Ice-Age Equator.* New Haven, CT: Yale University Press, 2007.

32. Bonaccorso E, Koch I, Peterson AT. Pleistocene fragmentation of Amazon species' ranges. *Diversity and Distributions* 2006; 12 (2): 157–164.

33. Wells G. Observing earth's environment from space. In: Friday L, Laskey R (eds.), *The Fragile Environment. The Darwin College Lectures.* Cambridge: Cambridge University Press, 1989: 148–192.

34. Loftis DG, Echelle AA, Koike H, Van Den Bussche RA, Minckley CO. Genetic structure of wild populations of the endangered desert pupfish complex (Cyprinodontidae: Cyprinodon). *Conservation Genetics* 2009; 10 (2): 453–463.

35. Ganopolski A, Rahmstorf S, Petoukhov V, Claussen M. Simulation of modern and glacial climates with a coupled global model of intermediate complexity. *Nature* 1998; 391 (6665): 351–356.

36. Hays JD, Imbrie J, Shackleton NJ. Variations in the Earth's orbit: pacemaker of the ice ages. *Science* 1976; 194: 1121–1132.

37. Petit J-R, Jouzel J, Raynaud D, *et al.* Climate and atmospheric history of the past 420,000 years from the Vostok ice core, Antarctica. *Nature* 1999; 399 (6735): 429–436.

38. Sigman DM, Boyle EA. Glacial/interglacial variations in atmospheric carbon dioxide. *Nature* 2000; 407 (6806): 859–869.

39. Galbraith ED, Jaccard SL, Pedersen TF, *et al.* Carbon dioxide release from the North Pacific abyss during the last deglaciation. *Nature* 2007; 449 (7164): 890–893.

40. Woodward FI. Stomatal numbers are sensitive to increases in CO_2 from pre-industrial levels. *Nature* 1987; 327: 617–618.

41. McElwain JC. Do fossil plants signal palaeoatmospheric CO_2 concentration in the geological past? *Discussion. Philosophical Transactions of the Royal Society* 1998; 353: 83–96.

42. Broecker WS, Denton GH. What drives glacial cycles? *Scientific American* 1990; 262 (1): 42–50.

43. Bray J. Volcanic triggering of glaciation. *Nature* 1976; 260: 414–415.

44. Houghton J. *Global Warming: The Complete Briefing.* Cambridge: Cambridge University Press, 2009.

45. Benson L, Burdett J, Lund S, Kashgarian M, Mensing S. Nearly synchronous climate change in the Northern Hemisphere during the last glacial termination. *Nature* 1997; 388 (6639): 263–265.

46. Moreno PI, Jacobson GL, Jr, Lowell TV, Denton GH. Interhemispheric climate links revealed by a late-glacial

cooling episode in southern Chile. *Nature* 2001; 409 (6822): 804–808.

47. Broncltnr WS, Kennett JP. Routing of meltwater from the Laurentide Ice Sheet during the Younger Dryas cold episode. *Nature* 1989; 341: 28.

48. Fairbanks RG. A 17,000-year glacio-eustatic sea level record: influence of glacial melting rates on the Younger Dryas event and deep-ocean circulation. *Nature* 1989; 342 (6250): 637–642.

49. Smith JE, Risk MJ, Schwarcz HP, McConnaughey TA. Rapid climate change in the North Atlantic during the Younger Dryas recorded by deep-sea corals. *Nature* 1997; 386: 818–820.

50. McManus J, François R, Gherardi J-M, Keigwin LD, Brown-Leger S. Collapse and rapid resumption of Atlantic meridional circulation linked to deglacial climate changes. *Nature* 2004; 428 (6985): 834–837.

51. Clark PU, Pisias NG, Stocker TF, Weaver AJ. The role of the thermohaline circulation in abrupt climate change. *Nature* 2002; 415 (6874): 863–869.

52. Camill P, Umbanhower C, Teed R, *et al.* Late-glacial and Holocene climatic effects on fire and vegetation dynamics at the prairie–forest ecotone in south-central Minnesota. *Journal of Ecology* 2003; 91 (5): 822–836.

53. Bennett KD. Postglacial population expansion of forest trees in Norfolk, UK. *Nature* 1983; 163: 164–167.

54. Jacobson GL, Jr, Webb T, III, Grimm EC. Patterns and rates of vegetation change during the deglaciation of eastern North America. In: Ruddiman WF, Wright HE (eds.), *The Geology of North America*. Boulder, CO: Geological Society of America, 1987: 277–288.

55. Huntley B, Birks HJB. *An Atlas of Past and Present Pollen Maps for Europe: 0–13000 Years Ago*. Cambridge: Cambridge University Press, 1983.

56. Schefuß, E, Schouten S, Schneider RR. Climatic controls on central African hydrology during the past 20,000 years. *Nature* 2005; 437 (7061): 1003–1006.

57. Rossignol-Strick M, Nesteroff W, Olive P, Vergnaud-Grazzini C. After the deluge: Mediterranean stagnation and sapropel formation. *Nature* 1982; 295: 105–110.

58. Sancetta C. The mystery of the sapropels. *Nature* 1999; 398 (6722): 27–29.

59. Lézine A-M. Late Quaternary vegetation and climate of the Sahel. *Quaternary Research* 1989; 32 (3): 317–334.

60. Singh G, Joshi RD, Chopra SK, Singh AB. Late Quaternary history of vegetation and climate of the Rajasthan Desert, India. *Philosophical Transactions of the Royal Society B: Biological Sciences* 1974; 267 (889): 467–501.

61. Gibbard P. Europe cut adrift. *Nature* 2007; 448 (7151): 259–260.

62. Beatty GE, Provan J. Post-glacial dispersal, rather than in situ glacial survival, best explains the disjunct distribution of the Lusitanian plant species *Daboecia cantabrica* (Ericaceae). *Journal of Biogeography* 2013; 40: 335–344.

63. Moore PD. The origin of blanket mire, revisited. In: Chambers FM (ed.), *Climate Change and Human Impact on the Landscape*. London: Chapman & Hall, 1993: 217–224.

64. Bond W, Van Wilgen B. *Fire and Plants*. London: Chapman & Hall, 1996.

65. Delcourt PA, Delcourt HR. *Prehistoric Native Americans and Ecological Change: Human Ecosystems in Eastern North America since the Pleistocene*. Cambridge: Cambridge University Press, 2004.

66. Moore PD. Down to the woods yesterday. *Nature* 2005; 433 (7026): 588–589.

67. Hailer F, Helander B, Folkestad AO, *et al.* Phylogeography of the white-tailed eagle, a generalist with large dispersal capacity. *Journal of Biogeography* 2007; 34 (7): 1193–1206.

68. De Toledo MB, Bush MB. A mid-Holocene environmental change in Amazonian savannas. *Journal of Biogeography* 2007; 34 (8): 1313–1326.

69. Brncic TM, Willis KJ, Harris DJ, Washington R. Culture or climate? The relative influences of past processes on the composition of the lowland Congo rainforest. *Philosophical Transactions of the Royal Society B: Biological Sciences* 2007; 362 (1478): 229–242.

70. Barber K, Maddy D, Rose N, Stevenson AC, Stoneman R, Thompson R. Replicated proxy-climate signals over the last 2000 yr from two distant UK peat bogs: new evidence for regional palaeoclimate teleconnections. *Quaternary Science Reviews* 2000; 19 (6): 481–487.

71. Blackford J. Palaeoclimatic records from peat bogs. *Trends in Ecology and Evolution* 2000; 15 (5): 193–198.

72. Payette S, Filion L, Delwaide A, Bégin C. Reconstruction of tree-line vegetation response to long-term climate change. *Nature* 1989; 341 (6241): 429–432.

73. Wanner H, Beer J, Bütikofer J, *et al.* Mid- to Late Holocene climate change: an overview. *Quaternary Science Reviews* 2008; 27 (19): 1791–1828.

74. Laird KR, Fritz SC, Maasch KA, Cumming BF. Greater drought intensity and frequency before AD 1200 in the Northern Great Plains, USA. *Nature* 1996; 384: 552–554.

75. Moberg A, Sonechkin DM, Holmgren K, Datsenko NM, Karlén W. Highly variable Northern Hemisphere temperatures reconstructed from low-and high-resolution proxy data. *Nature* 2005; 433 (7026): 613–617.

76. Cobb KM, Charles CD, Cheng H, Edwards RL. El Nino/Southern Oscillation and tropical Pacific climate during the last millennium. *Nature* 2003; 424 (6946): 271–276.

77. Abram NJ, Gagan MK, Liu Z, Hantoro WS, McCulloch MT, Suwargadi BW. Seasonal characteristics of the Indian Ocean Dipole during the Holocene epoch. *Nature* 2007; 445 (7125): 299–302.

78. Stocker T, Qin D, Plattner G-K, *et al. Climate Change 2013: The Physical Science Basis*. Cambridge: Cambridge University Press, 2014.

79. Rapp D. Future prospects. In: RappD (ed.), *Ice Ages and Interglacials*. Springer: Heidelberg, 2012: 327–375.

80. Williams JW, Jackson ST, Kutzbach JE. Projected distributions of novel and disappearing climates by 2100 AD. *Proceedings of the National Academy of Sciences* 2007; 104 (14): 5738–5742.

81. Williams JW, Jackson ST. Novel climates, no-analog communities, and ecological surprises. *Frontiers in Ecology and the Environment* 2007; 5 (9): 475–482.

82. Malhi Y, Aragao LEOC, Galbraith D, *et al.* Exploring the likelihood and mechanism of a climate-change-induced dieback of the Amazon rainforest. *Proceedings of the National Academy of Sciences* 2009; 106 (49): 20610–20615.

83. Willis KJ, Araújo MB, Bennett KD, Figueroa-Rangel B, Froyd CA Myers N. How can a knowledge of the past help to conserve the future? Biodiversity conservation and the relevance of long-term ecological studies. *Philosophical Transactions of the Royal Society B: Biological Sciences* 2007; 362 (1478): 175–187.

84. Brown JH. The desert pupfish. *Scientific American* 1971; 225: 104–110.

People and Problems

Section V

The Human Intrusion

Chapter **13**

The Pleistocene was a time of climatic instability, which had considerable impact on the distribution patterns of organisms over the face of the Earth. It was a time of extinctions, but also one of diversification for some types of organisms. There has been much debate concerning whether speciation became faster or slower during the Quaternary Ice Age, and the general conclusion is that extinction rates within the Pleistocene exceeded speciation rates [1]. For mammals, it was a time of extensive evolution, and most living species of mammal evolved during Quaternary times, driven by climatically unstable Quaternary environments [2]. One of the species that evolved at this time was our own species, Homo sapiens, which has had an even greater impact on the biogeography of the Earth than even the Ice Ages. It has therefore been suggested that this period of time should be known as the 'Anthropocene' [3,4].

The Emergence of Humans

The fossil history of humans is still very incomplete, but each year brings new material to light, helping to fill in the gaps and providing a more detailed picture of how anatomically modern humans emerged. The primates of the New World and the Old World became separated from each other some 40 million years ago (mya), and it is the Old World branch that is ancestral to humans. We are very closely related to the great apes, which include the orangutan *Pongo*, and especially to the gorilla *Gorilla* and the chimpanzees *Pan* (Figure 13.1). The separation of the human ancestral branch (the **hominins**) from the great apes (the two groups are known jointly as **hominids**) is thought to have taken place about 7 mya. A major source of evidence on which this estimate is based is the genetic similarity between humans and chimpanzees; almost 99% of human genetic makeup is shared with the chimpanzee, so their evolutionary divergence must have been relatively recent in geological terms. Palaeontological research into this relationship has been hampered by a lack of fossil material of chimpanzees: the earliest chimpanzee fossil, found in East Africa, dates back only 0.5 mya [5].

Trying to establish the biogeography of the early hominids, which lived during the Miocene over 5 mya, is extremely difficult because of the lack of fossils. However, we can instead study the fossil record of other, larger and more common mammal groups, such as the hyaenids (hyenas) and the proboscideans (mammoths and elephants), which are often associated with hominids [6]. These share a common set of patterns involving speciation in Africa in the Early Miocene and expansion into Europe, Asia and North America during the Middle Miocene, followed by a movement back into Africa. It is very likely that hominids (which include the ancestors of apes and humans) followed similar changing patterns of distribution in the Miocene.

There is still controversy about whether a single line of development emerged from the common ancestor with the chimpanzee, or whether a number of individual lineages interbred, eventually leading to the evolution of the human line.

Biogeography: An Ecological and Evolutionary Approach, Ninth Edition. Edited by C. Barry Cox, Peter D. Moore, Richard J. Ladle.
© 2016 John Wiley & Sons, Ltd. Published 2016 by John Wiley & Sons, Ltd.

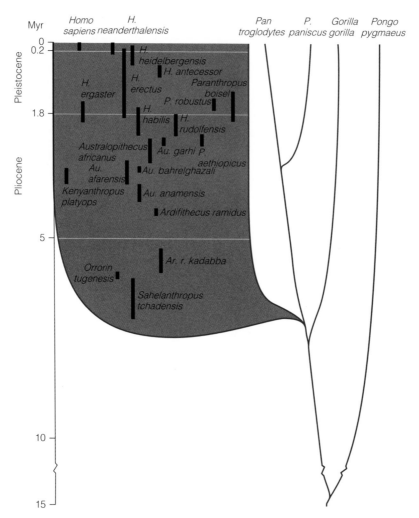

Figure 13.1 The relationships between the hominins and the great apes. From Carroll [4].

Whichever is the case, the hominin line became more bipedal and developed a larger brain and greater manual dexterity. While chimpanzees remained in the lower canopy of the forest, the ancestors of humans took to the savanna woodlands and grasslands.

One of the great problems underlying any discussion of human evolution results directly from our natural, intense interest in the subject. It has therefore been relatively easy to get grants to search for fossils related to our ancestry. It has also encouraged scientists to give a new scientific name (a new genus or a new species) to anything that is found. As a result, many more fossils related to our ancestry have been discovered and named than in any other group.

As is normal in palaeontology, most of the fossils are incomplete and fail to answer some of the many questions on the significance of their structure and adaptations. It seems very likely that our evolution, like that of other groups, has not been a simple, linear progress through time, but has involved a number of parallel lineages and branches. Fitting all the fossils into this complex pattern of evolution inevitably becomes difficult, and has often caused vigorous disagreement and arguments. The recognition of different fossil species is further complicated by variation within the species due to factors such as age, disease or sexual dimorphism.

In most groups, there are sufficient gaps in time and/or space between the different fossils that there is no difficulty in recognizing them as separate species. However, because of the comparative wealth of human fossils and the intensity of interest, this too becomes a problem. A recent practical solution has been to recognize six informal 'grades' rather than try to establish the detailed relationship of the different species. These grades are: early, archaic, megadont ('large-toothed') and transitional hominins, plus pre-modern *Homo* and modern *Homo* (Figure 13.2) [7]. 'Species' are placed in a particular grade on the basis of having a similar role in the ecosystem, similar posture and adaptations of the limbs, feet and hands, and similar diet. The grades are as follows.

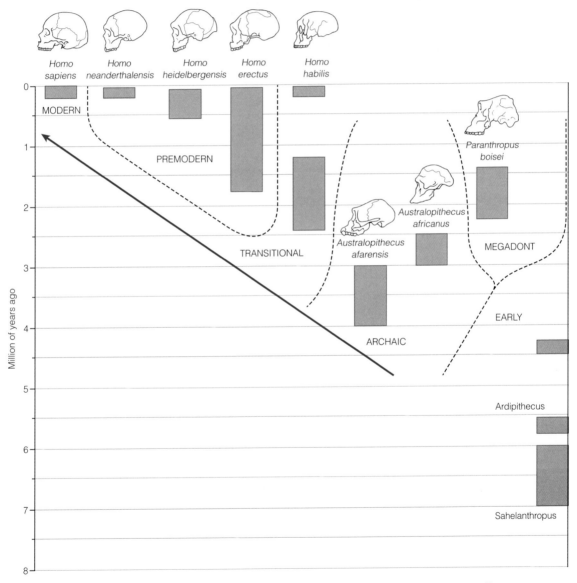

Figure 13.2 A scheme showing the possible interrelationships between hominins over the last 7 million years. Adapted from Wood and Lonergan [7].

Early Hominins

In 2002, Michel Brunet and his fellow researchers [8] discovered six fossil bones (a cranium and some lower jaws) that had hominin similarities. They were assigned to a new genus on the human line of evolution and called *Sahelanthropus tchadensis*. Until then, most finds associated with early human evolution had been discovered in East Africa, but this set of fossils came from further west, in the African state of Chad, in the Sahel region, south of the Sahara Desert. The fossil fauna associated with these finds suggests a date of Late Miocene, around 6–7 mya. So, if the genetic estimate of the separation from the ape line of evolution is correct, *Sahelanthropus* could be one of the very first organisms on the human line of development, as opposed to the great ape line. The skull of *Sahelanthropus* is remarkable in that it has the features of a chimpanzee when viewed from behind, but the front view looks very like the genus *Australopithecus*, a genus of hominins that was to become very important some 3 million years later. Its discovery has given strong support to those who prefer to think of human evolution as stemming from a single line, for it has precisely the combination of characters one might expect of such a stem.

Archaic Hominins

The fossil remains of the **australopithecines** (the name given to members of the genus *Australopithecus*) that succeeded *Sahelanthropus* are widely recorded from eastern and southern Africa; the earliest finds were in Tanzania and Ethiopia and dated from around 4 mya. Among the fossils of this age is the partial skeleton known as 'Lucy,' which has supplied a great deal of anatomical information about early hominins [9]. These fossils have been named *Australopithecus afarensis*. The extraordinary discovery of obviously bipedal fossil human footprints in volcanic ash from Tanzania [10] shows that they walked upright, on their hind legs, though their short legs suggest that they were not adapted to running. Until recently, it was thought that the ability to make stone tools only appeared with the genus *Homo* [11], the discovery of stone tools 3.3 million years old shows that it was, instead, the australopithecines that made this revolutionary invention. The habitat in which

they lived was open woodland and savanna, but very little is known of their precise way of life and the ecological niche they occupied. *A. afarensis* was probably merely one species within the ecosystem's complex food web, no more important than any other species. Molecular studies are helping to sort out the diet and ecological role of the australopithecines (Box 13.1).

Megadont Hominins

A short-lived side-line of hominin evolution lived in East Africa 2.3–1.4 mya. Characterized by heavy, strong jaws and large molar teeth with a thick coating of enamel, this hominin was originally named *Zinjanthropus* but is now known as *Paranthropus*. The males of this genus were much larger than the females. It was clearly adapted to feeding on more resistant food such as large seeds, nuts and C4 grasses and sedges.

Transitional Hominins

This group includes *Homo habilis*, a species that lived nearly 2 mya. It is different from the australopithecines in having an upright posture and larger brain. However, the structure of its arms and hands suggests it was still quite adept at climbing, and its ankle has australopithecine characteristics. It is therefore transitional, showing a mixture of characteristics, some advanced but others still primitive. It was accepted as a member of the genus *Homo* largely because there was evidence that it could make stone tools and, at the time at which it was discovered, it was assumed that this ability was confined to our own genus. Knowledge of its diet and ecological role is still fragmentary, but it is believed that the meat content of the diet increased (see Box 13.1).

Although bipedalism has advantages, such as freeing the hands for other tasks and elevating the head above the ground vegetation, it would not have allowed early members of the genus *Homo* to outrun the larger mammalian quadrupeds around them – including animals they hunted as well as those that hunted them. On the other hand, bipedal striding locomotion would have allowed them to run long distances [12]. It is possible that the loss of body hair took place at this time, in order to allow the body to cool down more readily by sweating.

Australopithecine diet

It is difficult to be sure about the diet of the australopithecines. It is possible that they were largely vegetarian but also took occasional animal prey, just like the chimpanzee. One attempt to reconstruct the australopithecine diet has used the enamel from their fossil teeth, which is extremely hard-wearing and can survive intact over millions of years. The chemistry of the enamel may reflect diet, especially in the isotopes of carbon that are contained within it. The technique is based on the fact that the two photosynthetic systems operating in plants, C3 and C4 (plus CAM), accumulate carbon from the atmosphere as a result of the activity of two different enzymes (see Chapter 2). The two enzymes have different capacities for discriminating between the two isotopes of carbon, 13C and 12C, with 13C being enriched by the C4–CAM system. So the organic products of photosynthesis differ depending on which type of plant has constructed them. The different proportions of carbon isotopes are also transmitted to the animals that consume them, so analysis of plant or animal organic matter can help determine their photosynthetic origins. When tooth enamel of australopithecine fossils was analysed [74], it was found to be rich in 13C, suggesting that the australopithecines had a diet that was rich either in C4 plant species (such as the tropical grasses, including their roots or seeds) or in the meat of herbivores that consumed C4 plant species. The lack of wear on the tooth enamel and the lack of any tools that would be needed for grinding fibrous grasses suggest that the carbon isotope ratios in australopithecines' teeth were because they were eating meat.

So the evidence implies that these hominins lived in open, savanna environments and that animals formed an important part of their diet. The volcanism of East Africa increased the fertility of its soils and therefore of its vegetation, which in turn encouraged the evolution of grazing herbivores. These herbivores may have included large mammals, such as elephants, rhinos, giraffes, gazelles and bovids, together with insects that fed on grasses. It is important to remember, however, that the australopithecines were small in stature, only about the size of chimpanzees (1–1.5 m tall and weighing 30–50 kg), so their ability to prey on very large mammals must have been limited. However, the carcasses of these mammals, killed by large carnivores, may have been quite abundant, and so facilitated the transition of humans from largely plant-based food to carnivory via scavenging.

Pre-modern *Homo*

A new hominin, *Homo ergaster*, appeared around 1.9 mya, closely followed *Homo erectus*. There is every reason to believe that these new species were the direct descendants of *H. habilis*. Analysis of Kenyan fossil footprints dated to 1.5 mya and believed to belong to *Homo ergaster/erectus* suggests that the feet of this species were essentially the same as those of modern humans [13].

Although *H. habilis* is believed to have spread into the Eurasian region of modern Georgia [14], the first species of our genus to be found far beyond Africa is that known as *Homo erectus*. It left Africa about 1.7 mya, and had spread into eastern Asia by 100 000 years later [15]. Some excavations in Java indicate that *H. erectus* may have survived in South-East Asia as late as the last Ice Age (50 000 years ago), in which case it would have overlapped with our own species in that area [16].

By 1.5 mya, *H. erectus* had developed much more sophisticated stone tools, such as hand axes. Even more significantly, there is evidence of the use of fire. Although at first it may have been used as a means of food preparation, the potential of fire as an aid in hunting must surely have been appreciated by this intelligent species. Both the fauna and the flora of the grasslands must have been altered by this new phenomenon in the environment.

The populations of *H. erectus* in Africa gradually evolved into a new hominin, *Homo heidelbergensis*, which lived between 600 000 and 100 000 years ago. A unique insight into the way of life of this species has been provided by the discovery of hunting spears, buried in compressed peat deposits in north Germany [17], dated to 400 000 years ago. This suggests that the ancestors of modern humans who occupied the northern regions of Europe were big-game hunters, and supports the argument that the hunting and butchering of animals using tools extend far back into human ancestry. *H. heidelbergensis* also reached Britain by about 500 000 years ago, although stone tools dating to about 900 000 years ago show that other hominins had reached the island even earlier.

The British palaeontologist Chris Stringer [18] suggests that an evolutionary split occurred in *H. heidelbergensis* between 400 000 and 300 000 years ago. The first evidence of this split is the appearance of *Homo neanderthalensis* about 200 000 years ago with evidence of *Homo sapiens* appearing later, in 160 000-year-old deposits from Ethiopia [19]; these latter finds support the view that our species evolved in Africa. The earliest reliable date for fossil *H. sapiens* outside Africa comes from Israel, with a date of 115 000 years. So it is likely that the human population of Africa began to expand and spread into other parts of the world at around that time.

Neanderthals were the first of these two species to enter Europe, about 45 000 years ago, followed by *H. sapiens* about 10 000 years later. DNA from bones dating 38 000 years ago from Uzbekistan in Central Asia has Neanderthal affinities, suggesting that the species may well have spread extensively through Asia [20]. There were no sharp breaks between these successive species in the history of human evolution: *Australopithecus*, *Homo habilis*, *H. erectus*, *H. heidelbergensis*, *H. neanderthalensis* and finally *H. sapiens*. These 'species' should be regarded as stages that palaeontologists find it convenient to recognize and name for ease of reference in what was really a gradual process of evolutionary change.

H. neanderthalensis and *H. sapiens* co-existed in Europe and Asia Minor 40 000 to 35 000 years ago [21]. So it is not surprising to find that there was some interbreeding between these two closely related species in the last half million years, as shown by the fact that analysis also shows that 1–4% of our own DNA comes from the Neanderthals. Recent studies suggest that such interbreeding may have been adaptive, helping modern humans to adjust to non-African environments [22].

Neanderthals disappeared from the fossil record about 28 000 years ago, though some would claim survival to 24 000 years ago in Gibraltar [23]. But why did they die out? It is possible that active competition, or even conflict, between the two species played a part. But it may be significant that the Neanderthals disappeared at a time that coincides with a major expansion in global ice volume. This climatic change may have placed an additional strain on Neanderthal survival. One fact is clear however, only *Homo sapiens* remained in Europe at the beginning of the Holocene.

Until comparatively recently, it seemed as though *H. sapiens* was also the only member of our genus present world-wide during the Holocene. This changed in 2003 when a skeleton of an adult hominin only about 1 m tall was unearthed during the excavation of cave sediments dating back only 18 000 years on the island of Flores, Indonesia [24]. Additional fossil bones of other members of the population were found in the cave in 2004, so the original discovery was not, as many at first thought, just a single aberrant individual. It was recognized as a new species, *Homo floresiensis*, which appears to be a dwarf form of its genus, like many other examples of animals or reduced size living in islands, with their limited supplies of nourishment (see Chapter 7). Using a basic energetics model, biogeographers have recently calculated that a greater number of small-bodied hominins could persist on Flores than larger bodied hominins, partly explaining how they could persist for so long on such a small island [25]. In fact, the fauna of the island also included a pygmy form of the elephant *Stegodon*, which *H. floresiensis* may well have hunted.

Though it was also suggested that it could be a pygmy form of our own species, the cranial structure of *H. floresiensis* does not support this view. The feet of these small people were unusually long and, though they were bipedal, the feet are in some ways more ape-like than human. This raises the possibility that this hominin may even not be a direct descendant of *H. erectus*, but may be derived from some other line of primate development [26]. The hunt continues for further samples, but the debate about human ancestry in this part of the world remains very active.

Modern *Homo*

Even before the last glaciation had begun, *H. sapiens* was spreading into the Middle East and also further south into the African continent (Figure 13.3). By the middle of the glaciation, our ancestors had reached the Asian interior, north of the Caspian

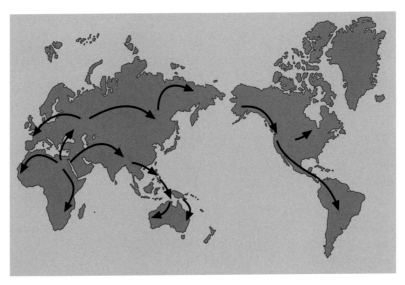

Figure 13.3 Map showing the probable routes taken during the spread of *Homo sapiens* out of Africa over the past 100 000 years.

Sea, and had also spread across the Tibetan Plateau into South-East Asia. Australia became populated by about 50 000 years ago, and humans had reached northern Europe and eastern Asia by the time the glaciation was at its peak, around 20 000 years ago. Only the Americas remained empty of our species.

When the last glaciation was at its height, a very large volume of the world's water was locked up in the ice of the expanded ice caps and glaciers (see Chapter 12), which meant that the sea level of the world's oceans was considerably lower, perhaps by around 100 m. Regions that are now beneath the sea were then exposed as land, and a land bridge formed across the Bering Straits, linking Siberia to Alaska. This is the most likely route by which North America was colonized. Fossil plant material dating from about 24 000 years ago, the time of the maximum glacial extent, has been found in Yukon and has revealed much about the nature of the environment on this land bridge. Grasses and prairie sage (*Artemisia frigida*) were abundant [27], so the tundra steppe vegetation would have supported herds of large herbivores, including woolly mammoth, horses and bison. The hunting peoples of eastern Asia probably followed these herds across into the New World. Stone spear-points known as Clovis points, dated to about 13 000 years ago, have been found widely across North

America, and it was logical to conclude that these were the traces of hunting by the recently arrived first human hunters. A human skeleton dated to 13 000–12 000 years ago, found in Mexico, that shows a mixture of Asian and Native American features also seems compatible with that scenario. But a range of other discoveries over the last 15 years suggests that our species arrived in the New World much earlier (reviewed in [28]). These include traces of human occupation about 15 500 years ago north of Austin in Texas, and also at Monte Verde in southern Chile about 14 000 years ago [29]. Human bones dated to about 13 000 years ago found in the Channel Islands off California also suggest that some, at least, of the early Americans may have arrived by boat, and this is supported by slightly later evidence of a maritime culture, similar to that on the Asian Pacific coast, on those same islands. The abundant seafood, ranging from fish to marine mammals, down the American Pacific coast, would have provided sustenance for them, and all this might be a clue to the origin of the people who lived in southern Chile. One problem is that most of the evidence for the presence of a sea-coast culture has been covered up by the rise in sea levels since that time. All in all, there is still a great deal to learn about the early history of our species in the Americas.

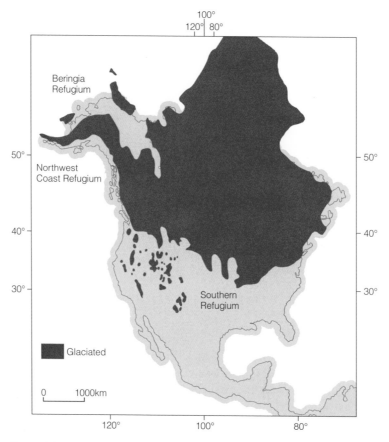

Figure 13.4 Maximum extent of the last (Wisconsin) glacial in North America, showing the three ice-free areas (refugia) which correspond to the language groups of Native Americans. From Rogers *et al.* [30].

A completely different line of enquiry on this problem comes from the study of the geography of human languages, which suggests that the colonization of the New World must have taken place before the major advance of the last glaciation about 22 000 years ago. The linguistic research of R.A. Rogers [30] has revealed three distinct groups of Native American languages, and these are centred on the three ice-free refugial areas of North America during the height of the last glaciation (Figure 13.4). This suggests that human populations were isolated in these areas during the glacial maximum and subsequently spread to other regions. The extinct language Beothuk, once spoken in Newfoundland, could belong to another population isolated in an eastern refugium. An alternative explanation is that there have been three separate invasions of North America from

Asia, thus accounting for the three language groups [31]. Genetic analysis of people from a wide range of Native American races, however, has indicated a very close similarity, suggesting that there was just one founding population of immigrants, who have subsequently diverged into communities as widely separated as the Aleuts of Alaska and the Yanomami people of Brazil. The current view of most researchers is that there was a single origin for the all the native peoples of America [32].

Modern Humans and the Megafaunal Extinctions

The spread of the human species during the last glaciation was accompanied by the extinction of many species of large mammals, known as the megafauna – first in Australia, then in Eurasia and finally in North America. It was long assumed that these extinctions were the result of the climatic changes, but the American anthropologist Paul Martin suggested that humans may have been the culprits [33]. He pointed out that most of the animals that became extinct were large herbivorous mammals or flightless birds, weighing over 50 kg body weight—in other words, precisely the fauna humans might have been expected to hunt. He also observed that similar extinctions had taken place in other, more southern areas, and suggested that the timing of these extinctions varied in significant fashion. In each case, the time corresponded with the evolution, or arrival, of a race of humans with relatively advanced hunting techniques. In Africa, for example, where *H. sapiens* evolved, the extinctions at this time were far less severe, probably because of the long history of ecological coexistence between herbivores and hominids, so that any changes in herbivore size in response to human hunting pressure had already taken place.

The record of extinctions has been studied in the greatest detail in North America. Martin suggests that 35 genera of large mammal (55 species) became extinct in North America at the end of the last (Wisconsin) glaciation, over twice as many as had taken place during all the earlier glaciations, and this at a time when the climate was already becoming warmer. This combination of features seems to support the idea that some agent other than climate was responsible, and it is reasonable to suspect the hunting activities of humans. But the American anthropologist J.E. Grayson [34] has shown that there was a similar rise in the level of extinctions of North American birds (ranging from blackbirds to eagles) at that same time. Since it is unlikely that early humans were responsible for the extinction of these birds, this observation throws doubt on the whole hypothesis of the dominant role played by humans in Pleistocene extinctions in general.

On the other hand, the fact that so many North American species became extinct at the same time (12 000–11 000 years ago) as hunting peoples arrived on the continent, whereas in Europe the extinctions were spread over a longer period, provides a strong body of circumstantial evidence to support Martin's theory [35]. The debate continues, and the extinction of so many different species of

mammals may not have been caused by one single factor. However, other studies show that the time of extinction can be precisely correlated with the arrival, or intensification, of settlement by human populations. Dale Guthrie of the University of Alaska has obtained radiocarbon dates for the Late Pleistocene and Early Holocene remains of many large mammal fossils from Alaska and the Yukon Territory [36] (Figure 13.5). Although the horse (*Equus ferus*) and the mammoth (*Mammuthus primigenius*) became extinct at about the time that humans were settling the area, other large mammals, such as wapiti or elk (*Cervus canadensis*) and bison (*Bison priscus*, which later evolved into *Bison bison*), had begun to expand their populations before the human invasion. The implication is that environmental changes, involving both

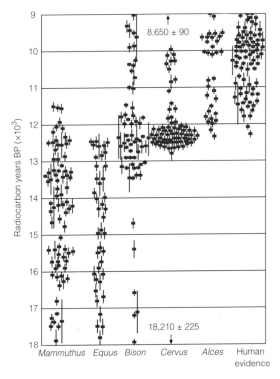

Figure 13.5 Radiocarbon dates of fossil mammal bones from North America showing the loss of mammoth and horse around the time of the Pleistocene–Holocene transition. The expansion of bison, moose and wapiti populations is also shown, as are dates for human occupations, mainly based on charcoal from hearths. From Guthrie [36].

climate and vegetation, had already created new conditions under which the balance of competition between the large herbivores had changed. It is always possible, of course, that human predation added to pressures on the declining species, but the situation is clearly complex, with many species already extinct or in decline when they came into contact with palaeo-Indians [37].

In Europe, the mammoth survived into the Younger Dryas stadial, but had vanished by the opening of the Holocene. The same was true of the giant Irish 'elk' (*Megaloceros giganteus*), which stood 2 m high at its shoulder and bore antlers with a 3.5-m span. Both species survived much longer in Asia: the giant elk was present in the Ural Mountains until 7700 years ago, and the mammoth survived in western Siberia until 3600 years ago [38]. In the case of the mammoth, climatic change led to the loss of 90% of its former geographical range between 42 000 and 6000 years ago, with suitable habitat mainly restricted to Arctic Siberia [39]. This suggests that there were complex interactions between climate, vegetation, competition and human predation.

In Australia, where human populations had maintained their hunting activities in the midlatitude regions through the global cooling of the last glacial episode, megafaunal extinction came earlier than in Europe and America. All 19 marsupial species that exceeded 100 kg in weight, and 85% of all animal species greater than 44 kg, became extinct during the Late Pleistocene (Box 13.2). Some of these, such as the ostrich-sized flightless bird *Genyornis newtoni*, have been studied in detail. By radiocarbon dating 700 of the fossil eggshells of *Genyornis*, it has proved possible to trace its decline from widespread and common 100 000 years ago to a sudden disappearance about 50 000 years ago, corresponding to the arrival of humans [40]. However, once again, there is evidence of climatic change at this same time [41].

So the case for human involvement in the megafaunal extinction process is complicated by the fact that it took place at a time of rapid climatic and environmental change. It has been proposed that some form of planetary impact took place at around 12 900 years ago that could have affected the climate and the megafauna [42]. The evidence

Does size matter?

Box 13.2 — Concept

One intriguing observation concerning the megafaunal extinction at the close of the last glaciation could prove helpful in interpreting the event. Fossil evidence suggests that several of the species heading for extinction underwent a period of size reduction prior to their final disappearance. In North America, for example, about 70% of the large mammals became extinct about 13 000 to 11 000 years ago, as determined by radiocarbon dating. Of these, the horses comprise an important group, and these have been studied intensively. Dale Guthrie of the University of Alaska has examined fossil metacarpal bones of two horse species in Alaska at the close of the last glaciation [75] and has found a 14% decline in the size of the bones between 20 000 and 12 000 years ago. He considers this finding to be due to the climatic changes taking place at that time leading to a decline in the availability of forage for the grazers. It is possible, therefore, that the climate change could have finally resulted in the extinction of these species. A similar decline in size has been found in the kangaroos of Australia around this time.

But is the size decline in herbivorous mammals necessarily proof of the climatic cause of megafaunal extinction? One could look at the data in another way. Human populations of the time may well have concentrated their hunting on the larger individual animals, and this could have resulted in an evolutionary selection in favour of smaller individuals. Modern trophy hunting is known to have an impact on the overall size of a population of prey animals, as in the case of bighorn sheep [76]. So human hunting in the Late Pleistocene could have selected against large individuals. A further consideration is the rate of reproduction among the prey species. On the whole, it was the slower breeding animals that became extinct at this time, again suggesting that they were unable to recover from constant harvesting by humans [77].

is based on the discovery of tiny diamonds in sediments marking the commencement of the Younger Dryas cold stadial. These roughly spherical nano-diamonds (less than 300 nm in diameter) have been found throughout North America, from Canada to Arizona, and in Germany. The layer is reminiscent of the iridium layer at the KT (end of Cretaceous) boundary, also associated with mass extinction. However, the presence of the diamonds does not fully confirm the impact of a shower of comet fragments, and certainly cannot explain all aspects of the megafaunal extinction data [43].

Plant Domestication and Agriculture

The success of *Homo sapiens* can be explained in many different ways, including high brain capacity, manual dexterity, toolmaking and social organization. Their adaptability was another feature that enabled the survival of our species through times of fluctuating climate. When prey animals became scarce, humans readily turned to alternative sources of food. By about 164 000 years ago, some human groups in South Africa had resorted to marine habitats for their main food supply, harvesting shellfish and other intertidal organisms [44]. Hunting as a means of subsistence was supplemented by gathering available resources, which included both animal and plant products.

By the end of the last cold period, people living in the Middle East were experimenting with a new technique for enhancing their food supplies. In the fertile region of Palestine and Syria grew a number of annual grasses with edible seeds, the ancestors of our wheat and barley. The people occupying these regions at the time were hunters and gatherers of the Upper Palaeolithic, feeding on a rich variety of animals, including gazelles, birds, rodents, fish and molluscs. Some settlements have been discovered in Israel dating from about 23 000 years ago, and their hearths contain residues of charred dough made from the seeds of wild barley, wheat and other grass species. Recognizing the value of these plants, successive generations of people must have encouraged the growth of such useful plants by removing shade-casting trees and shrubs, and disturbing soils so that their seeds germinated more effectively. It was then a simple step to retain

some of the seeds from one season to the next, and to select those strains that were richest in edible grains.

Tracing the ancestry of our modern species of cereals is difficult, but Figure 13.6 represents a possible scheme for the evolution of modern wheat, based on a study of the chromosome numbers of the various wild and cultivated species. The original wild wheat species had a total of 14 chromosomes (seven pairs) in each cell. *Triticum monococcum* (einkorn) was the first wild wheat to be extensively used as a crop plant. It probably hybridized with other wild species, but the hybrids would have proved infertile because the chromosomes could not pair up prior to the formation of gametes. Faulty cell division in one of these hybrids, however, could solve this problem because, once the chromosome number had doubled (**polyploidy**; see Chapter 6), chromosome pairing could take place and the species would become fertile. Fertile polyploid hybrid species formed in this way included another important crop species, emmer (*Triticum turgidum*, sometimes called *Triticum dicoccoides*) with 28 chromosomes. This evolutionary development probably took place naturally, without any human intervention, because emmer wheat was one of the plants found in association with Upper Palaeolithic gatherers [45]. However, modern bread wheat (*Triticum aestivum*) has 42 chromosomes, and this probably arose as a result of polyploidy following the hybridization of emmer with another wild species, *Triticum tauschii*. This species comes from Iran, and it probably interbred with emmer as a result of the transport of that wheat species by migrating human populations. Thus, early agriculturalists began not only to modify their environment, but also to manipulate the genetics of their domestic species. So the genetic modification of food-producing domesticated species is a process as old as agriculture itself.

More sophisticated studies of wheat seeds have used DNA fingerprinting. Manfred Heun, from the Agricultural University of Norway, and his colleagues [46] analysed 338 samples of the most primitive wheat species, einkorn, which still grows wild in the Near East. They were able to locate wild einkorn populations in a region of the Karacadag Mountains in southeastern Turkey and close to the Euphrates River, which provide genetic evidence of being the progenitors of domesticated

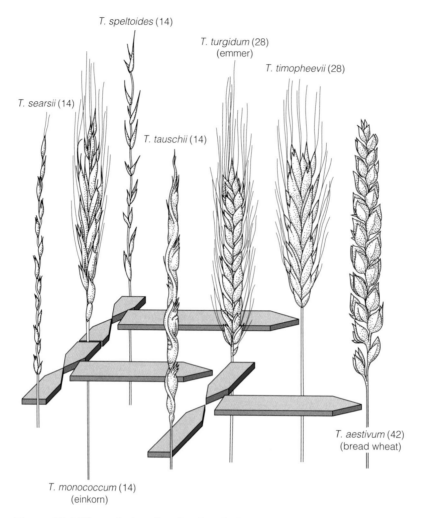

Figure 13.6 The evolution of modern bread wheat. This reconstruction is tentative, but it represents the probable course of crossings among the wild wheat species that led to the early domesticated forms of the genus *Triticum*, and the subsequent crossings of domesticated wheats with wild species and chromosome doubling that led to bread wheat. Figures in parentheses after names represent chromosome numbers.

einkorn. The application of molecular techniques in the study of domestication and of the biogeographical origins of many plants and animals is becoming increasingly important in solving some important and long-unanswered biogeographical and anthropological questions.

Molecular techniques have also been used to investigate whether plant domestication was a rapid process involving a limited number of ancestral forms, or a protracted series of trial-and-error domestications complicated by a constant genetic influx from wild species [47]. This process of mixing would have been constantly modified as the farmers selected for specific traits and brought in new genetic material to improve their crops [48]. As a result of the complete analysis of the genome of baker's yeast (*Saccharomyces cerevisiae*) in 1996, it has been found that even the domestication of this fungus involved a complex series of mixing wild strains and selecting the required characteristics [49].

Humans also grew and cultivated other wild plants in the Middle East (Figure 13.7), including barley, rye, oat, flax, alfalfa, plum and carrot. Further west, in the Mediterranean Basin, yet more native plants were domesticated, including pea, lentil, bean and mangelwurzel. The idea of domestication and organized agriculture resulted in rapid changes in the diet of early Neolithic people [50]. This generated a steady spread of agricultural techniques from the Near East across the continent of Europe during the Holocene (Figure 13.8), and was accompanied by the transport of domesticated plants from their native regions. It is still disputed whether this process involved the movement of peoples or only the cultural diffusion of new techniques [51]. If agriculture was an efficient means of stabilizing food supplies and avoiding some of the chance catastrophes of hunting and gathering, then it could have led to population expansion and the need to move to new areas.

Some archaeologists have investigated the association between agricultural spread and the dominance by certain language groups. For example, 144 of the languages spoken in Europe and Asia belong to a group known as Indo-European, and the reason for this dominance could be its spread with the populations that were developing agriculture in the Early Holocene. There are two variants of this idea [52]. It was at first suggested that basic Indo-European was spoken by the early agriculturalists when they spread out of Anatolia (in modern Turkey) 9500 to 8000 years ago. But a later suggestion points out that the reconstruction of this language includes words for wheeled vehicles, which only appeared following the domestication of the horse by herders on the steppes of western Eurasia, called the Yamnaya, more recently, 5000–6000 years ago. The DNA of their remains is a close match to that of individuals from Northern Europe who made a type of pottery known as 'Corded Ware'. This suggests that there was a massive westward migration of these early herders, who took with them their

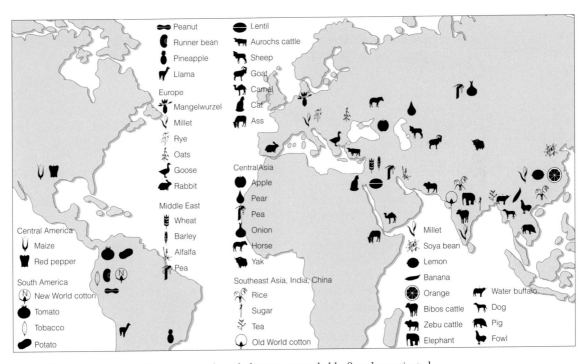

Figure 13.7 Areas where different animals and plants were probably first domesticated.

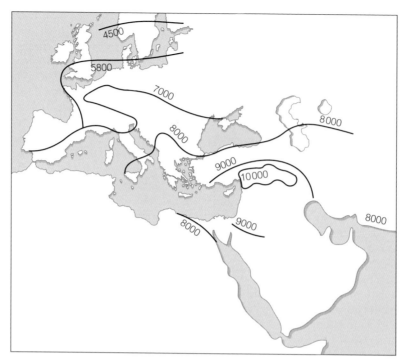

Figure 13.8 Map of Europe showing the spread of agriculture from the area of the Fertile Crescent. Dates are in radiocarbon years before present. Radiocarbon timescales diverge increasingly from calendar or solar timescales as one moves further into the past. A radiocarbon date of 10 000 years ago is approximately equivalent to 11 500 calendar years ago (9500 BC). The pattern is greatly simplified here, and there are problems concerning the precise time and direction of agricultural spread in such regions as the Balkans. From Willis and Bennett [51].

horses, wheeled vehicles and at least an early branch of the Indo-European language group (Figure 13.9).

The idea of plant domestication seems to have evolved independently in many different parts of the globe (Figure 13.10) and at many different times. In each area, appropriate local species were exploited: in southwest Asia, there were millet, soybean, radish, tea, peach, apricot, orange and lemon; central Asia had spinach, onion, garlic, almond, pear and apple; and in India and South-East Asia, there were rice, sugarcane, cotton and banana. Rice cultivation, for example, began in the coastal wetlands of eastern China over 7500 years ago [53]; settlers used fire to clear alder swamps ready for rice growing. Maize, New World cotton, sisal and red pepper were originally found in Mexico and the rest of Central America, while tomato, potato, common tobacco, peanut and pineapple

first grew in South America. In some cases there may have been independent cultivations of the same or similar species in different parts of the world: for example, emmer wheat may well have originated quite independently in the Middle East and in Ethiopia.

New World agriculture is thought to have begun in Central America with the cultivation of three major crops, maize (*Zea mays*), bean (*Phaseolus vulgaris*) and squash (*Cucurbita pepo*). Some argument has surrounded the time of this independent agricultural development, especially in relation to agricultural origins in other parts of the world. Radiocarbon dates of squash seeds from caves in Oaxaca, Mexico, place their cultivation at about 10 000 years ago, so an early origin for New World agriculture is now well established [54]. The origins of maize have long been debated, as explained in Box 13.3.

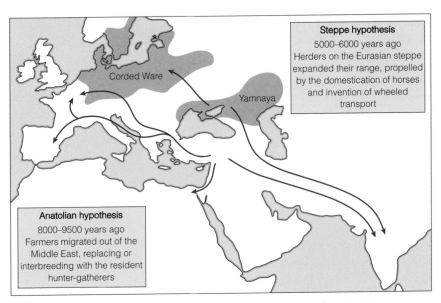

Figure 13.9 Two hypotheses on the origin of the Indo-European family of languages.

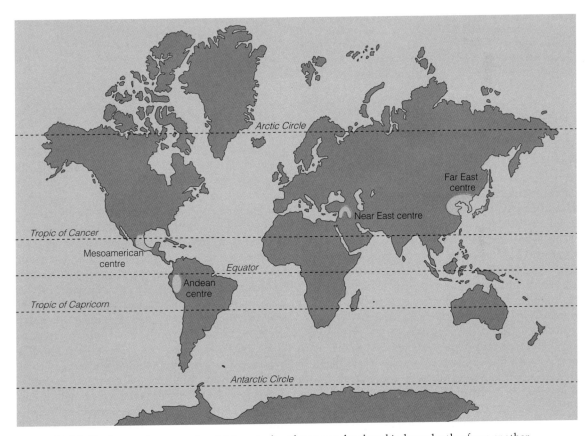

Figure 13.10 The four main centres in which agricultural systems developed independently of one another.

One of the most puzzling problems in the study of plant domestication has been the origin of maize. Most research workers studying this subject now agree that the most likely ancestor is the Mexican annual grass, teosinte. Indeed, both maize and teosinte are now regarded as subspecies of *Zea mays*, but structurally they are very different, especially in their flower and fruit structure. Teosinte has its fruits arranged in just two rows along the axis of the inflorescence, and each seed is surrounded by a persistent woody case that is shed with the seed. Its value as a food plant is therefore limited because it is difficult to separate the nutritious seed from its unpalatable case. Corn, or maize, has many rows of fruits in its inflorescence and, most important, the seeds are easily separated from their cases. The real problem is that there are no intermediate types between teosinte and maize, and neither are there any archaeological records of wild teosinte grains being collected and used for food. Very often, maize appears quite suddenly in the archaeological record of the diet of early human populations, and it may then become the dominant food resource. Its arrival at a site can be traced by isotopic changes in fossil human bone collagen. Maize is a C4 plant (see Chapter 3), and one of the characteristics of these plants is that the ratio of 13C to 12C in the sugars produced by photosynthesis differs from that found in C3 plants. This difference is retained in animals feeding on the plant, so it is possible to trace the importance of C4 plants in an animal's diet. In this case, it can be used to record the adoption of maize cultivation during the history of a human population.

The massive structural transformation involved in the evolution of maize from teosinte, however, may not have required major genetic change. The loss of the persistent fruitcase, which is the most critical limitation on the use of the plant for food production, involves a change in a single gene [78]. This change probably arose as a random mutation but was effectively selected out by the early farmers who recognized the value of this particular variety. This discovery not only goes a long way toward solving the mystery of maize evolution, but also demonstrates that relatively small genetic changes can result in considerable alterations in an organism's form, and hence its value as a domesticated species.

Animal Domestication

The domestication of some animals may have preceded that of plants. There is some evidence, for instance, that early cultures had domesticated the wolf or, in some North African communities, the jackal. It is likely that such animals were of considerable use in driving and tracking game and hunting down wounded prey, but determining when the dog was domesticated has proved very difficult using conventional archaeological techniques. Bones of dog/wolf associated with human settlements (found as far back as 400 000 years) could merely be a result of people eating the animal, rather than domesticating it. Joint burials of dogs with humans, which provide a reasonable indication of domestication, are known that date from the Early Holocene in the Near East, but little is known of earlier associations. Molecular studies may again give the best clue. Carles Vilà, from the University of California, Los Angeles, and his colleagues [55] have analysed mitochondrial DNA samples from 162 wolves and 140 dogs, representing 67 different breeds. Their studies support the idea that the dog evolved from the wolf, but the differences between the two groups suggest that the evolutionary separation (presumably associated with domestication and isolation of dogs from wolves) took place about 100 000 years ago, but later work suggests it could be as little as 30 000 years ago. As in the case of plant domestication, however, the combined processes of selection and backcrossing with wild races have probably confused the genetic record. This is clearly illustrated by the fact that none of the most ancient breeds of dog come from regions where the oldest archaeological remains have been found, and three of these breeds (basenjis, dingoes, and New Guinea singing dogs) actually derive from regions outside the natural range of wolves [56]. (The geographical regions from which many types of domestic animals originated are shown in Figure 13.7.)

It is likely, however, that many of the other animals that became associated with humans, such as sheep and goats, were domesticated much later, during the Early Neolithic period and soon after the first cultivation of plants. These were initially herded for their meat and hides but, once tame enough to handle, they would have also been a source of milk. The first traces of domesticated sheep come from Palestine around 8000 radiocarbon years ago. These may have originated from one of the three European and Asiatic sheep, or may have resulted from interbreeding among these species. The Soay sheep that has survived in the Outer Hebrides in Scotland almost certainly originated from the mouflon, either the European *Ovis musimon* or the Asiatic *Ovis orientalis*. Domestication of these animals may have resulted from the adoption of young animals orphaned as a consequence of hunting activity. In Israel, there is a marked shift in diet between 10 000 and 8000 years ago, gazelle and deer being replaced by goat and sheep. Almost certainly this was a consequence of domestication.

The aurochs, *Bos primigenius*, a species of wild cow, were frequent inhabitants of the mixed deciduous woodland that was spreading north over Europe during the postglacial period. The bones of this animal have been found in many of the sites where fossil remains of these forests have been preserved, such as in buried peats and submerged areas. The aurochs became extinct in the 17th century, but it was long thought that European cattle resulted from the domestication of this wild bovid. The analysis of DNA in modern cattle, however, has led to a very different conclusion [57]. The domestic cattle of Europe (which were subsequently introduced into North America) have DNA that is very distinct from that of fossil aurochs bones and is more closely related to *Bos taurus* of the Middle East. The cattle of Africa and India, on the other hand, seem to have arisen from different stock, suggesting that cattle domestication took place in several different locations and was based on different bovid species.

Humans modified the genetic constitution of their domesticated animals by selecting for certain qualities, such as placid behaviour, as well as for high meat or milk production. Conversely, the close association of people with animals also led to genetic modifications in the human species, as in the case of milk digestion. It has proved dif-ficult to establish the earliest date of human use of milk from domesticated mammals, but molecular and isotopic evidence can be used to detect the presence of those fatty acids associated with milk fat. It is likely that the milking of cattle was being practiced in the Near East about 9000 years ago [58]. Milk evolved in mammals as a means of increasing the growth rate of the very young. Its major constituent is the carbohydrate lactose. In the young mammal, this is digested by the enzyme lactase, but the gene controlling the production of this enzyme is normally switched off in humans as they are weaned. European races of humans are exceptional in that lactase production continues through the life of the adult, so they can still digest milk. The same is true of many East African peoples, but not of those in West Africa. The work of Sarah Tishkoff of the University of Maryland on the genetics of African peoples has suggested that the mutation leading to lactase persistence in adults took place relatively recently, probably as late as 7000 years ago [59]. It seems likely, therefore, that the increasing use of domesticated cattle in East Africa during the Holocene led to selection for the maintenance of the activity of the lactase gene. So the process of domesticating cattle eventually led to evolutionary change in the people who had caused it.

Diversification of *Homo sapiens*

As humans spread around the world and developed their own cultures and food resources, they continued to diversify in response to the new environmental pressures placed on them. Some of the modifications they developed resulted from their choice of food, as in the case of the persistence of lactase production in populations that consumed milk. However, some of the diversity found among human races is less easy to explain, as in the case of blood groups. The blood group Rh, for example, is found in 30% of all Caucasians and yet is rare in the peoples of the Far East and among Native Americans. The A and B blood groups are totally absent in Native Americans, yet are relatively common among Caucasians. As yet, it has proved impossible to provide a general explanation of these differences in terms of their selective value under different environmental circumstances. Some blood groups are associated with higher or lower incidence of particular diseases. For example, individuals with blood group O are more likely to suffer from stomach ulcers, and those with group A have a higher incidence of stomach cancer. But these are conditions normally found in older people, so one could argue that they are unlikely to have a strongly deleterious effect in evolutionary terms. On the other hand, the survival of a post-breeding, senior cohort in a population could have other advantages, such as assistance with rearing the young, or the transmission of tribal experience and wisdom, so the health of the old could still have advantages for the population as a whole.

One of the most visually obvious sources of diversity among humans is the colour of their skin. There is a great range of skin colours and shades within the human species, and there is an evident geographical correlation between skin types. On the whole, the peoples of the equatorial regions have darker skins than those of the high latitudes, and the most obvious explanation for this is protection from the intense light of the tropics, especially the harmful **ultraviolet radiation** (UVR). UVR can cause skin cancer and is also responsible for the destruction of certain B vitamins, such as folic acid, in the skin. The intensity of UVR is greater near the equator because light from the sun passes along a shorter path through the atmosphere, as explained in Chapter 3. At higher latitudes, light passes at a shallow angle through a greater depth of atmosphere, where more energy is reflected or absorbed. However, there are complications, such as the effect of increasing altitude, when UVR also increases with altitude, so the relationship is not a simple latitudinal one. Recent research shows that between 70 and 77% of the variation in skin colour can be accounted for by UVR level [60]. This is quite a high level of correlation and seems to account for the predominance of dark skin types in the tropics, so our African ancestors almost certainly had black skin. On the other hand, there is a positive advantage for lighter skin colour in the peoples of higher latitudes because, like everyone, they need vitamin D, which is manufactured in the skin from biochemical precursors when exposed to UVR. This vitamin is required for calcium metabolism and bone growth, so lack of UVR exposure

can lead to softening of the bones, collapse of the pelvis, death of the foetus and an increased susceptibility to tuberculosis. There is, therefore, a clear advantage for the people of the higher latitudes to have a pale skin colour that allows the penetration of ultraviolet radiation. In the course of evolution, different races have developed the most efficient level of skin pigmentation to assure the optimal level of vitamin D production while protecting tissues from excessive UVR [61]. However, recent research on fossil DNA shows that this happened slowly, only beginning in about 6000 BC, and was brought into Europe in the genome of the early agriculturalists as they spread into the continent from the Near East.

In excessive amounts, vitamin D can be toxic, so dark skin prevents this from occurring. Vitamin D balance may also explain one of the most obvious exceptions to the general latitudinal variation in skin colour, namely, the relatively dark colour of the Inuit people of the far north. These tundra-dwelling people feed mainly on fish, walruses, seals and polar bears, and the livers of these animals contain very high levels of vitamin D. The Inuit avoid eating large quantities of liver from their prey, but this source of vitamin D means that there is no need for weak skin pigmentation in order to boost its production. Another apparent exception to the UVR correlation with skin colour is the Bantu people of southern Africa, who have a darker skin than might be expected for the southern temperate latitudes they now occupy. But these people migrated south into these regions only within the last 2000 years, giving little time for evolutionary change, so once again the lack of correlation can be explained.

The recent movements of people around the world have obviously complicated any studies of human biogeography and adaptation, but there are examples where such movements have themselves produced new evolutionary developments. One good example is the blood condition called *sickle-cell anaemia* that, as its name suggests, causes anaemia and other malfunctions of the blood system. One would expect evolution to select strongly against this condition, but in West Africa the gene for sickle-cell anaemia occurs in over 20% of the population. The reason for this retention of a potentially harmful gene in the population is that it also provides the carrier with a high degree of resistance to the malarial parasite. As in the case of skin colour, there is a trade-off between positive and negative effects. When people were forcibly transported from West Africa to North America as slaves, they encountered conditions where malaria was less common and thus operated less strongly as a selective factor in population survival. Under these circumstances, the gene for sickle-cell anaemia became distinctly less advantageous, and its incidence among North Americans of African descent has now dropped to below 5%. In Central America, where malaria remains a greater risk, the gene is still found in 20% of the people of West African origin.

The Biogeography of Human Parasitic Diseases

When our species first evolved, as hunter-gatherers on the plains of Africa, the people would, like any other species, have been subject to a variety of diseases, some caused by viruses and bacteria, and others by parasitic organisms. Some of these diseases may have infected the ancestors of early humans, such as *Homo habilis* and *Homo erectus*. So our early African ancestors were probably already liable to such widespread infections as roundworms (*Ascaris*), hookworms (*Necator*) and amoebic dysentery (*Entamoeba histolytica*). All of these have infective stages that are passed out with the faeces of the infected individual and then wait in the soil or water until ingested by the next individual. As hunter-gatherers on the plains of Africa, however, our ancestors probably suffered less from such diseases than do the people who live there today in more sedentary settlements. Their habit of continually moving on from temporary campsites would have ensured that they did not remain for long near their own faeces, which might otherwise have acted as infective agents for the eggs or larvae of parasites. Furthermore, the members of each small, independent group would have been closely related to one another, and therefore all would have had a similar amount of immunity to any viral or bacterial infections. So though any such infection might rapidly lead to the death of most of the individuals in a particular group, any survivors would be immune to future infections. As a result, our ancestors would not have suffered from epidemics that spread from group to group.

The life cycle of most parasites involves not only the final, **definitive host** (such as a human being) but also an **intermediate host**, or **vector**, within whose body the parasite multiplies and is transformed into a stage that can infect a new definitive host. Flying, blood-sucking insects are particularly well suited to the role of intermediate host. They were probably quick to take advantage of the thin skin and reduced covering of body hair of this new species of hominid, even if it was as yet only present in low-population densities. The best-known of these diseases today is malaria, which is transmitted by the mosquito *Anopheles* and is caused by a protozoan (*Plasmodium*) that lives in the bloodstream of humans. The mosquito *Aedes* similarly transmits the virus that causes yellow fever and, along with other genera of mosquito, transmits the infective stage of the nematode worms (*Brugia* and *Wuchereria*) that cause elephantiasis. But mosquitoes are not the only culprits in such insect-transmitted diseases. Sandflies (*Phlebotomus*) carry the infective stage of the African disease known as leishmaniasis, the blackfly *Simulium* carries the infective stage of the nematode *Onchocerca* that causes river blindness, and the tsetse fly *Glossina* carries the infective protozoan *Trypanosoma* that is responsible for sleeping sickness. Finally, the vectors of the disease known as bilharzia or schistosomiasis are snails that live in the streams, rivers and lakes, in the aquatic infective stage during which they bore into humans when they enter the water.

Studies of the distribution patterns of parasitic diseases have shown a strong latitudinal gradient in the frequency of the diseases associated with protozoan parasites [62], with higher concentrations of such diseases in the tropics. The distribution pattern of any parasitic disease that requires an intermediate host is naturally limited by the environmental needs of both the final host and the vector. The year-round warmth of the tropics and subtropics provides a genial environment for all of them, and so it is not surprising that such diseases are prevalent there. It is also worth noting that the tropics also contain more species of bird and mammal [63], which therefore provide a varied reservoir of organisms that share our warm-blooded physiology and from which a transfer of host by 'spillover' (discussed further in this section) may be relatively easy.

As our ancestors spread northward from Africa, they were thus accompanied by most of these parasitic diseases. Only sleeping sickness did not succeed in spreading to Asia, apparently because the tsetse fly is restricted to Africa and the Arabian Peninsula. The other diseases are all prevalent in southern Asia, including the Indian subcontinent, while leishmaniasis is also found in southern Europe, and malaria has occurred as far north as southern England. (It is also possible that all these diseases were already present in Eurasia when *H. sapiens* arrived there, having been taken there earlier when our ancestor, *H. erectus*, spread to that continent.)

In time, people changed their way of life from nomadic hunter-gathering into more permanent settlements, surrounded by the animals and plants that they had domesticated. But their new closeness to animals brought with it a greater variety of disease exposure. The tapeworm *Taenia* finds its intermediate host in cattle and pigs, while the human diseases smallpox, tuberculosis and measles are all closely related to similar diseases of cattle. Similarly, the nematode worm *Trichinella*, which encysts in muscle cells, infects humans when they eat inadequately cooked pork (which may be why pigs are considered unfit for human consumption in the Middle East). At the same time, the irrigation systems that early farmers constructed in the Fertile Crescent of the Middle East may, by placing water courses permanently near their villages, have made it more likely that they would be infected by diseases transmitted by mosquitoes (whose larvae live in water). At the same time, homes and storehouses would have provided shelter and food for rats, from which they could have caught typhus. Finally, the higher population densities that accompanied all these changes would have made these early human communities more vulnerable to epidemics of disease. So there were certainly drawbacks, as well as advantages, to the development of domestication.

Just as humans gradually modified their domesticated animals to fit with their new environments, so the diseases and parasites the animals carried evolved to exploit the new opportunities offered by close human contact [64]. Many diseases of animals, whether prey species or domesticated, cannot be transferred to human beings. The characteristics of the disease organisms are so closely tuned to the nature of their host that they cannot pass through the barriers to enter and infect a different species. For example, most malarial parasites are species specific, so humans can only be infected by the human variety. Other diseases, however, such as rabies or bird influenza can be transmitted from animals to humans. Often such diseases are caused by agents that are relatively inefficient in terms of persistence, either because they prove fatal to the new host or because they are unlikely or unable to be transmitted between individuals of the new host. Thus, any epidemic among humans is unlikely.

In the next evolutionary stage, transmission from human to human becomes possible, as in the case of the Ebola virus and dengue fever. With human-to-human transmission, the disease becomes much more serious because outbreaks in human populations can persist. In later evolutionary developments, the agent may become increasingly adapted to its new human host, eventually being transmissible only between human beings, as in the case of the HIV/AIDS virus, which probably originated in wild populations of chimpanzees, but is now a specifically human disease. This is one example of what is called *spillover*, in which a pathogen passes from members of one host species (the chimpanzee) into members of another (humans). Malaria is another example of this, *Plasmodium falciparum* having originally been a parasite of African gorillas, but by spillover it infected humans. In Asia, there has been a recent spillover: *Plasmodium knowlesi*, a common parasite of monkeys, now also infects humans. Events such as this highlight an increasing threat for humans, as we increasingly move into close proximity to new possible sources of spillover, as we clear the tropical rainforests for logging or for their replacement by cash crops such as oil palms.

One biogeographical problem with diseases that have become exclusively human is tracing their origins unless, like the spillover of the HIV/AIDS virus, this took place relatively recently. A large number of such diseases seem to have originated in the Old World, but tracing the precise source can be difficult. For example, the bacterium *Helicobacter pylori* is present in approximately 50% of all human stomachs, where it can cause peptic ulcers and may even be a causative agent of stomach cancer. An extensive genetic survey [65] has shown that its genetic diversity decreases with distance from East Africa, suggesting that this was the original centre of infection and evolutionary development.

When the first people migrated through the colder lands of northern Asia and across the Bering Straits into the New World, they left behind their domesticated animals and, with them, their associated diseases. Perhaps the only parasite that is found naturally in both the Old World and the New World is the hookworm *Ancylostoma*. However, in South America a New World version of leishmaniasis evolved independently of that of the Old World. It is caused by a different species of the parasite and carried by a different sandfly. South America is also the home of Chagas disease, caused by a species of *Trypanosoma* related to that which causes sleeping sickness in Africa; it is present in South American mummies dating from 2000 BC. (Interestingly, the disease is endemic in the marsupial opossum of South America, but the animal is hardly affected by the infection, perhaps because this ancient inhabitant of the continent has, unlike humans, had many millions of years in which to adapt to the infective relationship and mitigate its effects.)

Another factor that reduced the prevalence of disease in the pre-Columbian peoples of North America was that they did not succeed in domesticating the large mammals of that continent. The reasons for this are discussed by the American biogeographer-physiologist Jared Diamond [66] of the University of California, Los Angeles, who surveyed the reasons for the fact that the Western version of civilization came to conquer or dominate the rest of the world. Biogeography, at a worldwide scale, was important. Eurasia is the world's largest landmass, extending widely from west to east, and so has a great variety of temperate, subtropical and tropical environments. As a result, a greater number of large mammals, possible candidates for domestication, evolved there than in any other continent. Furthermore, by chance, many of those of Eurasia could be domesticated (e.g. cattle, sheep, goat, pig and horse), while this was not true of any of the large mammals of North America, and is true only of the llama of South America. Even today, none of the other large mammals of North America, Africa or Australia have been domesticated. But the absence of domesticated animals in the New World also meant that their peoples were not exposed to the many diseases that had accompanied domestication in the Old World.

This advantage for the peoples of the New World was accompanied by a corresponding disadvantage, for they had no immunological defences when they were confronted by the spread of the peoples of the Old World. So the diseases that Eurasians had caught from their domesticated animals (smallpox, measles, influenza and typhus) ravaged the pre-Columbian Native Americans of the New World, killing 95% of those of North America and 50% of the Aztecs of Mexico and of the Incas of Peru. Smallpox, in particular, similarly devastated the peoples of southern Africa, Australia and the Pacific islands. On the other hand, because the Europeans who colonized the rest of the world came from higher latitudes, where tropical diseases were rare or absent, these diseases for many centuries acted as a barrier to large-scale European colonization of tropical Africa: West Africa was once known as the White Man's Grave. So geography, climate and mammalian evolution have all played important roles in controlling the variety and incidence of the diseases that plague our species.

Inevitably, the incidence of some of these diseases has been affected by the increasing environmental impact of humanity and its activities. For example, human operations such as mining, deforestation and road building have led to increases in the predominance of *P. falciparum* (the more virulent species of malaria, which causes cerebral malaria), at the expense of *Plasmodium vivax*, which causes a less serious variant of the disease. This has also been aided by the evolution of drug-resistant strains of *P. falciparum* and by the construction of dams and large-scale irrigation projects. These increase the area of water in which the mosquitoes can breed, as well as placing the water close to areas in which people live. The breakdown of control measures in the highlands of East Africa and Madagascar after the 1950s also led to an increase in malaria in these regions. In Madagascar, this was exacerbated by climatic changes involving an increase in temperature during the period of maximum mosquito abundance, December–January.

All these human activities, however, have also increased the size and widespread distribution of human populations available as hosts to *Leishmania* and *Trypanosoma*. Increasing aridity in parts of southern Africa has caused the movement of tsetse flies and *Simulium* blackflies into new areas, leading to increases in sleeping sickness and river blindness. On the other hand, the loss of forest in some parts of Africa has led to the loss of these vectors and a consequent reduction in these illnesses [67]. Environmental degradation, as well as human population pressures, resource exhaustion and disease, can thus combine to cause the collapse of human societies [68]. When societies do collapse, the environment often recovers remarkably rapidly, as is evidenced by the effects of the Black Death pandemic in Europe (1347–1352). When this plague spread into Europe from Asia, it resulted in 30–60% mortality among the human population. Analysis of contemporaneous pollen profiles from lakes has shown that arable farming was abandoned and pastoral activity was greatly reduced, resulting in the regrowth of many forest areas that had previously been cleared for agriculture [69].

Environmental Impact of Early Human Cultures

Environmental modification was an essential consequence of domestication and subsequent human spread. As the postglacial climatic optimum passed and conditions in the temperate regions became generally cooler and wetter, the agricultural concept continued to spread into higher latitudes. The incentive to modify the environment to make it more suitable for enhanced productivity of domestic animals and plants became a major driving force for the expanding human populations. Temperate forests are unsuitable for the growth of domesticated plants, because most of them have a southern origin and need a high light intensity. Similarly, domestic animals such as sheep and goats are not at their most efficient in a woodland habitat, preferring open grassland conditions. Cattle and pigs, on the other hand, can be herded within forests, but even they can be managed more efficiently in a habitat that is more open. In pre-agricultural times, the Mesolithic people of northern Europe discovered that the opening of forest and burning to retain open glades provided a higher productivity of red deer (*Cervus elaphus*). Like some of the Native American tribes of North America, who became closely dependent on the bison, the European people of the Middle Stone Age were often reliant on the red deer.

The intensification of forest clearance with the coming of agriculture in northern Europe is very apparent from pollen diagrams, where the pollen of open habitat species (e.g. grasses, plantains and heathers) rises and the proportion of tree pollen falls. The precise pattern of forest clearance and the development of heathland, grassland, moorland and blanket bog as consequences of this activity vary depending on local conditions and the pattern of human settlement. By 2000 years ago, the impact was severe through much of central and western Europe, although the forests of the far north had been little influenced at that time. Some of the most severe deforestation, judging from the pollen record, took place in the northwest of Europe, including the British Isles. Perhaps it was in this region that the forest was least able to recover from human impact, and the maintenance of heavy grazing kept the area relatively open.

In North America, the hunting and foraging strategies of the Native Americans produced a patchwork of seasonal settlements and camps that involved opening the forest. The use of fire led to the development of sharp divisions between habitats, such as prairie grasslands and woodland. Many groups subsisted on the gathering of nuts from trees, and it is likely that they managed habitats, clearing unwanted species and opening the canopy to enhance nut production [70]. Agriculture in the temperate zone in pre-European times was confined largely to the growing of maize and other open-habitat, weedy species, including purslane (*Portulaca oleracea*). This involved the

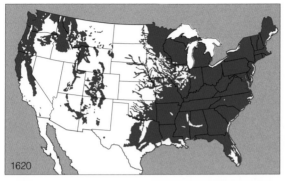

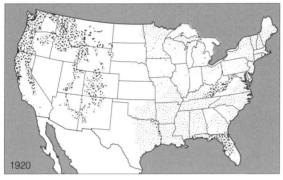

Figure 13.11 The extent of virgin forest in what is now the United States of America in 1620, 1850 and 1920, showing the progressive destruction of American old-growth forests. From Linz *et al.* [65].

clearance of small areas of forest, and the effect of these clearings can be detected in pollen diagrams [71]. On abandonment, these clearings seemed to recover, and there is little evidence for large-scale forest destruction of the European type. However, some changes in the composition of the North American forests may well have resulted from the activities of agricultural peoples. The burning and cutting of forests are often associated with a loss of some species, such as sugar maple and beech, and an increase in the abundance of fire-resistant pines and oaks, together with a general increase in the frequency of birch. Intensive clearance in the eastern United States and Canada was usually delayed until the arrival of European settlers in the 18th and 19th centuries. It is often marked in the pollen diagrams by a marked rise in the ragweed (*Ambrosia*).

There is still considerable debate about the impact of aboriginal peoples on the development of Australian vegetation. It is likely that the first invaders were mainly associated with the savanna regions, and undoubtedly they used fire as a means of managing and hunting game. Like the Native Americans, this practice would have resulted in sharp boundaries between savanna and forest habitats. Fire-resistant species, such as the eucalypts, would have been favoured by the use of fire [72].

Environmental change has accelerated in the last few centuries, and the face of the Earth is changing ever faster. Figure 13.11 illustrates the rate of destruction of primary forest in the United States over the past 400 years [73]. The destruction that temperate forests have experienced during these recent centuries is now being repeated in the tropical forests.

The evolution of human beings changed the environment for many other species. Some found new opportunities for extending their ranges as they adapted to the new circumstances, but many experienced new stresses, either because of direct destruction by humans or because their habitats were increasingly modified as human populations spread and became more numerous. But the enormous extent of that increase now threatens to overwhelm the resources of our planet. The recognition of that problem, and possible ways in which to address it, are the subjects of the next, final, chapter of this book.

Summary

1 The human line of evolution separated from the great apes around 7 mya and led to the development of many types of hominin (members of the human line of evolution rather than that of the great apes).
2 Africa is the likely location for the origin of the genus *Homo* and of our own species, *Homo sapiens*.
3 Of the various types of hominin, only *H. sapiens* survived beyond the last glaciation, although the discovery of *Homo floresiensis* fossils could indicate the local persistence of a separate line of evolution.
4 The extinction of many large vertebrate animals at the end of the last glacial had not occurred in former climatic cycles, and circumstantial evidence suggests that humans were involved in the extinction of many of them.
5 The domestication of animals and plants provided new opportunities for human food production and hence population expansion. It took place in a number of different locations around the world.
6 Many parasites and disease organisms evolved alongside humans and adopted this primate as their host. Other parasites and diseases of animals were favoured by domestication, and some of these adapted to enable them to move from domestic animals to humans.
7 Human spread, population increase and the need for agricultural development have meant that humans have increasingly manipulated and modified their environment.

Further Reading

Delcourt HR, Delcourt PA. *Prehistoric Native Americans and Ecological Change*. Cambridge: Cambridge University Press, 2004.

Diamond J. *Collapse: How Societies Choose to Fail or Succeed*. New York: Viking, 2005.

McIntosh RJ, Tainter, JA, McIntosh SK. *The Way the Wind Blows: Climate, History, and Human Action*. New York: Columbia University Press, 2000.

Pielou EC. *After the Ice Age: The Return of Life to Glaciated North America*. Chicago: University of Chicago Press, 1991.

Roberts N. *The Holocene: An Environmental History*. 2nd ed. Oxford: Blackwell, 1998.

Russell EWB. *People and Land through Time: Linking Ecology and History*. New Haven, CT: Yale University Press, 1997.

Stringer CB. *The Origin of Our Species*. London: Penguin Books, 2012.

References

1. Willis KJ, Niklas KJ. The role of Quaternary environmental change in plant macroevolution: the exception or the rule? *Philosophical Transactions of the Royal Society of London. Series B: Biological Sciences* 2004; 359 (1442): 159–172.

2. Lister AM. The impact of Quaternary Ice Ages on mammalian evolution. *Philosophical Transactions of the Royal Society of London B: Biological Sciences* 2004; 359 (1442): 221–241.

3. Crutzen PJ. *The 'Anthropocene'*. Berlin: Springer, 2006.

4. Carroll SB. Genetics and the making of *Homo sapiens*. *Nature* 2003; 422: 849–857.

5. McBrearty, S, Jablonski NG. First fossil chimpanzee. *Nature* 2005; 437 (7055): 105–108.

6. Folinsbee KE, Brooks DR. Miocene hominoid biogeography: pulses of dispersal and differentiation. *Journal of Biogeography* 2007; 34 (3): 383–397.

7. Wood B, Lonergan N. The hominin fossil record: taxa, grades and clades. *Journal of Anatomy* 2008; 212 (4): 354–376.

8. Brunet, M, Guy F, Pilbeam D, *et al. A new hominid from the Upper Miocene of Chad, Central Africa*. Nature, 2002; 418 (6894): 145–151.

9. Johanson, D, Edey M. *Lucy: The Beginnings of Humankind*. London: Simon & Schuster, 1990.

10. Hay RL, Leakey MD. The fossil footprints of Laetoli. *Scientific American* 1982; 246: 50–57.

11. Gibbons A. Paleoanthropology: tracing the identity of the first toolmakers. *Science* 1997; 276 (5309): 32–32.

12. Bramble DM, Lieberman DE. Endurance running and the evolution of *Homo*. *Nature* 2004; 432 (7015): 345–352.

13. Bennett MR, Harris JWK, Richmond BG, *et al.* Early hominin foot morphology based on 1.5-million-year-old footprints from Ileret, Kenya. *Science* 2009; 323 (5918): 1197–1201.

14. Lordkipanidze D, Jashashvili T, Vekua A, *et al.* Postcranial evidence from early *Homo* from Dmanisi, Georgia. *Nature* 2007; 449 (7160): 305–310.

15. Ciochon RL, Bettis EA III. Palaeoanthropology: Asian *Homo erectus* converges in time. *Nature* 2009; 458 (7235): 153–154.

16. Swisher CC III, Rink WJ, Antón SC, Schwarcz HP, Curtis GH, Suprijo Widiasmoro A. Latest Homo erectus of Java: potential contemporaneity with Homo sapiens in southeast Asia. *Science* 1996; 274 (5294): 1870–1874.

17. Thieme H. Lower Palaeolithic hunting spears from Germany. *Nature* 1997; 385: 807–810.

18. Stringer C. Modern human origins: progress and prospects. *Philosophical Transactions of the Royal Society B: Biological Sciences* 2002; 357 (1420): 563–579.

19. White TD, Asfaw B, DeGusta D, Gilbert H, Richards GD, Suwa G, Clark Howell F. Pleistocene *Homo sapiens* from Middle Awash, Ethiopia. *Nature* 2003; 423 (6941): 742–747.

20. Krause J, Orlando L, Serre D, *et al.* Neanderthals in central Asia and Siberia. *Nature* 2007; 449 (7164): 902–904.

21. Mellars P. Neanderthals and the modern human colonization of Europe. *Nature* 2004; 432 (7016): 461–465.

22. Sankararaman S, Mallick S, Dannemann M, *et al.* The genomic landscape of Neanderthal ancestry in present-day humans. *Nature* 2014; 507 (7492): 354–357.

23. Finlayson C, Pacheco FG, Rodríguez-Vidal J, *et al.* Late survival of Neanderthals at the southernmost extreme of Europe. *Nature* 2006; 443. (7113): 850–853.

24. Brown P, Sutikna T, Morwood MJ, Soejono RP, Jatmiko, Saptomo DW, Due RA. A new small-bodied hominin from the Late Pleistocene of Flores, Indonesia. *Nature* 2004; 431 (7012): 1055–1061.

25. Dennell RW, Louys J, O'Regan HJ, Wilkinson DM. The origins and persistence of *Homo floresiensis* on Flores: biogeographical and ecological perspectives. *Quaternary Science Reviews* 2014; 96: 98–107.

26. Jungers W, Harcourt-Smith WEH, Wunderlich RE, *et al.* The foot of *Homo floresiensis*. *Nature* 2009; 459 (7243): 81–84.

27. Zazula GD, Froese DG, Schweger CE, *et al.* Palaeobotany: ice-age steppe vegetation in east Beringia. *Nature* 2003; 423 (6940): 603.

28. Curry A. Coming to America. *Nature* 2012; 485 (7396): 30–32.

29. Meltzer DJ. Monte Verde and the Pleistocene peopling of the Americas. *Science* 1997; 276 (5313): 754–755.

30. Rogers RA, Rogers LA, Hoffmann RS, Martin LD. Native American biological diversity and the biogeographic influence of Ice Age refugia. *Journal of Biogeography* 1991; 18: 623–630.

31. Gibbons A. *The peopling of the Americas. Science* 1996; 274 (5284): 31–33.

32. Raff JA, Bolnick DA. Palaeogenomics: genetic roots of the first Americans. *Nature* 2014; 506 (7487): 162–163.

33. Martin PS, Wright HE. *Pleistocene Extinctions: The Search for a Cause.* New Haven: Yale University Press, 1967.

34. Grayson DK. Pleistocene avifaunas and the overkill hypothesis. *Science* 1977; 195 (4279): 691–693.

35. Stuart AJ. Mammalian extinctions in the Late Pleistocene of northern Eurasia and North America. *Biological Reviews* 1991; 66 (4): 453–562.

36. Guthrie RD. New carbon dates link climatic change with human colonization and Pleistocene extinctions. *Nature* 2006; 441 (7090): 207–209.

37. Boulanger MT, Lyman RL. Northeastern North American Pleistocene megafauna chronologically overlapped minimally with Paleoindians. *Quaternary Science Reviews* 2014; 85: 35–46.

38. Stuart AJ, Kosintsev PA, Higham TFG, Lister AM. Pleistocene to Holocene extinction dynamics in giant deer and woolly mammoth. *Nature* 2004; 431 (7009): 684–689.

39. Nogués-Bravo D, Rodríguez J, Hortal J, Batra P, Araújo MB. Climate change, humans, and the extinction of the woolly mammoth. *PLoS Biology* 2008; 6 (4): e79.

40. Miller GH, Magee JW, Johnson BJ, Fogel ML, Spooner NA, McCulloch MT, Ayliffe LK. Pleistocene extinction of *Genyornis newtoni*: human impact on Australian megafauna. *Science* 1999; 283 (5399): 205–208.

41. Cohen TJ, Larsen J, Gliganic LA, Larsen J, Nanson GD, May J-H. Hydrological transformation coincided with megafaunal extinction in central Australia. *Geology* 2015; G36346: 1.

42. Kennett DJ, Kennett JP, West A, *et al.* Nanodiamonds in the Younger Dryas boundary sediment layer. *Science* 2009; 323 (5910): 94.

43. Kerr RA. Did the mammoth slayer leave a diamond calling card? *Science* 2009; 323 (5910): 26.

44. Marean CW, Bar-Matthews M, Bernatchez J, *et al.* Early human use of marine resources and pigment in South Africa during the Middle Pleistocene. *Nature* 2007; 449 (7164): 905–908.

45. Piperno DR, Weiss E, Holst I, Nadel D. Processing of wild cereal grains in the Upper Palaeolithic revealed by starch grain analysis. *Nature* 2004; 430 (7000): 670–673.

46. Heun M, Schäfer-Pregl R, Klawan D, Castagna R, Accerbi M, Borghi B, Salamini F. Site of einkorn wheat

domestication identified by DNA fingerprinting. *Science* 1997; 278 (5341): 1312–1314.

47. Allaby RG. The rise of plant domestication: life in the slow lane. *Biologist* 2008; 55 (2): 94–99.

48. Purugganan MD, Fuller DQ. The nature of selection during plant domestication. *Nature* 2009; 457 (7231): 843–848.

49. Liti G, Carter DM, Moses AM, *et al.* Population genomics of domestic and wild yeasts. *Nature* 2009; 458 (7236): 337–341.

50. Richards MP, Schulting RJ, Hedges RE. Sharp shift in diet at onset of Neolithic. *Nature* 2003; 425 (6956): 366–366.

51. Willis KJ, Bennett KD. The Neolithic transition-fact or fiction? Palaeoecological evidence from the Balkans. *The Holocene* 1994; 4 (3): 326–330.

52. Callaway E. Language origin debate rekindled. *Nature* 2015; 518: 284–285.

53. Zong, Y, Chen Z, Innes JB, Chen C, Wang Z, Wang H. Fire and flood management of coastal swamp enabled first rice paddy cultivation in east China. *Nature* 2007; 449 (7161): 459–462.

54. Smith BD. The initial domestication of Cucurbita pepo in the Americas 10,000 years ago. *Science* 1997; 276 (5314): 932–934.

55. Vilà C, Savolainen P, Maldonado JE. Multiple and ancient origins of the domestic dog. *Science* 1997; 276 (5319): 1687–1689.

56. Larson G, Karlsson EK, Perri A, *et al.* Rethinking dog domestication by integrating genetics, archeology, and biogeography. *Proceedings of the National Academy of Sciences* 2012; 109 (23): 8878–8883.

57. Troy CS, MacHugh DE, Bailey JF, *et al.* Genetic evidence for Near-Eastern origins of European cattle. *Nature* 2001; 410 (6832): 1088–1091.

58. Evershed RP, Payne S, Sherratt AG, *et al.* Earliest date for milk use in the Near East and southeastern Europe linked to cattle herding. *Nature* 2008; 455 (7212): 528–531.

59. Check E. How Africa learned to love the cow. *Nature* 2006; 444 (7122): 994–996.

60. Diamond J. Geography and skin colour. *Nature* 2005; 435 (7040): 283–284.

61. Rees JL, Harding RM. Understanding the evolution of human pigmentation: recent contributions from population genetics. *Journal of Investigative Dermatology* 2012; 132: 846–853.

62. Nunn CL, Altizer SM, Sechrest W, Cunningham AA. Latitudinal gradients of parasite species richness in primates. *Diversity and Distributions* 2005; 11 (3): 249–256.

63. Dunn RR, Davies TJ, Harris NC, Gavin MC. Global drivers of human pathogen richness and prevalence. *Proceedings of the Royal Society B: Biological Sciences* 2010; 277 (1694): 2587–2595.

64. Wolfe ND, Dunavan CP, Diamond J. Origins of major human infectious diseases. *Nature* 2007; 447 (7142): 279–283.

65. Linz B, Balloux F, Moodley Y, *et al.* An African origin for the intimate association between humans and *Helicobacter pylori*. *Nature* 2007; 445 (7130): 915–918.

66. Diamond JM. *Guns, Germs, and Steel: The Fates of Human Societies.* New York: W.W. Norton, 1997.

67. Molyneux DH. Common themes in changing vector-borne disease scenarios. *Transactions of the Royal Society of Tropical Medicine and Hygiene* 2003; 97 (2): 129–132.

68. Diamond J. *Collapse: How Societies Choose to Fail or Succeed.* New York: Penguin, 2005.

69. Yeloff D, Van Geel B. Abandonment of farmland and vegetation succession following the Eurasian plague pandemic of AD 1347–52. *Journal of Biogeography* 2007; 34 (4): 575–582.

70. Delcourt PA, Delcourt HR. *Prehistoric Native Americans and Ecological Change: Human Ecosystems in Eastern North America since the Pleistocene.* Cambridge: Cambridge University Press, 2004.

71. Mcandrews JH. Human disturbance of north american forests and grasslands: the fossil pollen record. In: HuntlyB, WebbT III (eds.), *Vegetation History.* Dordrecht: Kluwer, 1998: 673–697.

72. Bowman DM. *Australian Rainforests: Islands of Green in a Land of Fire.* Cambridge: Cambridge University Press, 2000.

73. Williams M. *Americans and Their Forests: A Historical Geography.* Cambridge: Cambridge University Press, 1992.

74. Sponheimer M, Lee-Thorp JA. Isotopic evidence for the diet of an early hominid, *Australopithecus africanus. Science* 1999; 283: 368–370.

75. Guthrie RD. New carbon dates link climatic change with human colonization and Pleistocene extinctions. *Nature* 2006; 441: 207–209.

76. Coltman DW, O'Donoghue, Jorgenson IT, Hogg JT, Strobeck C, Festa-Bianchet M. Undesirable evolutionary consequences of trophy hunting. *Nature* 2003; 426: 655–657.

77. Cardillo, Lister A. Death in the slow lane. *Nature* 2002; 419: 440–441.

78. Wang H, Nussbaum-Wagler T, Li B, *et al.* The origin of the naked grains of maize. *Nature* 2005; 436: 714–715.

Conservation Biogeography

As we have seen, the study of biogeography has deep roots with much of the foundational work completed by the end of the 19th century. Nevertheless, the contemporary relevance of biogeographical research has never been greater. The current scientific consensus is that we are entering a unique period in the history of the Earth, a dramatic transformation of life on Earth reminiscent of some of the events of the far past that led to mass extinctions. Ultimately, the degree to which human action alters the diversity and distribution of life on Earth depends on the willingness and capacity of societies, organizations and individuals to conserve what is left of the natural world [1]. However, conservation resources are limited, and it is imperative that we make rational, empirically grounded decisions about where to invest these limited resources (both taxonomically and geographically). Biogeography has an essential role to play in this endeavour, providing tools and concepts to identify key processes and to make realistic predictions of what may happen to species and ecosystems under different scenarios of human development [2].

Welcome to the Anthropocene

The expansion of that singular and peculiar animal *Homo sapiens* out of Africa, marked the beginnings of a remarkable period of change and re-sorting for the world's biotas. The enormous impacts that humans have had on biological communities, landscapes and even the global climate have led some scientists to dub the current geological epoch the **Anthropocene** (from the Greek *anthropos* 'human being' and *kainos* 'new') in recognition of the pervasiveness, diversity and enormous magnitude of the various impacts that humans have had on the natural environment [3]. Although controversial, there is ample justification for a new geological epoch given that a wide range of environmental variables are now far outside their typical ranges during the majority of the Holocene – currently, the 'official' geological epoch stretching from the end of the Pleistocene (nearly 12 000 years ago) to the present day.

Specifically, in a blink of an eye from a geological perspective, humans have managed to significantly alter atmospheric chemistry, make the world's oceans far more acid, re-arrange and transform river systems, appropriate a huge proportion of global net primary productivity, accelerate rates of extinction, break down biogeographical barriers causing the homogenization of biotas and create an enormous variety of novel ecosystems and species assemblages without historical precedents. Many of these effects (e.g. extinctions and biological invasions) are irreversible, while others, such as the increase in the atmospheric concentration of greenhouse gases, may be beyond the limited capacity of the global community to address.

The main impact of the concept of the Anthropocene is symbolic, recognizing that the world has entered a distinct, undeniable period of human-induced environmental change. Perhaps the most controversial aspect of this proposed epoch is when it started, given the huge variation in the magnitude and geography of how humans have affected

Biogeography: An Ecological and Evolutionary Approach, Ninth Edition. Edited by C. Barry Cox, Peter D. Moore, Richard J. Ladle.
© 2016 John Wiley & Sons, Ltd. Published 2016 by John Wiley & Sons, Ltd.

their environment. For this reason, some scientists have suggested that the start of the Anthropocene should depend on when human impacts became regionally significant. In this context, New Zealand was the last major area with an intact biota to enter the Anthropocene since it was first colonized about 750 years ago. Other scientists have suggested that the start of the Industrial Revolution in the late 18th century is a more appropriate start date for the new epoch. There is also support for placing the start of the Anthropocene at the turn of the second millennium in the year 2000, in recognition of when the term was first coined [4].

From a biogeographical perspective, the Anthropocene is most notable for the accelerating loss of **biodiversity** and the remarkable re-sorting and restructuring of communities due to human-assisted migrations. The loss of biodiversity has occurred at every level of organization, from genes through to ecosystems [5], though it is perhaps the loss (extinction) of species that has been most studied and, importantly, has probably played the biggest role in alerting societies to the often catastrophic impacts of human actions on the natural world [6].

As we saw in Chapter 1, the reality of extinction only became widely accepted in the early 1800s, partly driven by George Cuvier's remarkable reconstructions of fossil elephants. Such huge and distinctive creatures were clearly very different from any living species and, given their size and dramatic appearance, it seemed very unlikely that they still existed. Nevertheless, for the Victorian scientists, it was one thing to accept that species might have gone extinct, and another thing entirely to attribute the cause of such extinctions to human action. This reluctance to point the finger at our own species is clearly illustrated by contemporary accounts of high-profile extinctions, such as that of the great auk (*Pinguinus impennis*), an impressive seabird that was hunted into extinction by European fishermen. Nineteenth-century accounts indicate that the last documented specimens were collected in 1844 and that the species was extinct by 1852 [7]. From present-day perspectives, overhunting was clearly to blame for the demise of the great auk, but contemporary writers had great difficulty accepting the pivotal role of human action. As the English naturalist James Orton expressed it: 'The upheaval or subsidence of strata, the encroachments of other animals, and climatal revolutions—by which of

these great causes of extinction now slowly but incessantly at work in the organic world, the Great Auk departed this life, we cannot say' [8, p. 540].

The strange (from a modern-day perspective) reluctance to attribute human causes to the contemporary extinctions lasted into the early twentieth century. As more and more evidence accumulated, the scientific community slowly coalesced around the idea that the wave of high-profile extinctions (e.g. the great auk, passenger pigeon, Carolina parakeet, quagga, etc.) was most likely being driven by human action. However, the scale of the problem was largely unknown, and attempts to quantify extinction rates and to compare them to background rates (from the fossil record) only began in earnest in the 1970s. It was during this decade that the word *crisis* became attached to the extinction concept, as scientists and conservationists began to uncover the full magnitude of human effects on the environment.

The belief that the current rate of extinction is many times greater than normal background rates is one of the fundamental tenets of the modern conservation movement [9]. The evidence for this is now undeniable, especially the rates of documented extinctions on oceanic islands, and comes from two main sources: (i) historically documented extinctions, and (ii) models, simulations and frameworks that relate environmental change (e.g. habitat loss and transformation) to probabilities of the extinction of individual species or to rates of extinction within specified areas and time frames (reviewed in [10]).

Biogeographical theories have played a central role in estimating extinction rates: the most widely used (and misused) method is based on the observation that the relationship between the size of an oceanic island and the number of species it contains can be effectively captured by a simple mathematical relationship, known as the *species–area curve* (see Chapter 7). This can be used to calculate how many fewer species should be found in ecosystems such as tropical forests after large areas have been deforested. The most high-profile example of such estimates is E.O. Wilson's 'conservative' prediction from 1992 that approximately 27 000 species are going extinct every year, based on the rate of tropical forest loss. Such enormous figures certainly capture the attention of the public and politicians, but have been criticized because they are based on a number of critical assumptions that are rarely resolved. Specifically, to calculate extinctions based on the *species–area relationship* (SAR), the total number of species and the proportion of endemic species prior to habitat destruction need to be accurately and precisely estimated. The slope of the species–area curve should also be known. Moreover, there is an underlying assumption that terrestrial islands such as rainforest fragments act like oceanic islands in a 'sea' of agricultural land. Finally, the number of species in the original habitat should be in equilibrium (see Chapter 7). Added to this, species do not go extinct immediately when habitat area is reduced, but are slowly lost due to a range of demographic, genetic and environmental effects. Of these assumptions, the estimated total number of species has the most scope for influencing the figure for global extinctions. There may be anywhere between 1 and 100 million species on Earth (see the Linnean shortfall, discussed in the 'Uncertainties and Shortfalls' section), many of which are arthropods in tropical forests. Choosing a higher number gives a higher number of total extinctions and thus a higher rate [11].

The American ecologist Stuart Pimm and his colleagues have recently developed an alternative way to analyse extinctions that circumvents the uncertainties involved in estimating species numbers or estimating extinction risk [12]. Rather than use absolute numbers as many previous estimates have done, their approach is to express extinction rates as fractions of known species going extinct over time, in this case extinctions per million species-years (E/MSY). They calculate recent extinctions by following cohorts from the dates of their scientific description – this approach excludes some of the most famous recent extinctions, such as the dodo, that went extinct before they were formally described. Taking birds as an example, at the time of writing taxonomists had described 1230 species of birds after 1900, with 13 of these subsequently becoming extinct. The 'bird cohort' has accumulated 98 334 species-years and, on average, a species has been known for 80 years. Based on these data, Pimm and his colleagues estimate the contemporary extinction rate for birds as $(13/98\ 334) \times 10^6 = 132$ E/MSY.

The advantage of this system of measuring global extinctions is that these estimates can be directly compared with estimates of baseline extinctions – the rate of extinction that would naturally occur in the absence of human influence. In an earlier article [13], Pimm had estimated that the background rate was approximately 1 E/MSY, although subsequent empirical studies on a range of fossil groups suggest that this is probably a considerable overestimate. For example, a recent study on marine taxa reported variations in extinction rates for fossil genera (which should be roughly similar to those for species) from 0.06 genera extinctions per million genera-years for cetaceans to as low as 0.001 genera E/MSY for brachiopods [14].

The current rate of extinction is not simply high; it is catastrophically high. Throughout the history of life on Earth, extinction is normally more or less balanced by speciation. But this is not the first time that extinctions have decimated life on Earth: there are at least five occasions in the last 600 million years where the world lost more than three-quarters of its species over a geologically short time period. Clearly, we have not reached this state yet, although evidence is accumulating that a sixth mass extinction, this time caused by human action, may occur within the next few centuries [15]. Mass extinctions in the past were characterized by a conjunction of unusual conditions, such as abnormal climate dynamics, atmospheric composition and very high-intensity levels of ecological stress. Most ecologists would agree that this 'perfect storm' of biophysical factors is a major characteristic of the Anthropocene, and it is unlikely that existing ecosystems, moulded in the absence of humans by the glacial–interglacial cycles that began 2.6 million years ago, will be able to resist the multiple onslaught of warming temperatures, habitat loss and fragmentation, pollution, overexploitation and invasive species.

Of course, the loss of species is only one aspect of the wide-scale biotic changes occurring across the planet. Unfortunately, the prognosis is not much better however we try to measure the state of the natural environment (Figure 14.1). In 2010, a collaboration of scientists from the world's leading research centres reviewed the state of global biodiversity using 31 separate indicators, including population trends, extinction risk, habitat extent and condition, and community composition [5]. The study was unusual in that it also attempted to measure trends in pressures on biodiversity such as resource consumption, invasive species, nitrogen pollution, overexploitation and climate change. Predictably, while

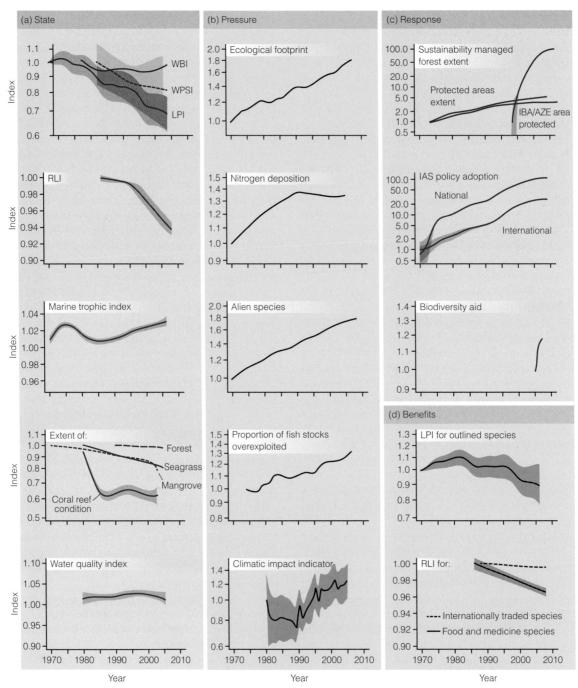

Figure 14.1 Indicator trends for (a) the state of biodiversity, (b) the pressures upon it, (c) the responses to address its loss, and (d) the benefits humans derive from it. Data are scaled to 1 in 1970 (or for the first year of data if before 1970), modelled (if >13 data points) and plotted on a logarithmic ordinate axis. Shading shows 95% confidence intervals except where unavailable (i.e. mangrove, seagrass and forest extent, nitrogen deposition, and biodiversity aid). AZE, Alliance for Zero Extinction site; IAS, invasive index species; IBA, Important Bird Area; LPI, Living Planet Index; RLI, Red List Index; WBI, Wild Bird Index; WPSI, Waterbird Population Status Index. From Butchart *et al.* [6].

indicators related to biodiversity declined, pressures on biodiversity showed inexorable increases. Perhaps the only bright spot on an almost unremittingly bleak picture was that investments in ways to alleviate biodiversity loss were increasing.

Less, and Less Interesting

We have seen that there is undeniable evidence that biodiversity, in all its aspects, is decreasing. However, in addition to extinctions, biological communities are also changing their composition as new species arrive. Some of these new arrivals are *climate migrants*, moving into new areas as anthropogenic climate change transforms the biophysical characteristics of habitats and ecosystems (see the 'Novel Climates and "No-Analogue" Communities' section). However, the 'winners' in the climate change lottery are likely to be in the minority given the extraordinary levels of fragmentation that compromise natural dispersal in most areas of the world. In many habitats, especially islands, the majority of the new arrivals have been transported by humans. Moreover, these species are only a narrow subset of the world's biota, leading to an increasing homogenization of life on the planet. An obvious example can be seen in almost every airport in the world, where the first birds a visitor sees are the Eurasian house sparrow (*Passer domesticus*) and the feral pigeon (*Colomba livia*).

Human-assisted dispersal of non-native species (sometimes referred to as *alien species* or *invasive species*) differs from natural dispersal events in the number and frequency of events [16]. Specifically, natural dispersal events, especially between major biogeographical regions, are rare in terms of the number of species and the frequency with which these species disperse. In stark contrast, humans are constantly moving a wide variety of species, both intentionally and unintentionally. Moreover, unlike naturally dispersing species that tend to have certain characteristic ecological traits (e.g. animals with the ability to fly, wind-dispersed seeds, etc.), human-dispersed species are ecologically diverse, with many having very limited natural dispersal ability.

A clear example of these dramatic human-induced increases in dispersal can be seen on the Hawaiian Islands, where contemporary rates of biological invasions are almost 1 million times higher than the rates before human colonization. Before the Polynesians arrived (AD 300–600), the invasion rate of terrestrial species was in line with that of other island systems – approximately 30 species per million years (0.00003 per year). This jumped to 20 000 species per million years (0.02 per year) after the arrival of the Polynesians, and then jumped again when Europeans arrived to approximately 20 per year [17] (Table 14.1).

It is also important to note that human-assisted dispersal routes follow economic and social ties rather than biophysical connections (e.g. ocean currents, air flows, etc.). This is creating a completely new type of biogeography that can only be understood through a detailed understanding of human behaviours [16]. Perhaps the clearest example of this is the transportation of aquatic organisms in the ballast water of ocean-going ships. These enormous vessels struggle to achieve stability to minimize drag with empty cargo holds, and it is now

Table 14.1 A comparison of prehistoric versus human-assisted biological invasions. Adapted from Ricciardi [17].

Characteristics	Prehistoric invasions	Human-assisted invasions
Long-distance dispersal events	Very rare	Common
Species transported per event	Few	Few to many
Propagule size	Typically small	Small to large
Mechanisms of dispersal	Few	Many
Temporal and spatial dynamics	Few, episodic short-distance events	Many, continuous long-distance events
Biotic homogenization	Weak and local	Strong and global
Potential for interacting with other stressors	Low	Very high

common practice to fill ballast compartments by pumping water into ballast tanks. The quantities are enormous: a typical commercial vessel might carry over 30 000 metric tonnes of water. The problem for conservation is that the ballast water taken from the port of departure is discharged into the arrival port, along with any animals that unintentionally came along for the ride. Current estimates suggest that the global fleet (about 35 000 commercial vessels) is transporting 7000–10 000 species at any given time [18].

From a biogeographical perspective, one of the most interesting aspects of all of this human-assisted dispersal is that changes in species diversity are frequently caused by the invasion of a relatively limited subset of ubiquitous non-native species into areas containing a unique subset of native species, some of which may be endemic. Since the same non-native species have been introduced into multiple sites, the net effect is for these ecologically diverse regions to become more similar (known as *biotic homogenization*). More generally, biotic homogenization can be thought of as the process by which former biotas lose their biological distinctiveness at any level of biological organization, including genetic and functional characteristics [16].

The enormous influx of non-native species into a wide variety of ecosystems has led to a dramatic and unique re-sorting and reconfiguration of many communities, especially those near major human settlements. However, the ecological consequences of such wide-scale biotic invasions are diverse and, often, counterintuitive. For example, while human density is associated with a decrease in species richness at small spatial scales, at larger spatial scales it is associated with increased richness [19]. These results indicate that at the local level, humans are driving species extinct through the effects of habitat loss, fragmentation and over-exploitation, but over larger areas relatively few species are being lost while many new (non-native) species are being added.

What is behind the Biodiversity Crisis?

As we have seen, one of the defining characteristics of the Anthropocene is the enormous and possibly unparalleled rates of species loss and habitat degradation. Clearly, human actions are to blame. Specifically, biodiversity decreases when habitat is damaged, fragmented, restructured or completely destroyed; when exotic species replace native species; when the biophysical conditions (e.g. the climate) change more rapidly than ecological communities can effectively adapt and when natural resources are exploited unsustainably. These factors act singly or in concert to drive down the abundance of populations. Once a population has been driven to very low numbers, stochastic factors such as demographic shifts, genetic degradation and chance environmental events such as disease outbreaks or unusual weather events can wipe out the last few individuals, causing extinctions [20].

The single factor responsible for the greatest reduction of biodiversity during the Anthropocene is undoubtedly habitat loss [21]. The history of large-scale habitat loss is complex, with rates accelerating and decelerating depending on economic and technological advancements which, in turn, are frequently connected with the history of colonization. In Europe, large-scale habitat loss has been going on for millennia, as is beautifully illustrated by Oliver Rackham's classic account of the history of British woodland [22]. British forests began to appear with the advent of post-glacial climate conditions about 12 000 years ago and, within 5000 years, mature forests covered most of the island. At about the same time, Neolithic farmers began to chop down forest for small-scale agriculture and, by time the Romans arrived about 2000 years ago, Britain's once vast forests had been largely reduced to small fragments. The loss of Britain's forests must have been catastrophic for populations of woodland species such as bears, wolves and beavers, which all eventually succumbed to the pressures of small population size and overexploitation in the Middle Ages [23].

In North America, deforestation rapidly followed upon the footsteps of the first European colonists. Until the beginning of the eighteenth century, deforestation was relatively small scale and mainly a consequence of clearance for subsistence farming by both indigenous peoples and the growing population of European immigrants and their descendants. This was followed by a period of more extensive deforestation for more intensive and specialized agriculture. Finally, the early twentieth century saw the creation of major logging for the timber industry.

Although habitat loss through deforestation has almost ceased in North America and northern Europe, it is still ongoing in most parts of the tropics, the very place where biodiversity is highest (see Chapter 8). Tropical forests originally covered between 14 and 18 million km^2, but by the late 1980s only about half that area remained. Rates of loss have dropped since then, but the numbers are still impressively large. A high-resolution satellite study from 2013 [24] indicated that 2.3 million km^2 of forest was lost between 2000 and 2012 – most of this loss was attributable to tropical regions where the area of forests was reduced, on average, by 2101 km^2 per year. As in Europe and North America, the major increases in tropical deforestation were initially attributable to land conversion for small-scale agriculture. However, towards the end of the 20th century, population growth rates in the tropics began to decrease and there was a demographic shift towards urbanization. However, there was no accompanying decline in deforestation as large-scale agriculture (e.g. palm oil, soybeans and beef) and the global demand for timber and paper have continued to drive forest loss.

Habitat loss is frequently associated with habitat fragmentation, although the effects of this process on biodiversity are less easily understood and quantified. It is thought that fragmentation has four main effects on habitat pattern with potential consequences for biodiversity [25]: (i) reduction in habitat amount, (ii) increase in number of habitat patches, (iii) decrease in size of habitat patches and (iv) increase in isolation of patches. It is important to distinguish the different ways in which these changes in habitat pattern might influence biodiversity. For instance, the **Theory of Island Biogeography (TIB)** suggests that habitat loss has a strong and consistently negative effect on biodiversity, whereas simply breaking up a habitat (fragmentation without appreciable habitat loss) has much weaker impacts that may be positive or negative.

The nature of the habitat between the remaining habitat fragments (known as the *habitat or landscape matrix*) may influence how species react to the impacts of fragmentation. For instance, Australian ecologist James Watson investigated woodland bird species in habitat patches in three fragmented landscapes in the Canberra area of New South Wales [26]. Watson found big differences between species and populations in how they responded to different types of matrices in which their woodland habitat was embedded (illustrated schematically in Figure 14.2). This study and others like it call into question the common paradigm of viewing terrestrial habitat fragments as islands in a sea of uninhabitable land.

Another problem in interpreting the effects of habitat fragmentation is that it is often conflated with habitat loss. Indeed, many studies have come to the conclusion that it is habitat loss rather than the degree to which a given amount of habitat is broken apart that is crucial in driving biodiversity loss. This is not to say that fragmentation effects are trivial. Recent research suggests that habitat configuration (the spatial arrangement of habitat at a given time) can exert a strong effect on fragment connectivity, edge and matrix effects (reviewed in [27]), which in turn affect which species will be retained within individual fragments and the landscape as a whole. This can be easily understood by considering hypothetical landscapes with exactly the same amount of habitat and exactly the same

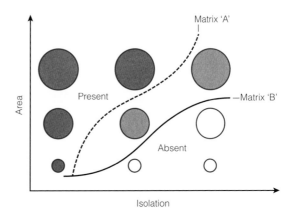

Figure 14.2 A modified species incidence function for a hypothetical species in a series of habitat islands. The occupancy of the species depends primarily on the area and isolation of the habitat island but also varies between Landscape A and Landscape B as a function of the quality of the matrix habitat. Dark blue circles indicate occupied habitat islands and white cells unoccupied habitat islands. The pale blue remnants and solid line indicate that a species would inhabit these remnants when in a landscape with matrix composition 'B' (favourable) but would not in matrix composition 'A' (less favourable; dashed line). Based on original ideas developed by Mark V. Lomolino and James E. Watson. From Whittaker *et al.* [39]. (Reproduced with permission of John Wiley & Sons.)

number of patches of the same size, but where the spatial arrangement of fragments is different (Figure 14.3). When the fragments are close together, there may be high ecological connectivity allowing free movement among patches. Conversely, when the patches are spread further apart, each patch may act as an isolated 'island' where the populations within are demographically stranded. The degree of connectivity will also be influenced by the permeability of the habitat matrix, meaning that an 'isolated' configuration in one landscape may be ecologically connected to other fragments in a landscape with a different matrix.

After habitat loss, human exploitation (over-hunting and over-harvesting) may be the second most important cause of species extinctions. This is especially problematic in many developing countries, where the recent adoption of modern hunting techniques and technologies has dramatically increased hunting efficiency [28]. A closely related problem is the over-reliance of many rainforest communities on bushmeat (meat from

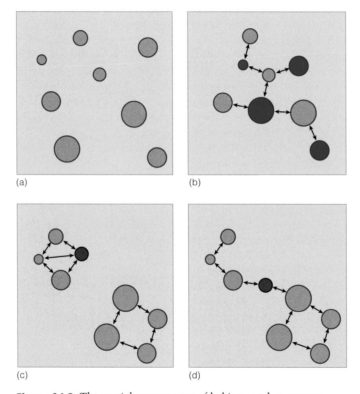

(a)

(b)

(c)

(d)

Figure 14.3 The spatial arrangement of habitat patches matters. Landscapes with the same amount of habitat and also with the same number of patches of the same size, but in different spatial locations, may result in situations where functional connectivity and habitat reachability are completely different for a hypothetical species. In scenario (a), the patches are too isolated and there are no biological fluxes among them (similar to a 'metapopulation in non-equilibrium'), while in scenario (b), the displacement of four patches allows free movement among all the patches (like in a 'patchy metapopulation'). In scenarios (c) and (d), the insertion (or restoration) of a small patch (in grey) in different locations may have very different effects on functional connectivity and the corresponding habitat network. From Villard and Metzger [27]. (Reproduced with permission of John Wiley & Sons.)

wild-caught animals). In many countries, this is exacerbated by a widespread preference for wild-caught meat and the high status associated with the consumption of species such as chimpanzees and gorillas. A recent review of bushmeat studies in central and west Africa [29] found that 177 wild species from 25 orders are hunted for meat, with 31 species classified as threatened by the International Union for Conservation of Nature (IUCN).

Bushmeat consumption may garner a lot of headlines, but it is in the marine environment that the effects of overexploitation are most visible. Fishing is one of the oldest human practices, and it has been having negative impacts on marine life for almost its entire history. A survey of the archaeological literature [30] shows that ancient human fishing frequently caused gradual shifts in the size of the fish caught and the serial depletion of species – unmistakable characteristics of overfishing. However, the depletion of the world's oceans really started to get going when the fishing process became industrialized in the early 19th century [31]. The initial technological breakthrough was the use of steam trawlers with power winches to gather nets. These were replaced by diesel engines in the 1920s, and, after World War II, fishing boats began to use the accoutrements of modern industrialized fishing, such as freezer trawlers, radar and acoustic fish finders [31].

Fishermen have always preferentially targeted larger species or species high in the food web, only turning to smaller species in lower trophic levels once their favoured species are no longer commercially profitable – a process the French marine biologist Daniel Pauly refers to as 'fishing down the food web' [32]. Such practices have seen the abundance of large predatory fishes in the ocean drop to 10% of their baseline numbers in the last 50 years and, in the case of sharks, to ~1% of their carrying capacity [33]. Likewise, tuna and billfish species have declined between 10% and 50% in all oceans [33].

Unsustainable exploitation can be seen as a special type of habitat degradation in which species of special value to humans are selectively removed from ecosystems. A more pervasive form of degradation is when human actions alter the biophysical conditions within a site or ecosystem, creating conditions that are incompatible with the continued survival of some of the native species. The clearest example of such degradation

is undoubtedly chemical pollution. Indeed, the impact of pollutants (especially pesticides) on organisms and ecosystems was instrumental in the birth of the modern environmental movement, inspiring books such as Rachael Carson's (1962) classic, *Silent Spring* [34].

The threat of pesticides and other toxicants is perhaps less visible than it once was. However, there is an increasing recognition that less toxic, but more widely used, compounds may be having considerable impacts on ecosystems. In particular, nitrogen pollution has recently been dubbed the 'third major threat to our planet after biodiversity loss and climate change' [35]. Europe and North America are currently the biggest sources of reactive nitrogen, but by 2020 half of anthropogenic nitrogen pollution will be produced by the developing world with potentially catastrophic consequences for these biodiverse regions. Nitrogen (and phosphorous) pollution are mainly caused by agriculture and urban activities, especially the use of fertilizers, although atmospheric deposition also contributes nitrogen. Their effects are most clearly seen in aquatic ecosystems, where super-abundance of these nutrients causes problems such as toxic algal blooms, loss of oxygen, fish kills, loss of biodiversity, loss of aquatic plant beds and coral reefs, and other problems [36]. Nutrient enrichment also causes practical problems for humans, reducing the quality of water for drinking, industry, agriculture, recreation and other purposes.

Concerns about pollutants have been largely replaced by the spectre of an even greater threat to ecosystems and species – the impact of anthropogenic climate change. Here, the effects are pervasive and global, and have the potential to cause an even greater impact on global biodiversity than habitat destruction and invasive species. Even under the unlikely scenario that greenhouse gas emissions are brought under swift control, global warming is unavoidable. The latest projections from the Intergovernmental Panel on Climate Change (IPCC) suggest that the global temperature will rise by 1.8–4°C this century as compared with late 20th-century baselines. Moreover, this increase will be accompanied by significant changes in precipitation patterns (rainfall and snow) and the seasonality of weather. The potential impacts on species and ecosystems are enormous, but they are not at all easy to predict.

Climate is a crucial factor for almost every aspect of an organism's ecology, physiology and behaviour, so the implications of changing climate are inherently complex to predict. This is an enormous challenge to biogeographers wishing to predict how individual organisms and ecosystems will respond. So far, most of the focus has been on answering two key questions: (i) how will the current geographical range of species be affected under different climate change scenarios? And (ii) how many species, and which ones, will be unable to adjust their geographical range in alignment with changing climate and therefore become threatened with extinction? Two general approaches have emerged to address these questions: mechanistic models and species distribution models.

Mechanistic models quantify the relationships between key physiological or behavioural processes and the external environment. For example, many freshwater fish such as trout or salmon are adapted to fast-flowing 'cool' rivers and are physiologically intolerant of higher water temperatures. Such critical temperature thresholds can be experimentally assessed, and the future range of the species can be forecast under different climate change scenarios. One of the key limitations of mechanistic models is that detailed physiological information is not available for many species, especially those that are already rare and may be at most risk from climate change.

A more flexible and widely used method for forecasting climate-induced range changes is a family of models known as *species distribution models* (SDMs) [37]. These relate the presence (or absence) of a species to some aspect of the environment, typically a number of standard climate variables. A basic species distribution model has three components (Figure 14.4). First, the climate and habitat within the observed geographical distribution of a species are analysed statistically. This produces a unique bioclimatic envelope (also known as *climate space*) representing the physical conditions that the species needs to survive. Second, the ability of the species to reach new habitats (dispersal) is quantified. Third, one or more climate change scenarios are chosen as the basis for forecasting the geographical distribution of the species in the future based on the new geographical location of the bioclimatic envelope. These scenarios typically contain a set of high, medium and low-impact forecasts which are applied to one or more future dates – typically 'round number' years, such as 2050 or 2100.

By comparing the current and future (predicted) ranges of species, it is possible to determine how ranges will contract or expand, how much overlap there is between current and future distributions, and whether a species has the capacity to move between these areas. If there is no geographical overlap between current and future ranges and dispersal is unlikely, the species may be 'trapped'

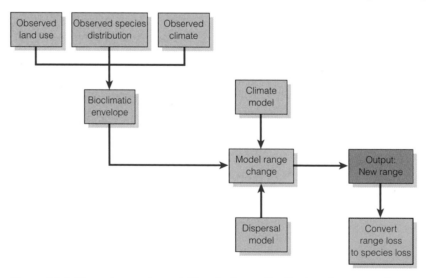

Figure 14.4 The main steps in modelling the future distribution of species under climate change.

within a climate that is unable to sustain a viable population and may be destined for eventual extinction. In the language of climate modelling, these species are *committed to extinction*. When these procedures are repeated for whole sets of species, the results can potentially be translated into overall patterns of changing diversity.

This was the approach taken in 2004 by the British biologist Chris Thomas and colleagues from a wide range of research institutes and conservation organizations [38]. Thomas' study, published in the journal *Nature*, used an SDM approach to model the potential effects of global warming on the distributional ranges of 1103 species of strategically chosen land animals and plants. The results suggested that under 'moderate' climate change scenarios, between 15 and 37% of the species in the study would be 'committed to extinction' by 2050. It is important to remember that decades (or even centuries) may elapse between the reduction of appropriate habitat through climate change and the eventual extinction. If, and this is a very big 'if', these results could be extrapolated to all of the (known and unknown) species in the world, this could mean that as many as a million species may find themselves living in bioclimatically unsuitable conditions by 2050.

SDMs have been strongly criticized because of the wide number of assumptions and uncertainty implicit in such complicated models [39]. For example, the data used by SDMs normally take the form of species range maps. These maps are necessarily generalizations, as sampling has not occurred at every point on the map. This means that the envelope drawn around the data points reporting species presence will inevitably contain numerous places where the species is actually absent. To increase consistency, scientists standardize the mapping of species ranges by first dividing the landscape into grids of cells of a fixed size. A grid cell will be considered as containing the species if the species is reported anywhere within that cell. However, if the cell size is large, it may only occur in a small part of that cell, leading to overestimations of the total area occupied. Conversely, the use of very small grid cell sizes can provide more precise and accurate representations of the range, but at the cost of enormous increases in sampling effort, not to mention the cost and time invested in acquiring the data.

The presence of a species in a grid square is ultimately based on scientific records which, depending on who made the observation, have varying degrees of certainty attached to them. Expert surveys or voucher specimens lodged in herbaria and museums have a high degree of certainty. However, such surveys are costly and relatively infrequent, so they are less likely to cover the entire potential range of a species. If the data were collected over a long time period, there is a risk of recording a species as being present in areas where it has long been locally extinct [40]. Thus, ranges are often inaccurate, being overestimated, underestimated or displaced from their true locations. Furthermore, some species may still be in the process of redistributing themselves after the major climatic shock of the last Ice Age. In this case, the observed distribution doesn't really reflect the climate tolerance of the species, and any resulting SDM will be inaccurate.

Despite their many limitations, SDMs provide a useful tool for exploring the potential effects of climate change on biodiversity. Perhaps the main take-home point of these models is that there will be climate change 'winners' and 'losers'. The biggest losers of all will be the species which no longer have any suitable climate and habitat within their dispersal range. Such a case could occur on mountains where the bioclimatic envelope moves upwards and, eventually, may even disappear entirely off the top of the mountain. This is referred to as the *escalator effect* [41], and it has had the positive effect of renewing scientific interest in mountain fauna and flora.

Crisis Management: Responding to Biodiversity Loss

The desire to preserve other forms of life in the face of human development is a social value with a relatively recent origin. The modern conservation movement emerged in the late 19th century in response to fundamental changes in world-views concerning the nature of the relationship between humans and the natural world [42]. These changes were prompted by a series of influential discoveries, events and circumstances, most notably the publication of Darwin's seminal book on evolution by natural selection (1859), a sudden and well-publicised rash of extinctions, the rapid demise of the vast forests of the American Great Lakes region and the discovery of the great apes (the first scientific description of a gorilla came in 1847).

The early conservation movement was concentrated in the large urban areas of Europe and North America and was motivated by two main values: (i) to preserve natural areas for the intellectual and aesthetic contemplation of nature; and (ii) the acceptance that human dominion over nature goes hand in hand with a moral responsibility to preserve species from extinction. In the latter half of 20th-century conservation, these foundational values began to change and new values were added. For example, Rachel Carson's seminal book *Silent Spring*, published in 1962, revealed the horrific impacts of agricultural toxins on wildlife and human health, prompting the emergence of values relating to environmental justice: the idea that both environmental risks (e.g. pollution) and environmental benefits (e.g. clean water) should be equally distributed throughout society. Likewise, growing concerns about unchecked population growth and resource use highlighted the importance of intergenerational equity: the responsibility of humanity to protect and improve the environment for present and future generations.

Since the late 1980s, the main focus of conservation has been the protection of the world's biological diversity, typically shortened to the more user-friendly term *biodiversity*. Although biodiversity is defined as the variation of life on Earth at all levels of biological organization, for practical reasons this is normally expressed in terms of genes, species or ecosystems. *Biodiversity* was coined as a way of making politicians and bureaucrats aware of looming extinction crises, particularly in the tropics. It quickly rose in prominence and was cemented in international policy through the **Convention on Biological Diversity** (**CBD**), formulated at the 1992 Earth Summit in Rio de Janeiro.

The CBD emphasized the economic and social benefits that efforts to save genes, species and ecosystems could bring. This approach appealed to technocrats and led to considerable increases in government funds to support biodiversity and multi-million-dollar projects to integrate conservation and development goals. On the 10th anniversary of

the Earth Summit, governments committed to 'a more effective and coherent implementation of the Convention and to achieve, by 2010, a significant reduction of the current rate of biodiversity loss'. Sadly, virtually none of the 2010 targets were met [5], with progress being restricted to very small geographical areas and specific interventions.

The CBD, along with Agenda 21 on sustainable development, provides a comprehensive framework for sovereign nations to develop a national legislative response to the biodiversity crisis. This should outline strategies, plans and programmes that respond to the changing circumstances of biodiversity in particular nations. Many countries have thus developed national biodiversity strategies (identifying strategic needs) or action plans (identifying technical responses and practical steps) as ways of fulfilling their obligations to the CBD (Table 14.2).

Table 14.2 Main technical responses to threats to biodiversity. Adapted from Ladle and Malhado [21].

Main threats	Responses
Habitat loss	Protected areas
	Ecological restoration projects
	Systems of quotas and fines
Habitat fragmentation	Protected areas
	Protected area networks
Habitat degradation	Protected areas
	Remediation and restoration measures
	Stricter emission controls on contaminants
	System of quotas and fines
Invasive species	Eradication
	Biocontrol
	Containment
	Invasion prevention measures
Over-exploitation	Anti-poaching measures
	Systems of quotas and fines
Climate change	Protected area networks
	Improved forecasting
	Translocations and reintroductions

The CBD lays great emphasis on *in situ* conservation of viable populations (Article 8) and calls for the establishment of protected area networks, the rehabilitation of degraded areas and the protection of habitats and species in natural surroundings. Governments worldwide have responded to this challenge by adding to (sometimes extensive) existing protected areas. There are currently more than 100 000 protected areas that cover 14.6% of terrestrial land area and 2.8% of the marine environment [43]. These are huge figures, but how effective are these protected areas and networks going to be at protecting the Earth's remaining biodiversity? This critical conservation question is fundamentally biogeographical. In the 'The Birth of Conservation Biogeography' section, we will see how biogeographers have contributed to contemporary conservation practice, drawing on concepts and tools developed over the previous two centuries to address the urgent contemporary crisis of biodiversity loss.

The Birth of Conservation Biogeography

As we have seen, humans have had a dramatic influence on the physical and biological components of ecosystems at every scale from local habitats to the entire Earth system – the dramatic effect of a single species has led to the widespread use of the informal geological term *Anthropocene* to refer to the current epoch during which human activities had a significant impact on the environment [44]. Scientists have played an essential role in documenting and modelling these changes in ecology and biogeography. Moreover, starting in the late 1800s the modern conservation movement has evolved in response to these threats, with the most prominent contemporary organizations (e.g. the World Wildlife Fund (WWF), Conservation International, The Nature Conservancy and the Wildlife Conservation Society) relying heavily on scientific evidence and reasoning to target resources for conserving the aspects of biodiversity they value most highly.

Even though conservation organizations have existed for well over 100 years, scientific interest in conservation problems arose much later. Before the 1970s, conservation practice was informed by a heterogeneous literature drawn from forestry and agricultural and biological sciences. A distinct

science of conservation only started to take shape in the late 1970s and early 1980s, eventually leading to 'conservation biology' emerging as a sub-discipline 'worthy' of academic study in its own right, with dedicated journals, textbooks and university courses [42]. Conservation biology drew heavily on concepts and theories from ecology, population biology and natural resource management with a strong focus on population or landscape-scale processes. The first international conference dedicated to conservation biology took place as recently as 1978 at the University of California, San Diego. However, conservation biology really took off in 1986 with the founding of the Society for Conservation Biology (SCB), followed closely by the first edition of their influential journal, also titled *Conservation Biology*, in 1987.

As a scientific discipline, conservation biology is primarily concerned with the application of population biology, taxonomy and genetics to conservation problems (e.g. extinction and population decline). In a now classic article from 1994 [20], the British ecologist Graham Caughley divided conservation biology research into two overarching paradigms – studies that seek an understanding of the proximate causes of population decline (the declining population paradigm) and those that are concerned with the consequences of small population size (the small population paradigm). Most research in conservation biology still falls clearly within these paradigms, though there has been an increasing realization that characteristics such as rarity and endangerment also need to be understood and that, more broadly, conservation biology needs to expand beyond its core conceptual background of ecology and systematics. Modern conservation biology textbooks now typically include a wide range of other academic disciplines, including those as diverse as anthropology, biogeography, environmental economics, environmental ethics, sociology and environmental law [e.g. 45].

Although not central to the development of conservation biology, the potential utility of biogeographical concepts for conservation planning was quickly recognized by the scientists of the day. Most notably, Jared Diamond drew attention to the similarity of protected areas in a 'ocean' of degraded or agricultural land and oceanic islands [46,47]. Based on the principles of MacArthur and Wilson's TIB, Diamond argued that: (i) the number of species a protected area can maintain (at equilibrium) will be a function of its geographical area and its degree of isolation. It follows that larger reserves that are near extensive areas of natural habitat will contain a greater number of species. (ii) If most of the natural habitat surrounding a reserve is destroyed, the protected area will maintain many more species than it can hold at equilibrium. These 'excess' species will slowly go extinct as the protected area relaxes towards its equilibrium level of species richness. The TIB has continued to play a central role in conservation planning at landscape and regional scales (reviewed in [48]), although the simplistic notion that protected areas (and habitat 'islands' in general) behave as oceanic islands has increasingly been questioned [2]. There is also a long history of using biogeographical principles as a means to identify priority areas or species for conservation action. For example, range size criteria are an important part of the IUCN's evaluation system for its Red List of Endangered Species. For example, the 'Endangered' category includes species with an estimated extent of occurrence of less than 5000 km^2 or an area of occupancy of less than 500 km^2 [49].

The first decade of the 21st century saw an increasing use of biogeographical principles for addressing conservation problems [39], the rise of SDMs to predict the impact of climate change on diverse species and communities [37] and a renewed interest in the ability of long-term ecology to provide insights into contemporary events [50]. These approaches were clearly differentiated from tradition conservation biology in their emphasis on broader geographical (landscape scale or above) and temporal scales. In 2005, Robert Whittaker (Professor of Biogeography at the University of Oxford) and his colleagues suggested that these diverse research strands could be grouped under a new subdiscipline of 'Conservation Biogeography', which they defined as 'the application of biogeographical principles, theories, and analyses to problems concerning the conservation of biodiversity ' [39].

The Scope of Conservation Biogeography

Given the breadth of biogeography and the multiple threats to the natural world, the scope of conservation biogeography is large and growing [51]. Broadly speaking, conservation biogeographers are interested in biophysical processes that are predominantly operating at coarse geographical scales (landscape scale and above) [52]. Clearly, this encompasses a huge range of issues and has spawned an enormous amount of theory and tools that are often applicable to limited scales of analysis – the difficulty of scaling up or scaling down is a constant problem in conservation biogeography, frequently making it difficult to extrapolate patterns (scaling up) or to determine site-level conservation actions (scaling down).

At finer scales of analysis (e.g. the landscape scale), biogeographical concepts such as the Equilibrium Theory of Island Biogeography (see Chapter 7 and above) and metapopulation theory are used to address issues such as the impacts of the number, size, configuration and connectivity of habitat fragments on biodiversity. By extension, these concepts can also be used to plan protected area networks that optimize species diversity. At coarser spatial scales, the accurate mapping of geographical patterns of species richness and centres of endemism, uncovering phylogeographical structure, and the precise identification of biogeographical regions are invaluable for prioritizing the allocation of conservation resources at regional and global levels. Indeed, influential schemes such as Conservation International's biodiversity hotspots [53] and WWF's ecoregions [54] are firmly grounded in biogeographical analysis.

It is important to remember that while biogeography provides much of the scientific underpinnings of conservation prioritization, the decision of *what* to prioritize depends on social values rather than scientific rationale – there is an important distinction between the processes leading to the adoption of a set of values and the process of developing scientific guidelines to implement these values [52]. Any system of conservation prioritization reflects to a greater or lesser degree how society values different biophysical features (e.g. endemism, species diversity, phylogenetic diversity, carbon storage, watershed protection, etc.). Inevitably, protecting areas on the basis of one feature (e.g. species diversity) will divert conservation resources from other areas with other characteristics of value for conservation (e.g. high endemism).

In summary, the tools and concepts of biogeography are of fundamental importance to conservation, providing insights and guidance for a wide range of activities. Unlike conservation biology, whose focus is primarily on population-level processes, conservation biogeography is concerned with coarse-scale patterns through time and space. South African biogeographer Dave Richardson and Rob Whittaker recently identified six core areas of research in conservation biogeography [51]: (i) the biogeography of habitat degradation (e.g. habitat fragmentation, homogenization, urbanization and other human-induced impacts); (ii) fundamental processes that influence rates and extent of biodiversity loss and recovery (e.g. colonization, climate as a fundamental determinant of distribution, dispersal, disturbance, extinction, persistence, range expansion, resilience and speciation); (iii) biodiversity inventory, mapping and data issues (e.g. atlas data, breeding bird surveys, citizen science, detectability and discovery probabilities, herbaria and other collections, and sampling intensity and biases); (iv) species distribution modelling (e.g. bioclimatic modelling, habitat suitability analysis, model performance, niche-based models, dispersal kernel analysis and presence-only data versus presence–absence data); (v) characterization of biotas (e.g. conservation threat status, diversity indices and patterns, ecoregions, endemism, rarity, range size, SARs, threatened species and identification of alternative baselines from long-term ecological data); and (vi) conservation planning (e.g. complementarity, congruence, conservation units, ecosystem services, gap analysis, global conservation assessments, irreplaceability, reserve networks and surrogates).

These diverse themes draw upon a wide variety of biogeographical methods and intersect with various other fields of enquiry, notably global change biology, molecular phylogenetics, invasion biology, bioinformatics and behavioural ecology. Moreover, they are united by a common set or overarching themes related to uncertainties and shortfalls in global biodiversity data, scale dependency, measuring niches and the impact of novel climates and ecosystems.

Uncertainties and Shortfalls

There are fundamental and practical limits on biodiversity knowledge (Table 14.3). This means that ecologists and conservationists have to work with incomplete and often unrepresentative data on a

Table 14.3 Definitions and original references for the seven main shortfalls of biodiversity knowledge. Adapted from Hortal *et al.* [74].

Seven main shortfalls of biodiversity knowledge
Linnean shortfall – Most of the species on Earth have not been described and catalogued; this concept can be extended to extinct species
Wallacean shortfall – The knowledge on the geographical distribution of most species is incomplete, being most times inadequate at all scales
Prestonian shortfall – Lack of data on species abundances and their dynamics in space and time are often scarce
Darwinian shortfall – Lack of knowledge about the tree of life and evolution of species and their traits
Raunkiæran shortfall – Lack of knowledge on species' traits and their ecological functions (this review)
Hutchinsonian shortfall – Lack of knowledge about the responses and tolerances of species to abiotic conditions (i.e. their scenopoetic niche)
Eltonian shortfall – Lack of enough knowledge on species' interactions and their effects on individual survival and fitness (this review)

limited number of organism characteristics. These gaps (known as *shortfalls*) in knowledge about the identity, distribution, evolution and dynamics of global biodiversity need to be carefully recognized, quantified and factored into conservation biogeography research. The inability to produce geographically unbiased and representative knowledge about biodiversity compromises our capacity to describe its existing state or to make accurate predictions about how it might change in the future. Biased data can also lead to misidentification of biogeographical processes [55] and inefficient use of limited conservation resources [56].

The two most important data shortfalls for conservation biogeography are the Linnean shortfall and the Wallacean shortfall (Table 14.3). The *Linnean shortfall* is named after the discrepancy between formally described species and the number of species that actually exist [57]. The size of the Linnean shortfall is unknown for two reasons. First, the number of formally described species is constantly changing due to new descriptions, revisions and unresolved synonyms as well as difficulties in establishing a unified species concept or agreement on operational tools to delimitate different taxa. The most comprehensive and authoritative global index of species currently available is the Catalogue of Life, which has records for more than 1.5 million species. Second, the predicted number of species is highly sensitive to the estimation method adopted and the parameter estimates used, leading to estimates ranging from 2 million to as many as 100 million eukaryotic species [58], with more recent global species richness estimates converging on a narrower band of 2–10 million species [59].

It is important to note that the Linnean shortfall is made up of two distinct categories: species yet to be sampled and collected species that have not yet been described. Species in the former category are most frequent in the poorly researched areas of the world, such as the forests of the south-central Amazon. The category of collected but as yet unidentified species may run into the thousands and are partly a consequence of a lack of funding and capacity in global taxonomy.

The Linnean shortfall also contains clear taxonomic and geographical biases. This is because certain taxa and regions have inevitably received far more attention than others, to the point that the proportions between known and estimated numbers of species vary between about 7% for terrestrial fungi and marine animals to over 70% for terrestrial plants [60]. Terrestrial vertebrates and vascular plants are orders of magnitude better known than almost all invertebrates (and certainly better known than unicellular organisms). Similar size-based patterns can be discerned in individual taxa, with larger, conspicuous and easily detectable species typically being recorded earlier and more extensively [e.g. 61]. Conversely, taxonomists have an odd habit of preferentially collecting rare species and, as a consequence, disregarding or underrepresenting more prosaic members of their special taxon of expertise [62]. These biases propagate to data on all other aspects of biodiversity: there are far more data on ecological interactions and functions of crop pollinators and economic pests, compared to those on their natural enemies and wild relatives. Likewise, game species and emblematic taxa are much better known than less popular groups.

The Linnean shortfall affects the extent and distribution of every other type of biodiversity information shortfall (Table 13.1) because, understandably, we typically have no data on the characteristics of unknown species. The exceptions are a limited number of characteristics that can be estimated in models fitted to ecological and evolutionary data about related species [e.g. 63], or attributed to Operational Taxonomic Units (OTUs) using next-generation sequencing techniques [64]. Beyond these very minor contributions to hidden biodiversity knowledge, incremental improvements of any aspect of biodiversity must necessarily be preceded, or at least accompanied, by filling in the Linnean shortfall.

The *Wallacean shortfall* refers to the lack of knowledge about the geographical distribution of species [65]. The lack of knowledge about species distributions is closely connected with temporal and spatial variation in surveying effort. That some regions are better sampled than others is inevitable given the wide country-level differences in survey capacity allied to broad-scale variability in accessibility. Like the Linnean shortfall, the Wallacean shortfall is particularly acute in remote and difficult-to-access regions of the developing world such as the forests of southwest Amazonia and the Congo basin. Approximately 40% of Amazonia has never been surveyed, and we do not have an accurate geographical distribution for any of the plant species that occur in this region [66]. More generally, the quality of distribution data typically varies in relation to political rather than ecological units, and may therefore be heavily biased due to diverse historical trends that have influenced the trajectory of collecting, analysing and collating biogeographical data within a given country or geopolitical unit.

The spatial distribution of information on the occurrence of species is biased towards certain regions, biomes and habitats. This is due to varying investment in surveys, the behavioural preferences of researchers [67] and strong historical patterns of colonization and inventory [68]. Biodiversity inventories therefore tend to be more comprehensive near residences or workplaces of collectors and taxonomists, field stations or, in general, any location with convenient access, infrastructure and logistics [69]. These geographical biases add strongly to the uncertainty of observed species distributions and have led to major errors in the known distribution of endangered species and conservation targets [70]. Furthermore, temporal shifts in the spatial coverage of surveys result in spurious changes in the known distribution over time [71], affecting our ability to identify past range shifts and discriminate actual patterns of extinction [72,73].

Both Linnean and Wallacean shortfalls are scale dependent in terms of both their resolution and the extent of data coverage and analysis [74] (see Table 14.3). At the largest possible grain size (the entire Earth), we have perfect knowledge of the distributions of any species that has been described. However, at smaller grain sizes, the Wallacean shortfall begins to grow as increasingly precise information on distributions is required. Finally, at very small grain sizes it becomes progressively harder to define the presence and absence of a species, especially for highly mobile animals that range over wide areas and habitat types. There is strong temporal variation at smaller grain sizes, with distributions fluctuating in relation to the ecological characteristics of the species in question.

From an applied perspective, the Linnean and Wallacean shortfalls have far-reaching influences because data on the identity and distribution of species are vital for assessing and identifying broad-scale patterns in biodiversity and the processes that create biodiversity (e.g. extinction) and, by extension, any form of conservation prioritization based on species diversity and endemism. For example, global extinction estimates are highly sensitive to assumptions about the number of extant species, especially those calculated on the basis of backward extrapolation from the SARs [10]. It has recently been argued that backward SAR calculations are only valid for the rare case of randomly distributed species, and therefore should not really be used to calculate extinction rates [75].

The Wallacean shortfall can also have profound impacts on the estimates of biodiversity threat. The range size of a species is often used in conservation planning, with small ranges frequently used as a prioritization criterion or as a proxy for threat. Indeed, geographical range restriction is an integral part of IUCN criteria to identify and classify species in danger of global extinction [49]. Attribution to the most highly threatened IUCN Red List categories defaults to range size if other data are lacking – with thresholds of 100 km^2 for 'Critically Endangered', 5000 km^2 for 'Endangered' and 20 000 km^2 for 'Vulnerable'. Several conservation prioritization methods [e.g. 76] use an arbitrary criterion of <50 000 km^2 to define range restriction or

local endemism (originally suggested by Terborgh and Winter in their classic article from 1983 [77]). Apart from the obvious problem that such a broad geographical category necessarily captures nearly all island endemics, many of which are clearly not under threat, the Wallacean shortfall means that prioritization based on many taxa or for certain regions will be highly uncertain.

Data bias is arguably an even bigger problem for the development of robust tools to support conservation practice [40]. The potentially corrosive effects of such data bias are clearly illustrated in the use of distributional ranges for systematic conservation planning. SDMs are perhaps the most widely used analytical and predictive tool in conservation. Almost paradoxically, biases in the Wallacean shortfall have the potential to strongly influence the performance of SDMs – which were originally designed to account for the lack of distributional knowledge. SDMs typically relate field observations of species occurrence (and sometimes absence) to environmental (often climatic) predictors using statistically or theoretically derived response surfaces that are supposed to account for the tolerances of species to abiotic conditions [10]. SDMs are routinely used for rare species where accurate data on their distributions are missing. However, if the representation of the niche provided by the occurrence data is biased, then the SDM results will consistently fail for these rare species [37], reducing the representativeness of any reserve networks that are identified by these models [78].

Scale Dependency

Scale issues have a clear relevance to the development of strategic conservation frameworks (e.g. hotspots) because they are dependent upon mapping diversity patterns. For example, Lennon *et al.* [79] found that the spatial patterns of species richness for British birds using a 10 km grain grid system were statistically unrelated to those using a 90 km grain system. This could lead to different conservation decisions depending on the scale of analysis used.

Scale may also influence interpretations of rarity. *Rarity* can mean that a species occurs at low density, and it can mean that a species occupies a small geographical range. However, all endemic species on oceanic islands (e.g. the Canaries) occupy 'small' geographical ranges and would be considered as threatened if a commonly used threshold of <50 000 km^2 (see above) was adopted [77]. Thus, there is a need to refine 'range-restricted' for conservation. However, to do this, more data are needed to link range size estimates with population estimates for large numbers of species in different ecological contexts (e.g. mainland versus island systems).

Finally, conservation biogeography also needs to deal with temporal scale dependency. Because conservation aims to ensure the survival of species in the long term, it is important to understand the temporal dynamics of biogeographical processes [50]. However, habitats and assemblages are changing so fast that most models that attempt to project changes into anything but the immediate future are based on 'snapshot' data or very short temporal series.

Measuring Niches

In his classic study, G. Evelyn Hutchinson [80] conceptualized two kinds of niches: the fundamental niche (that reflects the underlying physiological tolerances of a species to environmental conditions) and the realized niche (the more limited range of environmental conditions in which a species can exist due to the influence of biotic interactions). These concepts can also be applied to species distributions: a species' realized distribution often does not fully occupy the geographical space of its potential distribution. This distinction is particularly important for conservation biogeography because one of its fundamental goals is to understand and predict the impact of the rapidly changing biophysical habitat on the distribution (and survival) of species [81]. This is typically done through models that try to quantify the habitat preferences and tolerances of the species of interest by inferring them from either field-derived occurrence data or laboratory-derived physiological data.

The simplest approach is to quantify the association between a species' distribution and one or more external variables (e.g. temperature, presence of a competitor species, etc.) within that distribution. Such habitat preferences derived from localities where the study species is known to be present can be used to map the complete distribution of

potentially suitable habitat; the assumption is that the species of interest will occur wherever there is suitable habitat. Simple habitat models have been widely used to estimate the potential response of species to habitat loss, using habitat extent as a surrogate for population size.

The main advantage of these types of models is that they can incorporate a wide range of ecologically important variables. However, simple correlative approaches work less well when species' distributions are limited by non-habitat-related factors, such as competition, predation, food availability or dispersal limitation (i.e. when the realized niche is a very limited subset of the fundamental niche). Furthermore, even under the best of circumstances, many species have highly specific microhabitat requirements that cannot be easily detected on the basis of (necessarily) broad-scale habitat data.

A more sophisticated family of models of realized distributions are based on samples from across an environmental gradient, determining whether the species of interest is present (or absent) at sites with different environmental conditions. By doing this, each geographical location at which a species has been found has an equivalent position in *environmental space* (based on the subset of environmental variables selected for the study). This information can then be used to predict the potential distribution of a species within the known limits of its occurrence – very useful information for future surveys. Because these models are solely based on association between presence–absence data and environment variables, they tend to work best when the effects of historical factors and sampling biases are weak and where dispersal is not strongly constrained.

An alternative approach is to predict changes in species distributions based upon the fundamental niche. Such data are normally derived from physiological measurements and performance curves obtained from laboratory experiments. Although physiologically derived niches are generally more precise than occurrence-derived niches, they are not necessarily more realistic. This is because different subpopulations often show a lot of variability in their response to physiological conditions due to genetic differences, phenotypic plasticity and acclimation [82]. Thus, niche data obtained under laboratory conditions provide an incomplete representation of the fundamental niche. Nevertheless, even if incomplete, such data can be used cautiously to predict the occurrence pattern of a species, sometimes with remarkably impressive results.

In 2004, Kearney and Porter [83] modelled the fundamental niche of the nocturnal lizard *Heteronotia binoei* across the whole of Australia based on laboratory-derived physiological measurements (thermal requirements for egg development, thermal preferences and tolerances, and metabolic and evaporative water loss rates) and high-resolution climatic data (air temperature, cloud cover, wind speed, humidity and radiation). In other words, they calculated the climatic component of the fundamental niche of this lizard. These data were then used to produce high-resolution maps of the lizard's potential distribution.

Novel Climates and 'No-Analogue' Communities

How do you study a community you have never seen before? Biogeography is definitely entering uncharted territory due to the combined effects of climate change, invasive species and habitat loss, fragmentation and degradation. The rapidity, intensity and sheer scale of these processes mean that all over the world, new ecological communities are coming into existence that are made up of species that have so far survived the extinction crisis, climate migrants exploiting newly expanded fundamental niches and invasive species that have, courtesy of humans, recently jumped over previously insurmountable biogeographical boundaries. Climates and communities with no modern-day analogues are a problem for biogeographers because the most commonly used models to predict the ecological effects of climate change are (at least partially) parameterized from modern observations. Consequently, our models may fail to accurately predict ecological responses to these novel climates [84].

One of the main problems is that, unlike many aspects of biogeography, studying the past may be of little help. This is not to say that so-called **no-analogue communities** didn't occur in the past. In fact, the opposite is true: many past ecological communities were compositionally totally unlike any modern communities, even before the impacts of human actions began to take effect. The fossil record suggests that the creation and demise of

past no-analogue communities were climatically driven and, significantly, were linked to climates that are also without modern analogues [84] If, as appears inevitable, human-induced climate change continues unabated, many areas will have no-analogue climates and communities in the near future.

Despite the complexities and uncertainties involved, the latest biogeographical models are able to give some idea about what these new communities might look like. For example, the American ecologist Mark Urban and his colleagues recently modelled multiple competing species along a warming climatic gradient, including the effects of temperature-dependent competition, differences in niche breadth and interspecific differences in dispersal ability [85]. They found that the combined effects of competition and differences in dispersal ability decreased diversity and produced no-analogue assemblages. Moreover, no-analogue communities were more likely to form when thermal performance and competitive niche breadths were narrow. While the predictive value of such models is debatable, the identification of factors that are likely to give rise to no-analogue communities provides a valuable start point for discussions of long-term conservation planning.

Conservation Biogeography in Action

Perhaps the central goal of conservation biogeography is to help societies to maximize the use of very limited conservation funds. Clearly, we cannot provide the same level of protection for all of the remaining areas of the natural world, not all of the species that currently exist can be saved, and the impacts of climate change on species and ecosystems cannot be completely avoided. However, biogeography does provide some of the tools and concepts to maximize the benefits of conservation, allowing societies to concentrate their efforts on the places that will provide the greatest conservation benefits or which have the best chance of retaining their unique biological characteristics into the future.

The key to achieving these goals and the most important weapon in the fight against biodiversity are protected areas (PAs). Of course, it is important to note that the designation and placement of PAs do not simply depend on scientific principles (Figure 14.5), being influenced by a range of historic, cultural and legal factors. Nevertheless, within these constraints, biogeography has an important role in ensuring that conservation maximizes the impact of these irreplaceable conservation interventions.

PAs have a long and distinguished history (reviewed in [86]). However, it was not until the 1960s that serious thought was given to the establishment of a worldwide network of natural reserves that included examples from all of the world's ecosystems. In a series of classic papers, Dasmann [87,88] and Udvardy [89] put this idea of biogeographical representation into practice. Building upon earlier maps of faunal regions and vegetation zones, they created a nested hierarchy of biological regions and thereby provided a framework for the massive and rapid expansion of PAs that was shortly to follow.

What has become known as the *Dasmann–Udvardy framework* follows greats such as Alfred Russel Wallace by dividing the world into biotic realms. They then applied a biome classification system within each of these realms, distinguishing smaller biotic provinces on the basis of faunal

differences in birds and mammals: areas with less than 65% of their species in common were designated as separate provinces. The framework provides a clear methodology, and in principle the analysis can be repeated to assess the implications of using updated distributional data, different systems of biogeographical regions or alternative thresholds for faunal similarity.

With the advent of electronic databases and powerful computers, the Dasmann–Udvardy approach was quickly superseded by global schemes developed by international conservation non-governmental organizations (NGOs). This makes perfect sense in that these organizations have the resources and political power to realize the global ambitions of this approach to conservation prioritization. Perhaps the most closely related contemporary framework to the original Dasmann–Udvardy framework is WWF's Ecoregions scheme [90]. This scheme uses data on biogeography, habitat type and elevation to identify fine-scale biogeographical units with the aim of identifying natural units within which ecological processes are maintained. In this sense, the Ecoregions scheme is very much focussed on conserving ecological function (as opposed to composition), and its success is predicated on effectively identifying the most important ecological processes, flows and linkages and,

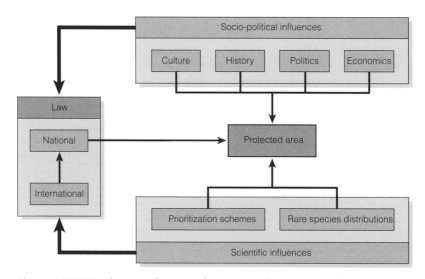

Figure 14.5 Main factors influencing the creation of new protected areas. The creation of a major new protected area is primarily achieved through national legislation but is heavily influenced by socio-political factors and is informed by science. From Ladle and Malhado [21].

critically, defining the geographical boundaries between these interlinked units [86].

In contrast to ecoregional approaches, the recently formed (1987) NGO Conservation International (CI) developed a 'hotspots' scheme [53,91] which was specifically designed to maximize the number of species 'saved' given the resources available for conservation. Rather than systematic representation, hotspots approaches use principles of irreplaceability and vulnerability to guide conservation planning on a global scale. Perhaps the first global application of these principles was by Oxford-based conservationist Norman Myers who identified 10 tropical forest 'hotspots' on the basis of very high levels of plant endemism and correspondingly high levels of habitat loss [92]. CI was quick to see the potential in Myers' approach and worked closely with him throughout the 1990s to develop quantitative criteria to identify hotspots. They eventually decided upon two strict criteria to qualify as a hotspot: a region had to contain at least 1500 endemic vascular plants (>0.5% of the world's total), and it had to have lost >70% of its original habitat [53]. Based on these criteria, 25 hotspots were identified that contained a staggering 44% of vascular plants and 35% of species in four major vertebrate groups (amphibians, mammals, birds and reptiles), despite covering a combined area of only 1.4% of the terrestrial surface of the Earth [53]. An updated analysis of distribution patterns based on the same criteria has identified 34 biodiversity hotspots that contain 50% of vascular plants and 42% of terrestrial vertebrates as endemics [91]. It is interesting to note that the hotspots scheme necessarily places higher conservation value on species that co-occur with lots of other range-restricted plant species on a regional scale [39]. This is not so much a weakness of this approach, but an inevitable consequence of almost all **prioritization scheme**s – in a world of limited conservation resources, there will always be winners and losers.

The marine environment has always been somewhat of a poor relative within global conservation. For example, a 2008 review revealed that only 0.7% (2.59 million km^2) of the world's oceans were within PAs [93]. Likewise, global prioritization schemes have only recently been developed, possibly also due to the conceptual difficulties of dividing and classifying the oceans (see Chapter 9). Using the success of terrestrial prioritization schemes as a spur, the Nature Conservancy's Mark Spalding and colleagues recently proposed a new global biogeographical classification system for coastal and shelf areas known as the Marine Ecoregions of the World, or MEOW [94]. The scheme, drawing heavily on previous studies, used a nested system of 12 realms, 62 provinces and 232 ecoregions – the latter defined as areas of relatively homogeneous species composition, clearly distinct from adjacent systems. The contributions of such prioritization schemes to investment in marine conservation are difficult to judge. Nevertheless, marine PAs are currently increasing far more rapidly than their terrestrial counterparts [95], and, correspondingly, informed and accurate prioritization could generate enormous benefits for biodiversity and sustainable development.

Biogeographical principles have also been used to prioritize conservation actions at lower spatial scales, particularly in the design of PA networks [96]. The first PAs were placed in an almost haphazard fashion depending on factors such as land availability and political motivation. This led to a very biased distribution of reserves that tended to be placed on dry, infertile and inaccessible areas – basically, places that weren't seen as having a lot of economic value. It was not until the 1970s that a more scientific approach was adopted, as some scientists began to apply the recently articulated TIB [97] to PA design [46].

Unfortunately, beyond some very general principles, the TIB proved to be largely inadequate as a tool to guide reserve selection. This is well illustrated by the ultimately pointless **SLOSS** (single large reserve or several small reserves) debate that eventually petered out when scientists finally came to the inconclusive answer, 'it depends' [98].

Scientists started to extend the basic principles of reserve design in the early 1980s by incorporating simple scoring systems (based on species richness or number of endemic species) to help guide choice of new PAs [99]. However, these early prioritization systems – though based on clear scientific principles and providing clear guidance – were also of limited practical utility. For example, choosing the biggest available reserve or the area with the greatest number of endemic species may not coincide with the wider goals for conservation of the landscape, and may actively conflict with demands for human use. In other words, the real world is complex and messy, and conservation planning needs to take this into account. Thus, the focus of PA network design has shifted towards identifying a range of scenarios, creating alternative proposed networks based on biogeographical algorithms that take into consideration the complexity of spatial planning across landscapes [98].

Contemporary conservation planning is based on five basic principles [96]: complementarity, representativeness, persistence, efficiency and flexibility. *Complementarity* simply refers to choosing areas with the aim of collectively achieving objectives (e.g. maximizing biodiversity). Complementary areas might, for example, contain different species or habitat types. *Representativeness* refers to how well reserve networks contain examples of every target feature of biodiversity (e.g. habitat types). *Persistence* refers to the ability of the network to ensure the continued existence of the biodiversity feature of interest (e.g. viable populations of target species). *Efficiency* is a measure of the impacts (e.g. economic costs) of conservation on society – the lower these impacts, the more 'efficient' the network. Finally, *flexibility* refers to the existence of alternative broadly equivalent network 'solutions' that land-use planners and politicians can use to make real-world decisions. As mentioned in this chapter, flexibility is a key facet of modern conservation planning, greatly enhancing the probability of a reserve network being implemented in reality.

One or more of these principles can be codified into selection algorithms depending on the objective of the network, providing a systematic way for scientists to identify credible conservation options for decision makers. Thus, modern conservation planning typically involves identifying a list of important conservation features, setting targets for each feature and using sophisticated statistical tools to identify priority areas for meeting these targets. Perhaps the most widely used decision support tool is Marxan (www.uq.edu.au/marxan; last accessed December 2015), which can be used to identify a subset of locations that meet the required (user-defined) conservation targets [100]. For example, Marxan could be used to identify reserves that contain 25% of each habitat type and 70% of threatened species in the regional species pool. Significantly, Marxan can also be used to cost different options, allowing scientists to factor in social factors (e.g. cost of banning natural resource extraction activities) and generate minimal 'cost' options.

For all the advances in biogeographically informed conservation planning (reviewed in [96]), the answer to one of the most critical questions has remained elusive – how much is enough? In other words, how do we ensure persistence of the resources we have managed to capture for conservation? Unfortunately the question is, ultimately, unanswerable given the unavoidable trade-offs involved in acquiring land for conservation. For sure, more is always better, but this always comes with additional costs for societies [96]. Moreover, these costs and their associated trade-offs are constantly changing due to the rapidly changing biophysical environment and social circumstances. As a consequence, the case for PAs needs to be restated. Their current and future value(s) to society needs to be explicitly stated and measured, providing incentives for expansion of the PA network and justifying the continued investment and development of existing PAs. In other words, as competition between land uses increases, the value of retaining areas of natural vegetation needs to be explicitly identified and quantified, justifying the continued investment of public and private organizations.

The Future is Digital

We are in the midst of an 'information revolution' with profound impacts on culture, science, politics and commerce [101]. The venerable discipline of biogeography will be in the front line of these changes, as new technologies transform the quantity and quality of biogeographical data available and our ability to access and analyse it. This, in turn, will allow scientists to better prioritize and protect wild nature in the face of environmental change.

The base unit of biogeographical knowledge is a record of a species (or, less frequently, another taxonomic unit) at a precisely defined point in time and space. Traditionally, collecting such data requires an experienced and knowledgeable observer who can accurately identify the species in question, collate the information and place it where other scientists can access it. Since Victorian times, this process has been performed in more or less the same way – a professional scientist would enter the field, collect specimens, bring them back to the laboratory for identification, place the preserved specimens in museum collections and publish the data in scholarly reports or scientific articles. Clearly, the rate of data collection is severely compromised by the number of experts available to: (i) collect records and (ii) identify species. Unfortunately, resources for curating museum collections are being cut around the world [102], and the last decades have also seen a global decline in taxonomists [103].

Taxonomists will always be needed. However, advances in mobile technology may eventually provide a way to circumvent the 'taxonomic impediment' and generate enormous quantities of data on species occurrences. Specifically, there have been four interlinked technological innovations with the potential to forever change the ways in which biogeographical data are collected. Perhaps the core component is the development of *apps* (an abbreviation of *software applications*), which are able to link the technological forces of cloud and mobile computing, social networking and 'big data' to transform mobile devices into sophisticated biodiversity sensors and powerful computers [104].

Although we are still a long way from replacing traditional field biologists and taxonomists, there are already some 'biogeography apps' that offer startling glimpses into how data collection may be transformed in the future. For example, the iBat app (Indicator Bats Programme, created by the Zoological Society of London) uses a handy ultrasonic bat detector that can be plugged into a standard smartphone, enabling users to record bat calls and upload the geo-referenced data to an online database. This database then uses an open-access classification tool (iBatsID) that deploys ensembles of artificial neural networks to classify time-expanded recordings of bat echolocation calls from 34 European bat species. Recordings not suited for machine identification are submitted to a Zooniverse Real Science online project (www.batdetective.org; last accessed December 2015) for crowd-sourced identification and discussion. Originally developed by astronomers to help manage 'data deluge', Zooniverse is a collection of online citizen science projects with over half a million registered users [105]. Using the remarkable pattern recognition capacities of the human brain, citizen volunteers classify, extract and discuss science data from photos of galaxies to ancient documents.

In the future, it may even be possible to automatically identify bird songs or camera-trap photos by using a new generation of evolutionary algorithms [106]. The underlying principle is that identification software could be trained to recognize species from images and, by extension, identify possible new species for which no records exist. When combined with advances in other fields such as DNA barcoding [107], the future may be far more data-rich than we can currently imagine. However, data collection is only one stage of the process. To be useful to biogeographers and conservationists, we also need tools to access and process new and existing records. Once again, advances in information technology mean that such tools (known generally as Biodiversity Information Systems) are becoming a reality. The most ambitious and biogeographically focussed of these is probably the 'Map of Life' project, the ultimate aim of which is a public, online, quality-vetted distribution map for every species on Earth that integrates and visualizes available distributional knowledge and which facilitates user feedback and dynamic biodiversity analyses [108] (Figure 14.6).

The Map of Life is not the only ambitious, global bioinformatics project currently being developed. The Encyclopedia of Life (www.eol.org; last accessed December 2015), a project inspired by biogeographer E.O. Wilson, aims to 'make available via the Internet virtually all information about life present on Earth' [109]. The plan is to have a website for every species that has been formally described (the list of which is available on another bioinformatics megaproject, the Catalogue of Life – www.catalogueoflife.org; last accessed December 2015). Like the Map of Life, the Encyclopedia is flexible and constantly evolving, so that it can easily incorporate new information on ecology, genetics and conservation. At the time of writing (April 2015), the Encyclopedia of Life had over 1.7 million pages with data and was growing rapidly.

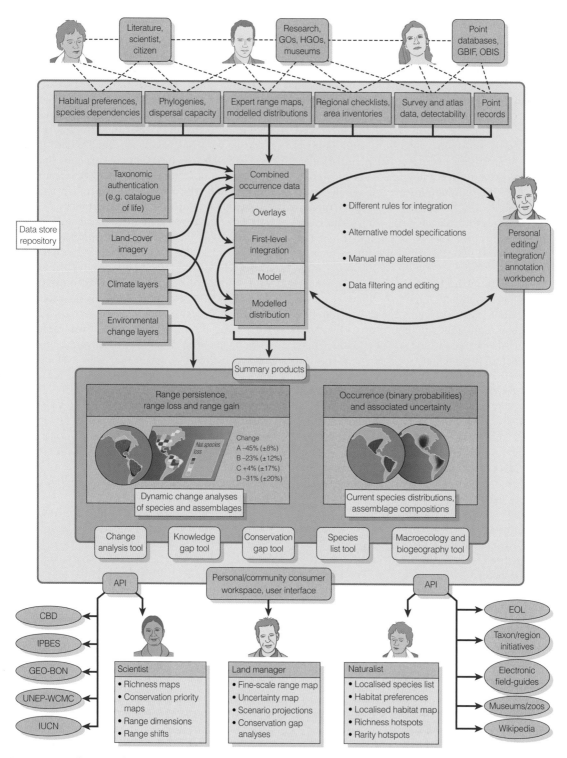

Figure 14.6 Schematic diagram showing how producers and consumers of species distribution information interact with the envisioned infrastructure, currently under implementation as a Map of Life. The planned web

Conclusions

There is an old Chinese curse, almost certainly apocryphal: 'May you live in interesting times'. These are certainly 'interesting times' to be a biogeographer – the information revolution, advances in molecular technology and an evolving culture of international collaboration are generating enormous quantities of biogeographical data and a new generation of sophisticated tools to uncover the hidden patterns of nature. All this is taking place during an unparalleled period of environmental change which will result in a re-sorting and readjustment of life on Earth, the magnitude of which has not been seen since the last mass extinction. These changes will have consequences at every scale, from local to global, and will have lasting impacts on the future evolution of life on our small, blue planet.

Conservation biogeography cannot provide all the solutions to the problems of biodiversity loss, though it can provide many of the tools and concepts to make scientifically informed choices about what and where to protect, and the potential consequences of future environmental change (e.g. climate change, habitat loss and invasive species). In this respect, the greatest challenge facing scientists may be turning their concepts and predictions into practice (guidelines, protocols, tools and applications that are useful at every level of conservation decision making). In many respects, the science is the 'easy' bit – influencing politicians and convincing sceptical publics are the real challenge. To effectively do this we may need to investment more in end-products, such as freely available and user-friendly tools and applications to analyse and visualize biogeographical data [110]. More generally, biodiversity needs to be fully integrated into other sectors of environmental governance. This will only occur when high-quality biodiversity data are available to key decision makers in forms that they can actually use. Conservation biogeographers can contribute by providing improved frameworks and concepts that give meaning and structure to biodiversity information, and by producing models and visualization tools that are of genuine practical importance.

As the Anthropocene continues, it will become increasingly difficult to find a balance between feeding the world's ever increasing population, adapting to the escalating impacts of man-made climate change and protecting the rapidly diminishing stocks of global biodiversity. Biogeographical tools and concepts will be essential for both predicting the consequences of these enormous changes to life on Earth and for formulating policies and practices that will allow humanity to hold on to at least some of the world's remaining natural areas. Our ability to stem the loss of biodiversity will be partly dependent on further advances in information technology, especially those related to remote sensing, data storage and mobile computing. These have the power to significantly enhance the predictive power of biogeography, opening new vistas of research and filling some of the enormous shortfalls in information about the identity and distribution of life on this small blue planet.

(*Figure 14.6 caption continued*) platform facilitates the uploading of species distribution information from many different organizations and sources, including data on habitat preferences, point occurrences and expert range maps. The infrastructure stores these data and provides a workbench for integrating them for one or many species. The data compiled, resulting summary information such as binary and probabilistic occurrence maps, and products from analysis tools can be provided to individual consumers, or served via Application Programming Interfaces (APIs) to other services or institutions such as the Encyclopedia of Life (EOL, http://www.eol.org), the GEO Biodiversity Observation Network (GEO-BON, http://www.earthobservations.org/geobon.shtml), initiatives connected to the Convention on Biological Diversity (CBD, http://www.cbd.int) or the Intergovernmental Platform on Biodiversity and Ecosystem Services (IPBES, http://www.ipbes.net). GBIF, Global Biodiversity Information Facility; GO, government organization; HGO, health and government operation; IUCN, Internetational Union for Conservation Network; OBIS, Ocean Biogeographical Information System; UNEP-WCMC, United Nations Environment Programme–Wildlife Conservation Mangement Committee. (All websites last accessed December 2015.) From Jetz *et al.* [108]. (Reproduced with permission of Elsevier.)

Summary

1 The enormous impacts that *Homo sapiens* have had on biological communities and landscapes since their emergence from Africa have led many scientists to dub the current geological epoch the 'Anthropocene': a distinct, undeniable period of human-induced environmental change.

2 There is strong evidence that we are entering a sixth mass extinction caused by human action that may wipe out a high proportion of the world's species in the next few centuries.

3 Biodiversity loss, range contraction and biotic homogenization are caused by several interlinked factors, including habitat loss and fragmentation, invasive species, habitat degradation, unsustainable exploitation and climate change.

4 Protecting the world's biodiversity will require action at multiple scales and needs to be supported by the best possible data and science.

5 Biogeography is uniquely placed to contribute to the development of biodiversity conservation through the development of concepts and tools to predict the consequences of human actions on the distribution of life and to identify effective strategies to slow or halt biodiversity loss.

6 Advances in technology have the potential to revolutionize the collection of biogeographical data in the coming decades, allowing scientists to construct much more realistic and dynamic maps of species distributions for use in conservation planning.

References

1. Ladle RJ, Jepson P. Toward a biocultural theory of avoided extinction. *Conservation Letters* 2008; 1 (3): 111–118.

2. Ladle RJ, Whittaker RJ. *Conservation Biogeography*. Oxford: John Wiley & Sons, 2011.

3. Corlett RT. The Anthropocene concept in ecology and conservation. *Trends in Ecology and Evolution* 2015; 30 (1): 36–41.

4. Crutzen PJ, Stoermer EF. The 'Anthropocene'. *IGBP Newsletter* 2006; 14: 17–18.

5. Butchart SHM, Walpole M, Collen B, *et al.* Global biodiversity: indicators of recent declines. *Science* 2010; 328 (5982): 1164–1168.

6. Ladle RJ, Jepson P. Origins, uses, and transformation of extinction rhetoric. *Environment and Society: Advances in Research* 2010; 1 (1): 96–115.

7. Bengtson S. Breeding ecology and extinction of the great auk (*Pinguinus impennis*): anecdotal evidence and conjectures. *The Auk* 1984; 101: 1–12.

8. Orton J. The great auk. *American Naturalist* 1869; 3: 539–542.

9. Sarkar S. *Biodiversity and Environmental Philosophy: An Introduction*. Cambridge: Cambridge University Press, 2005.

10. Ladle RJ. Forecasting extinctions: uncertainties and limitations. *Diversity* 2009; 1 (2): 133–150.

11. Pimm SL. The dodo went extinct (and other ecological myths). *Annals of the Missouri Botanical Garden* 2002; 89: 190–198.

12. Pimm SL, Jenkins CN, Abell R, *et al.* The biodiversity of species and their rates of extinction, distribution, and protection. *Science* 2014; 344 (6187).

13. Pimm SL, Russell GJ, Gittleman JL, Brooks TM. The future of biodiversity. *Science* 1995; 269 (5222): 347–349.

14. Harnik PG, Lotze HK, Anderson SC, *et al.* Extinctions in ancient and modern seas. *Trends in Ecology and Evolution* 2012; 27 (11): 608–617.

15. Barnosky AD, Matzke N, Tomiya S, *et al.* Has the Earth's sixth mass extinction already arrived? *Nature* 2011; 471 (7336): 51–57.

16. Olden JD, Lockwood JL, Parr CL. Biological invasions and the homogenization of faunas and floras. In: LadleRJ, Whittaker RJ (eds.), *Conservation Biogeography*. Oxford: Wiley-Blackwell, 2011: 224–244.

17. Ricciardi A. Are modern biological invasions an unprecedented form of global change? *Conservation Biology* 2007; 21 (2): 329–336.

18. Wonham MJ, Carlton JT. Trends in marine biological invasions at local and regional scales: the Northeast Pacific Ocean as a model system. *Biological Invasions* 2005; 7 (3): 369–392.

19. Pautasso M. Scale dependence of the correlation between human population presence and vertebrate and plant species richness. *Ecology Letters* 2007; 10 (1): 16–24.

20. Caughley G. Directions in conservation biology. *Journal of Animal Ecology* 1994; 63: 215–244.

21. Ladle RJ, Malhado AC. Responding to biodiversity loss. In: Douglas I, Hugget R, Perkins C (eds.), *Companion Encyclopedia of Geography: From Local to Global*. New York: Routledge, 2007: 821–834.

22. Rackham O. *Ancient Woodland: Its History, Vegetation and Uses in England*. London: Edward Arnold, 1980.

23. Mitchell-Jones A. The status and distribution of mammals in Britain. *Hystrix, Italian Journal of Mammalogy* 1996: 8 (1–2).

24. Hansen MC, Potapov PV, Moore R, *et al.* High-resolution global maps of 21st-century forest cover change. *Science* 2013; 342 (6160): 850–853.

25. Fahrig L. Effects of habitat fragmentation on biodiversity. *Annual Review of Ecology, Evolution, and Systematics* 2003; 34: 487–515.

26. Watson JE, Whittaker RJ, Freudenberger D. Bird community responses to habitat fragmentation: how consistent are they across landscapes? *Journal of Biogeography* 2005; 32 (8): 1353–1370.

27. Villard M-A, Metzger JP. Beyond the fragmentation debate: a conceptual model to predict when habitat configuration really matters. *Journal of Applied Ecology* 2014; 51 (2): 309–318.

28. Robinson JG, Redford KH. *Neotropical Wildlife Use and Conservation.* Chicago: Chicago University Press, 1991.

29. Taylor G, Scharlemann JPW, Rowcliffe M, *et al.* Synthesising bushmeat research effort in West and Central Africa: a new regional database. *Biological Conservation* 2015; 181: 199–205.

30. Jackson JBC, Kirby MX, Berger WH, *et al.* Historical overfishing and the recent collapse of coastal ecosystems. *Science* 2001; 293 (5530): 629–637.

31. Pauly D, Christensen V, Guénette S, *et al.* Towards sustainability in world fisheries. *Nature* 2002; 418 (6898): 689–695.

32. Pauly D, Christensen V, Dalsgaard J, Froese R, Torres F, Jr. Fishing down marine food webs. *Science* 1998; 279 (5352): 860–863.

33. Sala E, Knowlton N. Global marine biodiversity trends. *Annual Review of Environment and Resources* 2006; 31: 93–122.

34. Carson R. *Silent Spring.* New York: Houghton Mifflin, 1962.

35. Giles J. Nitrogen study fertilizes fears of pollution. *Nature* 2005; 433 (7028): 791–791.

36. Carpenter SR, Caraco NF, Correll DL, Howarth RW, Sharpley AN, Smith VH. Nonpoint pollution of surface waters with phosphorus and nitrogen. *Ecological Applications* 1998; 8 (3): 559–568.

37. Elith J, Leathwick JR. Species distribution models: ecological explanation and prediction across space and time. *Annual Review of Ecology, Evolution, and Systematics* 2009; 40 (1): 677.

38. Thomas CD, Cameron A, Green RE, *et al.* Extinction risk from climate change. *Nature* 2004; 427 (6970): 145–148.

39. Whittaker RJ, Araújo MB, Jepson P, Ladle RJ, Watson JEM, Willis KJ. Conservation biogeography: assessment and prospect. *Diversity and Distributions* 2005; 11 (1): 3–23.

40. Ladle R, Hortal J. Mapping species distributions: living with uncertainty. *Frontiers of Biogeography* 2013; 5: 8–9.

41. Marris E. The escalator effect. *Nature Reports Climate Change* 2007; 1: 94–96.

42. Jepson PR, Ladle RJ. *Conservation: A Beginner's Guide.* Oxford: Oneworld Publications, 2010.

43. Butchart SHM, Clarke M, Smith RJ, *et al.* Shortfalls and solutions for meeting national and global conservation area targets. *Conservation Letters* 2015; 8 (5): 329–337.

44. Crutzen PJ. *The 'Anthropocene'.* Berlin: Springer, 2006.

45. Primack RB. *Essentials of Conservation Biology.* London: Palgrave Macmillan, 2010.

46. Diamond JM. The island dilemma: lessons of modern biogeographic studies for the design of natural reserves. *Biological Conservation* 1975; 7 (2): 129–146.

47. Diamond JM. Island biogeography and conservation: strategy and limitations. *Comptes Rendus des Seances de la Societe de Biologie et de ses Filiales (Paris)* 1976; 160 (3): 1966.

48. Triantis KA, Bhagwat SA. Applied island biogeography. In: Ladle RJ, Whittaker RJ (eds.), *Conservation Biogeography.* Oxford: Wiley-Blackwell, 2011: 190–223.

49. International Union for Conservation of Nature (IUCN). *IUCN Red List Categories and Criteria: Version 3.1.* Gland: IUCN, 2001.

50. Willis KJ, Araújo MB, Bennett KD, Figueroa-Rangel B, Froyd CA, Myers N. How can a knowledge of the past help to conserve the future? Biodiversity conservation and the relevance of long-term ecological studies. *Philosophical Transactions of the Royal Society B: Biological Sciences* 2007; 362 (1478): 175–187.

51. Richardson DM, Whittaker RJ. Conservation biogeography – foundations, concepts and challenges. *Diversity and Distributions* 2010; 16 (3): 313–320.

52. Whittaker RJ, Ladle RJ. The roots of conservation biogeography. In: Ladle RJ, Whittaker RJ (eds.), *Conservation Biogeography.* Oxford: Wiley-Blackwell, 2011: 1–12.

53. Myers N, Mittermeier RA, Mittermeier CG, da Fonseca GAB, Kent J. Biodiversity hotspots for conservation priorities. *Nature* 2000; 403 (6772): 853–858.

54. Olson DM, Dinerstein E, Wikramanayake ED, *et al.* Terrestrial ecoregions of the world: a new map of life on Earth. *BioScience* 2001; 51 (11): 933–938.

55. Nelson BW, Ferreira CAC, da Silva MF, Kawasaki ML. Endemism centres, refugia and botanical collection density in Brazilian Amazonia. *Nature* 1990; 345 (6277): 714–716.

56. Mace GM. The role of taxonomy in species conservation. *Philosophical Transactions of the Royal Society of London. Series B: Biological Sciences* 2004; 359 (1444): 711–719.

57. Lomolino MV, Riddle BR, Whittaker RJ. *Biogeography*. 4th ed. Sunderland, MA: Sinauer, 2010.

58. May RM. Tropical arthropod species, more or less? *Science* 2010; 329 (5987): 41–42.

59. Caley MJ, Fisher R, Mengersen K. Global species richness estimates have not converged. *Trends in Ecology and Evolution* 2014; 29 (4): 187–188.

60. Mora C, Tittensor DP, Adl S, Simpson AGB, Worm B. How many species are there on Earth and in the ocean? *PLoS Biology* 2011; 9 (8): e1001127.

61. Gaston KJ. Body size and probability of description: the beetle fauna of Britain. *Ecological Entomology* 1991; 16: 505–508.

62. Garcillán PP, Ezcurra E. Sampling procedures and species estimation: testing the effectiveness of herbarium data against vegetation sampling in an oceanic island. *Journal of Vegetation Science* 2011; 22 (2): 273–280.

63. Raxworthy CJ, Martinez-Meyer E, Horning N, Nussbaum RA, Schneider GE, Ortega-Huerta MA, Peterson AT. Predicting distributions of known and unknown reptile species in Madagascar. *Nature* 2003; 426: 837–841.

64. Yahara T, Donoghue M, Zardoya R, Faith DP, Cracraft J. Genetic diversity assessments in the century of genome science. *Current Opinion in Environmental Sustainability* 2010; 2 (1–2): 43–49.

65. Riddle BR, Ladle RJ, Lourie S, Whittaker RJ. Basic biogeography: estimating biodiversity and mapping nature. In: Ladle RJ, Whittaker RJ (eds.), *Conservation Biogeography*. Oxford: Wiley-Blackwell, 2011: 45–92.

66. Bush MB, Lovejoy TE. Amazonian conservation: pushing the limits of biogeographical knowledge. *Journal of Biogeography* 2007; 34 (8): 1291–1293.

67. Sastre P, Lobo JM. Taxonomist survey biases and the unveiling of biodiversity patterns. *Biological Conservation* 2009; 142 (2): 462–467.

68. Diniz-Filho JAF, Bini LM, Hawkins BA. Spatial autocorrelation and red herrings in geographical ecology. *Global Ecology and Biogeography* 2003; 12: 53–64.

69. Hortal J, Lobo JM, Jimenez-Valverde A. Limitations of biodiversity databases: case study on seed-plant diversity in Tenerife, Canary Islands. *Conservation Biology* 2007; 21 (3): 853–863.

70. Dennis R, Thomas C. Bias in butterfly distribution maps: the influence of hot spots and recorder's home range. *Journal of Insect Conservation* 2000; 4 (2): 73–77.

71. Lobo JM, Baselga A, Hortal J, Jiménez-Valverde A, Gomez JF. How does the knowledge about the spatial distribution of Iberian dung beetle species accumulate over time? *Diversity and Distributions* 2007; 13 (6): 772–780.

72. Lobo JM. Decline of roller dung beetle (Scarabaeinae) populations in the Iberian peninsula during the 20th century. *Biological Conservation* 2001; 97: 43–50.

73. Huisman JM, Millar AJK. Australian seaweed collections: use and misuse. *Phycologia* 2013; 52 (1): 2–5.

74. Hortal J, de Bello F, Diniz-Filho JAF, Lewinsohn TM, Lobo JM, Ladle RJ. The seven fundamental shortfalls in large-scale knowledge for ecological and evolutionary research. *Annual Review of Ecology, Evolution, and Systematics* 2015; 46 (1): 523–549.

75. He F, Hubbell S. Estimating extinction from species–area relationships: why the numbers do not add up. *Ecology* 2013; 94 (9): 1905–1912.

76. Rodrigues AS, Akçakaya HR, Andelman SJ, *et al.* Global gap analysis: priority regions for expanding the global protected-area network. *BioScience* 2004; 54 (12): 1092–1100.

77. Terborgh J, Winter B. A method for siting parks and reserves with special reference to Columbia and Ecuador. *Biological Conservation* 1983; 27 (1): 45–58.

78. Rondinini C, Wilson KA, Boitani L, Grantham H, Possingham HP. Tradeoffs of different types of species occurrence data for use in systematic conservation planning. *Ecology Letters* 2006; 9 (10): 1136–1145.

79. Lennon JJ, Koleff P, Greenwood JJD, Gaston KJ. The geographical structure of British bird distributions: diversity, spatial turnover and scale. *Journal of Animal Ecology* 2001; 70 (6): 966–979.

80. Hutchinson GE. *A Treatise on Limnology*. New York: Wiley, 1957.

81. Fuller RA. Planning for persistence in a changing world. In: Ladle RJ, Whittaker RJ (eds.), *Conservation Biogeography*. Oxford: Wiley-Blackwell, 2011: 161–189.

82. McCann S, Greenlees MJ, Newell DA, Shine R. Rapid acclimation to cold allows the cane toad to invade montane areas within its Australian range. *Functional Ecology* 2014; 28 (5): 1166–1174.

83. Kearney M, Porter WP. Mapping the fundamental niche: physiology, climate, and the distribution of a nocturnal lizard. *Ecology* 2004; 85 (11): 3119–3131.

84. Williams JW, Jackson ST. Novel climates, no-analog communities, and ecological surprises. *Frontiers in Ecology and the Environment* 2007; 5 (9): 475–482.

85. Urban MC, Tewksbury JJ, Sheldon KS. On a collision course: competition and dispersal differences create no-analogue communities and cause extinctions during climate change. *Proceedings of the Royal Society B: Biological Sciences* 2012; 279 (1735): 2072–2080.

86. Jepson P, Whittaker RJ, Lourie SA. The shaping of the global protected area estate. In: Ladle RJ, Whittaker RJ (eds.), *Conservation Biogeography*. Oxford: Wiley-Blackwell, 2011: 93–135.

87. Dasmann RF. Towards a system for classifying natural regions of the world and their representation by national parks and reserves. *Biological Conservation* 1972; 4 (4): 247–255.

88. Dasmann RF. *System for Defining and Classifying Natural Regions for Purposes of Conservations: A Progress Report*. Morges: IUCN, 1973.

89. Udvardy MD, Udvardy M. *A Classification of the Bio-geographical Provinces of the World*. Morges: IUCN, 1975.

90. Dinerstein E, Olson DM. *A Conservation Assessment of the Terrestrial Ecoregions of Latin America and the Caribbean*. Washington, DC: World Bank, 1995.

91. Mittermeier RA, Turner WR, Larsen FW, Brooks RM, Gascon C. Global biodiversity conservation: the critical role of hotspots. In: *Biodiversity Hotspots*. Berlin: Springer, 2011: 3–22.

92. Myers N. Threatened biotas: 'hot spots' in tropical forests. *Environmentalist* 1988; 8 (3): 187–208.

93. Spalding MD, Fish L, Wood LJ. Toward representative protection of the world's coasts and oceans – progress, gaps, and opportunities. *Conservation Letters* 2008; 1 (5): 217–226.

94. Spalding MD, Fox HE, Allen GR, *et al*. Marine ecoregions of the world: a bioregionalization of coastal and shelf areas. *BioScience* 2007; 57 (7): 573–583.

95. Watson JE, Dudley N, Segan DB, Hockings M. The performance and potential of protected areas. *Nature* 2014; 515 (7525): 67–73.

96. Watson JE, Grantham HS, Wilson KA, Possingham HP. Systematic conservation planning: past, present and future. In: Ladle RJ, Whittaker RJ (eds.), *Conservation Biogeography*. Oxford: Wiley-Blackwell, 2011: 136–160.

97. MacArthur RH. *The Theory of Island Biogeography*. Princeton: Princeton University Press, 1967.

98. Possingham H, Ball I, Andelman S. Mathematical methods for identifying representative reserve networks. In: Ferson S, Burgman M (eds.), *Quantitative Methods for Conservation Biology*. Berlin: Springer, 2000: 291–306.

99. Margules C, Usher M. Criteria used in assessing wildlife conservation potential: a review. *Biological Conservation* 1981; 21 (2): 79–109.

100. Ball IR, Possingham HP, Watts ME. Marxan and relatives: software for spatial conservation prioritization. In: Moilanen A, Wilson KA, Possingham HP (eds.), *Spatial Conservation Prioritisation: Quantitative Methods and Computational Tools*. Oxford: Oxford University Press, 2009: 185–195.

101. Saylor M. *The Mobile Wave: How Mobile Intelligence Will Change Everything*. Philadelphia: Vanguard Press, 2012.

102. Kemp C. The endangered dead. *Nature* 2015; 518: 292–294.

103. Giangrande A. Biodiversity, conservation, and the 'taxonomic impediment'. *Aquatic Conservation: Marine and Freshwater Ecosystems* 2003; 13 (5): 451–459.

104. Jepson P, Ladle RJ. Nature apps: waiting for the revolution. *Ambio* 2015; 44 (8): 827–832.

105. Borden KA, Kapadia A, Smith A, Whyte L. Educational exploration of the zooniverse: tools for formal and informal audience engagement. In: Barnes J, Shupla C, Manning JG, Gibbs MG (eds.), *Communicating Science*. San Francisco: Astronomical Society of the Pacific, 2013: 101–108.

106. Gaston KJ, O'Neill MA. Automated species identification: why not? *Philosophical Transactions of the Royal Society of London B: Biological Sciences* 2004; 359 (1444): 655–667.

107. Hebert PD, Gregory TR. The promise of DNA barcoding for taxonomy. *Systematic Biology* 2005; 54 (5): 852–859.

108. Jetz W, McPherson JM, Guralnick RP. Integrating biodiversity distribution knowledge: toward a global map of life. *Trends in Ecology and Evolution* 2012; 27 (3): 151–159.

109. Wilson EO. The encyclopedia of life. *Trends in Ecology and Evolution* 2003; 18 (2): 77–80.

110. Ladle RJ, Whittaker RJ. Prospects and challenges. In: Ladle RJ, Whittaker RJ (eds.), *Conservation Biogeography*. Oxford: Wiley-Blackwell, 2011: 245–258.

Glossary

The words and concepts listed in this glossary are shown in **bold type** in the text on the pages where the concept involved is defined; these pages are also indicated in bold in the index.

Abyssal plain The deep ocean floor, which lies between the continental shelves (q.v.).

Abyssal zone The marine life zone above the abyssal plain (q.v.).

Adaptive radiation The evolutionary radiation of a group, based on a novel set of characteristics, that allows it to adapt to a wide range of ways of life.

Albedo An index of the extent to which incoming radiation is reflected, rather than absorbed.

Allele One of sometimes several versions of a gene (q.v.), located at a single position on the chromosome.

Allerød interstadial The brief warm spell before the Younger Dryas (q.v.) stadial.

Allopatric speciation The evolution of a new species in isolation from its parent species.

Alpha diversity The biodiversity (usually measured as species richness) in a local area.

Amphitropical distribution See *bipolar distribution*.

Angiosperms Flowering plants.

Anthropocene A name for the current geological epoch in recognition of human impacts on the natural environment.

Antitropical distribution See *bipolar distribution*.

Apomorphic The derived state of a characteristic.

Arborescent Tree-like.

Archibenthal zone See *bathyal zone*.

Archipelago speciation Inter-island dispersal, leading to a more complex pattern of cladogenesis.

Area biogeography An approach to biogeography that begins with the identification of areas of endemism.

Area of endemicity Area within which one or more taxa are found exclusively. Also known as an area of endemism.

Arid Extremely dry, with an annual precipitation of less than 10 cm.

Asiamerica An area of land that included Asia and western North America during the Cretaceous.

Assembly rules The hypothesis that some species are only found on islands where another particular species is absent, or in islands containing a larger total assemblage of species. More generally in ecology, the principles that determine the aggregation of various species to make up a community.

Biogeography: An Ecological and Evolutionary Approach, Ninth Edition. Edited by C. Barry Cox, Peter D. Moore, Richard J. Ladle.
© 2016 John Wiley & Sons, Ltd. Published 2016 by John Wiley & Sons, Ltd.

Australasia The Australian continent, plus the islands on the Australian continental plate – New Guinea, Tasmania, Timor and New Caledonia, and other smaller islands.

Australopithecines Members of the genus *Australopithecus*, an early primate.

Barrier An obstruction to the passage of organisms.

Bathyal zone The marine life zone above the continental slope (q.v.); also known as the archibenthal zone.

Bathypelagic zone Zone of complete darkness in the sea, from below the mesopelagic zone (q.v.) to a depth of 6000 m.

Benthic Organisms that live on the seafloor.

Beta diversity The rate or amount of change in species composition between local areas. Also known as turnover or differentiation diversity, beta diversity translates the number of species in local areas (alpha diversity) into the number of species in larger regions (gamma diversity). Low beta diversity means similar sets of species between patches or local areas, such that the gamma diversity will not be much higher than the alpha diversity.

Biodiversity A census of the diversity of species in different parts of the planet and in the whole of the planet. It may include the genetic variation within species.

Biodiversity hotspots Areas of the world that are exceptionally rich in species.

Biogeography Study of the distribution, and of the patterns of distribution, of living organisms at all levels, ranging from genes to whole organisms and biomes, and of the evolution of these.

Biological control Using an introduced predator to control other, invasive species.

Biological species concept See *species*.

Biological spectrum The variety of component functional types found in a biome.

Biome A large-scale ecosystem, such as desert or tundra, found in different parts of the world and characterized by a similar life forms of animals and plants.

Biosphere The part of the Earth that is inhabitable by living organisms.

Biota The total of all organisms that inhabit a given region, that is, its fauna plus its flora.

Biotic resistance The pressures of predation and parasitism that an invasive species finds in its new environment.

Bipolar distribution Distribution in which related organisms are found in temperate or polar environments on either side of the equatorial region, but not in the equatorial region itself. Also known as antitropical or amphitropical distribution.

Bloom Rapid, seasonal increase in the mass of marine phytoplankton.

Boreal Found throughout the Northern Hemisphere; usually referring to plant distributions in the cool temperate zones of the north.

Boreotropical Confined to the tropics of the Northern Hemisphere.

Bottom water A mass of cold, dense, saline water that sinks down to the ocean floor along the eastern side of Greenland and near the Antarctic Peninsula.

Brooks parsimony analysis or **BPA** A technique of pattern-based historical biogeographical (q.v.) analysis.

Buffon's Law The observation that similar environments, in different parts of the world, contain different groupings of organisms.

C3 plants Those plants in which the first product of photosynthesis is a sugar containing three carbon atoms; this is the most common form of photoynthesis.

C4 plants Those plants in which the first product of photosynthesis is a sugar containing four carbon atoms; this is most advantageous in high light intensity and temperature.

Carrying capacity The number of species, or the biomass, or the highest population level of a species, that can be supported by a given area of a particular habitat or environment.

Chaos theory A method of analyzing a situation in which minor differences in initial conditions are highly influential in the ultimate outcome.

Chloroplast The green structures in plant cells in which the conversion of energy from sunlight takes place.

Chromosome The thread-like body within the nucleus of a cell that carries its DNA (q.v.).

Circumboreal A pattern of distribution that stretches around the northern regions.

Clade A group of taxa that includes their common ancestor and all its descendants.

Cladistic biogeography The analysis of patterns of endemism using cladistics (q.v.). Also known as vicariance biogeography.

Cladistics The system of analysis of the evolutionary relationships of taxa that views this as a series of dichotomously branching events, using the presence of shared derived characters to identify the location of each branch.

Cladogram A diagram portraying the results of a cladistic analysis.

Clements approach An interpretation of plant communities that suggests that they behave as integrated units, rather like an individual organism. See also *Gleason approach*.

Climate The whole range of weather conditions experienced within that area, including temperature, rainfall, evaporation, sunlight and wind through all the seasons of the year.

Climatic envelope The sum of all the climatic variables that limit the distribution of a species or a biome (q.v.).

Climatic relict A species that survives only in a few 'islands' of favourable climate.

Climax The final, stable, self-perpetuating assemblage of plants in a region.

Climax state A stable, self-perpetuating system.

Climax vegetation In which vegetation develops over the course of time, passing through several different assemblages of plants to finally reach this state.

Clinal Variation along a cline.

Cline Where there is gradual change in genetics and form along a gradient.

Cloud forest A highly humid forest type, in which plants living entirely in the canopy with no roots reaching the ground are unlikely to experience desiccation.

Coastal biome A biome of shallow seas.

Co-evolution When species not only tolerate one another, but evolve to become dependent on one another.

Community An assemblage of different species that live together in a particular habitat and interact with one another.

Competitive exclusion A situation in which the presence of one species prevents the presence of another.

Composition The species of organisms forming a community.

Conservation biogeography Biogeographical research that has a direct bearing on conservation.

Continental drift The process by which the continents separated over time. This is now known as plate tectonics (q.v.).

Continental rise The wedge of sediments, derived from the erosion of the continents, that lies at the foot of the continental shelf and extends onto the edge of the abyssal plain (q.v.).

Continental shelf That portion of the lower lying parts of a continent that are covered by shallow sea.

Continental slope That region in the sea, below the edge of the continental shelf, at which the seafloor descends steeply until it reaches the floor of the deep ocean that lies between the continental shelves.

Convention on Biological Diversity or **CBD** An international policy, formulated at the 1992 Earth Summit in Rio de Janeiro, that emphasizes saving genes, species and ecosystems.

Convergence Lines in the oceans along which water that is more dense and saline sinks below adjacent water that is lighter and fresher.

Corridor A pathway that permits the passage of most organisms.

Cosmopolitan group One that is widely distributed throughout the world.

Creation Science An antievolutionary interpretation of the Bible that holds that everything on the planet was created in a short burst of divine activity a few thousand years ago.

Crown group The earliest common ancestor of a clade (q.v.), plus all of its descendants, which therefore have all of the characteristics found in all of those descendants. Contrast with stem group (q.v.).

Dansgaard–Oeschger cycles The cycles of alternating warmth and cold during the last glacial.

Darwin's finches The finches that colonized the Galápagos Islands off the coast of western South America, where they underwent a radiation into a variety of forms.

De Geer route A former connection between northeastern Greenland and northwestern Europe.

Deciduous Perennial plants that shed their leaves during a cold or dry season.

Definitive host A final host in which a parasite settles.

Deoxyribonucleic acid A complex molecule, able to reduplicate itself, that lies at the heart of the genetic system; also known as DNA.

Diploid The genetic condition of most body cells, which have pairs of each chromosome, one having originated from each parent.

Disharmonic biota An island biota that lacks some of the normal components because these have been unable to reach, or survive in, the island.

Disjunct A pattern of distribution in which the areas occupied by a given organism are discontinuous and separated from one another.

Dispersal biogeography An approach to historical biogeography based on the assumption that related taxa that occupy ranges that are separate from one another arrived in them by crossing pre-existing barriers.

Dispersal–vicariance analysis or **DIVA** A type of parsimony-based tree fitting (q.v.), designed to cope with a reticulate pattern (q.v.) of relationship, in which no general area cladogram is assumed.

Dispersalism A concept assuming that, where a taxon or taxa are found on either side of a barrier to their spread, this is because they had been able to cross that barrier after it formed.

Dispersion When a species is able to extend its range into an area previously not available.

Divergence A region in the Atlantic or Pacific where there is a shear between currents going in different directions.

DNA See *deoxyribonucleic acid.*

Dominant Active, as in an allele.

Duplication The presence of two related species in an area, when it is uncertain whether this arose by dispersal or vicariance.

East Pacific Barrier The barrier to the dispersal of the short-lived larvae of organisms that live on the continental shelves, formed by the wide, deep, almost island-free expanse of the East Pacific Ocean.

Ecological biogeography The study of aspects of biogeographical phenomena that focuses on the interactions between organisms and their environments.

Ecological equivalents Species that have developed similar traits for similar environments.

Ecological network The sheer complexity of the interactions within an ecosystem.

Ecological species concept See *species.*

Ecophysiology A discipline that examines how plants and animals vary in their physiological processes in response to the environment.

Ecosystem The basic unit of ecology, taking into account an area's plants, animals and climatic and soil aspects.

El Niño Southern Oscillation or **ENSO** The cycle of changes in climate through much of the world that is caused by variations in wind and current intensity in the southern Pacific Ocean.

Endemic An organism that is found only in a particular region.

Energy flow The process by which solar energy is fixed by plants and then passes in turn to herbivores, carnivores, detritus feeders and decomposers.

Energy hypothesis The explanation of latitudinal gradients of species diversity as the result of variation in the amount of energy that is captured by vegetation.

Epicontinental sea Shallow sea that covers the lower lying parts of the continents.

Epipelagic zone The upper, warmer layers of the sea, containing a high concentration of living organisms; up to 200 m deep.

Epiphytes Plants that use other plants for support.

Equatorial submergence The phenomenon whereby the level of the transition between the cold waters of the bathyal marine life zone (q.v.) and the warmer waters above it is deeper at lower latitudes.

Euphotic zone The upper few tens of metres depth of the sea, in which there is enough sunlight for photosynthesis to take place.

Euramerica A large landmass, during the mid Palaeozoic 400 million years ago, made up of today's North America plus Europe, that lay across the equator. Also used for the land area that included eastern North America and Europe during the Cretaceous.

Eurytopic An organism that has wide ecological tolerance.

Eustatic A change in sea level caused by a change in the volume of water in the sea.

Evapotranspiration The sum total of water that evaporates directly from the surface of the ground, plus that lost by the uptake and loss of water by plants.

Evenness A similar population size.

Event-based methods Methods of cladistic biogeographical analysis that specify what event (vicariance, duplication, dispersal or extinction) took place at each point of branching of a biological area cladogram (q.v.).

Evolutionary relicts A pattern of distribution in which a formerly more dominant organism now inhabits only the scattered remains of a formerly continuous area, due to competition with another species or group.

Extant group One that is still alive today.

Facilitation A situation in which the presence of one species aids the addition of another to the community.

Family A group of genera (q.v.).

Filter An ecological barrier that prevents the passage of certain categories of organism.

Foraminifera Plankton that have robust outer cases that survive sedimentation to the ocean floor and there accumulate as fossil assemblages.

Formation Classification at a simpler level than that of the biome, based simply on vegetation.

Functional type Classification of species according to their physiological and ecological capacities.

Fundamental niche The theoretical or ideal type of niche; the sum of all the niche requirements under ideal conditions when the species is given unimpeded access to resources.

Gamma diversity The total species richness in a region comprising a number of smaller patches or habitats.

Gene A region of the DNA (q.v.) that is responsible for one or several characteristics of the organism.

General area cladogram or **GAC** See *pattern-based methods*.

General dynamic model The hypothesis that volcanic oceanic islands show a progression in which there is a linkage between area, altitude, erosion, habitat diversity, number of species and the proportion of them that are endemic to the island.

Generalized track See *track*.

Genetic drift A situation in which the frequency of an allele in a population is not controlled by selective pressures.

Genotype The total of all the genes of an organism, making up its total genetic inheritance.

Genus, plural **genera** A group of species (q.v.) that are closely related to each other.

Geodispersal The result of a geological rather than a biological event.

Geological area cladogram A cladogram (q.v.) that shows the sequence of separation of areas of land from one another.

Glacial relict A species whose distributions have been modified by the northward retreat of the great ice sheets during the Pleistocene Ice Ages.

Glaciation The spread of glacial ice into lower latitudes.

Gleason approach An individualistic interpretation of plant communities that emphasizes the varied ecological requirements of its component species. See also *Clements approach*.

Gondwana The supercontinent that, in past geological time, was formed of the southern continents (South America, Antarctica, Africa and Australia) plus India, before these split apart.

Gradualistic evolution In which evolutionary change normally takes place at a steady, gradual rate.

Greenhouse effect Where clouds and carbon dioxide absorb the heat re-radiated by the Earth, so causing an increase in its temperature.

Greenhouse gas A gas that contributes to the greenhouse effect (q.v.).

Guild A group of animals, not necessarily related taxonomically, that use the same resource or overlap significantly in their environmental requirements.

Gyre The huge mass of horizontally rotating water that fills a large part of an ocean basin.

Habitat The general type of environment within which an organism lives, for example forest or marsh.

Hadal zone The marine life zone at a depth of over 6 km (3.5 miles).

Hadopelagic zone Deepest part of the ocean, within the submarine trenches (q.v.).

Halocline The level within the sea at which there is a rapid change in the salinity of the water.

Haploid The genetic condition of a sperm or ovum, which has only one of each pair of chromosomes.

Haplotype A group of genes in an organism that are inherited together from a single parent.

Heinrich events When a cold phase is accompanied by massive discharges of icebergs into the North Atlantic Ocean.

Hekistotherm A plant that lives in polar regions.

Hermatypic corals The reef-forming corals in which there is a symbiotic relationship between the polyps of the coral and the algae, known as zooxanthellae, that live within them.

Historical biogeography The study of aspects of biogeographical phenomena that focuses on the origins and subsequent history of lineages and taxa.

Holarctic Found throughout the Northern Hemisphere; usually referring to mammals.

Holocene The current interglacial.

Hominids The group including humans and great apes.

Hominins The evolutionary lineage that branched away from that leading to the great apes, and that includes humans and their ancestors.

Hotspot A location, deep within the Earth, from which a plume of hot material rises to the surface. Where this occurs within an ocean, it leads to the formation of a volcano, which either reaches the surface as an island or remains as a submerged 'seamount' or 'guyot'. The term is also used of a region of the Earth with unusually high biodiversity.

Hybrid The result of a mating between two different species or divergently adapted genotypes.

Hydrological cycle The movement of water from oceans via water vapour to precipitation and back to the ocean via streams and rivers.

Hydrothermal vents Points in the mid-oceanic spreading ridges where cold seawater penetrates into the rocks surrounding the area where hot lava is emerging, and reacts chemically with them so that minerals are precipitated out.

Hydrozoan A type of colonial organism, such as coral.

Ice wedge Ice that forms in the ground by freezing and pushing up soil.

Incidence The pattern of occurrence of a species on islands and the factors that affect this pattern.

Indian Ocean Dipole A system of oceanic circulation that takes place in the Indian Ocean to the west of the Indonesian islands.

Individualistic concept In which the composition of species varies geographically as the physical limits of said species are encountered.

Indo-Pacific plant kingdom The region including India, South-East Asia and the islands of the Pacific Ocean, which contains a characteristic flowering plant flora.

Inertia Resistance to change.

Insolation The projected solar energy input.

Intelligent Design An antievolutionary belief that evolution by natural selection is incapable of producing current species' degree of adaptation, and therefore must have appeared by divine action.

Interglacial A period of time, between glacial events, that was warm enough for temperate vegetation to establish itself.

Intermediate host A host within whose body the parasite multiplies and is transformed into a stage that can infect a new definitive host.

Interstadial A period of warmth, between glacial events, that was too short or too cool for temperate vegetation to establish itself.

Intertropical convergence zone or **ITCZ** Where 'trade winds', found in both the Northern and Southern Hemispheres, meet in the region of the Equator.

Island arc The line of islands that form along a mid-oceanic region where old ocean crust is disappearing into the Earth.

Isolating mechanism Genetic systems that prevent mating between two different species, or that lead to any offspring having reduced fertility.

Isostatic A change in sea level relative to land level caused by a change in the level of the continental surface.

Jump dispersal When a species is able to disperse across a barrier; also called simple dispersal.

K-selected species Species that are slower to reproduce than r-selected species (q.v.), but are more able to sustain their population when this is close to the carrying capacity (q.v.). Characteristic of later colonists in a succession.

Keystone species A species that has an important influence on many other species in the ecosystem.

Kleptoparasitism Stealing food from another predator's catch.

Landscape ecology The study on the ecology of cultural landscapes.

Lapse rate The rate of a fall in atmospheric temperature with increasing altitude.

Laurasia The supercontinent that, in past geological time, was formed of the northern continents (North America and Eurasia), before these split apart.

Lessepsian exchange A link between marine faunas, through which they can exchange organisms; named after the link between the Mediterranean Sea and the Red Sea formed by the Suez Canal.

Life form A type of living creature characterized by an assemblage of structural and physiological features that adapt it to life in a particular type of environment.

Limiting factor One that is responsible for limiting the pattern of distribution of an organism.

Lithosphere A geological term referring to the surface of the Earth.

Little Ice Age A time (between about AD 1350 and 1850) of generally low temperature that occurred over a very wide area in the Northern Hemisphere.

Loess Blown sand.

Lusitanian species Species with a disjunct distribution pattern between Spain and Portugal and the west of Ireland.

Macroecology The study of the assembly and structure of biotas that focuses on their general, large-scale patterns and mechanisms.

Marsupial In this group, one of the two major groups of living mammals, the young leave the uterus at a very early stage and complete their development in the mother's pouch, in contrast with those of the placentals (q.v.).

Medieval Warm Period A peak in warmth around AD 1000 to 1150.

Mediterranean climate One with hot, dry summers and cool, wet winters.

Megafauna The large terrestrial vertebrate component of a fauna.

Megatherm A plant that prefers temperatures above 20°C.

Mesopelagic zone The zone of reduced light intensity in the sea, lying below the pycnocline (q.v.), and which extends down to a depth of about 1000 m.

Mesotherm A plant that prefers temperatures between 13°C and 20°C.

Metabolic theory The explanation of latitudinal gradients of species diversity as being the result of variation in the metabolic activity of organisms.

Metopopulation A series of separated subpopulations between which genetic exchange may be limited.

Microclimate The physical conditions of temperature, intensity of light, humidity, and so on, that are found in a particular small-scale environment.

Microhabitat The fine-scale environment within which an organism lives (e.g. forest floor).

Microtherm A plant that prefers temperatures below 13°C.

Mid-Atlantic Barrier Where the deep waters of the Atlantic create a barrier to the dispersal of shelf organisms between the African tropics and the South American tropics.

Mid-Continental Seaway A shallow seaway that at one time ran across North America from the Arctic Ocean to the Gulf of Mexico.

Mid-domain effect An explanation for gradients of diversity that suggests that it is due merely to variations in the range of individual species.

Migration When animals alter their distribution patterns in concert with the seasons.

Milankovich cycles An explanation of the sequence of alternating warm and cold climatic episodes, over the last 2 million years, as being the result of variations in the Earth's orbit and in the tilt of its axis.

Mitochondria Part of the cell that is responsible for the cell's control of respiration.

Model-based methods Methods of cladistic biogeographical analysis based on stochastic models, in which the biological processes involved are quantified and the principle of parsimony is not used in selecting the most likely explanation. Also known as parametric methods.

Monophyletic group One in which all its members are descended from a single common ancestor.

Monotypic species A species that exists in just one form.

Mutation Sudden alterations in the biochemical structure of a gene.

Natural selection The process whereby, due to differential survival and reproduction, the more advantageous characteristics persist into the next generation, while the less advantageous gradually disappear.

Nekton Organisms that swim in the waters of the sea.

Neoendemic A species that has only recently evolved and has not yet had time to spread from its centre of origin.

Neritic The shallow-sea realm.

Nested clade analysis A type of phylogeographical biogeography (q.v.) in which the various states of a given molecule in the taxa involved are arranged into a series of groups differing in only one mutational alteration, and that is continued until they are all included in a single clade.

Neutral theory of biodiversity The proposal that assemblages of species are merely a collection of randomly selected species.

Niche The set of physical and biological conditions and resources within which an organism is found, and the role that a species plays within the community.

Niche partitioning The subdivision of resources (e.g. between daytime and night-time predators).

No-analogue community A community that is compositionally unlike other current or past communities.

Nominate subspecies A subspecies that is given the same subspecific name as the specific name.

Null hypothesis A statistical technique that estimates how much similarity there would be between the results of the action of two sets of phenomena, assuming that there is no causal relationship between them. This can then be compared with the actual degree of similarity in order to find out whether or not this is the result of chance.

Nutrient cycling The process by which nutrients pass from organism to organism, and finally into the soil, from which they can be reused by plants.

Ocean baselines See *track*.

Oceanic circulation The worldwide pattern by which seawater is warmed in tropical latitudes and then circulated to higher latitudes before being returned to the tropics. Also known as the oceanic conveyor belt or thermohaline circulation.

Oceanic conveyor belt See *oceanic circulation*.

Oriental zoogeographical region India plus South-East Asia, a region that contains a characteristic mammalian fauna.

Oroboreal flora Flora found only in the mountainous areas of Asia and North America.

Oxygen isotopes The three isotopes of oxygen are ^{16}O, ^{17}O and ^{18}O.

Pachycaul Thick-stemmed plants bearing terminal clusters of tough, leathery leaves.

Palaeoendemic A type of endemism that results from a species having survived in an area for a long time, protected by physical barriers to dispersal.

Palaeomagnetism A technique that uses the presence of magnetized particles in rocks to deduce the movements of the rocks through time, and therefore of the landmasses in which they lay.

Palaeotemperature Historic temperature readings.

Palmae Province The Cretaceous equatorial region, containing a characteristic megathermal (q.v.) forest.

Palynology The analysis of pollen grain assemblages.

Panbiogeography An approach to historical biogeography based on the identification of generalized tracks (q.v.) and that relies on vicariance (q.v.) rather than dispersal (q.v.).

Pangaea The supercontinent that, in past geological times, was formed of all today's continents before these split apart.

Panthalassa The single, worldwide ocean that existed when all the continents were united in Pangaea (q.v.).

Paradigm A theory based on a great variety of independent lines of evidence.

Parametric methods See *model-based methods*.

Parapatric In which the distributions of the populations are adjacent to one another but only overlap very narrowly.

Parsimony A principle of analysis in which the explanation involves the minimum number of assumptions; also known as economy of hypothesis.

Parsimony analysis for comparing trees or **PACT** A form of analysis that looks for a common pattern of relationships among trees.

Parsimony-based tree fitting A type of event-based biogeographical (q.v.) analysis that uses the principle of parsimony (q.v.) in deciding on the most likely explanation.

Pattern-based methods Methods of cladistic biogeographical analysis that commence with an attempt to find a common pattern of relationships, known as a general area cladogram (GAC), which shows the physical history of the relationships between the areas of endemism that are involved.

Pelagic The organisms that swim or float in the sea.

Periglacial The regions of treeless tundra immediately outside areas of past glaciation.

Phenotype The total characteristics of an organism, resulting from the action of its genes.

Photoperiodism A process in which flowering in many plant species is triggered by a response to a particular day length.

Phylogenetic biogeography The analysis of patterns of distribution of organisms using groups whose interrelationships are analyzed using cladistics (q.v.).

Phylogeography A type of phylogenetic biogeography (q.v.) in which the interrelationships of the taxa are based on data from their DNA.

Phytodetritus The remains of phytoplankton, forming an important constituent of the muds and oozes that cover the seafloor.

Phytoplankton Tiny, single-celled organisms that carry out most of the photosynthesis in the sea.

Phytosociology A distinct branch of plant geography, in which the plant communities are classified and may be arranged in a hierarchy.

Picoplankton Tiny, single-celled planktonic organisms.

Placental In this group, one of the two major groups of living mammals, the whole period of development of the young takes place in the uterus, in contrast with the situation in the marsupials (q.v.).

Plankton Minute organisms that float in the waters of the sea.

Planktonic bloom An overgrowth of plankton due to excess heat and light, coupled with a low level of predation.

Plant form Aspects of the morphology, anatomy and physiology of plants that are related to their ability to cope with environmental stresses.

Plant formation A large-scale ecosystem, such as desert or forest, found in different parts of the world, and characterized by a similar set of life forms of plants. (If the fauna also is included, the result is known as a biome.)

Plant functional types Plants with different ways of coping with their environments.

Plateau A larger area caused by, and surrounding, a hotspot (q.v.).

Plate tectonics The explanation of the history of the continents and oceans as resulting from the movements of the tectonic plates (q.v.).

Plesiomorphic The original ancestral or primitive state of a characteristic.

Pluvials Times of wet climate and high water levels.

Pollen grains Plant remnants; one of the most valuable sources of fossil evidence to reconstruct past conditions and habitats.

Polyploidy The doubling or multiplication of the whole set of chromosomes within the cells of an organism.

Polytypic species A species that contains many races or subspecies.

Postglacial relict A species whose current distribution is a reflection of climatic changes that have taken place since the last glaciation ended.

Predictability Stability; biodiversity appears to render an ecosystem predictable.

Prey switching When predators turn to alternative food species if the numbers of their usual prey populations are reduced.

Prioritization scheme Any scheme that reflects conservation prioritization, e.g. to maximize the number of species saved.

Productivity The amount of plant material that accumulates in a given area in a given time.

Pseudocongruence In which the same area cladogram may have arisen more than once but at different times, as a result of a repetition of the same geographical change.

Punctuated equilibrium In which a comparatively large number of changes are seen to take place at the same time.

Pycnocline The level within the sea at which there is a rapid change in the density of the water.

r-selected species Species with a high potential rate of population increase. Characteristic of early colonists of a succession. See also *K-selected species*.

Race A genetically or morphologically distinct set of populations of a species, confined to a particular area.

Range The geographical area within which an organism is found.

Range extension When a species extends its area of distribution or range (q.v.) until it meets barriers to its further spread.

Rapoport's Rule The observation that organisms found in high latitudes tend to have broader geographical and altitudinal ranges and ecological tolerances than those found in lower latitudes.

Realized niche Where the species is found over a smaller range than would have been predicted (e.g. due to competition).

Recessive Inert, as in an allele.

Refugium, plural refugia A location in which some organisms have been able to survive a period of unfavourable conditions.

Relict An organism that now has a more limited distribution than it once had. In the case of a habitat relict or climatic relict (q.v.), this is because of climatic change; in the case of a glacial relict (q.v.), the organism has been left behind, in areas of cold climate, by the northward retreat of the Ice Age climates.

Rescue effect When local extinction of a species is prevented by the immigration of individuals of that species from elsewhere.

Resilience Hardiness, in which a species or ecosystem can rapidly return to its original state following a disturbance.

Reticulate pattern A network-like pattern of relationship between areas of endemism, caused by their having had more than one type of relationship with one another over time.

Ring species A species that lives in a circle, or ring, around a barrier.

Saporpel Black anoxic sediments.

Sclerophyll Having tough, thick, evergreen leaves. Also known as scleromorph.

Seafloor spreading See *spreading ridges*.

Seamount A submerged volcano that formed above a hotspot (q.v.); also known as a guyot.

Seismic waves Shock waves caused by earthquakes.

Sensory drive A type of evolutionary change, associated with differences in the sensory systems and behaviour.

Shelf break The edge of the continental shelf, at which point the seafloor starts to descend more steeply, forming the continental slope, until it reaches the deep ocean floor or abyssal plain.

Shelf zone The marine life zone above the continental shelf (q.v.).

SLOSS debate The argument as to the relative advantages of Single Large, or Several Small, nature reserves in retaining species and reducing extinctions.

Small island effect A size threshold after which the diversity of species drops dramatically.

Solar forcing In which variations in astronomical conditions determine global climate.

Spatial separation When a species is restricted, by some of its characteristics, to a specialized microhabitat within the area available to it.

Species The fundamental unit of the taxonomic system, which can be defined in a variety of ways. Using the biological species concept, the species is a group of natural populations whose members can all breed together to produce offspring that are fully fertile, but that in the wild do not do so with other such groups. Using the ecological species concept, the species is a group of natural populations whose members all possess a set of characteristics (morphological, behavioural, physiological, etc.) that adapt it to a particular ecological niche.

Species–energy theory The hypothesis that the number of species on an oceanic island is controlled by the amount of energy that falls upon it.

Species richness The number of species present within an ecosystem.

Spreading ridges The worldwide system of chains of submarine volcanic mountains, where new seafloor is being formed as the regions on either side move apart, a process known as seafloor spreading.

Stability Consistency; a stable ecosystem could be defined as one that rapidly returns to its original state following disturbance.

Stem group The ancestors of a crown group (q.v.), which may have had some of the characteristics of the crown group.

Stenotopic An organism that has limited ecological tolerance.

Stochastic A result or process produced by chance, that is, not caused by the action of a regulatory force.

Stone stripe When the water freezing around stones in the ground has the effect of forcing them to the surface.

Strategy The outcome of many generations of selection of individuals and genotypes, conserving those best fitted for prevailing conditions.

Structure Arrangement of the biomass of the vegetation into layered forms.

Subduction The process by which old ocean floor is drawn back into the Earth at the system of oceanic trenches (q.v.).

Subspecies A genetically or morphologically distinct set of populations of a species, confined to a particular area.

Succession A regular change in a community over time. When this begins with bare ground, it is sometimes called primary succession, as distinct from secondary succession, which refers to changes after the collapse or destruction of an existing community.

Sweepstakes route A dispersal route through which it is extremely difficult to pass, so that organisms can only do so by a chance combination of favourable circumstances, or by special adaptations to facilitate its passage.

Sympatric speciation The evolution of a new species within the same area as that of its parent species.

Taxon, plural taxa Any unit in the system of naming and classifying organisms, e.g. species (q.v.).

Taxon–area cladogram A cladogram (q.v.) in which the name of each taxon has been replaced by that of the area in which it is found.

Taxon cycle The hypothesis that the distribution and ranges of individual species in island communities go through a cycle of expansion and contraction.

Taxonomy The study of the naming of organisms and their placement into a hierarchical system of classification.

Tectonic plates The assemblage of regions of the Earth's surface, containing ocean floor with or without continents, and which move across the face of the Earth, fusing or subdividing.

Temporal niche A preferred stage in successional development when a species' attributes are most effective in competing and in establishing a sustainable population.

Temporal separation A situation in which two species occupy similar niches in the environment but at different times of the day.

Terminal Eocene Event The marked climatic cooling that took place at the end of the Eocene Epoch, involving a decrease in mean annual temperature and an increase in the mean annual temperature range.

Terrane A small area of originally marine or volcanic rocks, which have become scraped off against the edge of a continent as the ocean floor that bore them has become subducted below it, and which therefore totally differ from the other rocks that now surround them.

Tethys Ocean   An ocean that once separated the southern continents from the northern continents, now represented only by the Mediterranean Sea.

Theory of Island Biogeography or **TIB**   A theory that suggests that the changing, and interrelated, rates of colonization and extinction of organisms on oceanic islands eventually lead to an equilibrium between these two processes. The rate of replacement of species, or turnover rate, then becomes approximately constant, as does the number of species on the island. The theory also suggests that there is a strong correlation between the area of the island and the number of species that it contains at equilibrium.

Thermocline   The level within the sea at which there is a rapid change in the temperature of the water.

Thermohaline circulation   The worldwide pattern by which seawater is warmed in tropical latitudes and then circulated to higher latitudes, before being returned to the tropics. Also known as the oceanic conveyor belt or oceanic circulation.

Thulean route   A former land connection between Greenland and Europe across the area now occupied by Iceland.

TIB   See *Theory of Island Biogeography*.

Till   A clay-rich deposit, containing quantities of unsorted, rounded and scarred boulders and pebbles, left behind during the melting and retreat of a glacier.

Timberline   The elevation at which forest vegetation gives way to alpine scrub.

Track   A line that connects the separate ranges of a set of related taxa, used in the theory of panbiogeography (q.v.). Where a number of unrelated sets of taxa show identical tracks, this is known as a generalized track; where these run across ocean basins, they are known as ocean baselines.

Transform fault   A region of active earthquake activity, where different plates move past one another.

Trenches   The system of deep submarine canyons where old ocean floor is consumed, disappearing downward into the Earth.

Trophic level   One of a series of levels within an ecosystem through which energy passes from organism to organism.

Turgai Sea   A shallow seaway that once separated Europe from Asia; also known as the Obik Sea.

Turnover rate   The rate of replacement of species on oceanic islands. See *Theory of Island Biogeography*.

Ultraviolet radiation   Radiation from the sun that can cause skin cancer and also the destruction of certain B vitamins (e.g. folic acid) in the skin.

Ungulate   Hoofed.

Uniformitarianism   The idea that present-day conditions can be used as a key for understanding past processes with no need to interpret geology in the light of supposed past global catastrophes.

Vector   See *intermediate host*.

Vicariance biogeography   An approach to historical biogeography based on the assumption that related taxa, occupying ranges that are separate from one another, arrived in them before the appearance of the barriers that now separate them, rather than by dispersal across the barriers after they had formed.

Vicariant speciation   A process of speciation that arises after an organism has become isolated as a result of vicariance (q.v.).

Wallacea   The region, containing many islands, that lies between the continental shelves of South-East Asia and Australia.

Wallace's Line   The north–south line, running through Wallacea (q.v.), that separates the predominantly Asian fauna to the west from the predominantly Australian fauna to the east.

Xerophyte   A plant that can tolerate low levels of moisture.

Younger Dryas   A cold episode that took place between 12 700 and 11 500 solar (calendar) years ago.

Zonation A regular spatial sequence of replacement of species caused by a similar sequence of change in physical or chemical conditions.

Zooplankton Tiny animals in plankton that consume phytoplankton.

Zooxanthellae A type of algae whose photosynthetic activity provides energy and nourishment to other species.

Index

Page numbers in *italics* indicate figures or tables, those in **bold** indicate the pages on which the subject of the entry is defined. Plates are indexed as Plate 1, Plate 2, etc.

Biogeography: An Ecological and Evolutionary Approach, Ninth Edition. Edited by C. Barry Cox, Peter D. Moore, Richard J. Ladle.
© 2016 John Wiley & Sons, Ltd. Published 2016 by John Wiley & Sons, Ltd.

arachnids *93, 120,* 121, 335
Araucaria 295, 309, 310, 342, 363
    pollen diagrams *364–365*
arborescent **455**
Arbutus unedo (strawberry tree) 55–57, *56,* 383
Archaeochlus 233
archibenthal zone *see* bathyal zone
archipelago speciation **196,** *215,* **455**
Arctic tern *78,* 78
area biogeography 233–236, *234*
areas of endemicity **232, 455**
Argyroxiphium (silverswords) 205, 206
arid regions 344, 363, *367,* 381–382, *382,* **455,** Plate 1
    distribution *366, 367*
armadillo *301*
Armenian gull *34*
arrow-worms 267
Artemesia
    A. frigida (prairie sage) 405
    A. herba-alba 98, *99,* Plate 1
    A. norvegica (Norwegian mugwort) 53–54, *54*
    pollen diagrams *362*
artiodactyls *301*
Ascaris (roundworms) 417
ash *358*
Asiamerica **455**
assembly rules **216–218,** *217,* **455**
association 6
Asteraceae (daisies) 42–46, *42–45,* 201
    pollen diagrams *378*
Astralium rhodosteum 284
auroch 415
Austral Polar Current 262, *264*
Australasia **456**
Australia 331–334
    biomes *333*
    cold periods 363
    megafaunal extinctions 408
Australian bell miner 74
australopithecines/*Australopithecus* **402,** 404, **456**
    A. afarensis 401, 402
    diet 403
avocet 71

bacteria *120*
baker's yeast 410
Balanus balanoides 69
bamboo grass 220
bar-headed goose 37
barnacles 69
barriers **456**
    biological 37
    climatic 37
    East Pacific Barrier **277, 458**
    Mid-Atlantic Barrier **277, 462**
    ocean biomes 267–268
    Panama Isthmus 272, 277
    shallow-sea environment 276–278
bathyal zone *259,* **259,** *269,* **456,** Plate 4
bathypelagic zone *259,* **261, 456**
bats *301*
    altitudinal diversity 136
Bay-Area model 245
Bayesian Island Biogeographical model (BIB) 243, 244

beard-tongue 174, *174*
beech *see Nothofagus*
Bellamya unicolor 184
benthic organisms **260,** *274,* **456**
beta diversity **141, 456**
Betula 377
    B. alleghaniensis (yellow birch) 140
    pollen diagrams *357, 358, 378*
Bidens 203
biodiversity 5–7, **117–146,** 426, **456**
    and altitude 134–136, *135*
    dynamic 142–143
    and glaciation 132–133
    hotspots **136–139,** *137, 138*
    indicator trends *428*
    intermediate disturbance hypothesis 141–142
    knowledge shortfalls 439–441, *439*
    latitude and species range 133–134
    latitudinal gradients 123–131, *123–131*
    neutral theory **23, 38, 142–143, 462**
    scale dependency 441–442
    species numbers 118–123, *120–122*
    succession 139–141, *139, 140*
    tropical regions 131–132, *131*
biodiversity crisis 430–435, *431, 432, 434*
    response to 435–437, *436*
BioGeoBEARS model 245
biogeographical regions
    ancient 291–313
    history 323, *324*
    modern 315–352
biogeography **456**
    cladistic (pattern-based) **233–236,** *234,* **457**
    conservation *see* conservation biogeography
    and creation 4–5
    dispersal **458**
    ecological **3–4,** 6, 19–20, **458**
    historical **4, 460**
    history 1–30
    island 208–212, *209–212*
    ocean *see* ocean biogeography
    phylogenetic **232–233,** *233,* **464**
biological barriers 37
biological control **82, 456**
biological invasion 429–430, *429*
biological species concept **179, 456**
biological spectrum **101, 456**
BIOME 3 model 113
biomes 6, 20, **100–101,** *101, 103,* **456**
    climate diagrams *111*
    coastal **263, 457**
    early 305–311, *305, 307, 308*
    modelling 112–113
    mountain 103–106, *104, 105,* 134–136, *135*
    ocean 267–268
    polar 263, *264*
    trade winds 263, *264*
    westerly winds 263, *264*
    see also specific types
biosphere **120, 456,** Plate 2
biota **40, 456**
biotic assemblages 98–102, *99, 101*
biotic diversity *210, 211*
biotic homogenization 430